COMPREHENSIVE CHEMICAL KINETICS

COMPREHENSIVE

Section 1. THE PRACTICE AND THEORY OF KINETICS (3 volumes)

Section 2. HOMOGENEOUS DECOMPOSITION AND ISOMERISATION REACTIONS (2 volumes)

Section 3. INORGANIC REACTIONS (2 volumes)

Section 4. ORGANIC REACTIONS (6 volumes)

Section 5. POLYMERISATION REACTIONS (3 volumes)

Section 6. OXIDATION AND COMBUSTION REACTIONS (2 volumes)

Section 7. SELECTED ELEMENTARY REACTIONS (1 volume)

Section 8. HETEROGENEOUS REACTIONS (4 volumes)

Section 9. KINETICS AND CHEMICAL TECHNOLOGY (1 volume)

Section 10. MODERN METHODS, THEORY, AND DATA

CHEMICAL KINETICS

EDITED BY

R.G. COMPTON

M.A., D.Phil. (Oxon.)
University Lecturer in Physical Chemistry
and Fellow, St. John's College, Oxford

Co-editor for Vol. 29

A. HAMNETT

M.A., D.Phil. (Oxon.)
University Lecturer in Inorganic Chemistry
and Fellow, St. Catherine's College, Oxford

VOLUME 29

NEW TECHNIQUES FOR THE STUDY OF ELECTRODES AND THEIR REACTIONS

ELSEVIER
AMSTERDAM–OXFORD–NEW YORK–TOKYO
1989

ELSEVIER SCIENCE PUBLISHERS B.V.
Sara Burgerhartstraat 25
P.O. Box 211, 1000 AE Amsterdam, The Netherlands

Distributors for the United States and Canada

ELSEVIER SCIENCE PUBLISHING COMPANY INC.

655 Avenue of the Americas
New York, NY 10010
U.S.A.

ISBN 0-444-41631-5 (Series)
ISBN 0-444-42999-9 (Vol. 29)

with 340 illustrations and 24 tables

Printed in The Netherlands

COMPREHENSIVE CHEMICAL KINETICS

Volumes in the Series

Section 1. THE PRACTICE AND THEORY OF KINETICS (3 volumes)

Volume 1 The Practice of Kinetics
Volume 2 The Theory of Kinetics
Volume 3 The Formation and Decay of Excited Species

Section 2. HOMOGENEOUS DECOMPOSITION AND ISOMERISATION REACTIONS (2 volumes)

Volume 4 Decomposition of Inorganic and Organometallic Compounds
Volume 5 Decomposition and Isomerisation of Organic Compounds

Section 3. INORGANIC REACTIONS (2 volumes)

Volume 6 Reactions of Non-metallic Inorganic Compounds
Volume 7 Reactions of Metallic Salts and Complexes, and Organometallic Compounds

Section 4. ORGANIC REACTIONS (6 volumes)

Volume 8 Proton Transfer
Volume 9 Addition and Elimination Reactions of Aliphatic Compounds
Volume 10 Ester Formation and Hydrolysis and Related Reactions
Volume 12 Electrophilic Substitution at a Saturated Carbon Atom
Volume 13 Reactions of Aromatic Compounds

Section 5. POLYMERISATION REACTIONS (3 volumes)

Volume 14 Degradation of Polymers
Volume 14A Free-radical Polymerisation
Volume 15 Non-radical Polymerisation

Section 6. OXIDATION AND COMBUSTION REACTIONS (2 volumes)

Volume 16 Liquid-phase Oxidation
Volume 17 Gas-phase Combustion

Section 7. SELECTED ELEMENTARY REACTIONS (1 volume)

Volume 18 Selected Elementary Reactions

Section 8. HETEROGENEOUS REACTIONS (4 volumes)

Volume 19 Simple Processes at the Gas–Solid Interface
Volume 20 Complex Catalytic Processes
Volume 21 Reactions of Solids with Gases
Volume 22 Reactions in the Solid State

Section 9. KINETICS AND CHEMICAL TECHNOLOGY (1 volume)

Volume 23 Kinetics and Chemical Technology

Section 10. MODERN METHODS, THEORY, AND DATA

Volume 24 Modern Methods in Kinetics
Volume 25 Diffusion-limited Reactions
Volume 26 Electrode Kinetics: Principles and Methodology
Volume 27 Electrode Kinetics: Reactions
Volume 28 Reactions at the Liquid–Solid Interface
Volume 29 New Techniques for the Study of Electrodes and their Reactions
Volume 30 Electron Tunneling in Chemistry. Chemical Reactions over Large Distances

Contributors to Volume 29

W.J. ALBERY, F.R.S.	Department of Chemistry, Imperial College of Science and Technology, South Kensington, London SW7 2AY, Gt. Britain
P.A. CHRISTENSEN	Inorganic Chemistry Laboratory, University of Oxford, Oxford OX1 3QR, Gt. Britain
R.G. COMPTON	Physical Chemistry Laboratory, University of Oxford, Oxford OX1 3QZ, Gt. Britain
S. DENNISON	Inorganic Chemistry Laboratory, University of Oxford, Oxford OX1 3QR, Gt. Britain
R. GREEF	Department of Chemistry, The University, Southampton SO9 5NH, Gt. Britain
A. HAMNETT	Inorganic Chemistry Laboratory, University of Oxford, Oxford OX1 3QR, Gt. Britain
J.A. HARRISON	Chemistry Department, University of Newcastle-upon-Tyne, Newcastle-on-Tyne NE1 7RU, Gt. Britain
R.E. HESTER	Chemistry Department, University of York, York YO1 5DD, Gt. Britain
C.C. JONES	Unilever Research, Quarry Road East, Bebbington, Wirral, Merseyside L63 3JW, Gt. Britain
R.L. LANE	Inorganic Chemistry Laboratory, University of Oxford, Oxford OX1 3QR, Gt. Britain
A.R. MOUNT	Department of Chemistry, Imperial College of Science and Technology, South Kensington, London SW7 2AY, Gt. Britain

R. PARSONS, F.R.S. — Department of Chemistry,
The University,
Southampton SO9 5NH, Gt. Britain

L.M. PETER — Department of Chemistry,
The University,
Southampton SO9 5NH, Gt. Britain

J. ROBINSON — Department of Physics,
University of Warwick,
Coventry CV4 7AL, Gt. Britain

P.R. TREVELLICK — Inorganic Chemistry Laboratory,
University of Oxford,
Oxford OX1 3QR, Gt. Britain

P.R. UNWIN — Physical Chemistry Laboratory,
University of Oxford,
Oxford OX1 3QZ, Gt. Britain

A.M. WALLER — Physical Chemistry Laboratory,
University of Oxford,
Oxford OX1 3QZ, Gt. Britain

Preface

Volume 29 of *Comprehensive Chemical Kinetics* is intended to give an account of new techniques for the study of electrodes and their reactions. It should both extend and complement Volumes 26 and 27 of the series which provide an introductory treatment of modern electrochemical methodology and reactions. The volume covers the various branches of spectroelectrochemistry and also some recent purely electrochemical advances.

In-situ spectroelectrochemical techniques are covered by chapters on infrared, Raman EPR, ellipsometry, electroreflectance, and photocurrent spectroscopy. Ex-situ UHV experiments are treated in a separate chapter. New electrochemical directions are described in chapters on hydrodynamic methods, channel electrodes, and microelectrodes. A final chapter covers computing strategies for the on-line accumulation and processing of electrochemial data.

The editors thank Dr. Andrew Waller for his assistance in compiling the index.

Oxford R.G. Compton
August 1988 A. Hamnett

Contents

Preface x

Chapter 1 (P.A. Christensen and A. Hamnett)

In-situ infrared studies of the electrode–electrolyte interface 1
1. The theory of IR reflection and transmission 2
   1.1 Specular external reflection 6
   1.2 Internal reflection 10
2. Attenuated total reflectance (ATR) 13
   2.1 Ion radical intermediates 15
   2.2 Molecular adsorption 16
   2.3 Corrosion 17
   2.4 Electropolymerisation 19
   2.5 The double layer 20
3. Transmission IR 24
4. External reflectance 25
   4.1 An electron-transfer reaction in the thin-layer 28
   4.2 Adsorption 28
   4.3 Electro-oxidation of small organic molecules 29
   4.4 Adsorbed hydrogen and its effects on double-layer structure 36
   4.5 Adsorption of ethylene derivatives 39
5. Fourier-transform techniques 42
   5.1 Investigations of simple double layers 48
   5.2 Adsorption 51
   5.3 Ion radical intermediates 55
   5.4 Adsorption at non-metal electrodes 58
6. Polarization-modulation techniques 61
7. Transient techniques 67

References 74

Chapter 2 (R.E. Hester)

Raman spectroscopic studies of species in situ at electrode surfaces 79
1. Introduction 79
   1.1 Resonance Raman enhancement of sensitivity 80
   1.2 Surface-enhanced Raman spectroscopy (SERS) 81
   1.3 Surface-enhanced resonance Raman spectroscopy (SERRS) 85
2. Spectroelectrochemical cells 85
   2.1 Electrode surface activation for SERS 88
   2.2 Electrode potential dependence of SERS 91

3. Chemical species adsorbed on metal electrodes 94
3.1 Modified electrodes 99
4. Non-aqueous systems. 100
5. Semiconductor electrodes 101
Acknowledgements 103
References 103

Chapter 3 (R. Parsons)

The use of ex-situ UHV techniques to study electrode surfaces 105
1. Introduction. 105
2. Early work 106
3. Experimental techniques 106
3.1 Direct transfer systems 107
3.2 Transfer using a glove box. 112
3.3 Parallel experiments. 113
3.4 Transfer of an electrode in and out of an electrolyte: immersion and emersion 115
4. Some examples of the use of ex-situ techniques in electrochemical problems. . . 116
4.1 The nature of oxidized platinum electrodes 116
4.2 The adsorption of hydrogen on platinum 118
4.3 The deposition of silver on platinum electrodes protected by iodine 122
5. Conclusion 126
References 126

Chapter 4 (W.J. Albery, C.C. Jones and A.R. Mount)

New hydrodynamic methods 129
1. Introduction. 129
2. Different hydrodynamic systems 129
2.1 Rotating disc 129
2.2 Wall-jet. 130
2.3 Tube electrode 132
3. Ring–disc electrodes 133
3.1 The collection efficiency 133
3.2 Ring–disc simulation programs. 134
3.3 Ring–disc transients. 136
3.4 Ring–disc pH transients 136
4. Measurement of diffusion coefficient by rotation speed step 138
5. Electrochemical ESR. 139
6. Wall-jet systems 142
6.1 Packed-bed wall-jet electrodes 142
6.2 Colloidal deposition 142
7. Photoelectrochemistry 146
References 147

Chapter 5 (J. Robinson)

Microelectrodes 149
1. Introduction. 149
2. In vivo studies. 150
3. Finite size effects 152
4. Experimental details 155
4.1 Electrode design and construction 155
4.2 Electronic equipment 157

5. Application of microelectrodes 159
5.1 The study of kinetics by steady-state measurements 159
5.2 Studies in resistive media 163
5.3 Transient studies of electrode kinetics 167
5.4 Nucleation and phase growth 169
6. Closure 171
List of symbols 171
References 172

Chapter 6 (P.R. Unwin and R.G. Compton)

The use of channel electrodes in the investigation of interfacial reaction mechanisms 173
1. Introduction 173
1.1 Mass transport as a variable in the study of electrode processes 174
1.2 Channel and tubular electrodes 176
2. Mass transport to the channel electrode 181
2.1 The Levich equation 181
2.2 Numerical methods 184
2.3 The Singh and Dutt approximation 188
3. Voltammetry and related experiments at channel electrodes 193
3.1 Steady-state voltammetry 194
3.2 Linear sweep and cyclic voltammetry 196
3.3 Alternating current voltammetry 198
3.4 Chronopotentiometry 200
3.5 Chronoamperometry 200
3.6 Anodic stripping voltammetry (ASV) 204
3.7 Pulsed-flow voltammetry (PFV) 206
4. Electrode kinetics and coupled homogeneous reactions 206
4.1 ECE and related mechanisms 206
4.2 EC reactions 217
4.3 CE reactions 218
4.4 EC′ reactions 219
5. Design and fabrication of channel electrode cells and the associated flow system 220
5.1 Channel electrode cells 220
5.2 Flow systems 222
5.3 Apparatus for studies in turbulent flow regimes 224
6. Extension of channel electrode methodology: practical examples 224
6.1 The double channel electrode 225
6.1.1 The double channel electrode in the investigation of electrode reaction mechanisms 226
6.1.2 The double channel electrode in the investigation of anodic metal dissolution 236
6.2 The study of electrode reaction mechanisms in a turbulent flow regime at a micro-tubular electrode 244
6.2.1 Mass transport to channel and tubular electrodes under a turbulent flow regime 244
6.2.2 Current–potential relationship for steady-state electron transfer . . . 251
6.2.3 Electrode kinetics and coupled homogeneous kinetics: the measurement of the rate of fast chemical reactions 253
6.3 Channel electrodes in the study of pitting corrosion 256
6.3.1 Theoretical models 256
6.3.2 Experimental investigation of pitting corrosion at channel electrodes 264
6.3.3 Concluding remarks 268

6.4 Channel electrode methodology in the study of the dissolution kinetics of ionic solids . . . 268
6.4.1 Calcium carbonate dissolution and channel electrodes . . . 268
6.4.2 Dissolution with amperometric detection: the relationship between the shielding of the transport-limited detector electrode current and the rate constant for the heterogeneous process on the crystal surface. . 271
6.4.3 The dissolution of calcite at pH 3 . . . 279
6.4.4 The dissolution of calcite in aqueous polymaleic acid (PMA) solution at around pH 3 . . . 281
6.4.5 The dissolution of calcite in aqueous maleic acid solution (3 mM): a morphological study. . . . 283
Appendix 1 . . . 287
Appendix 2 . . . 288
Appendix 3 . . . 289
Acknowledgements . . . 290
References . . . 290

Chapter 7 (A.M. Waller and R.G. Compton)
In-situ electrochemical ESR . . . 297
1. Introduction . . . 297
2. Electron spin resonance: an introductory survey. . . . 298
2.1 ESR theory . . . 298
2.1.1 The *g* factor . . . 299
2.1.2 Hyperfine coupling . . . 300
2.1.3 The ESR linewidth . . . 303
2.2 ESR instrumentation . . . 305
3. Electrochemical ESR. . . . 305
3.1 Historical development . . . 305
3.1.1 Ex-situ methods . . . 309
3.1.2 In-situ methods . . . 310
4. Contemporary in-situ electrochemical ESR . . . 312
4.1 The Bard and Goldberg cell . . . 312
4.2 The Bond cell for in-situ ESR at low temperatures . . . 313
4.3 The Allendoerfer cell . . . 314
4.4 The Compton–Coles cell . . . 317
4.5 The Albery in-situ cell. . . . 322
4.6 The coaxial in-situ cell . . . 323
5. Applications . . . 329
5.1 The mechanism of the reduction of dicyanobenzene . . . 329
5.2 Triphenylacetic acid. . . . 330
5.3 Triphenylmethanol . . . 334
5.4 ECE/DISP1: the reduction of fluorescein . . . 336
5.5 Polymer-coated electrodes . . . 339
5.5.1 Poly(*N*-vinylcarbazole) . . . 340
5.5.2 Polypyrrole . . . 341
5.5.3 Molecular motion within polymer-coated electrodes. . . . 343
5.6 A spin-labelled electrode. . . . 344
5.7 Spin trapping . . . 346
6. Concluding remarks . . . 349
Acknowledgements . . . 349
References . . . 349

Chapter 8 (L.M. Peter)

Photocurrent spectroscopy 353
1. Introduction 353
2. Light absorption and charge carrier generation in solids 354
3. Collection of photogenerated carriers 357
4. Photocurrent generation in semiconducting films of finite thickness 360
5. Demonstration of photosensitivity of anodic films on metals 363
6. Experimental aspects of photocurrent spectroscopy 364
7. Derivation of absorption spectra from photocurrent spectra 366
   7.1 Bi_2S_3 films 366
   7.2 The anodic oxide on iron 369
   7.2 PbO films on lead electrodes 371
8. Determination of film thickness by photocurrent spectroscopy 371
   8.1 TiO_2 films on Ti 371
   8.2 Corrosion of Pb in H_2SO_4 373
9. Internal photoemission 375
10. An example of a p-type surface film: the oxide film on Cu 378
11. Organometallic films on metal electrodes 381
Acknowledgements 382
References 382

Chapter 9 (A. Hamnett, R.L. Lane, P.R. Trevellick and S. Dennison)

Electroreflectance at semiconductors 385
1. Introduction 385
2. Basic theoretical development 393
3. Low-field theories 402
4. Applications of the Franz–Keldysh theory to more heavily doped III/V semiconductors 405
   4.1 The effect of inhomogeneous electric field 405
   4.2 The effect of d.c. potential 406
   4.3 The effect of donor density 407
   4.4 The effect of thermal broadening 408
   4.5 The effect of a.c. amplitude 409
   4.6 The effect of optical cross-section 409
5. Experimental results 413
   5.1 Non-aqueous solvents 423
6. Conclusions 425
Acknowledgements 425
References 425

Chapter 10 (R. Greef)

Ellipsometry 427
1. Introduction. What is ellipsometry? 427
2. Theory. Three important parameters 427
   2.1 Phase 428
   2.2 Complex refractive index and the Fresnel equations 430
   2.3 Film thickness 432
3. Instrumentation 432
4. The ideal three-phase model. SiO_2 on Si 434
5. Calculations for the three-phase model 438
6. A "nearly perfect" three-phase system 439

7. Non-uniform and rough films. Aluminium corrosion 441
8. Aluminium in alkaline solution: non-uniform and rough films 442
9. Thin films. A biological example 444
10. Spectroscopic ellipsometry 446
11. Profiling of complex layers 448
12. Geometric roughness. An exact solution. 449
13. Reviews of ellipsometry applications 450
14. Outlook 450
15. Appendix 450
Acknowledgements 452
References 452

Chapter 11 (J.A. Harrison)

A computing strategy for the on-line accumulation and processing of electrochemical data 453
Abstract 453
1. Introduction. 453
2. On-line accumulation of electrochemical data 455
3. Choice of electrochemical methods and the processing of electrochemical data . 459
4. Current–potential curve for a single electrode reaction with single electron transfer: description by standard rate constant 460
5. Impedance–potential curve for a single electrode reaction with single electron transfer: description by standard rate constant 461
6. Current–potential curve for a single electrode reaction with single electron transfer and complexing of the reactant 463
7. Impedance–potential curve for a single electrode reaction with single electron transfer and complexing of the reactant. 464
8. Single electrode reaction with more than one electron transfer. 464
9. Two electrode reactions in parallel 466
10. Current–potential curve for two electrode reactions in series. 466
11. Impedance–potential curve for two electrode reactions in series 467
12. Single electrode reaction: comparison of experiment and theory 468
13. Dissolution and passivation of titanium in acid solution 470
14. Corrosion of stainless steel and its component metals in acid solution 475
14.1 Double layer capacity curves 475
15. Oxidation of chloride and bromide ions on ruthenium dioxide/titanium electrodes 477
16. Reduction of palladium chloro complexes 481
17. Corrosion of dental amalgams 489
18. Conclusions 496
List of symbols 496
References 497

Index 499

Chapter 1

In-situ Infrared Studies of the Electrode–Electrolyte Interface

P.A. CHRISTENSEN and A. HAMNETT

This chapter is not intended as a comprehensive review of in-situ studies of the near-electrode region by infrared spectroscopy, but rather the intention is to highlight the versatility of IR-based techniques (and particularly those involving external reflectance) and the wide range of application of such techniques to modern electrochemistry.

The study of electrochemical systems has always had a major contribution to make to the field of applied science. This has been facilitated by the ease and simplicity with which the rate of an electrochemical reaction can be controlled at the electrode–electrolyte interface by means of the electrode potential. However, it has long been known that many electrochemical reactions take place with complex mechanisms and the importance of techniques that are capable of yielding information on species at or near an electrode has been widely accepted.

Most studies that aim to improve the characterization of the electrode surface have been carried out using high-vacuum techniques such as Auger [1, 2], XPS [3, 4], SIMS [5, 6], etc. However, these techniques involve the removal of the electrode from the electrolyte and the information derived from them may not reflect the state of the electrode in-situ. In addition, many of these techniques lack the molecular specificity afforded by vibrational spectroscopy and it has long been realised that IR spectroscopy would be an ideal method if it could be applied to the in-situ study of the electrode surface. Information from IR would include, potentially, molecular composition and symmetry, bond lengths and force constants (perhaps allowing us to estimate the strength of a chemisorption bond), and molecular orientation.

However, there are two major problems in measuring the IR spectra of species in the near-electrode region. Firstly, all common electrochemical solvents absorb IR light strongly; this is especially true of water, since it has a broad strong absorption covering the whole of the mid-IR range. Secondly, there is a problem of sensitivity: how can we detect the very low absorptions from species on or near an electrode when conventional IR sources are weak and detectors very noisy? One possibility of circumventing the solvent absorption problem is to use Raman spectroscopy, which offers an alternative technique for the measurement of vibrational spectra. It has the immense advantage of using radiation in the visible region which is not absorbed by common electrochemical solvents, but one experimental dif-

References pp. 74–77

ficulty that might have been envisaged is that of the sensitivity of the technique to the very small amount of surface and near-surface species of interest; the problem was overcome by roughening the surface electrochemically to enhance the signal, though the exact mechanism is still a matter of controversy. Weaver and co-workers [7] have found that the effect is strongly dependent on surface morphology.

The first results in this area were reported in 1974 by Fleischmann et al. [8], who reported Raman spectra from pyridine on a roughened silver surface in contact with an electrode solution. This has led to a great deal of work using surface enhanced Raman spectroscopy (SERS) [9]. However, the technique is limited to a small number of substrates that give rise to signal enhancement on electrochemical roughening [9–11]. Another difficulty is the uncertainty as to whether the spectra arise from adsorbate molecules typical of those covering the electrode surface or only particular "atypical" structures. As a result, information on surface species derived from this technique is difficult to interpret quantitatively and it is evident that there is a need for a simpler, unenhanced technique employing more sensitive instruments. Encouraging results with *unenhanced* Raman spectroscopy have recently been reported by at least two groups [12, 13] and SERS is reviewed in detail elsewhere in this volume.

The difficulties with SERS and its restriction to a small number of electrode materials has refocussed attention on IR techniques and these are summarised in Fig. 1.

The elegant experiment, which apparently avoids the problem of solvent absorption, is attenuated total reflectance and this was the first in-situ IR technique to be developed. More recently, transmission and specular reflectance modes using IR radiation have also been carefully investigated and it has become evident that each of these three methods has both advantages and disadvantages. Before considering each technique in detail, it will be useful to develop a consistent theoretical framework.

1. The theory of IR reflection and transmission

The reflection and refraction of electromagnetic radiation at an interface can be discussed with relation to Fig. 2. The incident, reflected, and refracted rays are shown in this figure as I, R, and T and reflection and transmission coefficients can be derived straightforwardly from standard electromagnetic theory. In order to focus our ideas and to define the convention used in this section, we may write down the expression for the electric field, E, of the radiation, assuming that the light is propagating in the z direction (with unit vector $\hat{\mathbf{k}}$) and the electric field is oriented in the x direction (with unit vector $\hat{\mathbf{i}}$)

$$E = E_0 \hat{\mathbf{i}} \exp\{i(\omega t - kz)\} \qquad (1)$$

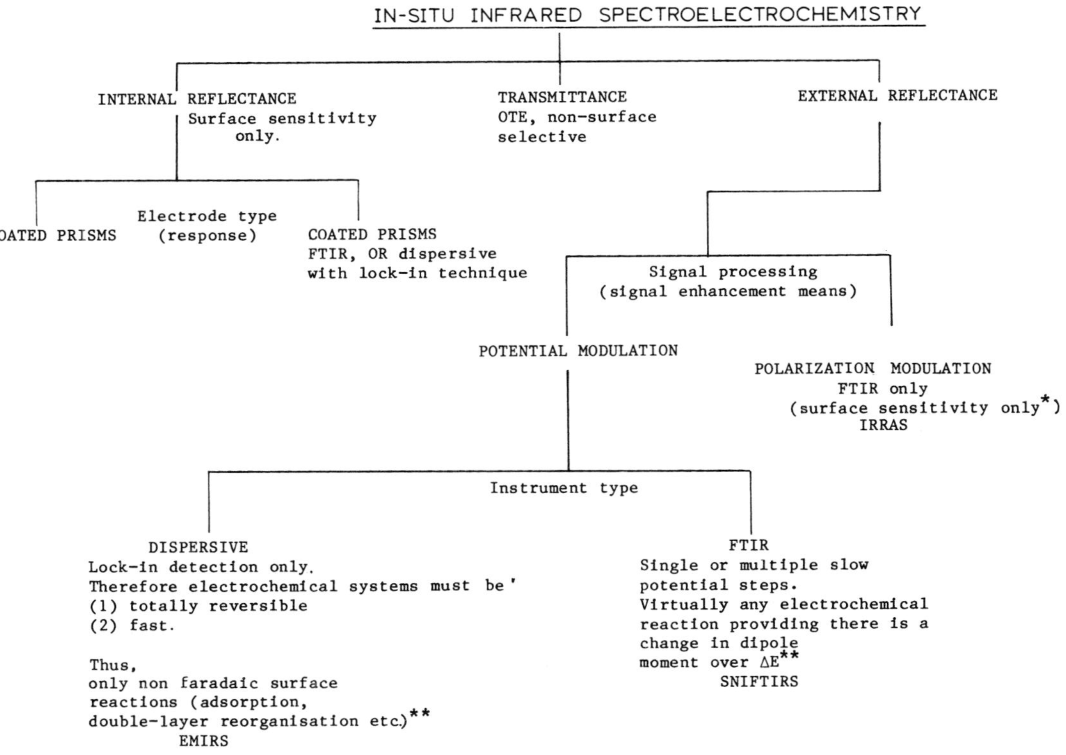

Fig. 1. Diagram showing the relationship between the various in-situ infrared electrochemical techniques.

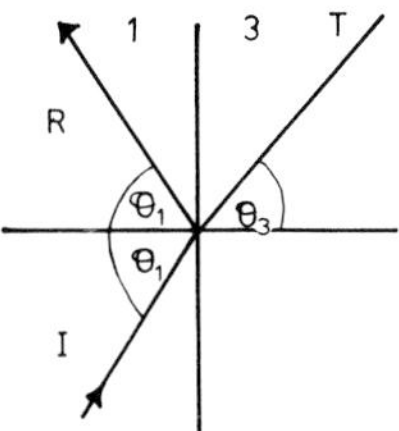

Fig. 2. Reflection and transmission of IR at the electrode–electrolyte interface. θ_1 is the angle of incidence and θ_3 the angle of refraction. The medium containing the incident and reflected light is referred to as 1 and that containing the transmitted light as 3.

and the magnetic induction, B, associated with the radiation is then given, from Maxwell's equations, by

$$B = \frac{E_o \hat{\mathbf{j}} \exp\{i(\omega t - kz)\}n}{c} \tag{2}$$

where c is the veolcity of light in vacuo, the vector $\mathbf{k}$ has *magnitude* $k = \omega/v$, and v is the phase velocity of the light in the medium. By recalling that $v = c/n$, where n is the refractive index, we see that $k = \omega n/c = 2\pi n/\lambda$. If the medium is absorbing, n and k become *complex*: under these circumstances, we will write the refractive index with a circumflex as $\hat{n}$, where

$$\hat{n} = n_R - in_I \tag{3}$$

and

$$E_x = E_o \exp\{i\omega(t - n_R z/c)\} \exp\{-n_I \omega z/c\} \tag{4}$$

from which it can be seen that n_I can be related to the absorption coefficient α, of the medium. Explicitly, since the rate of intensity decay $\langle E^2 \rangle \sim \exp(-\alpha z) \sim \exp(-2n_I\omega z/c)$, then

$$\alpha = \frac{4\pi n_I}{\lambda} \tag{5}$$

In addition to the refractive index, which is, of course, a function of the frequency of the incident radiation, we may also define the *dielectric function* of the medium: in fact

$$\varepsilon = n^2$$

and if n is complex, then ε will also be complex and written $\hat{\varepsilon}$

$$\hat{\varepsilon} \equiv \varepsilon' - i\varepsilon''$$

where

$$\varepsilon' \equiv n_I^2 - n_R^2$$

$$\varepsilon'' \equiv 2n_I n_R$$

In considering reflection or transmission of light at an interface, we can distinguish two cases: in the first, the light is *linearly* polarised with the electric vector *parallel* to the plane of reflection. In the second, the electric field is *perpendicular* to the plane of reflection and the electric-field vectors in the two cases are denoted by $\mathbf{E}^p$ and $\mathbf{E}^s$, respectively. The importance of these cases is that the reflected and transmitted electric-field vectors are also either both parallel or perpendicular to the plane of reflection, respectively: for light polarised at other angles, the transmitted and reflected waves will show more complex behaviour. If we refer to the medium containing incident and reflected light as 1 and the medium containing *transmitted* light as 3, as in Fig. 2, then it can be shown that the reflection coefficients r^p and r^s and the transmission coefficients t^p and t^s are given by [14, 15]

$$r^s = \frac{\xi_1 - \xi_3}{\xi_1 + \xi_3}; \quad r^p = \frac{\eta_1 - \eta_3}{\eta_1 + \eta_3}; \quad t^s = \frac{2\xi_1}{\xi_1 + \xi_3}; \quad t^p = \frac{2\eta_1(n_1/\hat{n}_3)}{\eta_1 = \eta_3} \tag{6}$$

where $\xi_j = \hat{n}_j \cos\phi_j = (\hat{\varepsilon}_j - \varepsilon_1 \sin^2\phi_j)^{1/2}$, $\eta_j = \xi_j/\hat{\varepsilon}_j$, $\hat{\varepsilon}_j$ is the (complex) dielectric function for medium j, and $\hat{n}_j$ the corresponding (complex) refractive index. We will assume for simplicity that medium 1 is non-absorbing, so that n_1 and ξ_1 are real. These coefficients are defined as the ratios of the reflected or transmitted electric field vector to the incident electric field vector and are, in general, complex numbers unless neither medium is absorbing. Our interest will be in the changes that occur in these quantities on formation of a thin film, where thin implies that the thickness, d, is much less than the incident wavelength λ.

The changes that occur are best understood in terms of the energy flow associated with the radiation. In general, this is given by the corresponding Poynting vector of the light, $\mathbf{S}$, defined by

$$\langle \mathbf{S} \rangle = \tfrac{1}{2}\,\mathrm{Re}(\mathbf{E} \times \mathbf{H}^*) \tag{7}$$

where $\mathbf{E}$ and $\mathbf{H}$ are the electric and magnetic field vectors associated with the electromagnetic radiation, $\mathbf{H}$ is related to $\mathbf{B}$, the magnetic induction, through the formula $\mathbf{B} = \mu_o \mathbf{H}$, where we have assumed that the medium through which propogation is taking place is non-magnetic, and $\mathbf{H}^*$ is the complex conjugate of $\mathbf{H}$. $\langle \mathbf{S} \rangle$ is the energy flux per unit area and is aligned along the direction of propagation of the radiation. For light in a non-absorbing medium, the expectation value of $\mathbf{S}$ is given by

$$\langle \mathbf{S} \rangle = \left\{ \frac{n\langle \mathbf{E}^2 \rangle}{\mu_o c} \right\} \cdot \hat{\mathbf{k}} \tag{8}$$

where $\hat{\mathbf{k}}$ is a unit vector in the direction of propagation of the radiation.

The *reflectivity* of the surface may now be interpreted in terms of the components of the Poynting vector perpendicular to the surface (which takes account of the fact that the magnitude of the Poynting vector refers to the energy flow per unit area of the incident or reflected beam, but we must refer to unit area of illuminated surface). Defining the reflectance, R, as

$$R = \frac{\langle \mathbf{S}_1^- \rangle \cdot \cos\theta_1}{\langle \mathbf{S}_1^+ \rangle \cdot \cos\theta_1} \tag{9}$$

where $\langle \mathbf{S}_1^- \rangle$ refers to the Poynting vector of the reflected ray and $\langle S_1^+ \rangle$ to that of the incident ray. By inserting eqn. (8) into eqn. (9), we obtain

$$R = \frac{(E_1^-)^2}{(E_1^+)^2} \equiv |r^2| \tag{10}$$

where E_1^- is the magnitude of the electric field vector of the reflected radiation and E_1^+ that of the incident radiation. In a similar way, the transmittance of an interface, T, is given by

$$T = \frac{\langle \mathbf{S}_3 \rangle \cdot \cos\theta_3}{\langle \mathbf{S}_1^+ \rangle \cdot \cos\theta_1} \tag{11}$$

where $\langle \mathbf{S}_3 \rangle$ is the Poynting vector of the transmitted ray, and from eqns. (8) and (11) we find, for light polarised perpendicular to the plane of reflection

$$T^{\mathrm{s}} = \frac{(E_3)^2 \cos\theta_3 \cdot \mathrm{Re}(n_3)}{(E_1^+)^2 \cos\theta \cdot n_1} \tag{12}$$

$$\equiv \frac{|t^{\mathrm{s}}|^2 \, \mathrm{Re}(\xi_3)}{\xi_1} \tag{13}$$

and a check using the appropriate forms of eqn. (6) for r and t shows that $R^{\mathrm{s}} + T^{\mathrm{s}} = 1$, as demanded by energy conservation. The transmittance for light polarised parallel to the plane of reflectance is complicated somewhat by the need to consider components both parallel to the y and z axes of Fig. 2, and it can be shown that

$$T^{\mathrm{p}} = \frac{|n_3 t^{\mathrm{p}}|^2 \, \mathrm{Re}(\xi_3/n_3^2)}{\xi_1} \tag{14}$$

Again, it is easy to show that $R^{\mathrm{p}} + T^{\mathrm{p}} = 1$.

The above theory may be developed in two different ways, depending on the experimental arrangement. We first consider the case of specular *external* reflection and then extend the theory to *internal* reflection.

1.1 SPECULAR EXTERNAL REFLECTION

In the presence of a thin film, as shown in Fig. 3, the situation becomes more complex. For the case of specular reflection, with the electric field polarised in the x direction (i.e. perpendicular to the plane of reflection), the continuity conditions at the (1, 2) interface demand that

$$E_1^+ + E_1^- = {}^{(12)}E_2^+ + {}^{(12)}E_2^- \tag{15}$$

and

$$n_1 E_1^+ \cos\theta_1 - n_1 E_1^- \cos\theta_1 = {}^{(12)}n_2 E_2^+ \cos\theta_2 - {}^{(12)}n_2 E_2^- \cos\theta_2 \tag{16}$$

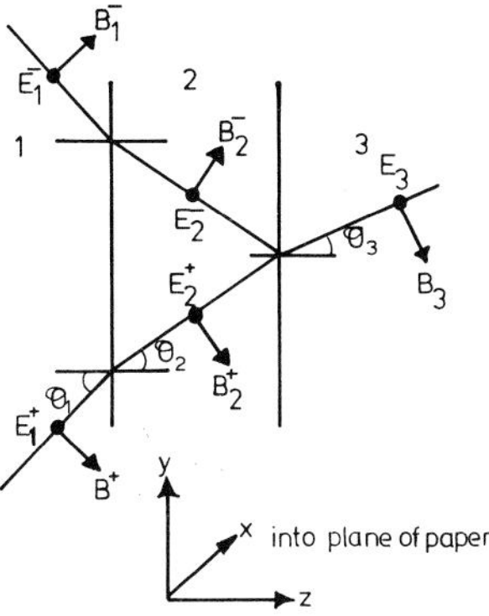

Fig. 3. Reflection and transmission at an interface at which there is a thin layer of (partially) transparent material, referred to as medium 2.

where the superscript (12) implies the value of E_2 in the film (medium 2) just inside the interface between media 1 and 2. Similar continuity equations may be written for the interface between media 2 and 3: in particular, continuity of the electric field implies that

$$^{(23)}E_2^+ + {}^{(23)}E_2^- = E_3 \tag{17}$$

where the absence of any reflected ray in medium 3 simplifies the analysis. We have seen that, in the absence of any second film

$$\frac{E_1^-}{E_1^+} = r_{13} \tag{18}$$

$$\frac{E_3}{E_1^+} = t_{13}$$

By using the continuity conditions and relating the electric field at (1, 2) and (2, 3) through eqn. (1), we find that

$$\frac{E_1^-}{E_1^+} = r_{123} = \frac{\{r_{12} + r_{23}e^{-2i\beta}\}}{1 + r_{12}r_{23}e^{-2i\beta}} \tag{19}$$

and

$$\frac{E_3}{E_1^+} = t_{123} = \frac{t_{12}t_{23}e^{-i\beta}}{1 + r_{12}r_{23}e^{-2i\beta}} \tag{20}$$

where $\beta = 2\pi d\xi_2/\lambda$. If $d/\lambda \ll 1$, then the exponents may be linearised; after some algebra we find that, for perpendicular polarised light

$$r_{123}^s \approx r_{13}^s\left[1 - \frac{2i\beta\xi(\xi_2^2 - \xi_3^2)}{\xi_2(\xi_1^2 - \xi_3^2)}\right] \tag{21}$$

and

$$t_{123}^2 \approx t_{13}^2\left[1 - \frac{i\beta(\xi_2^2 + \xi_1\xi_3)}{\xi_2(\xi_1 + \xi_3)}\right] \tag{22}$$

References pp. 74–77

From these two expressions, we can calculate the *changes* in reflectance and transmittance that occur on film formation. We find

$$\Delta R^s = \frac{8\pi d\xi_1 |r_{13}^s|^2}{\lambda} \operatorname{Im}[(\xi_2^2 - \xi_3^2)/(\xi_1^2 - \xi_3^2)] \tag{23}$$

$$\Delta T^s = \frac{4\pi d \operatorname{Re}(\xi_3)|t_{13}^s|^2}{\lambda \xi_1} \operatorname{Im}[(\xi_2^2 + \xi_1\xi_3)/(\xi_1 + \xi_3)] \tag{24}$$

Provided the film is non-absorbing, then, whilst both ΔR and ΔT are finite conservation of energy ensures that $\Delta R^s = -\Delta T^s$, as may be verified from eqns. (23) and (24). In addition, if $\xi_1^2 < \xi_2^2$, or, equivalently, $n_1 < n_2$, then it is evident that $\Delta R^s < 0$. In other words, *even if the film does not absorb in the IR region being examined, some changes in reflectance and transmittance will occur on film formation.*

If the film does absorb in the wavelength region of interest, then ξ_2 will be complex. Under these circumstances, $\Delta R^s + \Delta T^s < 0$, since some light is absorbed by the film and the overall energy balance is altered. From eqns. (23) and (24), we find that the total amount of energy absorbed by the film is

$$A^s = \frac{-8\pi d\xi_1 \varepsilon_2''}{\lambda|\xi_1 + \xi_3|^2} \tag{25}$$

It must be emphasised that the changes in R^s and T^s now consist of two different contributions: the first is the change associated with presence of a third phase at the interface, and will take place whether or not the film is absorbing, and the second is associated directly with the light absorbed by the film. In the latter case, the overall absorption is partitioned in a rather complex way between reflectance and transmittance: it is not, in general, true to say that all the changes in absorption will be manifested as a change in reflectance. However, this may be an acceptable approximation in those cases where there is a highly reflecting phase, such as a metal, that forms medium 3.

We can evaluate expression (25) by a different route. The energy dissipated in an absorbing medium may be calculated from the change in the magnitude of the Poynting vector. For an absorbing medium, the Poynting vector is given by

$$\langle \mathbf{S} \rangle = \left(\frac{n_R}{\mu_o c}\right) \exp(-2\omega n_I z/c) \langle E_o^2 \rangle \hat{\mathbf{k}} \tag{26}$$

and for a thin film, $d \ll \lambda$, we find that the energy dissipated per unit area is

$$A^s = \omega\varepsilon_o\varepsilon_2'' d \langle E_o^2 \rangle \tag{27}$$

For the film, the total electric field is given, at the (2, 3) interface, by eqn. (17) and it will be sufficient to take the magnitude of the field as uniform in the film. We may also assume, to the same level of accuracy, that the field E_3 can be calculated from the interface theory in the absence of the film, i.e. E_3 can be obtained from eqn. (18). By using eqn. (18) and (17) in eqn. (27), we obtain eqn. (25) once more.

For the case of parallel polarisation, rather similar considerations can be invoked, the only additional consideration being that the polarisation direction can be resolved along both y and z directions. The absorbances can then be calculated as [15]

$$A_y^{\mathrm{p}} = \frac{8\pi d\varepsilon_2'' \eta_1 |\eta_3|^2}{\lambda |\eta_1 + \eta_3|^2} \tag{28}$$

$$A_z^{\mathrm{p}} = \frac{8\pi d\xi_1 \varepsilon_2''}{\lambda |\eta_1 + \eta_3|^2} \tag{29}$$

Clearly, as far as the surface is concerned, A^{s} and A_y^{p} refer to the electric vector polarised *parallel* to the surface whereas A_z^{p} refers to the electric vector polarised *perpendicular* to the surface. For metals in the IR, where n_{I} and ε_3'' are very large, this distinction is important. An examination of the forms of eqns. (25), (28) and (29) shows that, as n_{I} increases, A^{s} and A_y^{p} decreases much more rapidly than A_z^{p}. Given the proviso above, the changes ΔR will become very small for s-polarisation and will only be significant for the component of parallel-polarised light in the z direction. This is a result of considerable importance since, regardless of ε_2'', it implies that the only absorptions making a significant contribution to ΔR are those associated with the electric vector oriented along the z axis, i.e. perpendicular to the surface. This is the *surface selection rule* [107]. It is more important in the IR than in the visible since values of ε_3'' are much larger for metals in the IR. We may insert some numbers to illustrate the extent of the surface selection rule: for copper at 5 μm wavelength, $n_{\mathrm{R}} \approx 3$ and $n_{\mathrm{I}} \approx 10$. In this case, A^{s} and A_y^{p} are both more than an order of magnitude smaller than A_z^{p}; ΔR^{p} is then dominated by the latter, whereas ΔR^{s} is an order of magnitude smaller.

It is of interest to pursue the physical reason for this effect. Consider first the case of perpendicularly polarised light: we examine the effect of increasing the imaginary part of $\hat{n}_3$ without limit, which corresponds, according to eqn. (5), to making the substrate highly optically absorbing. Under these circumstances, r^{s} will tend to, but will not reach, the value -1. Physically, this means that the *magnitude* of E_1^- approaches E_1^+ but with a phase change that approaches π or 180°. The net electric field vector in medium 1, $\langle E_1 \rangle$, is given by $\langle E_1^+ + E_1^- \rangle$ and if we write $r^{\mathrm{s}} \equiv (R^{\mathrm{s}})^{1/2}\exp(i\delta_{\mathrm{r}}^{\mathrm{s}})$, then the magnitude of the square of E_1^{s}, is given, from the equations above, by [14]

$$\frac{\langle (E_1^{\mathrm{s}})^2 \rangle}{\langle (E_1^+)^2 \rangle} = [(1 + R^{\mathrm{s}}) + 2(R^{\mathrm{s}})^{1/2}\cos\{\delta_{\mathrm{r}}^{\mathrm{s}} + 4\pi(z/\lambda)\xi_1\}] \tag{30}$$

if $R^{\mathrm{s}} \to 1$ and $\delta_{\mathrm{r}}^{\mathrm{s}} \to \pi$, then it is clear that, at $z = 0$, $E_1^{\mathrm{s}} \to 0$ and, indeed, within a distance comparable with the wavelength of the incident IR radiation, E_1^{s} will remain very small.

References pp. 74–77

Rather similar considerations apply to parallel reflection. From the equations given above, we find, for the components parallel to the z and y axes [15]

$$\frac{\langle (E^{\mathrm{p}}_{1y})^2 \rangle}{\langle (E^{+}_{1})^2 \rangle} = \cos^2\theta_1[(1 + R^{\mathrm{p}}) - 2(R^{\mathrm{p}})^{1/2}\cos\{\delta^{\mathrm{p}}_{\mathrm{r}} + 4\pi(z/\lambda)\xi_1\}] \tag{31}$$

$$\frac{\langle (E^{\mathrm{p}}_{1z})^2 \rangle}{\langle (E^{+}_{1})^2 \rangle} = \sin^2\theta_1[(1 + R^{\mathrm{p}}) + 2(R^{\mathrm{p}})^{1/2}\cos\{\delta^{\mathrm{p}}_{\mathrm{r}} + 4\pi(z/\lambda)\xi_1\}] \tag{32}$$

It is evident from eqn. (6) that, as n_{I} increases, $r^{\mathrm{p}} \rightarrow 1$ and so $\delta^{\mathrm{p}}_{\mathrm{r}} \rightarrow 0$. Thus, from above, the component of **E** parallel to the surface undergoes a phase change approaching 180°, but that component perpendicular to the surface does not. The net result is that at and near the surface (within a distance of the order of the wavelength of light), the total electric field parallel to the surface is very small.

If an ordered film now forms on the surface whose thickness is small compared with λ, then the result of the cancellation referred to in the previous paragraph is that those IR transitions associated with dipole moment changes parallel to the surface will effect the reflectance far less than those with dipole moment changes perpendicular to the surface. As a concrete example, 4-cyanopyridine will show strong IR absorption if it adsorbs perpendicular to the surface, but weak absorption if it adsorbs flat on the surface.

1.2 INTERNAL REFLECTION

Electrochemical application of IR internal reflection depends upon an arrangement of the form shown in Fig. 4, where the incident medium is 1, medium 2 is a thin metallic film that can act as an electrode, medium 3 is a film forming on the electrode whose properties we wish to investigate, and medium 4 is the electrolyte. Medium 1 is usually an IR-transparent material such as germanium.

The central feature of the experiment is the exploitation of the phenomenon of *critical internal reflection*. This is covered in more detail below, but we can consider briefly the nature of the effect in non-absorbing

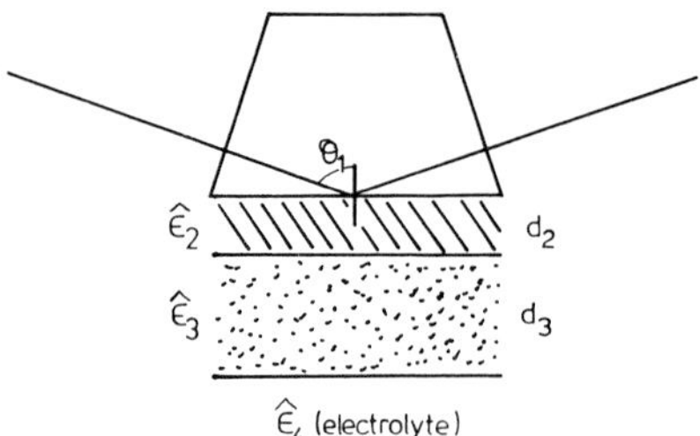

Fig. 4. Arrangement of the internal reflection crystal, electrode (medium 2), and electrolyte (medium 3) in an electrochemical cell based upon the internal reflection of infrared light.

media to clarify the approach. If light is reflected and refracted at the interface with a second medium of lower (real) refractive index than that through which it first passes, then for angles of incidence above a critical value, $\theta_c \equiv \sin^{-1}(n_3/n_1)$, a propagating refracted ray cannot exist in the second phase. Under these circumstances, provided the second medium is non-absorbing, all the energy of the incident ray is present in the reflected ray. However, although a propagating ray cannot form in the second phase, an examination of the continuity equations for the electromagnetic radiation at the interface shows that a damped evanescent wave does penetrate into the second medium, with a penetration depth related to the wavelength of the incident radiation. This wave plays a key role if the second medium is optically absorbing for, under these circumstances, energy may be lost to the second medium and less than 100% of the incident light energy is then reflected. It will be shown below that there is a clear, if rather complex, relationship between the energy loss and the imaginary part of the refractive index of the second medium.

For the arrangement shown in Fig. 4, the following equations for the fractional change in reflectance on formation of a thin film corresponding to medium 3 have been derived [15]

$$(\Delta R^s/R^s) = (8\pi d_3\xi_1/\lambda)\,\mathrm{Im}\left[\frac{(\hat{\xi}_3^2 - \hat{\xi}_4^2)F^s}{\xi_1^2 - \hat{\xi}_4^2}\right] \tag{33}$$

$$(\Delta R^p/R^p) = (8\pi d_3\xi_1/\lambda)\,\mathrm{Im}\left[\frac{\hat{\varepsilon}_3(\hat{\eta}_3^2 - \hat{\eta}_4^2)F^p}{\varepsilon_1(\eta_1^2 - \hat{\eta}_4^2)}\right] \tag{34}$$

where the functions F^s and F^p have the rather complex forms

$$F^s = 4\hat{\xi}_2^2\exp(2i\beta_2)\Bigg[\hat{\xi}_2^2\{(1 + \exp(2i\beta_2)\}^2 + (\xi_1^2\hat{\xi}_4^2 - \hat{\xi}_2^2) \times \{1 - \exp(2i\beta_2)\}^2/(\xi_1^2 - \hat{\xi}_4^2) + 2\hat{\xi}_2\hat{\xi}_4(\xi_1^2 - \hat{\xi}_2^2) \times \frac{\{1 + \exp(4i\beta_2)\}}{\xi_1^2 - \hat{\xi}_4^2}\Bigg]^{-1} \tag{35}$$

$$F^p = 4\hat{\eta}_2^2\exp(2i\beta_2)\Bigg[\hat{\eta}_2^2\{(1 + \exp(2i\beta_2)\}^2 + (\eta_1^2\hat{\eta}_4^2 - \hat{\eta}_2^2) \times \{1 - \exp(2i\beta_2)\}^2/(\eta_1^2 - \hat{\eta}_4^2) + 2\hat{\eta}_2\hat{\eta}_4(\eta_1^2 - \hat{\eta}_2^2) \times \frac{\{1 - \exp(4i\beta_2)\}}{\eta_1^2 - \hat{\eta}_4^2}\Bigg]^{-1} \tag{36}$$

where $\beta_2 = 2\pi\xi_2 d_2/\lambda$, and we *assumed* that $d_3 \ll \lambda$.

These equations are rather complex, but they can be simplified under certain circumstances. To understand some of the basic physics underlying the phenomenon of internal reflection, we consider the value of the square

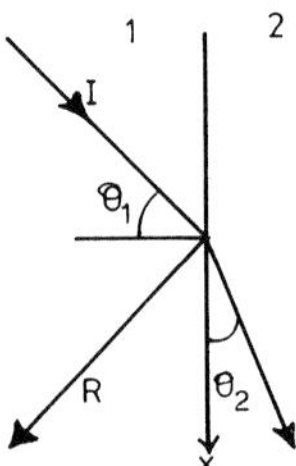

Fig. 5. Internal reflection and transmission of IR light at an interface.

of the electric field vector for the radiation transmitted into the second phase of the interface shown in Fig. 5. From eqn. (6), using the same algebra that led to eqns. (30)–(32), we find

$$\frac{\langle (E_2^s)^2 \rangle}{\langle (E_1^+)^2 \rangle} = |t^s|^2 \exp\{4\pi(z/\lambda) \cdot \mathrm{Im}(\hat{\xi}_2)\} \tag{37}$$

$$\frac{\langle (E_{2y}^p)^2 \rangle}{\langle (E_1^+)^2 \rangle} = |\hat{\xi}_2 t^p/\hat{n}_2|^2 \exp\{4\pi(z/\lambda) \cdot \mathrm{Im}(\hat{\xi}_2)\} \tag{38}$$

$$\frac{\langle (E_{2z}^p)^2 \rangle}{\langle (E_1^+)^2 \rangle} = |n_1 \sin\theta_1 \cdot t^p/\hat{n}_2|^2 \exp\{4\pi(z/\lambda) \cdot \mathrm{Im}(\hat{\xi}_2)\} \tag{39}$$

Even if there is an intervening layer, the *direction* of the refracted beam in medium 3, as shown in Fig. 6, is unaltered, though the magnitude of the electric field is reduced by the fact that the transmission through the layer must be taken into account. If the overall transmission functions are t^s and t^p, then, again, using the nomenclature of Fig. 6.

$$\langle (E_3^s)^2 \rangle = |t^s|^2 \exp[(4\pi/\lambda) \cdot \mathrm{Im}\{\hat{\xi}_3(z - d_2)\}] \tag{40}$$

$$\langle (E_3^p)^2 \rangle = (|\sin\theta_3|^2 + |\cos\theta_3|^2)|t^p|^2 \exp[(4\pi/\lambda) \cdot \mathrm{Im}\{\hat{\xi}_3(z - d_2)\}] \tag{41}$$

Clearly, if medium 3 is absorbing, then even at angles less than the critical angle, $\hat{\xi}_3$ we will have an imaginary component < 0 and the wave will be damped in this medium. Consider, however, the situation in which medium 3 is transparent but $n_3 < n_1\sin\theta_1$. Then $\hat{\xi}_3 \equiv (\hat{n}_3^2 - n_1^2\sin^2\theta_1)^{1/2}$ will be purely

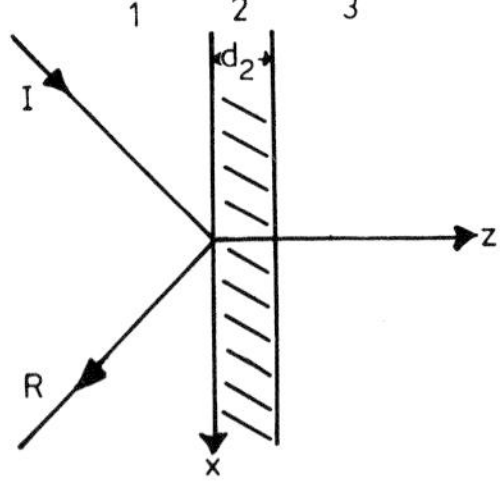

Fig. 6. Internal reflection and evanescent wave at an interface at which there is an intervening layer of medium 2.

imaginary. Under these circumstances, the wave is again damped in medium 3, with a penetration depth of $\lambda[4\pi(n_1^2\sin^2\theta_1 - \hat{n}_3^2)^{1/2}] \approx 0.1\lambda$. This is termed the *evanescent wave* and it becomes of considerable importance if medium 3 actually has a non-zero value of $\mathrm{Im}(\hat{n}_3)$, i.e. medium 3 is absorbing. If this is so, energy can be absorbed from the evanescent wave, which becomes further damped with an associated penetration depth.

$$\tau = \frac{\lambda}{-4\pi \cdot \mathrm{Im}(\hat{\xi}_3)} = \frac{\lambda \cdot \mathrm{Re}\,\hat{\xi}_3}{n_{3R}\,n_{3I}} \tag{42}$$

These equations govern the design of ATR cells. If, for the moment, we imagine a two-phase system as in Fig. 4, then we can calculate the thickness of metal needed to attenuate the light by a factor $1/e$. For a typical metal, such a copper, with $n_{3R} = 3$ and $n_{3I} = 10$, using eqn. (42) we find that this attenuation will take place within a 500 Å film at 5 μm wavelength. It follows that if we use a metal film as the electrode, we must not make it so thick that light does not penetrate into the surrounding film/electrolyte region and this thickness will be a few hundred angstroms.

It is evident that, if $\mathrm{Im}(\hat{\xi}_3) \neq 0$, then the concept of a true critical angle becomes blurred, though it should be emphasised that the optical constants of the thin metal layer, constituting medium 2 in Fig. 5, do *not* affect the existence of critical behaviour between media 1 and 3. Even if medium 2 is strongly absorbing, provided that $\mathrm{Im}(\hat{n}_3) \ll 1$, the critical angle will be a well-defined quantity. As emphasised above, the only effect of medium 2 is to attenuate the evanescent wave before it reaches medium 3. As medium 3 does become more absorbing, we lose any sharp angle at which internal reflection rises sharply in intensity; instead, as is evident from the form of eqn. (35) and (36), there will be an increasing change in ΔR at all angles reflecting the increasing size of $\mathrm{Im}(\hat{\xi}_3)$.

2. Attenuated total reflectance (ATR)

In this method, as indicated above, the light is passed through an IR-transparent crystal at such an angle that it suffers one or more internal reflections. Either the crystal itself or, more usually, a thin metallic layer deposited upon it, is the electrode. In the early experiments, where the IR-transparent crystal was also the working electrode, a compromise had to be reached between the electrical conductivity of the crystal and its IR transparency since good conductors are also good absorbers; thus, Ge prisms were used [18]. Subsequently, very thin layers of metal (Au, Pt, or Fe) were evaporated on to the IR crystal to form the working electrode.

However, there remains a problem with this arrangement. If too thin a layer of a metal such as gold is deposited, a purple layer of high resistance is formed owing to the deposition of the gold as islands rather than as a continuous film [19]: if too thick a layer of gold is deposited, the evanescent

References pp. 74–77

wave will be severely damped. It is possible to reduce the aggregation problem with very thin films by using an ultra-thin coating of chromium [20], which is initially evaporated on to the Ge prism and which apparently aids the spread of the gold film. However, as will be evident from the analysis above, the effect of additional layers of metal will be further to reduce the magnitude of the electric field vector in the double layer and sensitivity becomes a major headache [21]. Only with the signal enhancement associated with FTIR has further exploration of the use of chromium become profitable.

The sensitivity problem may be overcome by using multiple internal-reflection prisms and/or using multiple scans to increase the signal-to-noise ratio. If the process under investigation is suitably fast *and reversible*, e.g. potential-induced double-layer reorganization [22], phase-sensitive detection may be used. In this technique, the potential at the working electrode is modulated as a square wave and the detector is then "locked-in" to the frequency of this modulation. By this means, only the signal modulated at this frequency is measured, the detector being scanned across the wavelength scale until a full spectrum is collected. The advent of Fourier

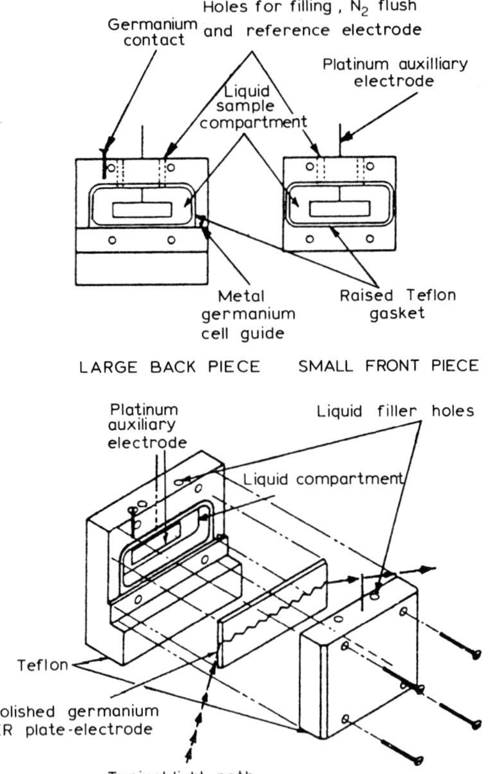

Fig. 7. Experimental arrangement of an internal reflectance spectroelectrochemical cell.

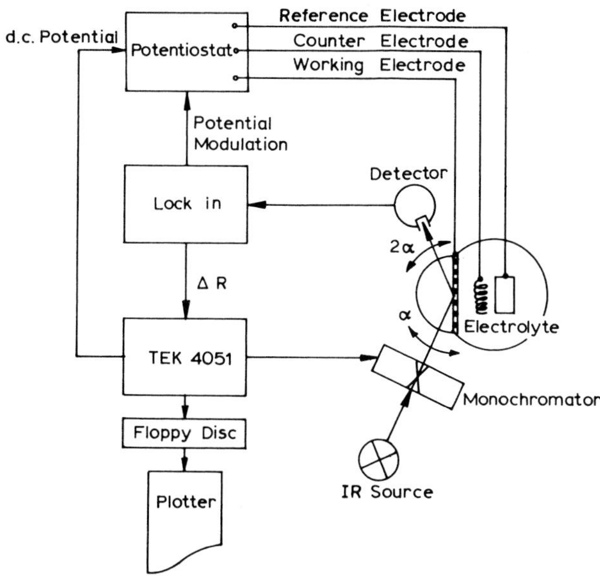

Fig. 8. Schematic representation of the experimental arrangement for attenuated total reflection of infrared radiation in an electrochemical cell.

transform infrared spectrometers with their Fellget and Jacquinot [23] advantages has increased the sensitivity of the technique with respect to single-step spectral subtraction experiments, but still cannot match the sensitivity of phase-sensitive methods (by the nature of FTIR instruments, such a "lock-in" detection system can find no simple analogue in FTIR spectroscopy [24]).

A typical multiple-reflection spectroelectrochemical ATR cell is shown in Fig. 7 and Fig. 8 shows a schematic diagram of a variable-angle single-reflection ATR cell for phase-sensitive detection.

Papers published on work carried out using spectroelectrochemical ATR techniques cover many topics [18, 20–22, 25–29], a selection of which are reviewed below.

2.1 ION RADICAL INTERMEDIATES

Figure 9 shows internal reflection spectra taken during the electrolysis of *p*-benzoquinone on a Ge prism electrode at potentials $\lesssim -0.35$ V vs. NHE showing the formation of the *p*-benzoquinone radical anion in DMSO, as reported by Tallant and Evans in 1969 [25 and references cited therein]. They also detected the anion radical of Benzil, as well as unassigned reduction products of acetophenone and benzophenone. Because of the relatively low sensitivity of the technique, rather high concentrations of reactants had to be used in order to be able to detect any products ($\gtrsim 10$ mM). An additional problem was the relatively long settling times before steady-state concentra-

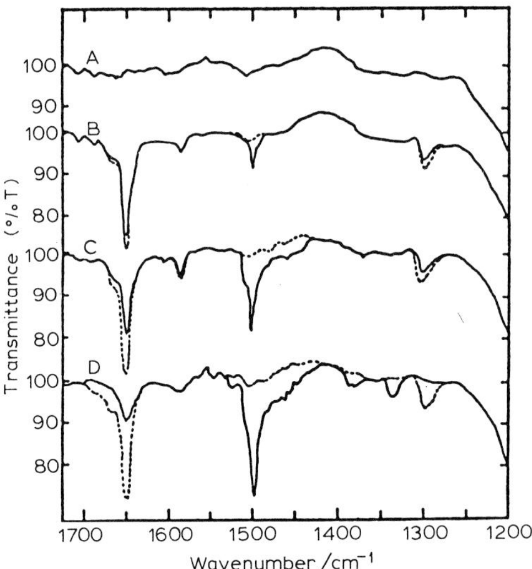

Fig. 9. Internal reflection spectra during electrolysis of *p*-benzoquinone. solid lines: A, Base line, 0.2 M TBAP in DMSO; B, 0.2 M *p*-benzoquinone during electrolysis at −0.7 V vs. PQRE; C, electrolysis at −1.1 V; D, electrolysis at −2.0 V. Broken lines: B, C, D, 0.2 M *p*-benzoquinone before electrolysis.

tions were reached (~60 s) owing to the resistance of the Ge place and the unequal potential distribution across its surface: the reference electrode was placed so as to measure the potential of the centre of the Ge electrode.

2.2 MOLECULAR ADSORPTION

Hatta et al. [29] used the enhancement of surface electromagnetic fields discussed above [30, 31] to study thiocyanate adsorbed at an Ag electrode/ aqueous electrolyte surface. The silver electrode was in the form of a 180 Å film deposited on to the base of a hemicylindrical Ge prism. The IR measurements were performed at an angle of incidence of 70°, at which the maximum enhancement is expected [32]. Throughout the experiments reported, they failed to observe the band due to solution-free $^-$SCN, indicating the very short penetration depth of the evanescent wave into solution, which can be calculated to be ~1000 Å. Figure 10 shows IR-ATR spectra of the C≡N stretching vibration of SCN^- measured at various potentials (vs. Ag/AgCl). Figure 10(a) was interpereted as desorption of the $^-$SCN as the potential of the Ag electrode is rendered more cathodic. (The pzc of Ag under the conditions used is ~ −0.9 V vs. Ag/AgCl and hence the electrode is positively charged at all the potentials shown in Fig. 10.) As the potential is stepped to the original more anodic value, the SCN^- is re-adsorbed. The spectra in Fig. 10(b) were obtained on a fresh electrode by stepping the potential from −0.2 to 0.1 V. They attributed the observed increase in band intensity and frequency between −0.2 and 0 V to the increase in positive charge on the Ag surface. When the potential passed through 0 V, a low-frequency tail app-

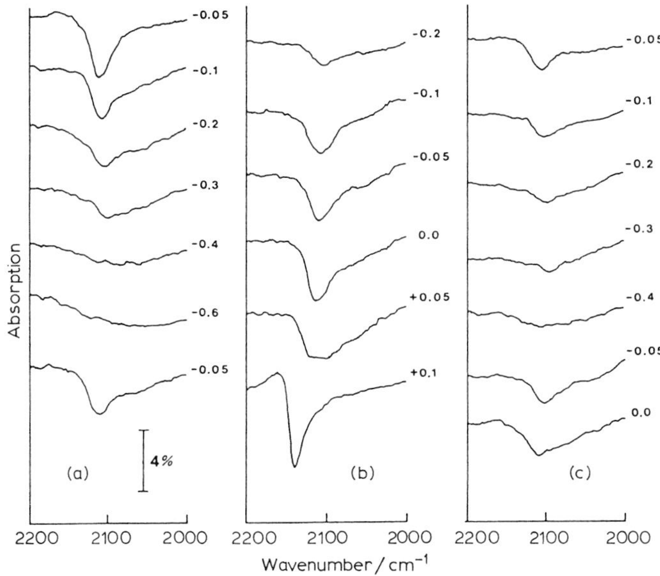

Fig. 10. ATR spectra of the CN stretching vibration of thiocyanate at an Ag/electrolyte interface measured at indicated electrode potentials (V) vs. Ag/AgCl. The angle of incidence of radiation is 70°. Electrolyte: (a), (b) 0.1 M $NaClO_4$ + 0.01 M NaSCN; (c) 0.1 M NaCl + 0.01 M NaSCN. Thickness of Ag electrode: (a), (c) 18 nm; (b) 17 nm. Measurement order is top to bottom in each case.

eared, whilst the peak frequency remained constant, while at 0.1 V, an intense band at 2141 cm^{-1} appeared which was assigned to the ν_1 vibration of AgSCN, formed as a relatively thick film on the silver electrode surface. Figure. 10(c) shows the potential dependence of the adsorption in the presence of Cl^- and shows the remarkable inhibition of the adsorption of SCN^- by the specific adsorption of Cl^-. An interesting point was made that the potential dependence of the weaker SCN^- feature in NaCl nearly paralleled that in $NaClO_4$, indicating the relative unimportance of specific interaction between the adsorbed SCN^- and Cl^- ions.

2.3 CORROSION

The oxidation products of a thin film of Fe electrode deposited on Ge (4–25 nm thick) in 5 M KOH were detected by Neugebauer et al. [27]. The experiment was carried out using an FTIR spectrometer. Figure 11 shows the potentials at which the spectra in Fig. 12 were taken; the base potential used by the authors at which the reference spectrum was taken was -1400 mV. As can be seen from the spectra, no real changes occur until a potential of ~ -900 mV. The authors interpreted the spectra in terms of the formation of the $HFeO_2^-$. Thus, at potentials of -860 to -850 mV, the $HFeO_2^-$ anion occurs as a soluble intermediate of the iron oxidation in alkaline solutions

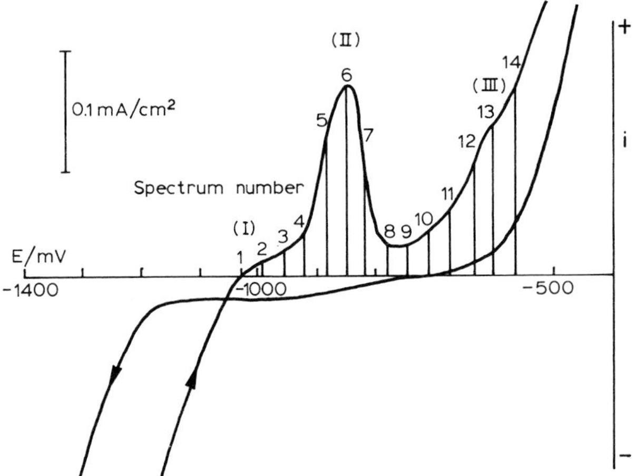

Fig. 11. Typical cyclic voltammogram of an iron layer on a germanium crystal in 5 M KOH. Scan rate, 2 mV s^{-1}. Limits of the sweep −1400 to −400mV, starting at −1400 mV. Reference electrode: Pt wire.

according to

$$Fe + 2OH^- \rightleftarrows HFeO_2^- + H^+ + 2e^- \qquad (43)$$

The hydrogen ions produced by reaction (43) react with carbonate, present as an impurity in aqueous KOH, to produce CO_2 (antisymmetric stretch at 2370 cm^{-1}). The $HFeO_2^-$ ion then hydrolyses to give solid $Fe(OH)_2$ at the

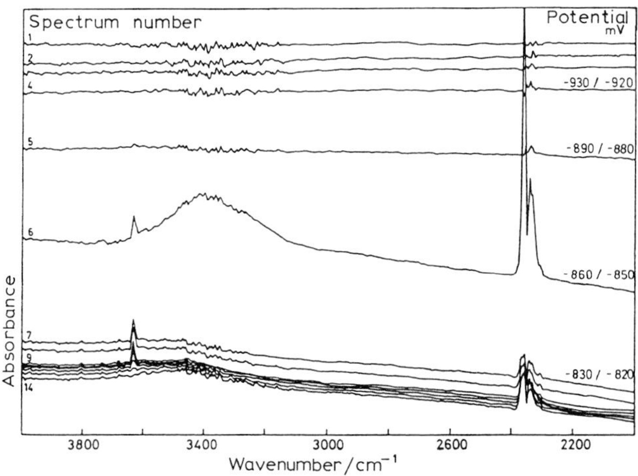

Fig. 12. Multi-internal reflection IR spectra of an iron layer on a germanium crystal at various potentials. Scan rate, 2 mV s^{-1}. All spectra are on the same absorbance scale. The spectra number indicates the potential at the end of the sampling of 10 single interferograms in a time range of approximately 5 s (corresponding to a potential range of 10 mV). The time between two samplings is necessary for the storage of the interferograms in the computer equipment.

surface (the former giving rise to the broad hydrogen-bonded OH stretch at 3400 cm^{-1} [33, 34], the latter to the "free" OH stretch at 3630 cm^{-1} [33]). In the course of the reaction, the authors noted that the optical transmission of the iron layer increases. They attribute this to the dissolution of the iron layer.

2.4 ELECTROPOLYMERISATION

In 1984, Neugebauer et al. [35] conducted an in-situ FTIR-ATR investigation into the electropolymerisation of 3-methylpolythiophene using, as a working electrode, a thin gold layer deposited on a germanium internal reflection crystal. The actual electropolymerisation process itself was not followed; instead, the IR spectra of the neutral and oxidised forms of the polymer were obtained. The 3-methylthiophene was polymerised as the oxidised form on to the gold-coated Ge crystal from a 0.4 M solution of the monomer in acetonitrile by holding the potential at -1.5 V vs. SCE. Once the polymer was sufficiently thick, it was reduced to its neutral state by switching the potential to 0 V. Poly(3-methylthiophene) is red in its neutral state and blue when oxidised to its metallic state; the two states are thought to be

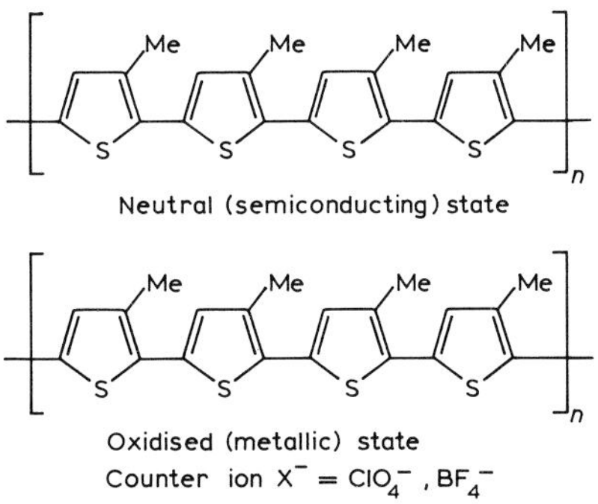

Neutral (semiconducting) state

Oxidised (metallic) state
Counter ion $X^- = ClO_4^-$, BF_4^-

Figure 13 shows ATR-FTIR spectra recorded during the oxidation of a 3-methylthiophene film, Fig. 14 a spectrum of the oxidised state, and Fig. 15 a current density curve of the polymer film on a germanium crystal with an evaporated gold layer showing the potential regions in which spectra were taken.

It was suggested that the strong intensities of the bands growing during the oxidation process were due to a vibronic intensity enhancement effect due to, for example, the coupling of the skeletal backbone vibrations with the π-electron charge oscillations along the chain [37]. The bands in the spectra of the two forms were largely unassigned, although features at 1100 or 1060 cm^{-1} could be assigned to the influx of counterion into the near-electrode region (ClO_4^- and BF_4^-, respectively) on oxidation and the extremely broad band beginning at 1600 cm^{-1} and extending to 4000 cm^{-1} was assigned to electronic transitions in the metallic state due to the presence of free carriers [36].

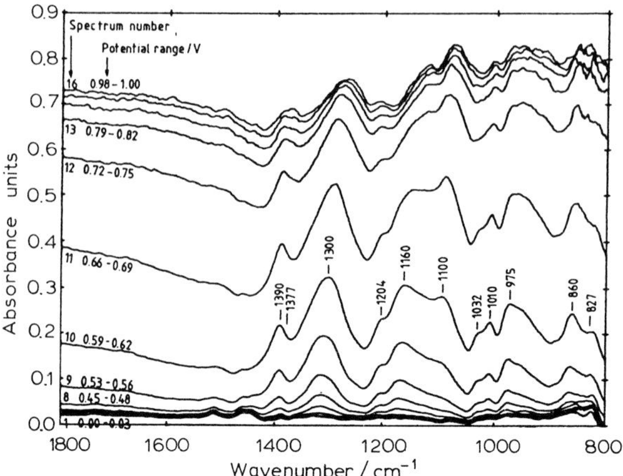

Fig. 13. ATR-FTIR spectra recorded during the oxidation of a 3-methylthiophene polymer in contact with an electrolyte solution containing tetrabutylammonium perchlorate. The spectrum numbers correspond to the numbers in Fig. 15. The spectra are smoothed but not baseline corrected. The spectrum of the electrolyte solution was subtracted.

2.5 THE DOUBLE LAYER

The enhancement of the electromagnetic fields at the surface of an internally reflecting crystal arise via the excitation of surface plasmon polaritons (SPP) [30] by the incident IR light. These SPPs are collective electronic excitations at metal surfaces and, theoretically, provide a sensitive probe of the optical properties of the interfacial region via resonance with the in-

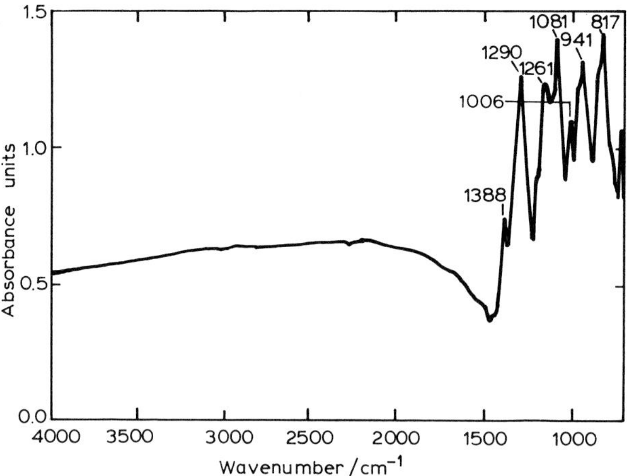

Fig. 14. ATR-FTIR spectrum of the oxidized form of 3-methylthiophene polymer on the surface of a reflection element in contact with an electrolyte solution containing tetrabutylammonium perchlorate. The smoothed spectrum originates from 200 added interferograms. The spectrum of the electrolyte solution was subtracted.

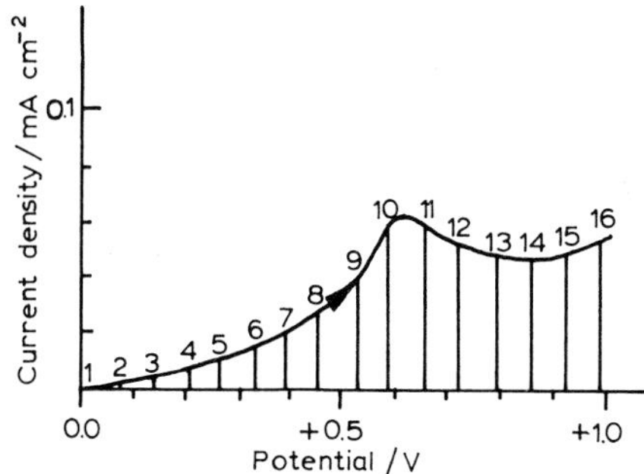

Fig. 15. Potentiodynamic current density/potential curve of a 3-methylthiophene polymer film on a germanium crystal with an evaporated gold layer. Electrolyte solution as in Fig. 13, sweep range 0.1–1.0 V, sweep rate 0.002 V s^{-1}. The numbered vertical bars indicate the potentials at which ATR-FTIR spectra (each spectrum originating from 20 interferograms) were recorded.

cident IR beam. In general, the use of SPP excitation in electrochemical systems is restricted to a spectral range where the electrolyte is optically transparent [38], since the presence of an absorbing medium is expected to dampen the polariton fields, and thus the optical resonance, and thereby reduce the sensitivity of the method. However, Neff et al. [22] and Lang et al. [20] have reported studies intended to assess the magnitude of the damping when a strongly absorbing aqueous electrolyte was employed in investigations aimed at using SPP excitation to probe the vibrational properties of the water molecules in the interfacial region. As reflectance changes of only a few units $\times$ 10^{-4} ($\Delta R/R$) were anticipated, a phase-sensitive detection system was employed (solvent re-organisation etc. was anticipated as being very fast and reversible).

In order to achieve SPP excitation, the light must be incident at the inner surface of the crystal near the critical angle and be *p*-polarised [39]. The applied potential was modulated by $\pm$ 50 mV either side of the point of zero charge of the evaporated/partially crystallised gold layer on a quartz internal reflectance crystal.

Both the intensity and the frequency of the –OH stretch in water are critically dependent upon the extent of hydrogen bonding to the immediate environment [40] and thus is a sensitive probe of the various types of water existing at the electrode–electrolyte interface. However, there are additional physical effects that may be detected by the phase-sensitive IR-ATR technique such as those derived from the variation of the electric field at the surface; these electroreflectance effects are primarily due to changes in the surface electron density.

The authors surmised that the infrared properties of water in the solvation shell of different anions and cations might be very different. Thus, they decided to use the proton as a direct means of monitoring ionic excess concentrations in the double layer via its strong hydrogen bonding to water. By using only a small potential modulation about the pzc of the gold layer, such that the electric field across the interface was very low, they deduced that a significant proportion of the reflectivity change arising from the potential modulation must be due to changes in proton concentration (and thus the double-layer composition) providing the electrolyte anion was chosen as having very small IR absorption. The electrolyte chosen was 0.5 M H_2SO_4 in D_2O, with small amounts of H_2O added for the investigation of the –OH stretch in DHO; H_2O gives problems due to Fermi resonance with the overtone of the HOH bending mode and dynamic dipole coupling [40(a)].

Figure 16 shows differential infrared ($\Delta R/R$) spectra taken for different concentrations of H_2O in D_2O.

The structure between 3000 and 3700 cm^{-1} was assigned by the authors to the –OH stretch of HDO molecules in the double layer and the non-linear growth of the intensity of this structure with the concentration of H_2O was ascribed to the increased damping of the optical resonance in the gold film due to the increasing amount of absorbing medium near its surface. The broad background in the figure was thought to be due to changes in the electron density at the metal surface and to changes in the double layer refractive index. Figure 17 shows difference spectra with this background subtracted.

The authors used these difference spectra as a basis for comparison with their calculations of the contribution to $\Delta R/R$ which may result from the variation of the proton excess in the double layer. Sulphate ions have only a minor influence on the –OH absorption spectrum of water [41] and the authors decided that the HSO_4^- ion present at the low pH employed in the experiment would not behave much differently. Since the same is true for the Cl^- ion, the authors used optical data which was available for different concentrations of HCl [42] in their model calculations. The imaginary part of the complex refractive index, k, was plotted [22] for HCl solutions in H_2O in the spectral range of interest ($\sim 3000\,cm^{-1}$). The effect of increasing

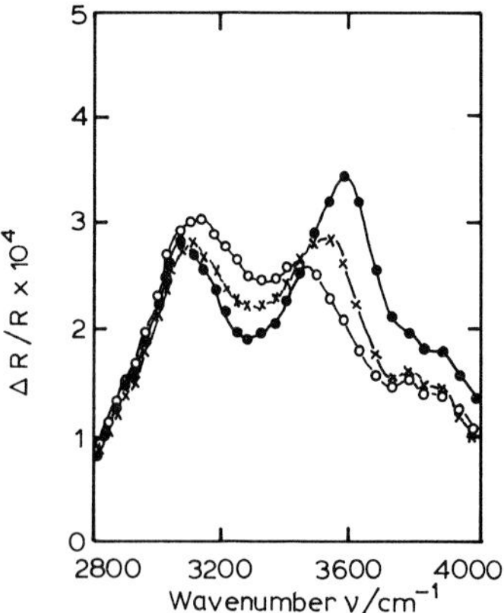

Fig. 16. Effect of the addition of 0.5% (□), 1.0% (×), and 3.0% (○) H_2O to 0.5 M D_2SO_4 on the differential reflectance spectrum of a gold film, modulated around the potential of zero charge.

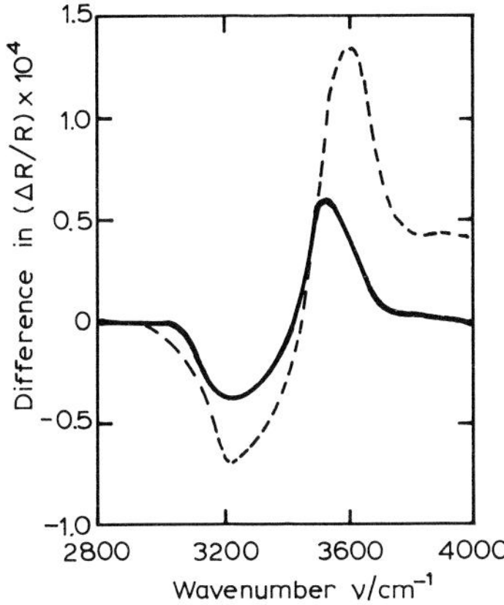

Fig. 17. Difference spectra of $\Delta R/R$ taken from the data of Fig. 16. ——, 1–0.5% H_2O; – – –, 3–0.5% H_2O.

proton concentration was to reduce absorption at higher wavenumbers and increase it at lower wavenumbers with respect to pure water. Qualitatively, the result is the same as in Fig. 17, assuming that the interfacial concentration of protons increases as the potential is taken cathodic of the pzc due to electrostatic attraction. The effect of increasing the proton concentration is to enhance the formation of strong hydrogen bonds to water molecules, which, in turn, increases the intra-molecular force constant for the – OH stretch.

Data on the influence of H^+ on HDO absorption in the double layer were not available, and the authors had to make several approximations: one was to reduce the k values for the H^+/H_2O data [22] according to the percentage of H_2O in solution and another was to assume that the thickness of the double layer was 4 Å. The (interfacial) acid concentration was assumed to vary between 0 and 2 M [43] as the potential was stepped from 450 to 350 mV (vs. SCE). Using these and several other approximations, a modulation ($\Delta R/R$) spectrum was calculated for the difference between a solution with 1 and 0.5% H_2O. The result is shown in Fig. 18. The calculated result was 5 times smaller than the experimental, but nevertheless bore some qualitative resemblance to the results in Fig. 17 and the authors were justifiably encouraged. This work does suggest that the IR-ATR technique may provide a probe of ion concentration in the double layer.

The slightly later paper [20] investigated the effect of increasing the solution absorption coefficient on the damping of the surface plasmon polariton and the effect of increasing the angle of incidence of the internally incident light on the differential reflectance spectra of the $H_2O/D_2O/H_2SO_4$ system. The authors conclude that low absorption coefficients in the liquid are required for the application of IR-ATR spectroscopy to the study of the metal–electrolyte interface; unfortunately, using isotopic mixtures of H_2O

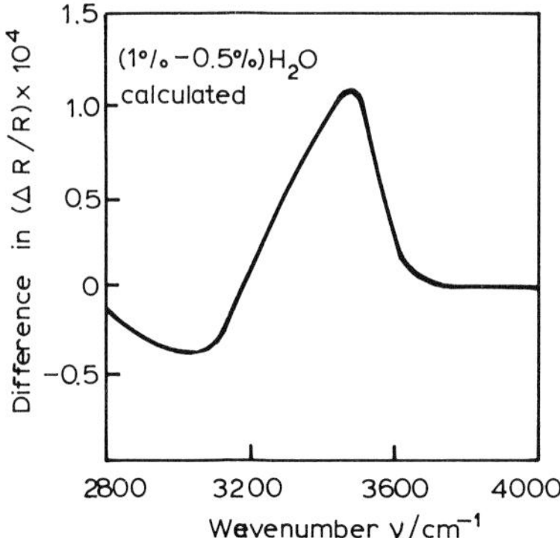

Fig. 18. Calculated difference spectra in $\Delta R/R$, based on the four-phase model. Angle of incidence, 70°. 1–0.5% H_2O.

and D_2O to fulfil this condition reduces the number of chromophores at the surface such that their contribution to $(\Delta R/R)$ is swamped by other effects.

Thus, the first experimental approach used to record in-situ IR spectra of the near-electrode region took advantage of total internal reflection to minimise the effective path length in the liquid phase and so reduce the large attenuation imposed on an IR beam by all commonly used solvents. The techniques has both advantages and disadvantages: one disadvantage is that, because of the high surface selectivity (i.e very low bulk penetration of the evanescent wave), high concentrations of absorber are required. In addition, if an uncoated prism is employed, the method of data collection is limited by the electrical characteristics of the electrode (i.e. it must take into account response times and IR effects across the crystal, etc.). However, the advent of high-throughput FTIR spectrometers and the development of methods to ensure the deposition of thin metallic coatings [21] has meant that metal-coated electrodes can be used almost routinely with sufficient sensitivity to obtain workable spectra. Thus, the technique of in-situ IR-ATR spectroscopy has much to offer.

3. Transmission IR

This method was also investigated early in the development of in-situ IR electrochemistry owing to its relative simplicity. However, the use of conventional transmittance IR methods is very restricted in electrochemical applications where most chemical change occurs within the Helmholtz layer, only a few molecules thick. Hence, the transmission approach is relatively insensitive to changes occurring in the near-electrode region. The detection of chemical changes in solution can be facilitated by using several optically transparent electrodes in multi-layer units. Such a method, in which gold minigrid electrodes [44] were used as spacers to give the necessary thin solvent layer, was used with some success to monitor the reduction of ninhydrin [45]. In the large majority of cases, however, in-situ IR electrochemical studies have employed either internal or external reflectance methods.

4. External reflectance

External reflectance techniques employing potential modulation were first developed by Bewick and Kunimatsu [46] using a dispersive IR spectrometer. Subsequently, Davidson et al. [47] reported similar experiments with Fourier transform IR. In both cases, a thin-layer cell was employed to overcome the strong solvent absorption. The remaining problem of detecting the small number of absorbing species in the near-electrode region was overcome by using potential modulation and signal processing regimes, though the detailed experimental set-up obviously depends on the type of IR instrument used. This in turn is chosen according to the degree of reversibility and relaxation time of the electrochemical system to be studied.

In 1974, Clark et al. [48] published an assessment into the feasibility of studying adsorbed species at metal electrodes in-situ by infrared spectroscopy, whether via internal or external reflectance or transmittance, "... would be impracticable unless improved measurement sensitivity could be obtained either by the use of air-cooled thermal and photo-electric detectors or lasers operable in the IR region". However, in 1979, Pons [49] in his doctoral thesis, reported that the UV–VIS specular external reflectance technique, which had been demonstrated to be effective in the study of a variety of electrochemical systems [50], could be utilised in the infrared. The first papers by Bewick et al. [46, 51], describing the new technique appeared in the literature in 1980. As had long been accepted, the problem of saturation absorption of infrared light by the solvent could be reduced by using a thin layer technique. Exact calculations using the Fresnel reflection equations [52] show that the radiation absorbed by an aqueous solution layer of the order of 1 μm thick in contact with a reflective electrode will not lead to solvent saturation effects; in addition, the change in absorbance brought about by forming a monolayer of adsorbate on the electrode is sufficient to be measured by the technique available. This effect is shown in Fig. 19.

The electrochemical cell design employed in thin-layer reflection-absorption studies is shown in Fig. 20.

The technique developed by Bewick for use with a dispersive IR spectrometer is termed electrically modulated infrared spectroscopy (EMIRS) and is essentially a direct development of the UV–VIS specular reflectance technique [50], modulated specular reflectance spectroscopy (MSRS). As in MSRS, radiation is specularly reflected form a polished electrode surface while the electrode potential is modulated with a square wave between a base potential and the working potential at which the process of interest occurs; the wavelength range of interest is then slowly scanned. Only that

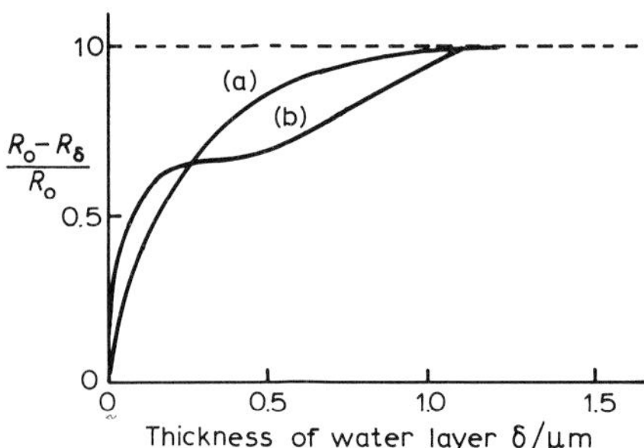

Fig. 19. Normalized absorbance vs. thickness plot for a layer of water in contact with a platinum reflector measured using an infrared beam at the wavelength of the maximum absorptin of the O–H stretching mode and incident at an angle of 45°, calculated using (a) Beer's Law and (b) the full Fresnel reflectance equations.

component of the optical signal having the same modulation frequency and phase as the applied potential square wave is amplified, hence increasing the sensitivity up to the level necessary for the detection of potential-induced absorption changes of submonolayer quantities of adsorbate (typically 10^{-4} absorbance units). A high-intensity scanning monochromator is employed for the spectral scan and averaging of successive scans is used to increase the S/N ratio [23] (see Fig. 21).

This intensity difference, ΔR, observed in an EMIRS experiment may result from several sources, e.g. electroreflectance effects arising from changing electron densities at the surface of an electrode under the influence of the applied potential, changes in the amount of adsorbed species, or re-organisation of the double layer, etc. Spectra arising from changes in the adsorbed layer may be derived from

(a) a change in the amount of species present with potential-induced absorption or faradaic effects,

(b) a potential-dependent shift of the IR band frequency, or

(c) a potential-dependent change in the orientation of the adsorbed species.

The spectra obtained are presented as plots of $(\Delta R/R)$ vs. $\tilde{\nu}$, the $(\Delta R/R)$ ratio removing the wavenumber dependence of the (unchanging) bulk solvent absorption, instrumental energy throughout, detector sensitivity, and electronic response. Thus, the signal processing causes the single-beam spectrometer to act as a "pseudo"-double beam instrument [53]; the sample and reference beams being separated in time rather than space. In an experimental paper, Bewick et al. [53] have given examples of possible IR reflectance spectra and the corresponding difference spectra aising from them*.

Because an EMIRS experiment involves a relatively fast potential mo-

* It must be remembered that both EMIRS and SNIFTIRS are *difference* techniques and thus the spectra they give rise to are not simple transmittance spectra.

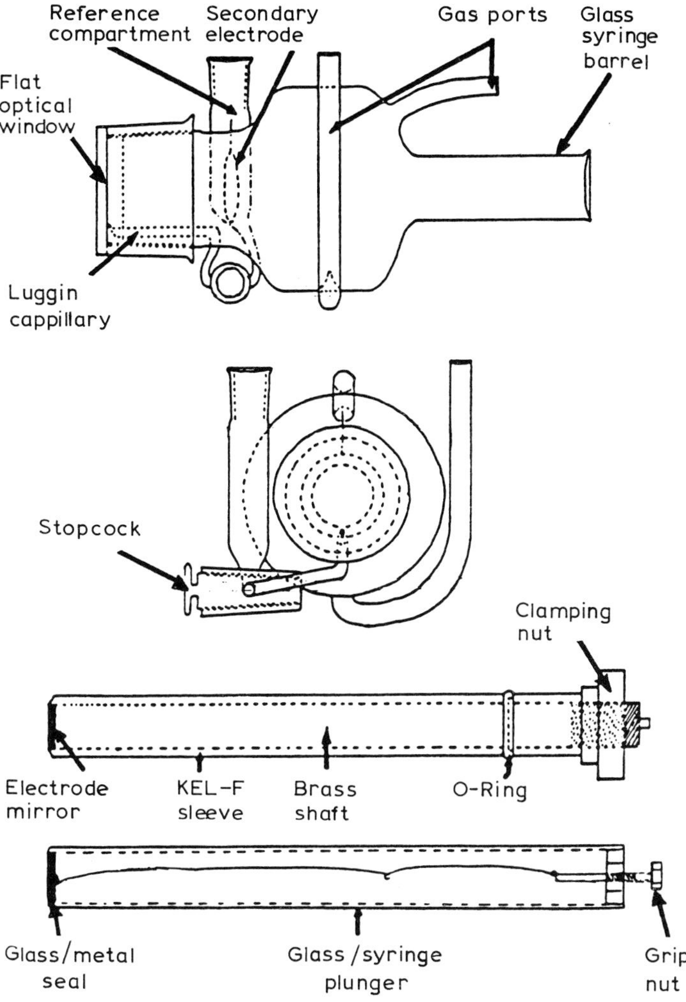

Fig. 20. Design of the spectroelectrochemical cell used for the EMIRS and SNIFTIRS measurements.

dulation (10 Hz [54]), the type of electrochemical systems that can be investigated by the technique are confined to those having very fast relaxation times and that are totally reversible, otherwise the spectra obtained may be extremely difficult to interpret. Thus, Bewick et al. and later Kunimatsu, confine their EMIRS experiments [55, 56] to the investigation of double-layer structure or adsorption. With apparently one exception [51], faradaic processes are effectively too slow for the EMIRS technique to be used with any advantage in their study. Examples taken from these applications are given below.

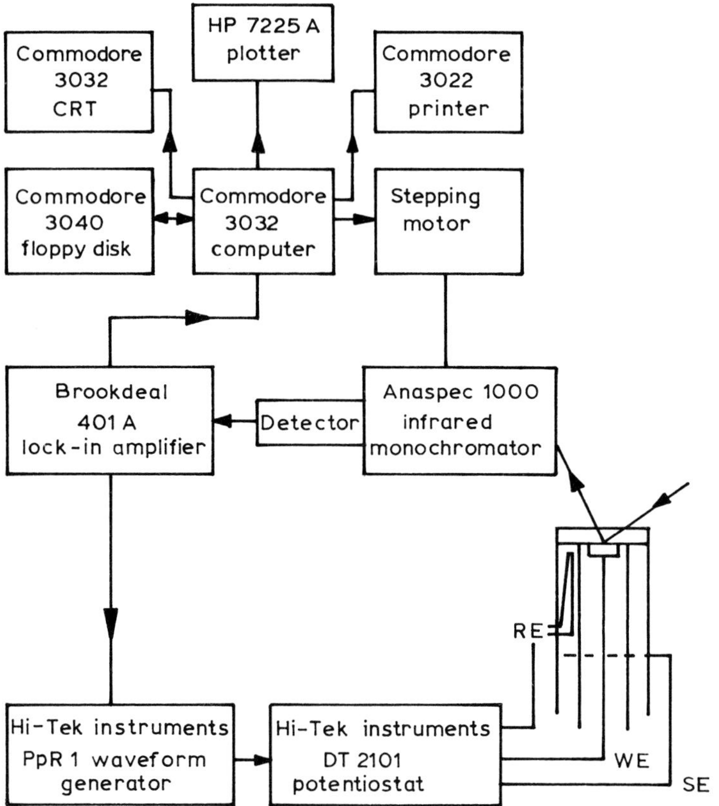

Fig. 21. Block diagram of the complete spectrometer system used in EMIRS.

4.1 AN ELECTRON-TRANSFER REACTION IN THE THIN-LAYER

In an early paper, Bewick et al. [51] reported the spectrum of the thianthrenonium cation radical obtained from the in-situ oxidation of thianthrene (0.01 M) in acetonitrile at 1.2 V vs. Ag/Ag^+. The spectrum was observed on modulating the potential between 0 and 0.2 V and was described as closely comparable with that of an authentic sample of thianthrenonium perchlorate obtained with a conventional spectrometer.

4.2 ADSORPTION

Bewick et al. [51] have also reported one of their first adsorption studies using EMIRS; the electrochemical adsorption of indole on to a platinum electrode. Modulating the potential of a platinum electrode between −1.1 and 0 V vs. Ag/Ag^+ (0.01 M) in contact with 0.05 M n-Bu_4NBF_4 in acetonitrile containing 15 mM indole causes the adsorption of the indole at 0 V, which is then stripped off at the −1.1 V base potential. The EMIRS spectrum obtained with a 10Hz modulation frequency is shown in Fig. 22.

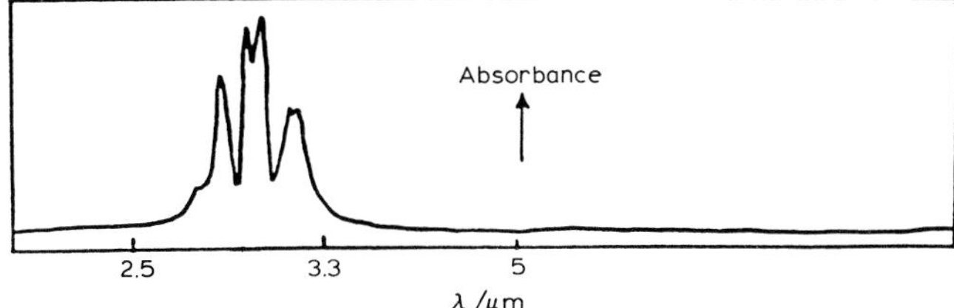

Fig. 22. EMIRS spectrum of 15 mM indole in 0.05 M n-Bu_4NBF_4/acetonitrile. Modulation limits − 1.1 to 0 V vs. Ag/Ag^+; modulation frequency 10 Hz.

The above spectrum was interpreted by the authors after first noting that the EMIRs spectrum yielded the difference between the absorbance of free indole in solution (at − 1.1 V) and adsorbed indole (at 0 V) since the modulation frequency was too high for appreciable amounts of additional indole to diffuse into, or out of, the thin layer during the adsorption/desorption step. Thus, since the adsorption was perturbing only the N–H stretching region without any visible effect upon the C–H regions of the spectrum, the authors concluded that the bonding to the Pt was via the nitrogen. Furthermore, given that the indole molecule is planar, it cannot be adsorbed flat upon the surface since the surface selection rule [57, 58] would not permit any absorption to occur.

4.3 ELECTRO-OXIDATION OF SMALL ORGANIC MOLECULES

A major focus of more recent studies on adsorption at metal electrodes has been the investigation of the mechanism of electro-oxidation of organic fuels (methanol, formic acid, formaldehyde, etc. [55, 56]) and the electro-reduction of carbon dioxide. The former type of reaction is important in the context of the development of fuel cells; a major problem has been the poisoning of the anode by carbon fragments and mechanistic insights are urgently needed. In the latter case, the development of CO_2 sensors has a high priority.

The electrocatalytic oxidation of small organic molecules at noble metal electrodes has been intensively investigated over a period of many years, both by conventional electrochemical methods [59], including radiotracer measurements, and by an increasing battery of non-conventional techniques. The metal anodes in such reactions are rapidly poisoned by the formation of strongly adsorbed species that block the catalytic surface, and the identification of these species has become a major objective.

Coulometric work suggested either a $\geqslant$(COH)ads surface species occupying three surface sites [59(c)] or a –(CO)ads species [60] as the most likely blocking agents. The same poison appears to be produced from a range of organic substrates and on a range of noble metal anodes*. EMIRS, IR–LSV, and IRRAS investigations of the adsorbed fragments formed in these reac-

* Except formic acid on Pd for which electrochemical measurements show that no strong poisons are formed [61].

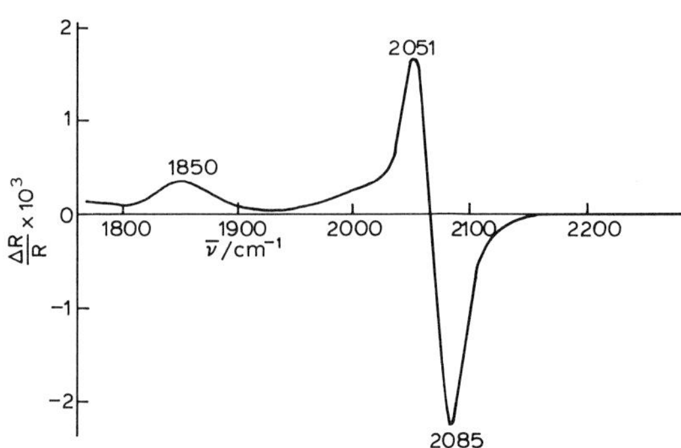

Fig. 23. Reflectance spectrum from Pt/1 M H_2SO_4, 0.5 M methanol. Modulation from 0.05 to 0.45 M (NHE) at 8.5 Hz. Scan rate 0.0127 μm s^{-1}.

tions have now begun to form a coherent picture, which we will review in outline.

The first report in the literature of spectra taken, using the EMIRS technique, of the electrosorbed fragments from the electrocatalytic oxidation of methanol on Pt was in 1981 by Beden et al. [62]. Figure 23 shows the spectrum obtained over the range 1800–2300 cm^{-1} by modulating the potential of the reflective platinum electrode, immersed in 0.5 M MeOH/1 M H_2SO_4, between +0.05 and +0.45 V vs. NHE. Figure 24 shows the dependence of the spectra on the modulation amplitude. As with almost all EMIRS (and SNIFTIRS) spectra, features having a positive $\Delta R/R$ correspond to more absorbing species present at the (usually lower) base potential.

The weak band at 1870 cm^{-1} was ascribed by the authors to a bridged (Pt)(Pt)>C=O species, the intense bipolar band centred at ~2070 cm^{-1} to linearly bonded –C≡O. The bipolarity of the band was identified as a potential-dependent frequency shift; the energy of the C≡O stretch was higher at more positive potentials. The spectra in Fig. 24 were reported to indicate a

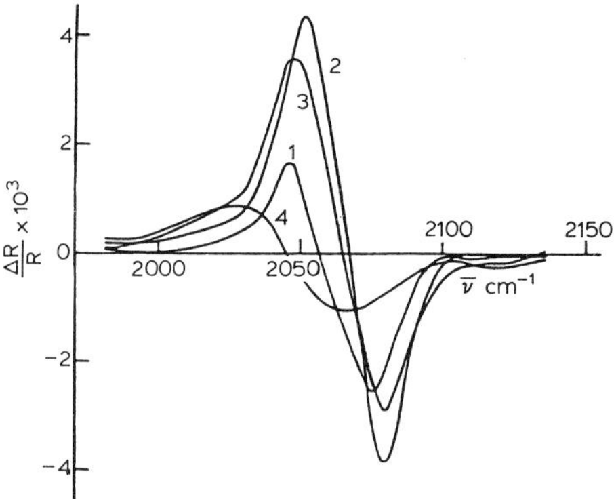

Fig. 24. Dependence of the spectra on the modulation amplitude. (1) 0.05–0.30 V, (2) 0.05–0.55 V, (3) 0.05–0.80 V, (4) 0.05–0.95 V. Scan rate 3.18 × 10^{-3} μm s^{-1}.

decrease in the coverage by adsorbed species on entering the region of sustained methanol oxidation (> 0.7 V), with the proviso that "the measurements were non-steady state, with a perturbation at 8.5 Hz, and the dissociative desorption of methanol is slow." The major conclusions of this first paper were that the dominant adsorbed species (poison) was Pt–C≡O, with also some $\begin{matrix}(Pt)\\ \end{matrix}\!\!>\!C{=}O$ [(Pt)/(Pt) bridged C=O], particularly at higher potentials, and that there was no spectroscopic evidence for $(Pt)_3{\equiv}C{-}OH$ under the conditions of the experiment. These conclusions stand up well, even when compared with those derived from the combined results of all the later in-situ IR studies [55, 56], differing only in the identity of the species absorbing at 1850 cm^{-1}. The conclusions of this paper were reinforced by a later report [63] of investigations using the IRRAS technique which established the dependence of the C≡O band wavenumber on electrode potential as the primary source of the bipolar band in the EMIRS spectrum, as will be seen below. This latter paper is itself of some note since it is the first to use the acronym EMIRS.

In 1982, Beden et al. [64] published an investigation into the nature of CO_2 reduced on Pt at potentials within the hydrogen region ($0 \leqslant E \leqslant 150$ mV vs. NHE, 1 M H_2SO_4). After saturating the solution with CO_2 for 30 min at open circuit, the cell was switched in at +350 mV (vs. NHE). The potential was then modulated between +350 and +550 mV and a spectrum collected. Several other spectra were collected at varying modulation widths by keeping the positive modulation limit constant at +550 mV whilst lowering the negative limit. The authors reported observing no spectra between 1700 and 2100 cm^{-1} until the lower limit was within the region of weakly adsorbed hydrogen [65]. Figure 25 shows spectra obtained at modulation limits of +550 → + 250 mV and +550 → + 150 mV; the former outside and the latter inside the hydrogen adsorption region. The authors observed that the spectra in Fig. 25 were consistent with the earlier-held view [66] that the reduction of CO_2 to the linearly bonded (CO)ads species takes place by reaction with adsorbed hydrogen. The spectrum taken between +550 and +150 mV shows the bipolar Pt–C≡O band centred at ~2060 cm^{-1}, but no feature at 1860 cm^{-1}. Switching in the cell at +50 mV after CO_2 saturation at open circuit and modulating to successively higher potentials gave the spectra shown in Fig. 26. These spectra showed both the features previously observed in the electro-oxidation of MeOH on Pt; however, the band centred at 1860 cm^{-1} was now bipolar and tentatively ascribed by the authors to a –CO species (*not* $Pt_3{\equiv}C{-}OH$) adsorbed at a *three*-coordinate surface site.

The variation in intensity and widths of the positive and negative peaks centred around 2060 cm^{-1} were interperted in terms of a potential-dependent shift in band position and energy of linearly adsorbed C≡O at constant

References pp. 74–77

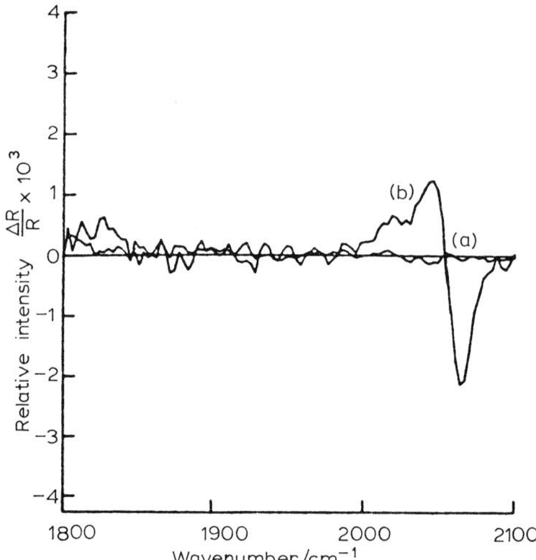

Fig. 25. EMIRS spectra for modulation between (a) + 250 and + 550 mV, (b) + 150 and + 550 mV. $\Delta R/R$ is proportional to absorption and a positive sign corresponds to greater absorption at the more negative potential.

coverage. The authors assumed that, over the time-period of the experiment (30 min to 10 h), the coverage would reach a steady-state and not show an appreciable modulation in step with the imposed 8.5 Hz square-wave modulation of the electrode potential. The paper concluded that

(a) the reduction of CO_2 involves reaction with adsorbed hydrogen and

(b) the strongly adsorbed species responsible for poisoning the electro-

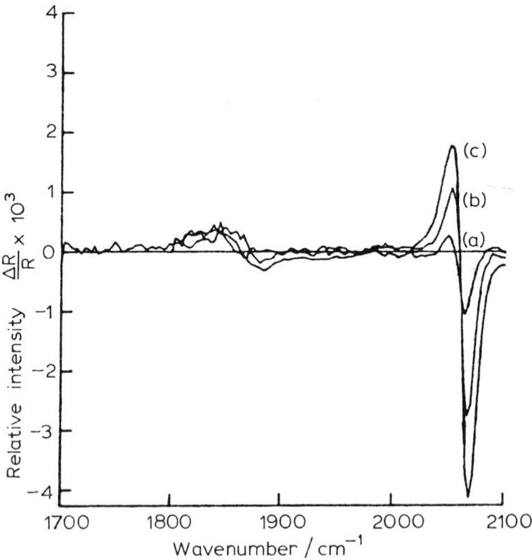

Fig. 26. EMIRS spectra for modulation between (a) + 50 and + 250 mV, (b) + 50 and + 350 mV, and (c) + 50 and + 450 mV.

catalytic oxidation of many small organic molecules can be formed via the reduction of CO_2 and is probably a CO species in a three-coordinated site on the metal surface, giving rise to an IR feature at or around 1860 cm^{-1}: it is not $\geqslant$C–OH.

(c) The latter feature appears to be formed more readily via CO_2 reduction than by direct electrosorption of CO, HCOOH, or HCHO under conditions that exclude formation of CO_2.

This latter point is an important one in view of the fact that adsorbed hydrogen is almost certainly one of the transient species formed by electrosorption of hydrogen-containing small organic (fuel) molecules, and CO_2 will also be present in the electrolyte after a few electrochemical measurements.

(d) Essentially, the same poisoning species are formed by the reduction of CO_2 as in the electro-oxidation of HCHO, CH_3OH etc.

The authors' conclusions were somewhat tentative with respect to the behaviour of the feature at 1860 cm^{-1} and perhaps in any detailed treatment of the potential dependence of adsorption.

This is, of necessity, the case since EMIRS is a difference technique and, as such, does not allow any quantitative investigation into the adsorption process, i.e. it is not possible to compare any integrated band intensities since the separate "absolute" spectra of the species present at the two potentials cannot be obtained. Nonetheless, the paper clearly does show the versatility of the technique in giving an insight into an electrochemical system, aided by the high degree of energetic control afforded by potential modulation.

As indicated above, the *quantitative* information that can be obtained from the EMIRS technique is somewhat restricted by its "difference" nature and it would be extremely useful to determine the "absolute" spectrum at a given potential. Kunimatsu [67, 68] has reported the development of a technique, which is a variation on EMIRS, called linear sweep voltammetry (IR–LSV), that indirectly provides just such an absolute spectrum. The method uses absorbance vs. potential measurements at a fixed wavenumber while linearly sweeping the electrode potential. Both of the IR–LSV papers to date report work on the electro-oxidation of methanol in $HClO_4$, concentrating upon only that area of the spectrum corresponding to linearly bound C$\equiv$O. Figure 27 shows examples of the change in $\Delta R/R$ with electrode potential under a fast linear cyclic voltammetric condition. The absorbance at a particular potential is taken with respect to the plateau region which lies either to a higher or lower potential of the peak depending upon the wavenumber concerned [i.e. (a)–(f) in Fig. 28]. The validity of the assumption that the absorption at this "floating" plateau was zero was tested by the author by comparing a calculated difference spectrum obtained by the IR–LSV method between two potentials with one obtained by the EMIRS technique modulating between the same two potential limits [41]. The agreement was good [67]. The reflection–absorption spectrum at a given potential

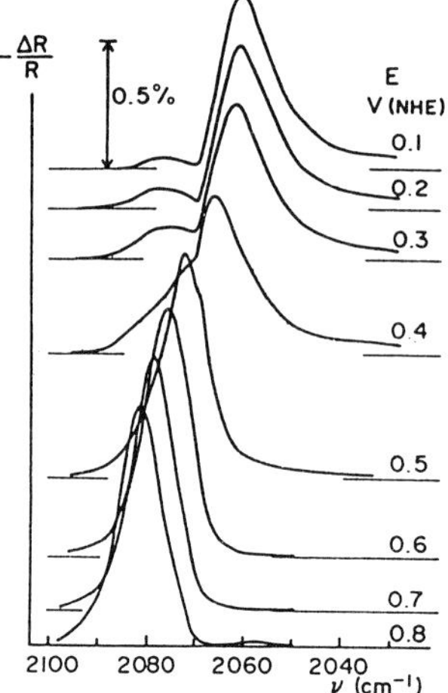

Fig. 27. Potential dependence of the IR reflection absorption spectrum of the linearly adsorbed CO on a platinum electrode in 5 M CH_3OH + 1 M $HClO_4$.

is calculated for as many wavenumbers as is necessary. Figure 27 shows the potential dependence of the "absolute" spectrum between 2050 and 2100 cm^{-1}. The author then used the assumption that the total integrated band intensity of a feature arising from absorption of IR light by an adsorbed species is proportional to the coverage dependence of the integrated band intensity and peak wavenumber so obtained (data manipulation not possible with a difference technique such as EMIRS).

The shift of the band maximum to higher wavenumbers with increasing anodic potential at constant coverage was explained by the author as being due to a change in the bond character between the adsorbed C≡O and the

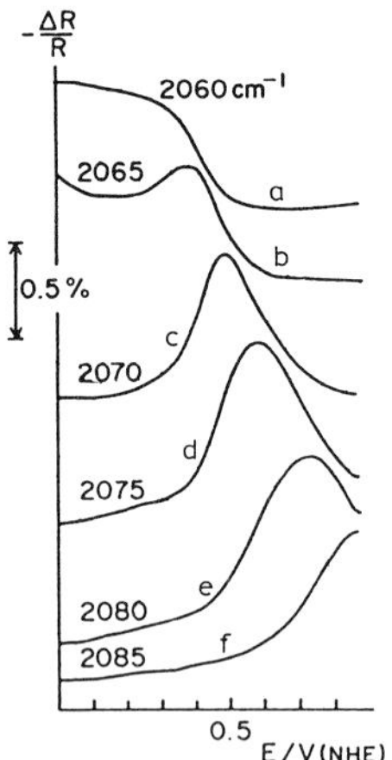

Fig. 28. Change with potential of the reflection absorption at constant wavenumbers from a platinum electrode under a fast cyclic voltammetric condition. 5 M CH_3OH + 1 M $HClO_4$; sweep rate 8.5 V s^{-1}, 256 scans.

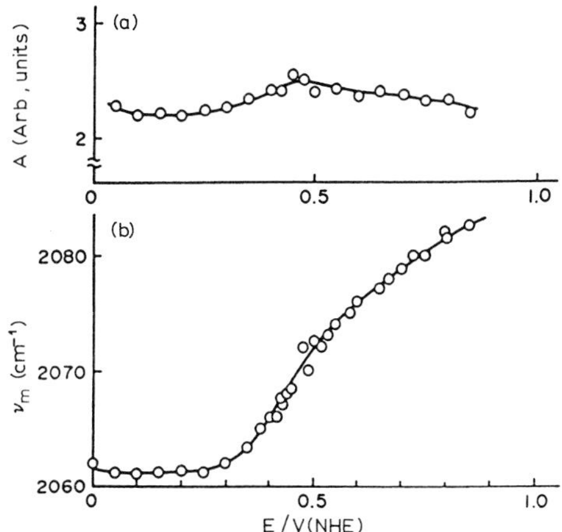

Fig. 29. (a) Potential dependence of the integrated band intensity A and (b) the wavenumber for the maximum absorption $\tilde{\nu}_T$ for the linearly adsorbed CO on a platinum electrode in 5 M CH_3OH + 1 M $HClO_4$.

surface Pt atoms. Thus, increasing the potential of the electrode increases its surface positive charge density, decreasing the back-donation of metal electrons into the antibonding $2\pi^*$ orbital of the C≡O, thus weakening the strength of the Pt–C bond and increasing that of the C≡O bond. This leads, in turn, to the observed increase in the frequency of the C≡O band. As can be seen in Fig. 27 there is some difference in the frequency and bandshape of the features that depends upon whether the electrode potential is in the hydrogen region (0.1–0.4 V vs. NHE) or double layer region (0.5–0.8 V vs. NHE) [65]. The finer detail of the change in the spectrum with potential in these regions can be seen in Fig. 30(a) and (b), These spectra were explained by the author in terms of co-adsorbed hydrogen atoms. Thus, a new band develops around 2075 cm^{-1} as the amount of co-adsorbed hydrogen atoms decreases with the more anodic potential, becoming very sharp at potentials where all the adsorbed hydrogen atoms have been oxidatively stripped off the electrode [65]. This suggests that the 2075 cm^{-1} band is (CO)ads on a hydrogen-free surface; whilst the 2062 cm^{-1} feature is (CO)ads with weakly co-adsorbed hydrogen atoms. Neither band shifts appreciably from 0 to = 0.3 V vs. NHE, the intensity of the 2075 cm^{-1} band merely increasing at the expense of the 2062 cm^{-1} feature. At higher potentials, the former band shifts to lower frequency and the latter to higher frequency. These results infer that adsorbed C≡O bound to a Pt surface partially covered with weakly bound hydrogen atoms has a band maximum at ~2062 cm^{-1}, its intensity simply decreasing with coverage of the weakly co-adsorbed hydrogen. Where the (CO)ads is on the surface with co-adsorbed strongly bound hydrogen, decreasing amounts of the latter cause a shift in the frequency of the feature to higher wavenumber and a decrease in its intensity. Kunimatsu did not try, at this point, to derive any further conclusion on the possible direct

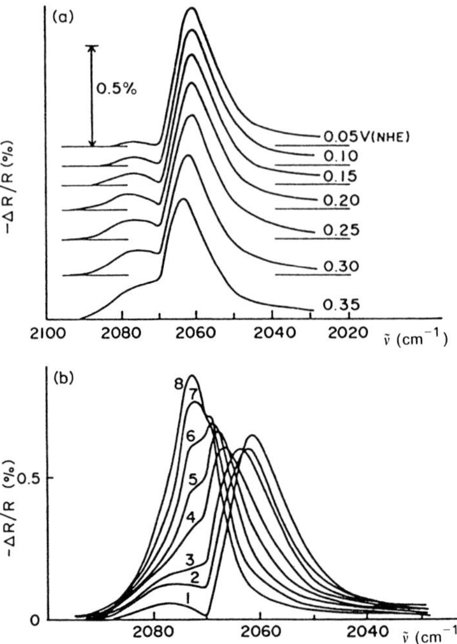

Fig. 30. Change of the IR reflection absorption spectrum of the linearly adsorbed CO on a platinum electrode with potential (a) in the hydrogen region between 0.5 and 0.35 V (NHE) and (b) across the hydrogen region into the double layer region. Curve 1, 0.15 V (NHE); curve 2, 0.30 V (NHE); curve 3, 0.35 V (NHE); curve 4, 0.40 V (NHE); curve 5, 0.425 V (NHE); curve 6, 0.450 V (NHE); curve 7, 0.475 V (NHE); curve 8, 0.50 V (NHE).

interaction between the co-adsorbed hydrogen atoms and the (CO)ads species.

Thus, the IR–LSV technique provides an elegant, if indirect, way to obtain the "absolute" spectra of species adsorbed at an electrode and, if used in conjunction with EMIRS, would nicely complement the latter technique via its ability to furnish quantitative information on adsorption.

To April 1987, the in-situ IR studies on the electro-oxidation of small organic molecules show that the strongly adsorbed fragments which act as poisons in these reactions are all CO species. There has been no spectroscopic evidence for the presence of $\geqslant$COH, even for conditions of less than saturation coverage by (CO)ads. In addition, the adsorbed CO is very stable, requiring fairly high potentials for its oxidation to CO_2 [55]. The reader is referred to the reviews by Bewick and Pons [55] and Foley et al. [56, 69] and the references cited therein for a more detailed treatment.

4.4 ADSORBED HYDROGEN AND ITS EFFECTS ON DOUBLE-LAYER STRUCTURE

Cyclic voltammetric studies of Pt and Rh electrodes in aqueous acid solution show two kinds of adsorbed hydrogen at potentials just anodic of hydrogen evolution. As the potential is made increasingly positive, the

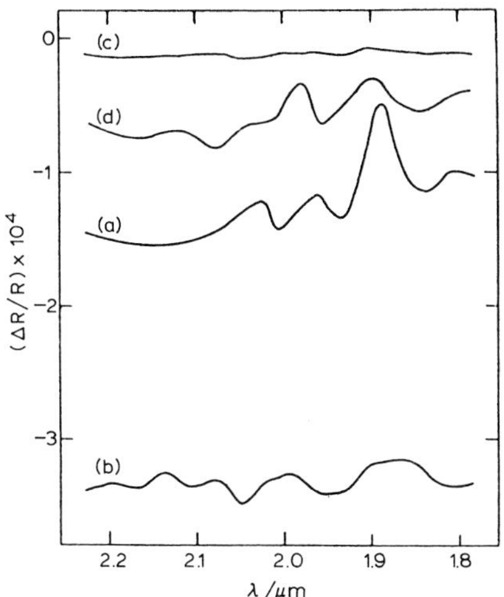

Fig. 31. Spectrum in the H_2O $\tilde{\nu}_2 + \tilde{\nu}_3$ region for 1 M H_2SO_4 using modulation limits (a) 50–170 mV (NHE); (b) 220–340 mV; (c) 450–570 mV; (d) 40–110 mV.

weakly adsorbed hydrogen is first oxidised off leaving the strongly bound form, which is then itself oxidised. The potential then passes into the double layer region between the oxidation of the strongly bound hydrogen and oxide formation. The first report to appear in the literature on the newly developed EMIRS technique [46] (to be followed by several subsequent papers [53–55, 65, 70]) was concerned with the investigation of these forms of surface hydrogen and the perturbations they imposed upon the surface water structure relative to that at the electrode in the reference state (i.e. the electrode held at a potential in the double-layer region). From this initial report, and the subsequent papers, it became apparent that modulation of the potential of a Pt or Rh electrode in aqueous acid solution between two limits, both in the double layer region, produces a difference spectrum consisting only of a small, flat, and relatively featureless baseline shift, attributed (by comparison with previous UV–VIS studies [54]) to the electroreflectance effects [as can be seen in Fig. 31(c)]. Modulation of the same magnitude between a potential in the double-layer region and a potential in the region where only strongly bound hydrogen is present gives rise to a featureless spectrum with a relatively large shift in baseline [Fig. 31(b)], the sign of which corresponds to increased reflectivity on forming the strongly adsorbed hydrogen. These results were entirely consistent with the hypothesis [71] that strongly adsorbed hydrogen consists of a proton buried just below the metal surface with its electron in the conduction band of the metal. Adsorption of such a species would thus increase the surface electron density of the electrode, causing a large increase in reflectivity, but would have little effect upon the water structure at the surface since the adsorbed hydrogen atoms are not available for bonding to the water molecules. Mod-

TABLE 1

Adsorption bands from the complex between water and weakly bound hydrogen (after ref. 55)

Band	Electrolyte	Band/cm^{-1}	Halfwidth/cm^{-1}	Intensity (10^4 $\Delta R/R$)
HDO bend	1 M $H_2[SO_4]$ (67%)	1460	120	1.1
H_2O bend	1 M $H[ClO_4]$	1610	200	1.2
OD stretch	1 M $H_2[SO_4]$ (67%)	2500	380	3.8
OH stretch	1 M $H_2[SO_4]$ (67%)	3480	490	5.7
$\tilde{\nu}_2 + \tilde{\nu}_3$ D_2O	1 M $D_2[SO_4]$	3790	215	0.5
$\tilde{\nu}_2 + \tilde{\nu}_3$ HDO	1 M $H_2[SO_4]$ (67%)	4980	305	0.82
$\tilde{\nu}_2 + \tilde{\nu}_3$ H_2O	1 M $H_2[SO_4]$	5240	130	0.80

The normal vibrational modes of water.

$\tilde{\nu}_1$ $\tilde{\nu}_2$ $\tilde{\nu}_3$

ulating between the double layer region and that of weakly bound hydrogen results in a number of adsorption bands appearing superimposed upon a smaller baseline shift [Fig. 31(a)] (the latter due to the overlap of the weakly and strongly bound hydrogen regions). The bands are all of the same sign, corresponding to increased absorption when the surface is covered with the weakly bound hydrogen, and at frequencies corresponding to the vibrational bands of water. By using mixtures of H_2O and D_2O, the authors assigned the observed bands, and several others, as shown in Table 1. They then applied the surface selection rule [57, 58] and deduced that water bonds to the electrode surface via oriented dimer units hydrogen-bonded to the weakly bound hydrogen atoms [see Fig. 32(a)]. The authors assumed that the dimer units were hydrogen-bonded to the water further away from the surface and that the reference state, water at an electrode held at a potential in the double-layer region, consisted of water molecules having an average orientation as in Fig. 32(b); (an orientation supported by other electrochemical data [72]).

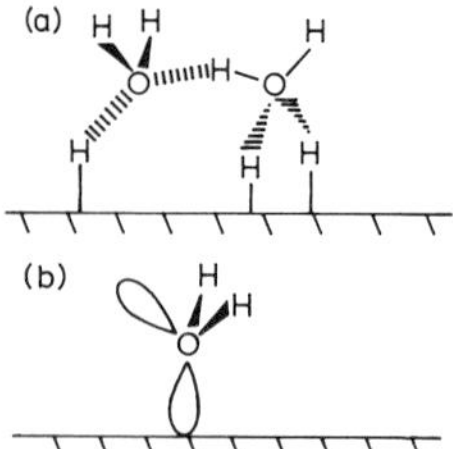

Fig. 32. The structure and orientation of (a) weakly bound hydrogen and its associated water on a Pt or Rh electrode, (b) water on the electrode surface at potentials in the double layer region.

Thus, EMIRS provided an elegant insight into the structure of the double-layer at platinum or rhodium electrodes. These papers are also of interest in that the authors also reported employing a *modulated* (cf. the later work of Kunimatsu [67, 68]) form of linear sweep voltammetry to study the reflectivity ($\Delta R/R$) change of an electrode at a particular wavenumber as a function of potential.

4.5 ADSORPTION OF ETHYLENE DERIVATIVES

The final example of the wide-ranging applications of EMIRS to electrochemical problems was chosen to illustrate the old adage that rules were meant to be broken. In a review paper in 1983, Bewick [70] reported the EMIRS spectrum of acrylonitrile on gold adsorbed from solution in 1 M H_2SO_4. This consisted of two features, as shown in Fig. 33(a) and (b), a noisy feature at 1525 cm^{-1} due to the C = C stretching mode and a weaker band at 2125 cm^{-1} assigned to the C ≡ N stretch. In the former case, such noise in this particular region of the spectrum is not an infrequent occurrence in in-situ IR spectroscopy, since this is one of the regions of the spectrum of very strong (liquid) water adsorption and, consequently, low energy throughput. The spectra obtained of the adsorbed acrylonitrile are extremely interesting for several reasons.

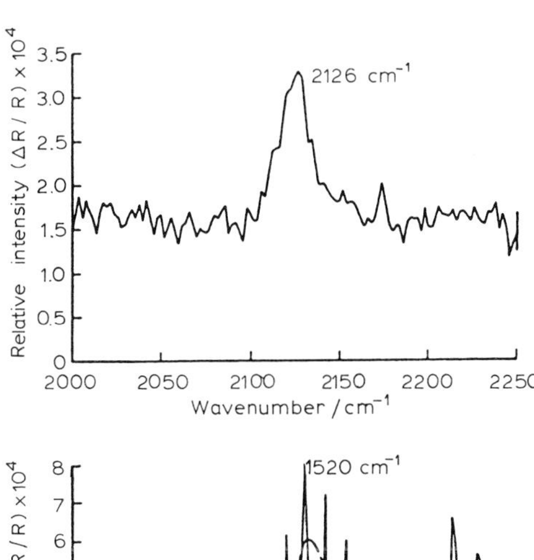

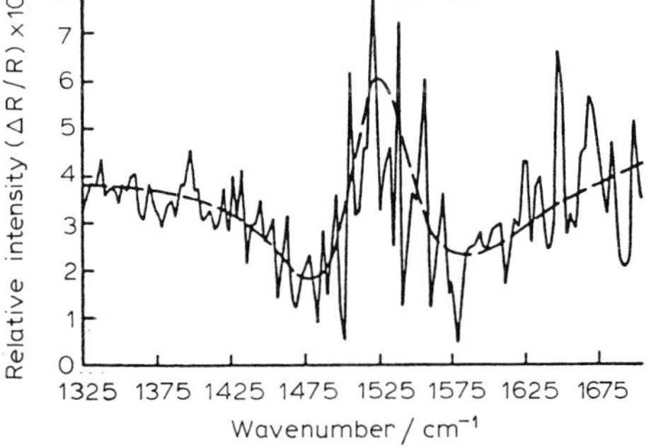

Fig. 33. EMIRS spectra of acrylonitrile adsorbed on an Au electrode in 0.05 M acrylonitrile/1 M H_2SO_4. Modulation 50–1000 mV (NHE).

References pp. 74–77

(a) Both bands are weak compared with the intensities expected for adsorption via the nitrogen of the CN group (i.e. the C≡N stretch is an order of magnitude weaker than that of adsorbed benzonitrile [55]).

(b) Both bands are red-shifted by about 100 cm^{-1} compared with free acrylonitrile in solution. If the molecule was adsorbed vertically via the nitrogen lone-pair electrons, a shift to higher energy would be expected [56]; however, the large red shifts imply that the molecule is adsorbed flat on the electrode surface to maximise interaction with the π orbitals. This is supported by (a) but directly contravenes the selection rule [57, 58] since a flat orientation would result in the C=C and C≡N features being absent from the IR spectrum.

(c) There are no bands that can be assigned to the loss of solution acrylonitrile, which suggests that the coverage of the electrode over the potential range studied remains constant; an observation supported by electrochemical data, [55]. However, the observed bands are both unipolar and positive, corresponding to increased absorption at the more cathodic potential, yet a potential step applied at constant coverage would be expected to give, if anything, a bipolar band (due to a potential-induced frequency shift); the technique is a difference technique requiring some form of a spectroscopic change between the two modulation limits. Thus, the required potential-induced change may be due to a change in oscillator strength brought about by the changing electric field at the surface. Further evidence for a mechanism dependent on the electric field arises from the potential dependence of the intensity of the C≡N stretch reported in one [56] of the several papers that have featured acrylonitrile subsequent to Bewick's initial report [53, 55, 56, 70, 73]. Figure 34 shows the dependence of the intensity of the 2125 cm^{-1} feature as a function of the modulation amplitude. If the fall in intensity at

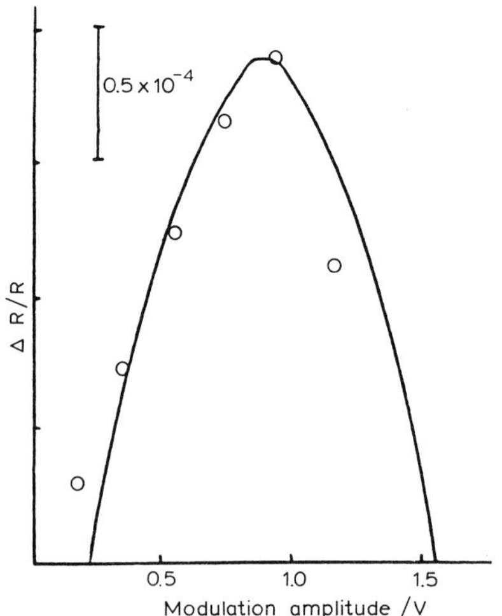

Fig. 34. Intensity of the C≡N stretch band of adsorbed acrylonitrile as a function of the amplitude of the potential modulation (base potential = − 0.195 V vs. SCE). The curve sketched in the figure is the theoretical squared electric field dependence.

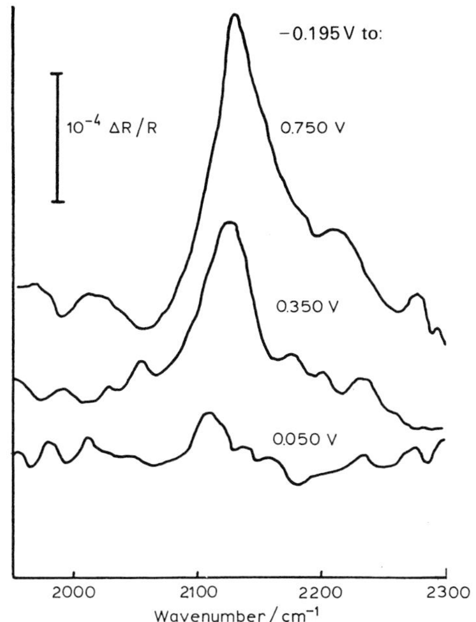

Fig. 35. EMIRS spectrum of 0.05 M acrylonitrile adsorbed at a gold electrode in 1.0 M H_2SO_4. The base potential is − 0.195 V vs. SCE. Modulation at 8.5 Hz.

higher modulation amplitudes was merely a result of the desorption of acrylonitrile, then a feature of opposite sign would be expected to appear due to the solution-free species. However, as can be seen from Fig. 35, this is not the case. Bewick and Pons [55] and Foley et al. [56] explained this intensity dependence in terms of the electric field at the electrode surface. Thus, typical electric field strengths in the near-electrode region are in the range of 10^6–10^8 V cm^{-1} [74]. This field induces a static dipole in the molecule proportional to its polarizability* and perpendicular to the electrode surface. During a vibration, the polarizability of the molecule will change causing a fluctuating dipole to be superimposed on the static field-induced dipole in step with the normal mode vibration, which may have an appreciable component perpendicular to the surface, thus allowing an otherwise "forbidden" vibration to appear in the IR spectrum. Such bands would be monopolar and their relative intensities would depend upon their relative polarizability changes and thus may be different from those in the free molecule. The intensity of such a band is expected to vary with the square of the field strength [56, 76]. An experimental test of this is difficult since the dependence of the electric field on the applied potential, and the potential at which the field strength is zero (the pzc) are not known. However, it would be expected that the C≡N peak intensity should reach a maximum when one

* Hence Korzeniewski and Pons [75] chose to study pyrene as a possible example of this "electrochemical Stark effect" [76] since it was known to adsorb flat on an electrode and was highly polarizable and it did, indeed, exhibit a band forbidden by the surface selection rule.

of the modulation limits is at the pzc. The full line in Fig. 34 is the theoretical curve predicted by the squared electric field strength dependence of intensity [56] (sic) and fits sufficiently well to the experimentally observed results. The maximum intensity observed corresponds to a modulation between − 0.2 and + 0.7 V vs. SCE. The anodic limit is not an unreasonable potential for the pzc*.

5. Fourier-transform techniques

Once allowance is made for its apparent surface-selective "limitation" and its requirement that the electrochemical system under investigation be fast and reversible, EMIRS is a versatile technique that can provide valuable mechanistic information for faradaic and non-faradaic processes. The surface selection rule [57, 58] can act as a double-edged weapon: it greatly simplifies the spectrum of an adsorbed species and aids assignment, but it may render a surface electrochemical process invisible to IR.

There are, however, some problems with the use of simple dispersive instruments and lock-in techniques. Figure 36 shows a simplified schematic representation of a conventional IR spectrometer.

Continuous radiation is dispersed by a grating (or prism), the wavelength of interest being selected by a slit. Scanning is achieved by rotating the grating or prism. These are two obvious disadvantages to this arrangement.

(a) The spectral resolution is determined by the slit width; the narrower the slit, the higher the resolution. However, slits attenuate the beam, resulting in a loss in intensity and also a loss of some percentage of the information the beam may be carrying as the signal-to-noise (S/N) ratio decreases.

(b) In-situ studies require a high S/N ratio, so each wavelength may have to be examined for a relatively long time. Long-term instrumental and experimental drift in these dispersive instruments may then become a problem.

An alternative spectroscopic approach is not to disperse polychromatic radiation with a prism or grating but rather to create an interference pattern or interferogram that becomes modified by the presence of an absorbing sample. The interferogram is then manipulated mathematically to yield the absorption spectrum of the sample.

The interferometer used in most common utility FTIR spectrometers was designed by Michelson in 1891 [78]. Although the phenomenon of the interference of light had been recognised much earlier, Michelson's interferometer allowed the two interfering beams to be well-separated in space so

---

* An alternative mechanism has also been suggested: vinyl derivatives bind to the surface by a Chatt–Duncanson form of σ-donor π^*-acceptor bonding [77]. Alteration of the ethylenic bond length will alter the σ-donor π^*-acceptor capacity of the ligand and charge may then flow between ethylene and metal giving rise to a dipole-moment change perpendicular to the surface. The dependence of this effect on electric field will undoubtedly be far from straightforward.

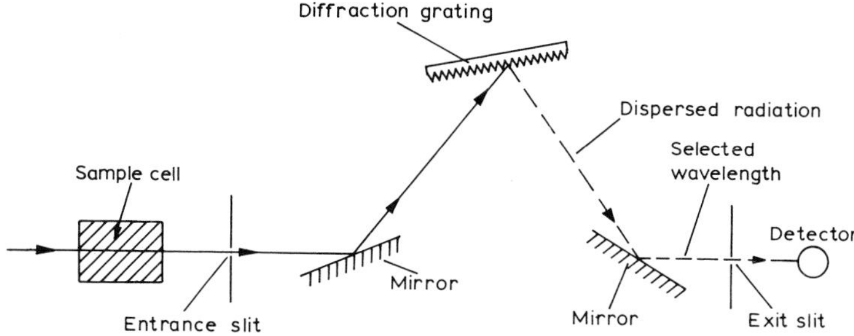

Fig. 36. Simplified schematic representation of a grating infrared spectrometer.

that their relative path differences could be conveniently and precisely varied. Figure 37 shows schematic diagrams of Michelson's interferometer and of the optical arrangement of an FTIR spectrometer employing such an interferometer.

The Michelson interferometer consists of two mutually perpendicular plane mirrors, one of which can move along the axis shown in Fig. 37(a) and

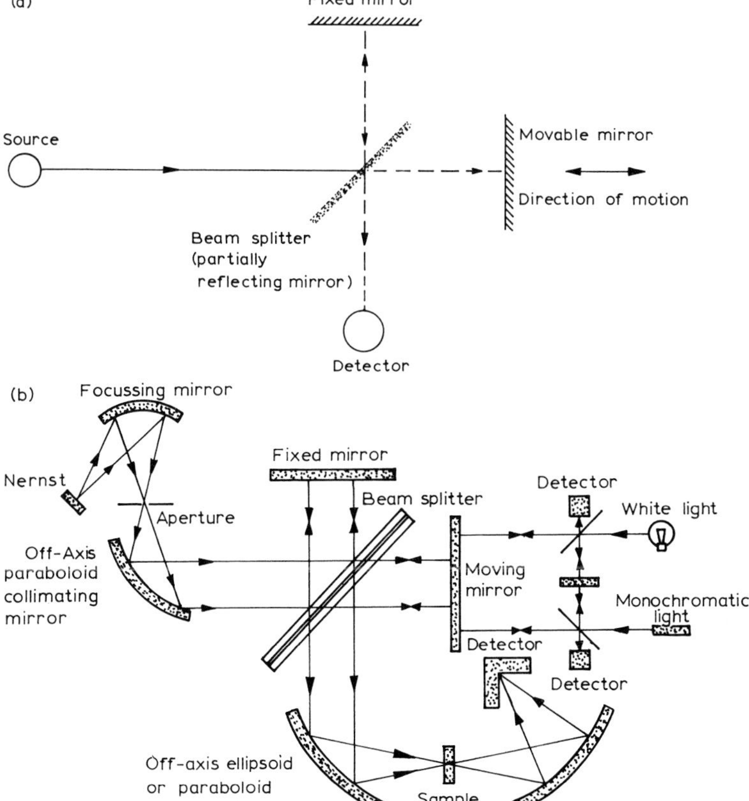

Fig. 37. Schematic diagrams of (a) the Michelson interferometer and (b) an FTIR spectrometer.

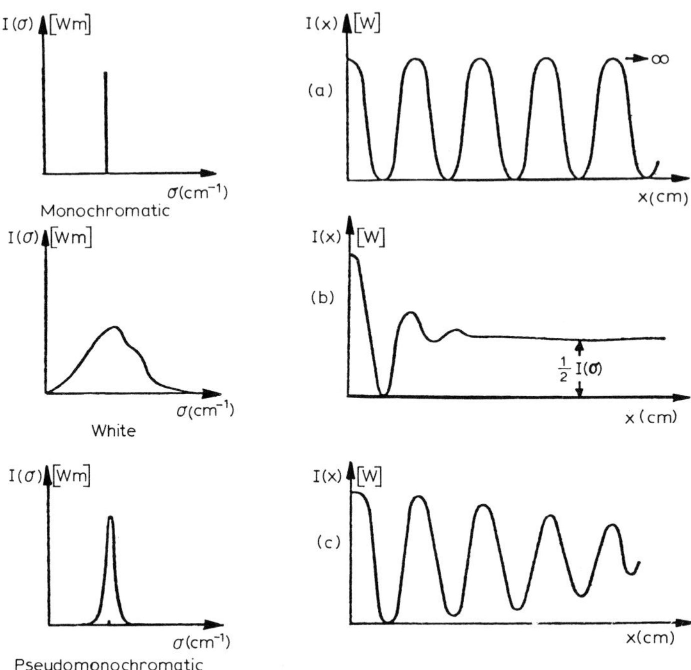

Fig. 38. Spectra and their interferograms.

the other is fixed. A 50% beam splitter is placed between the two mirrors at an angle of 45°. Collimated infrared radiation from a broad-band source (e.g. Nernst Glower or Globar), enters the interferometer and is split into two beams of equal intensity by the beam splitter. The beams are then recombined constructively or destructively depending upon the mirror displacement, x, from the position of zero path difference. The reconstructed beam (the interferogram) has a maximum intensity when the mirrors are equidistant ($x = 0$), the two beams interfere constructively for all wavelengths, and the intensity reaching the detector is the "centreburst" of the interferogram. As the movable mirror is displaced from $x = 0$, the amount of constructive interference decreases rapidly, and the intensity of the recombined beam is thus rapidly attenuated. The interferogram is the resultant intensity at the detector after interference and recombination at the beam splitter; it is normally plotted as function of the mirror retardation, $\delta = 2x$. Figure 38 shows some interferograms and the corresponding light sources that gave rise to them. Physically, the interferogram is built up at every mirror retardation of the superposition of waves of a range of amplitudes, frequencies, and phases. (Thus, it is important to note that, since all wavelengths enter the interferometer, the intensity at every point in the interferogram contains information concerning the entire spectrum.) This superposition corresponds exactly to the mathematical operation known as a Fourier transformation. Now, one of the properties of Fourier transforms is that the Fourier transform of a Fourier transform is the original function. Hence, in a FTIR instrument, physical interference effects produce the first transform (the interferogram) and a dedicated computer using modern fast algorithms then transforms this to give the spectrum of the source modified by the detector response (both constant) and the sample.

The laser in Fig. 37(b) is the source for a second interferometric system. The output of the detector of this second interferometer is a sine wave since the source is monochromatic (see Fig. 38). As this second interferometer is mechanically tied to the main interferometer and the frequency of the laser, and hences its output sine wave, are accurately known, the latter can be used as a very accurate measure of the displacement, x, of the mirror. Some FTIR instruments [23] employ a third interferometric system that is also tied to both the laser and main systems. This has a white tungsten source, which, having a high frequency than the infrared source, provides a very sharp centreburst at its detector with respect to the main IR centreburst (see Fig. 38). The fixed mirror of this third interferometer is adjusted so that its centreburst occurs just prior to that of the main interferogram. This provides a very accurate reference point for the measurement of x. The precise determination of x thus ensures the overlapping of spectra taken during signal averaging so that the resolution is not degraded.

Of the several advantages of FT over dispersive infrared spectrometers [23], two are particularly important in the in-situ study of the near-electrode region.

(a) The throughput (or Jacquinot) advantage. Due to the lack of energy-dissipating optical components in an FTIR spectrometer and the use of large surface area mirrors, the energy of the beam striking the sample in an FTIR spectrometer is greater than that in a dispersive infrared instrument (for spectra recorded at the same resolution).

(b) The multiplex (or Fellgett) advantage. This is the gain in speed afforded by using an FTIR spectrometer over a dispersive instrument to attain a similar signal-to-noise ratio. This made possible by the fact that, in an FTIR instrument, all the intensities of all the spectral frequencies are sampled concurrently (as was discussed above). Thus, the monochromator of a grating spectrometer permits only one resolution element of the spectrum, $\Delta\nu$, to be examined at a time. Frequencies above and below this small frequency band are masked off from the detector. If time, T, is available for observing the spectral region between $\tilde{\nu}_1$ and $\tilde{\nu}_2$ ($\tilde{\nu}_1 < \tilde{\nu}_2$) at resolution $\Delta\tilde{\nu}$, the time spent observing resolution element, $\Delta\nu$, is

$$T_{\Delta\nu} = \frac{T\Delta\tilde{\nu}}{(\tilde{\nu}_2 - \tilde{\nu}_1)}$$

For an FTIR spectrometer, all of the spectral elements are observed for time T. Thus, the time spent sampling the resolution $\Delta\tilde{\nu}$ is also equal to T. Since the signal-to-noise ratio improves as the square root of the observation time

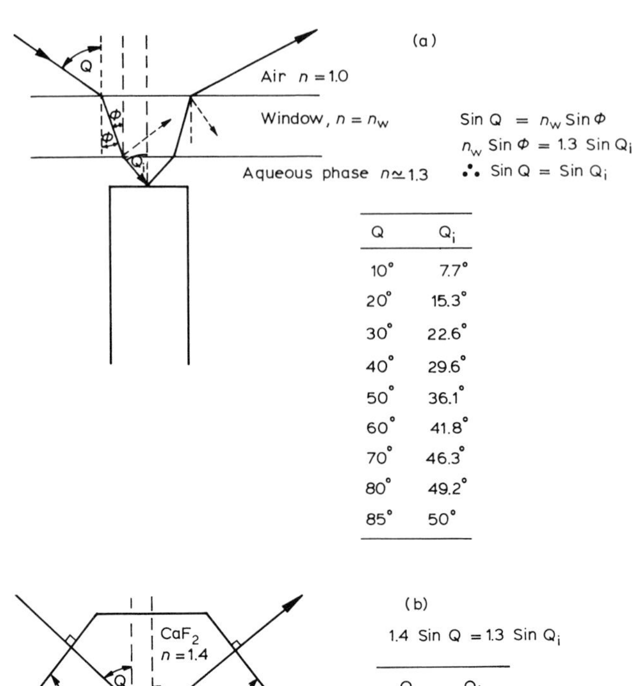

Q	Q_i
10°	7.7°
20°	15.3°
30°	22.6°
40°	29.6°
50°	36.1°
60°	41.8°
70°	46.3°
80°	49.2°
85°	50°

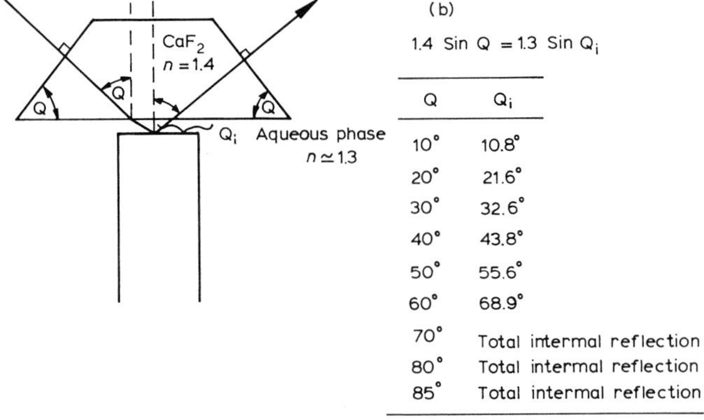

Q	Q_i
10°	10.8°
20°	21.6°
30°	32.6°
40°	43.8°
50°	55.6°
60°	68.9°
70°	Total internal reflection
80°	Total internal reflection
85°	Total internal reflection

Fig. 39. Angle of incidence of IR light on reflective electrode (Q_i) (a) for various angles of incidence on a plate window and (b) for various angles of the faces of a prismatic window at normal incidence.

[23], the multiplex advantage M_A is given by

$$M_A = \sqrt{\frac{T}{T\Delta\tilde{v}/(\tilde{v}_2 - \tilde{v}_1)}}$$

$$M_A = \sqrt{\frac{\tilde{v}_2 - \tilde{v}_1}{\Delta\tilde{v}}}$$

where $M_A = \sqrt{1000}$ for $4\,cm^{-1}$ resolution, range 400–$4400\,cm^{-1}$.

A third advantage is proffered by the laser referencing system that affords high wavenumber precision (the Connes advantage [23]), aiding the attainment of high S/N ratios by accurate co-adding and averaging. However, this may be partially degraded in those systems employing a N_2 purge since the refractive index of nitrogen changes with $\tilde{\nu}$ [23].

Pons and co-workers [79] reported the first potentially modulated in-situ FTIR studies of the near-electrode region and they then developed the technique [24, 55, 56, 69] and eventually coined the acronym SNIFTIRS (subtractively normalised interfacial Fourier transform infrared spectroscopy). Corrigan et al. [81–83] and Bockris and co-workers [80, 84–88] have also reported studies employing the technique, or variations on it. These techniques all employ some form of potential modulation regime; as with EMIRS, intended to cancel out all those absorptions that do not change with potential (bulk solvent, window, etc.), the spectra are again presented as $(\Delta R/R)\tilde{\nu}$ vs. $\tilde{\nu}$. However, the stepwidths (i.e. the time spent at each potential) in a SNIFTIRS experiment are much longer than those in EMIRS; several tens of seconds instead of a tenth of second [89, 90].

By the nature of FTIR, the intensity of the IR light hitting the detector in an FTIR spectrometer is modulated by the Michelson interferometer. The frequency of this modulation, $F_{(\tilde{\nu})}$, depends upon the frequency of the IR light, $\tilde{\nu}$ (cm^{-1}) and the velocity of the mirror in the interferometer, V (cm S^{-1}), as [23]

$$F_{(\tilde{\nu})} = 2\tilde{\nu} V$$

Now, in order to employ a locked-in detection system, as in EMIRS, the modulation frequency of the potential at the electrode would have to be at least an order of magnitude greater than $F_{(\tilde{\nu})}$. Thus, the potential modulation would have to be between 70 and 100 KHz, too great to allow sufficient relaxation time for most electrochemical processes to respond. As a consequence, lock-in detection has not been employed in in-situ FTIR studies and the sensitivity of SNIFTIRS is less than that of EMIRS. Nevertheless, the FT method does have the sensitivity necessary to detect monolayers, and submonolayers, of adsorbed species [55, 56]. This arises out of the very large improvements in S/N ratio available to FT (compared with dispersive) infrared spectrometry by the Jacquinot and Fellgett advantages.

The "slow" stepwidths employed in SNIFTIRS-type experiments means that such studies are not limited to fast, reversible (i.e. generally non-faradaic) processes. For totally reversible studies, fast or slow, multiple steps can be used to attain the necessary S/N ratio after co-adding and averaging; the potential is repeatedly stepped between the base potential, E_1, and working potential, E_2. The scans at each potential, E_1, are co-added and averaged and subtracted from the co-added and averaged scans taken at E_2. The stepwidth can be anything from one scan [91] up although, in order to cancel out long-term experimental and instrumental drift [23, 91], Corrigan and Weaver [82] recommend that stepwidths are less than 60 s long. For irreversible reactions, a single step employing many scans at each potential E_1 and E_2 [92], or multiple steps to successively higher potentials, ("staircasing" [93]), can be used.

References pp. 74–77

The thin-layer cell used in SNIFTIRS experiments is essentially the same as that used in EMIRS. However, Pons and co-workers [92] (SNIFTIRS), Bewick et al. [53] (EMIRS), and Kunimatsu and co-workers [94] (using the IRRAS technique) have all employed prismatic CaF_2 windows with 65° bevelled edges [see Fig. 39(b)]. There are two main advantages of this arrangement.

(a) A plate window only allows an experimental maximum angle of incidence on the electrode of 40–46° (allowing for the increasing percentage of reflected over refracted light at each face of the window with increasing θ), this being irrespective of the window material [see Fig. 39(a)]. The prismatic window allows the infrared light to be incident on the electrode at angles ~ 20° nearer to grazing incidence, an important advantage in surface studies involving adsorbed species.

(b) Because of the possibility of multiple internal reflections within the window, the light finally incident upon the detector may contain a percentage of spurious information, leading to the degrading of the S/N ratio, a possibility avoided by using a 60 or 65° bevelled CaF_2 window.

The SNIFTIRS technique has been used to investigate many processes in the near-electrode region [24, 55, 56], including adsorption [79, 91], electron-transfer [93, 95], solvent–electrolyte interactions [79], and processes occurring at semiconductor [84] and glassy carbon electrodes [96]. Several examples chosen from these applications are given below.

5.1 INVESTIGATIONS OF SIMPLE DOUBLE LAYERS

The first reports of the use of FTIR spectroscopy to record vibrational spectra of the electrode–electrolyte interface was by Pons and co-workers in 1981 [79] and 1982 [97]. The authors obtained in-situ spectra of the potential-dependent population of the species in the double layer formed at a Pt electrode immersed in CH_3CN/0.1 M $LiClO_4$/(n-butyl)$_4NBF_4$. Figure 40(a) shows the differential spectra of 0.1 M (n-butyl)$_4NBF_4$ in CH_3CN obtained between the base potential of -0.5 V vs. Ag/Ag$^+$ and successively higher working potentials. Figure 40(b) shows the differential spectra obtained for the same system except with added water. The potential regime employed was reported in detail in one [37] of the several reviews published on the technique [24, 55, 56, 89, 90]. In essence, the dedicated computer collected eight scans at each potential (base, E_1, and working, E_2), the step being repeated as many times as required for the S/N ratio desired. Only the last four scans in each batch of eight were co-added and saved in a reference or sample (depending upon whether taken at E_1 or E_2, respectively) file; this allowed about 4 s (the first four unused scans) for the cell to come to a steady state (see Fig. 41). The co-added and averaged reference file was then subtracted from the sample file to give ΔR which was then (presumably) normalised to give $(\Delta R/R)\tilde{v}$ vs. $\tilde{v}$.

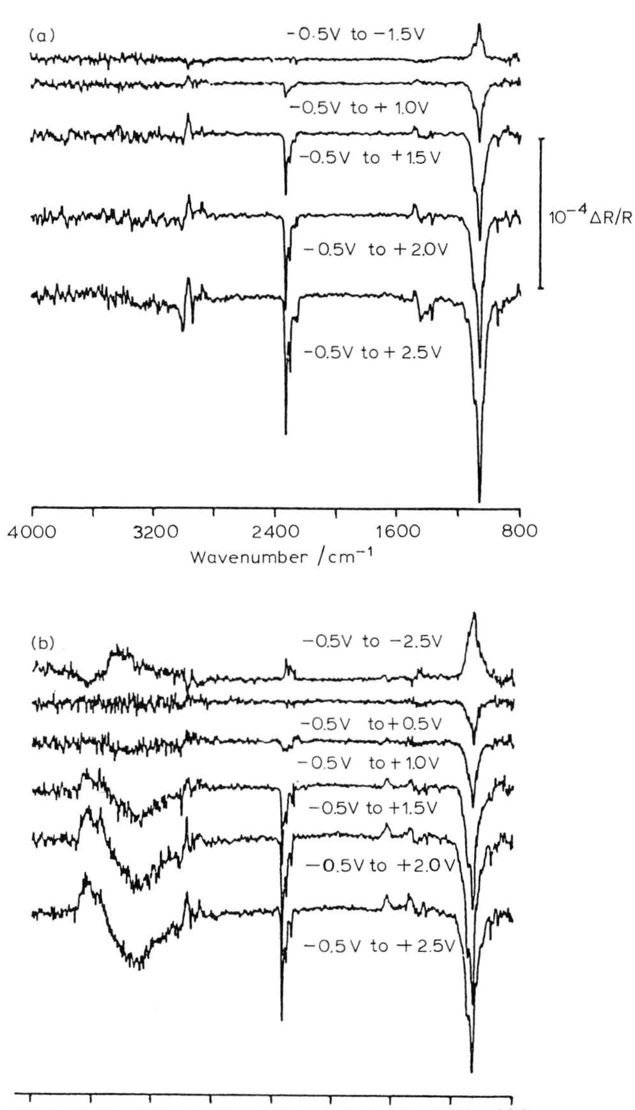

Fig. 40. (a) Difference spectra of anhydrous 0.10 M tetrabutylammonium fluoroborate in acetonitrile. (b) As for (a) except 0.1 M H_2O added.

The authors interpreted the spectra in terms of the adsorption of the acetonitrile and the movement of the anion into the double layer, on positive polarization of the platinum electrode. Thus, the increasing adsorption of acetonitrile at potentials positive of -0.5 V vs. Ag/Ag^+ (the base potential was chosen for its proximity to the pzc of platinum in the chosen electrolyte [98]) is shown by the positive feature at 2350 cm^{-1} whose intensity increases

References pp. 74–77

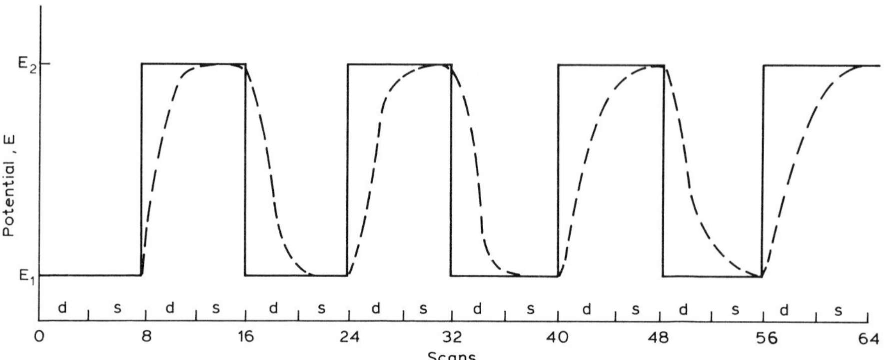

Fig. 41. Schematic representation of timing used for spectra acquisition (after ref. 82). Points d represent spectra that are discarded; points s spectra that are saved and co-added. – – –, Amount of absorbing species for the case in which the amount is greater at E_2 than at E_1.

with the applied anodic potential. This band is due to the $-C \equiv N$ fundamental, strongly blue-shifted due to adsorption [79] via the lone pair on the nitrogen. The increasing number of anions in the optical path is shown by the other relatively strong negative feature at 1060 cm^{-1} (1102 cm^{-1} for BF_4^-). The upward extending (positive) fine structure on the $C \equiv N$ stretch was assigned to the loss of "bulk" acetonitrile.

The broad negative feature at 3350 cm^{-1} was tentatively attributed to the gain of water hydrogen-bonded to the anion. The authors assigned the positive bands at 3625, 3550, and 1625 cm^{-1} to the loss, at the more anodic potentials, of water symmetrically hydrogen-bonded between two CH_3CN molecules. This is a well-characterised species for which the ν_1, ν_2 and ν_3 modes occur at the observed frequencies [99].

These results were used by the authors to postulate a model for the changing population of species in the optical path. Thus, rendering the electrode potential anodic of -0.5 V vs. Ag/Ag^+ causes the anions to move into the double layer and acetonitrile to adsorb. These species are in equilibrium with those in the thin layer, which are in turn replenished from the electrolyte solution outside the thin layer. Hence, the movement of anions into the double layer leads to a net increase in the total number of anions in the optical path and of any associated water. This leads to a displacement of some surface MeCN, both free and complex with water, particularly at the higher positive potentials.

Changes in the C–H stretching region were assigned to the $(n\text{-butyl})_4N^+$ cation, the authors confining their interpretation to assigning the form of the feature between 2900 and 3050 cm^{-1} to a change in band shape; the major part of the band extends upwards (positive) and was thus attributed to a decrease in the amount of cation in the double layer with increasingly positive potential.

These first papers boded well for the new technique, indicating the considerable amount of information on the composition of the double layer that could be gained by relatively simple experiments [55]. The SNIFTIRS method was not extended into aqueous solution until 1983 by Pons [89] and Habib and Bockris [85] and both these reports concerned the adsorption of inorganic anions.

5.2 ADSORPTION

The first paper by Pons [89] on the application of SNIFTIRS to aqueous studies concerned the adsorption of SCN^- at a polycrystalline Pt mirror electrode. Figure 42 shows SNIFTIRS spectra of the $-C \equiv N$ fundamental region taken at various potentials anodic of the base potential of -0.4 V vs. SCE. For working potentials up to $+0.6$ V, no features are observed in the chosen spectral region. However, at potentials $\geqslant 0.65$ V, a positive-going band at 2075 cm^{-1} appears, indicating the loss of bulk SCN^- corresponding to the adsorption of a species at potentials of $+0.65$ V or above. At higher potentials, little change is observed in the spectra until the oxidation of SCN^- at $+1.1$ V. The author attributed the feature at 2100 cm^{-1} to S-bonded SCN^-, on the basis of the correspondence between his results and those of Fleischman and co-workers [100] who employed Raman to study SCN^- adsorption at Ag electrodes. According to one of the published reviews of SNIFTIRS and EMIRS [55], the SERS measurements indicated that S-bonded adsorbed SCN^- is the major species formed at Ag, but detailed interpretation of the data was rendered difficult by effects arising out of the enhancement mechanism of the SERS technique [9–11] (as was discussed above).

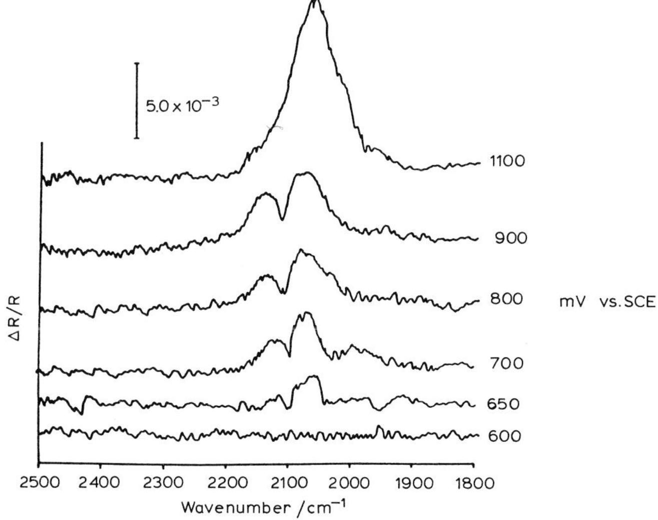

Fig. 42. SNIFTIRS difference spectra from a platinum electrode in 1 M NaF and 0.01 M KSCN. Measured at the voltages shown against a fixed base potential of $+400$ mV vs. SCE reference (300 scans at each potential).

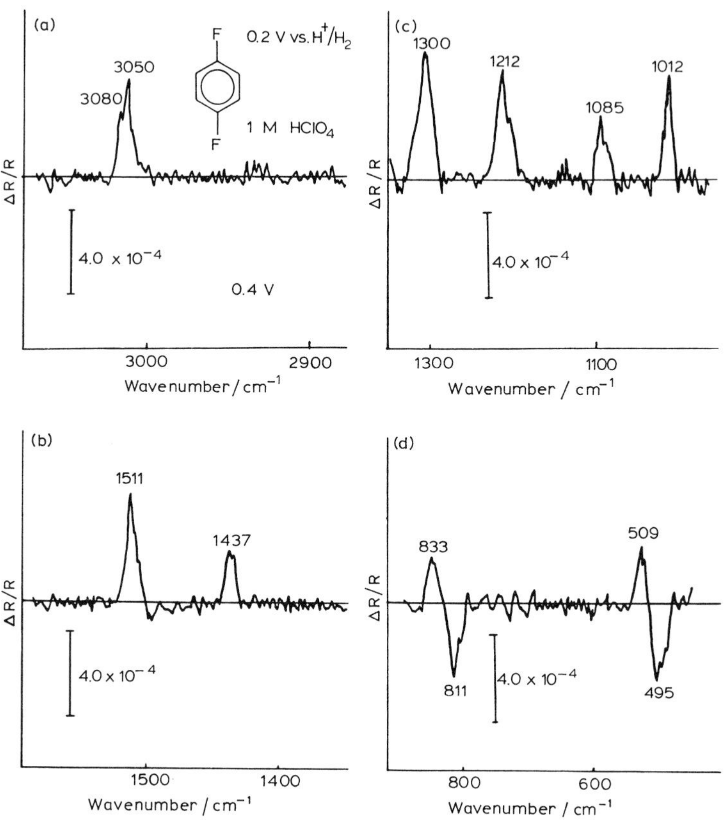

Fig. 43. Difference spectrum for *p*-difluorobenzene at a platinum mirror electrode in 1.0 M perchloric acid solution. Modulation limits are +0.200 V vs. NHE (base potential) and +0.400 V. (a) and (b) 1600 normalized scans, (c) and (d) 450 normalized scans.

A report in the literature by Pons and Bewick [91] in 1985 provided an elegant example of the powerful influence of the surface selection rule. The paper concerned the adsorption of difluorobenzenes and employed a hitherto unused technique in which alternate interferograms taken at the two different potentials (base and working) were phase-inverted. This resulted in these interferograms being automatically subtracted by the spectrometer instead of added. The resulting "co-added" and averaged interferogram was thus a difference interferogram of the two potential states and could be transformed in the usual manner.

Figure 43 shows SNIFTIRS spectra of *p*-difluorobenzene taken at a Pt mirror electrode in aqueous acid solution for modulation between the base potential of −0.2 V and +0.4 V (vs. NHE). Table 2 shows the IR-active normal vibrational modes of the substrate. Of these, only the last three (the b_{3u} modes) involve vibrations having a substantial component perpendicular to the electrode surface. From the work of Hubbard and co-workers [101], the difluorobenzenes are expected to adsorb flat for monolayer (or sub-monolay-

TABLE 2

p-Difluorobenzene infrared active modes

C–H stretch	3050 cm^{-1}
C–H stretch	1511 cm^{-1}
C–F stretch	1212 cm^{-1}
C–H bend	1012 cm^{-1}
C–C bend	737 cm^{-1}
C–H stretch	3080 cm^{-1}
C–C stretch	1437 cm^{-1}
C–C–C stretch	1300 cm^{-1}
C–H bend	1085 cm^{-1}
C–F bend	350 cm^{-1}
C–H umbrella	833 cm^{-1}
C–C–C bend	509 cm^{-1}
C–F in phase	186 cm^{-1}

er) coverage and the authors interpreted their results in terms of just such adsorption. Thus, since the concentration of the *p*-difluorobenzene is less than that required to form an adsorbed monolayer on the surface, absorptions due to the loss of solution species at the more anodic (working) potential will be positive whilst SSR-allowed absorptions by the difluorobenzene adsorbed at the working potential will be negative. In the frequency range investigated (3100–400 cm^{-1}), only two negative features would be expected due to SSR-allowed vibrations of the adsorbed species; those coresponding to the 833 and 509 cm^{-1} peaks of the free species. These are, indeed, observed in the SNIFTIRS spectra along with the positive features of the randomly orientated solution species, as bipolar bands. Their bipolarity is a result of the interaction of the electrode surface with the π-aromatic system which decreases the bond orders involved in these modes.

Figure 44 shows difference spectra taken in the region of the out-of-plane C–C–C bend of *p*-difluorobenzene for increasing values of the working potential with respect to the -0.2 V base potential. The negative "lobe" of this band can clearly be seen to be increasingly red-shifted as the potential is rendered more anodic, in keeping with flat adsorption. Under the same conditions as those for the *p*-difluorobenzene experiments, the *ortho* and *meta* isomers gave SNIFTIRS spectra indicative of flat adsorption. The authors remarked upon the absence of any SSR-forbidden features in the SNIFTIRS spectra of the difluorobenzene isomers that may have been rendered IR-active by the electrochemical Stark effect [76]. They concluded that the polarization changes during the vibrations of the adsorbed difluorobenzenes were insufficient to allow the Stark effect to be important.

Increasing the concentration of *p*-difluorobenzene in the thin layer up to 0.5 mM causes no change in the spectra. At solution concentrations $\geqslant 0.8$ mM, however, there is an appreciable increase in the intensities of the bands and negative bands appeared for all the modes. The authors attributed

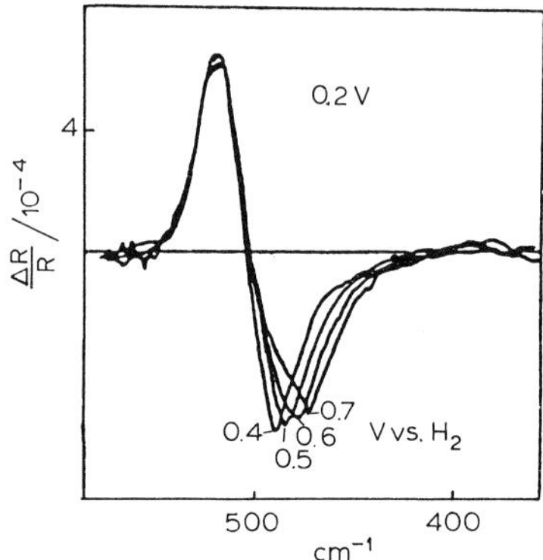

Fig. 44. Difference spectrum (690 normalized scans) of *p*-difluorobenzene as a function of the magnitude of the modulation potential. Other parameters are the same as those in Fig. 43.

these observations to an increase in the surface density of adsorbed species by forced edgewise packing.

This work ably illustrates the importance of the surface selection rule. Unfortunately, the phase inversion technique described in this paper has not been further developed by Pons and co-workers and there are some difficulties associated with it; it is evident, for example, that exact balancing of the positive and negative phases will lead to complete cancellation of the centreburst. This is of significance as the spectrometer software may well rely on the location of this centreburst to allow the Fourier transform to take place. It is, therefore, essential to build in mis-match into the phase-inversion amplifiers, though this in turn makes the technique very difficult to use quantitatively. Suffice to say that the authors of this report have not found it easy to use in practice and have relied on the subtraction of spectra already transformed as described above.

Two [80] of the several reports by Bockris and co-workers on investigations employing the SNIFTIRS technique [84–88] concerned the study of the adsorption of thiourea on a passivated iron electrode, a potentially important area as thiourea is reported [102] to inhibit electrochemical corrosion. In the first paper [88(a)], the IR-determined surface coverage was compared with radio-tracer data and showed good agreement between 0.4 and 1.1 V vs. NHE, with the adsorption of the thiourea reaching a maximum at 0.9 V. Below 0.4 V, the IR data differed from those of the radiotracer method, the authors postulating that this was due to the adsorbed urea changing its orientation at the more cathodic potentials (the radiotracer method is insensitive to such an orientation change). The extension of SNIFTIRS to non-noble metals is a particularly significant development, particularly in view of the fact that, to the authors's knowledge, it has not proven possible to measure a surface-enhanced Raman spectrum on iron.

5.3 ION RADICAL INTERMEDIATES

As was stated above, the slower time scale of a SNIFTIRS experiment compared with the EMIRS technique allows spectra to be obtained of the products of non-faradaic and faradaic reactions. This is because the relaxation time of any such potential system no longer determines whether or not it can be investigated. In addition, the far greater throughput of an FT over a dispersive instrument allows much greater solution thickness to be employed.

Up to a solution thickness of 50 μm, diffusion of reactants into, and products out of, the thin layer will be very slow. The long time scale of a SNIFTIRS experiment thus allows species in the thin layer to become equilibrated with the electrode, leading to a quick and simple means of investigating faradaic processes. Using such a thick layer results in an increased optical path length in which to detect the solution species (and thus increased S/N ratios). In addition, according to Pons and co-workers [92] referencing a report by Kunimatsu et al. [102], for solution layers greater than a few microns, adsorbed species are not detected since the electric field of the IR beam at the electrode surface is too low. Thus, a thick solution layer may be chosen if the investigation of continuous faradaic processes is of interest without any complicating spectra from adsorbed species; whereas investigations of the double-layer necessitate as thin a solution layer as possible. (The minimum solvent layer thickness that can be experimentally achieved appears to be $\sim 1\,\mu$m [79, 90, 95, 97]*.) Because of the Jacquinot advantage and the concurrent use of sensitive detectors, spectra can also be obtained with thin-layer widths of $\lesssim 50\,\mu$m [92] providing there is a sufficient "window" in the solvent absorptions. However, complications in the difference spectra obtained then arise due to diffusion to and from the thin layer**.

---

* Two methods of determining the thickness of the solution layer have been reported, depending upon whether CH_3CN or H_2O is the solvent. In the former case, the integrated band intensity of the $C \equiv N$ fundamental at 2220 cm^{-1} gives the solution thickness, d, via the simple equation [90].

$$d = \frac{2A \cos \theta}{\varepsilon C}$$

where A is the absorbance, θ the angle of incidence of the light on the electrode, ε the extinction coefficient of the band, C the concentration of the CH_3CN. However, the report in which this appears neglects to give ε or a suitable reference. In the latter case [92], the thickness of an aqueous solvent layer was determined from the charge under thin-layer voltammetric peaks (again unreferenced).

** The authors of this chapter have carried out investigations on the diffusion of species into aqueous thin layers [103] using a flow-through spectro-electrochemical cell. It was found that, for aqueous thin layers of greater than a few microns, the introduction of fresh species into the cell caused gross reflectivity changes. These were thought to be caused by the re-organisation of the double layer and electronic effects in the metal electrodes, these effects as a whole being magnified by the aqueous layer.

References pp. 74–77

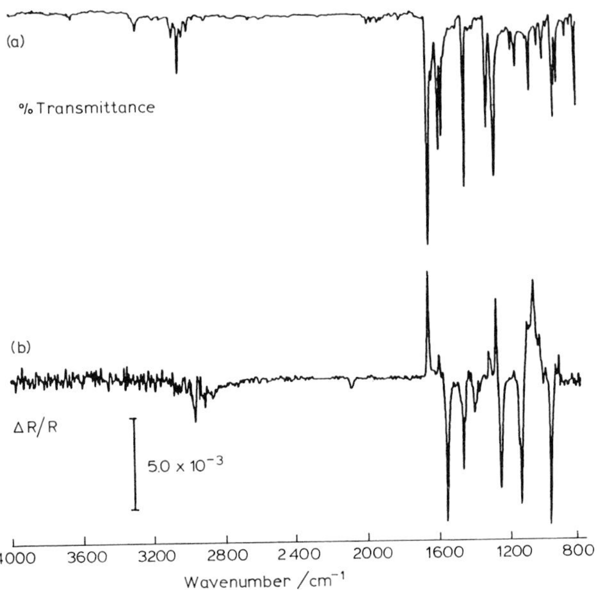

Fig. 45. (a) Transmission spectrum of benzophenone. (b) SNIFTIRS difference spectrum of benzophenone reduction to the ketyl anion radical.

In 1983, Pons et al. [95] reported the in-situ spectra of the anion radicals of benzophenone, anthracene and, tetracyanoethylene (TCNE). This study exemplifies the above reasoning with respect to determining the thickness of the solvent layer in a SNIFTIRS experiment. Thus, the thin layer was 1.0 μm in the benzophenone experiment, 14 μm in the anthracene, and 17 μm in the TCNE experiments. Figure 45(b) shows SNIFTIRS spectra taken of 10 mM benzophenone in 0.1 M tetra-*n*-butyl ammonium fluoroborate/acetonitrile electrolyte at a Pt mirror electrode for modulation between the base potential of -1.75 V vs. Ag/Ag^+ and the working potential of -2.50 V. The corresponding "transmission" spectrum of benzophenone in the same cell arrangement is shown in Fig. 45(a) for comparison. The radical anion spectra were obtained via the pulsed potential method and were the average of 40 scans. Agreement between the authors' results and a previous paper reporting the vibrational spectrum of the benzophenone ketyl radical was reported to be very good, slight differences being attributed to solvent effects. The results of this and later papers reporting further studies [55, 56, 90, 95] showed a large difference between the frequency of the C=O stretch in the ketyl radical (1555 cm^{-1}) and the benzophenone substrate (1661 cm^{-1}). This was attributed to the localising of the transferred electron on to the C=O bond. In the original paper [95], it was reported that the band at 1464 cm^{-1} in the radical anion spectrum, previously unobserved in its vibrational spectrum, was only present when using very thin solution layers. On increasing the thin layer from 1 to 200 μm and reducing 90% of the benzophenone in the thin layer to its ketyl radical yielded a SNIFTIRS spectrum with greatly enhanced peak intensities, but the band at 1464 cm^{-1} is no longer observed; it is similarly absent if the polarization of the incident light is changed from p to s. In the later papers [55, 56, 90], features at 1340 and 2120 cm^{-1} (in addition to the peak at 1464 cm^{-1}) were reported as exhibiting similar behaviour: unobserved through thick solvent layers and absent from conventional spectra of ketyl radical. In the first paper, the authors concluded that the benzophenone ketyl radical exists both in a solution-free and adsorbed form, the latter bonding to the surface through the C=O group. In a later review [56], the adsorbed species was postulated as a complex between the ketyl anion and adsorbed acetonitrile [104].

The anthracene radical anion was electrogenerated at -2.5 V vs. Ag/Ag^+, using a base potential of -1.5 V. The solution thickness of 14 μm indicated that the spectrum of the radical anion would not be complicated by absorptions from any adsorbed species. The SNIFTIRS spectrum obtained is shown in Fig. 46(b) and shows only those features due to the product radical, betraying a weakness of such differences techniques. As can be seen from Fig. 46(b) the spectrum is almost exclusively the negative bands corresponding to the anthracene radical anion; any positive bands due to the loss of the substrate are absent. The authors postulated that this was due to the extinction coefficients of the product features being a great deal larger than those of the substrate bands and/or that some anthracene bands were being cancelled out by radical anion features of the same frequency, bandshape, and intensity.

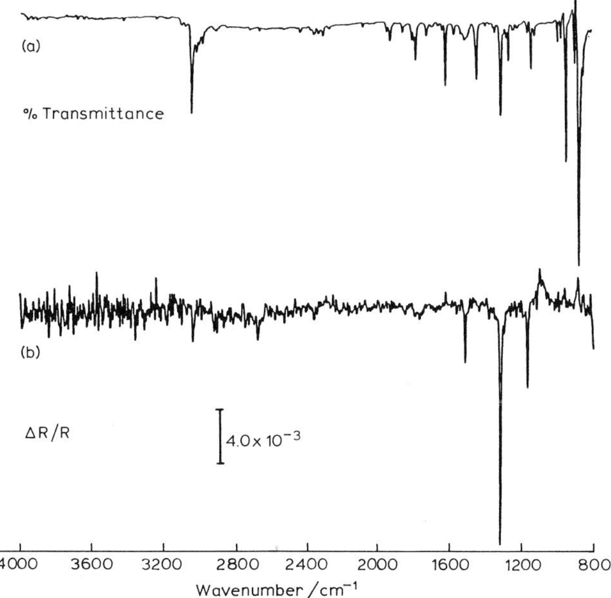

Fig. 46. (a) Transmission spectrum of anthracene, KBr pellet. (b) SNIFTIRS difference spectrum for anthracene reduction to the anion radical. Platinum electrode in 0.1 M TBAF in acetonitrile. Modulation potential -1.50 to -2.50 V vs. Ag/Ag^+ reference.

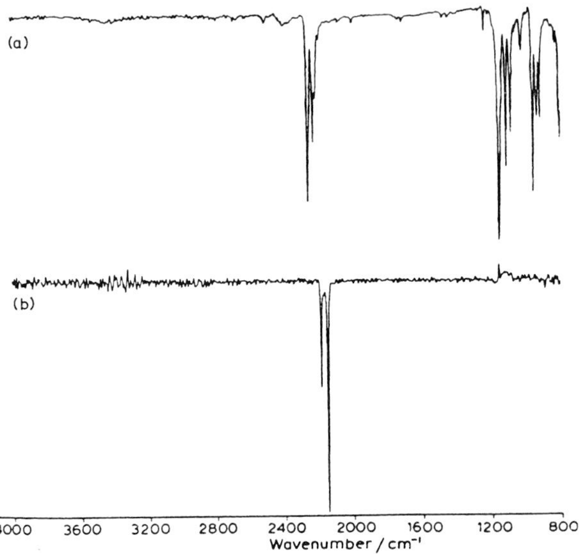

Fig. 47. (a) Transmission spectrum of TCNE. (b) Differential FTIR spectrum of TCNE between +0.25 and −0.25 V.

Figure 47(b) shows the difference spectrum of TCNE at Pt in TBAF/MeCN electrolyte for modulation between +0.25 V (vs. Ag/Ag^+) base potential and −0.25 V working potential. The spectrum is dominated by the two negative features at 2187 and 2148 cm^{-1}. The latter band was reported as 10^4 × greater than would be predicted for a simple Beer's Law calculation, using the concentration of substrate present [105]. The former feature was assigned by the authors, on the basis of work by Devlin and co-workers [106], to the C≡N stretch enhanced via the formation of a charge-transfer complex between the anion radical and a surface Pt atom. The more intense feature at 2148 cm^{-1} was attributed to the solution-free C≡N fundamental stretch enhanced by the formation of an electron donor/acceptor complex between the anion radical and neutral TCNE. Further studies were carried out on the TCNE system [56] and these demonstrated considerable complexity in the system.

5.4 ADSORPTION AT NON-METAL ELECTRODES

In 1983, Bockris and co-workers [84] reported obtaining spectra of adsorbed species during the photoassisted reduction of CO_2 at a semiconductor electrode. This was the first in-situ study in which an electrochemical reaction was induced via the illumination of a photoactive electrode and monitored in real time with an IR beam. The electrode material was a p-CdTe single crystal illuminated normal to its surface by a tungsten halogen lamp via a light guide; the spectroelectrochemical cell and optical arrangement was otherwise essentially the same as in a SNIFTIRS experiment. Figure 48

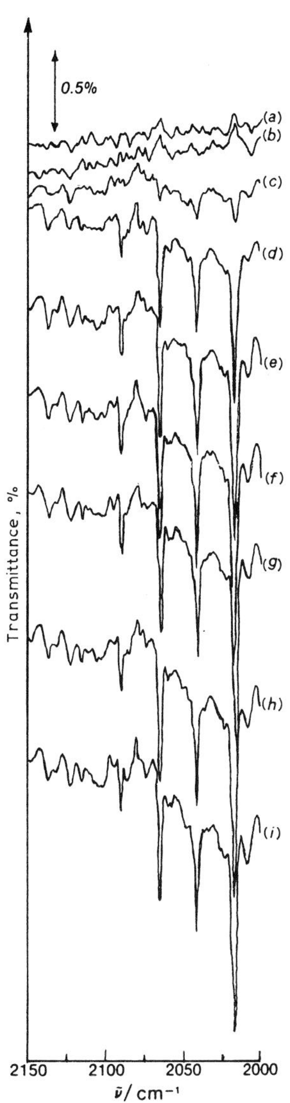

Fig. 48. IR spectra recorded at various values of the potential (vs. Ag/Ag) in MeCN + 0.1 M TABF. (a) −900, (b) −1100, (c) −1300, (d) −1500, (e) −1700, (f) −1900, (g) −2100, (h) −2300, (i) −2500 mV. (The potential at which the reference spectrum was taken, against which the subsequent spectra were ratioed in Fig. 48, is not stated.)

shows IR spectra (1000 normalised scans) reflected from the illuminated p-CdTe electrode in 0.1 M TBAF/MeCN saturated with CO_2 at various potentials between −900 and −24090 mV vs. Ag/Ag^+.

As can be seen from Fig. 48, four negative (product) features appeared in the chosen spectral range at potentials cathodic of −1100 mV vs. Ag/Ag^+; (i.e. at 2017.5, 2041.4, 2065, and 2090 cm^{-1}). These features were only observed when the solution was saturated with CO_2 and the electrode was under illumination. After photoelectrolysis, CO was detected as the product by gas chromatography. The authors deduced that the features were due to intermediates in the photoassisted reduction of CO_2 to CO. The ratios of the integrated band intensities of all four features remained constant at all potentials and electrolysis times, indicating that they were all due to the same species. In addition, the peaks were not observed when the polarization of the incident IR light was changed from p to s, which suggested that the species was adsorbed on to the electrode surface. The integrated band intensities of the four peaks increased with increasing cathodic potential. However, no frequency shift was observed (cf. the potential-dependent frequency shift of CO on Pt etc. [55, 56, 62] and the integrated band intensity of the features remained constant at longer electrolysis times. The authors thus suggested that the adsorbed species exhibited a potential-dependent maximum coverage and attributed the four features to the CO_2^- (ads) species on the basis of these results and their photocurrent vs. potential studies. In the latter studies, the relative height of the peaks increased almost in parallel with the photocurrent. They suggested

$$CO_2 + e^- \rightarrow CO_2^- \text{ (ads)}$$

as accounting for the observed features in the IR spectra followed by

$$CO_2 + CO_2 \rightarrow O{=}C{-}O{-}CO{-}O^-$$

$$O{=}C{-}O{-}CO{-}O^- + e \rightarrow O{=}C{-}O{-}CO{-}O$$

$$O{=}C{-}O{-}CO{-}O^- \rightarrow CO + CO_3^{2-}$$

(after Amatore and Savéant [107]) to account for the product CO. The increased stability of CO_2^-(ads) on p-CdTE was explained in terms of the crystallochemical model of Suchet [108]. In a recent paper, Chandrasekaran and Bockris [80] have reported the IR spectrum of CO_2 radical adsorbed on platinum in acetonitrile/$LiClO_4$ electrolyte, using the polarization-modulation IRRAS technique [94, 109–114].

In 1985, Datta et al. [96] reported in-situ FTIR studies of adsorption at a glassy carbon electrode. The paper was a preliminary report and the authors concluded that phosphate and sulphate are adsorbed at a glassy carbon electrode whilst perchlorate was apparently not, in keeping with the generally held view.

Thus, SNIFTIRS is an extremely versatile technique, especially when it is realised that virtually any electrode may be employed providing it is IR reflective, e.g. metals, doped semiconductors, basal plane and edge pyrolytic graphite, and glassy carbon, etc. As yet, its applications have only been touched upon, but it is expected that the technique is, and will continue to be, an increasingly powerful tool in the hands of the electrochemist.

6. Polarization-modulation techniques

The potential modulation methods of in-situ IR studies rely on switching the potential at a reflective electrode between "rest" and "active" states to generate "pseudo-double-beam" behaviour. In this manner, unchanging absorptions (window, solvent), the response of the detector, and the wavelength-dependent emission envelope of the source are all cancelled out. As a result, a spectrum is generated that represents a potential-dependent difference spectrum of the near-electrode region. The IR-LSV technique provides an indirect means of obtaining "absolute" spectra of the near-electrode region. At first sight, an alternative "direct" approach is that involving polarization modulation.

Many recent experimental studies on the first measurement of IR absorption by an adsorbed layer from the gas phase on a metal surface have utilised the polarization-modulation technique [109]. Golden and co-workers reported the development of instrumentation (using conventional dispersive optics) able to record detailed IR reflection–absorption spectra from molecules adsorbed on single-crystal surfaces without interference from other absorbers (e.g. atmospheric CO_2 and H_2O) in the optical path. The polarisation-modulation techniques takes advantage of the different response of adsorbed molecules to incident IR radiation polarized at different orientations to the surface (e.g. s and p). In such a technique, a photoelastic modulator is used to modulate the IR beam between the two polarizations. From Greenler's theory [57], such a polarization modulation gives rise to an a.c. voltage at the detector that is proportional to $I_p - I_s$, the difference in intensity of the two polarizations. Since I_p is adsorbed by surface-adsorbed molecules and I_s is not, this signal contains the absolute IR spectrum of the adsorbed species. The incident light is also chopped mechanically to determine the attenuation factor of the optical system, equal to $(I_p + I_s)^{-1}$ as a function of wavelength. The normalised ratio $(I_p - I_s)/(I_p + I_s)$ is then plotted as a function of frequency to give the spectrum of the absorbed species. In 1982, Russell et al. [63] applied this technique to in-situ studies of the electrode–electrolyte interface; referring to it as infrared reflection absorption spectroscopy (IRRAS). In order to enhance the intensity of IR absorption in an in- or ex-situ PM experiment grazing angles as near as possible to the "Greenler angle" of $\sim 88°$ are employed, where the absorption coefficient of p-polarised light is at a maximum: up to 5000 $\times$ that at normal incidence [112]. The electrochemical cell and optical arrangement required for an IRRAS experiment is similar to that used in EMIRS and SNIFTIRS and it has been suggested that IRRAS should be more sensitive than SNIFTIRS (69] in that the former can take full advantage of phase-sensitive detectors.

In practice, the situation is less clear cut. As indicated above, the surface selection rule arises because of a phase change of π on reflection in $\langle E_s \rangle$.

This reduces the magnitude of $\langle E_s \rangle^2$ not only at the surface but for an appreciable distance into the solvent layer. As a result, $(I_p - I_s)$ not only contains information on the surface layer, but also on a layer at least 0.1 μm thick of solution. Clearly, this hardly matters for gas-phase adsorption, but it is far more serious for highly absorbing aqueous layers and later papers have resorted to taking the *difference* between IRRAS spectra recorded at different potentials to annul contributions from the solvent. Obviously, such a procedure substantially reduces any initial advantage IRRAS might have had over SNIFTIRS.

Figure 49 shows the thin-layer spectroelectrochemical cell used for experiments and Fig. 50 a block diagram of the necessary instrumentation.

The large "effective range" of the SSR implies that the use of IRRAS as a means of obtaining an "absolute" spectrum of an adsorbed species is effectively restricted to those species having absorption in those areas of the IR free from strong solvent features, e.g. for water, this "window" is in the range $\sim$2000–3000 cm^{-1}. Figure 51 shows an FT-IRRAS spectrum of the near-electrode region for CO adsorbed on 1 N H_2SO_4 at 0.4 V vs. NHE; the very strong water absorption right across the spectral range can clearly be seen in comparison with the tiny (but relatively strong in terms of in-situ IR studies) CO absorption, ($\Delta R/R \simeq 4\%$) at 2140 cm^{-1}. Most studies in this area of in-situ PM have been carried out by Kunimatsu and co-workers [62, 94, 110–114]. The majority of their papers have been on the electro-oxidation of

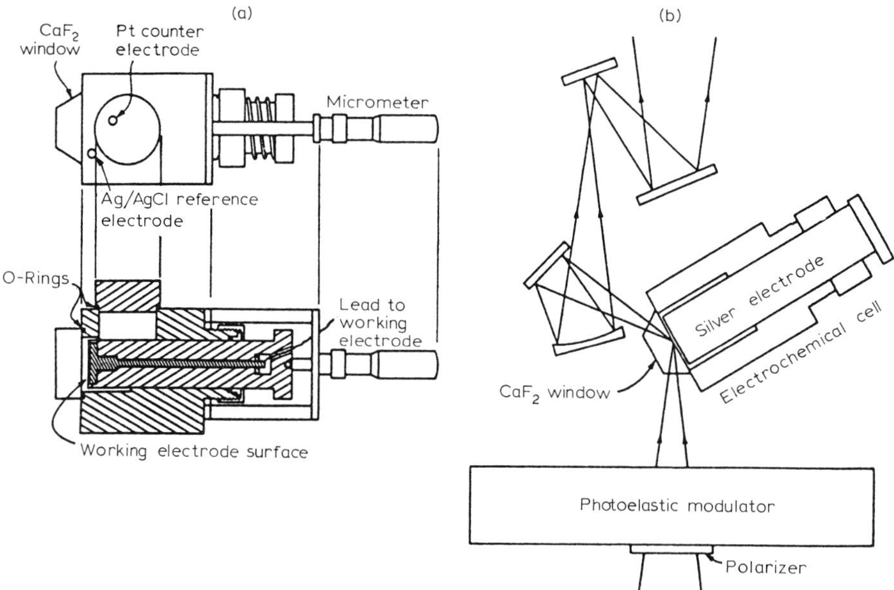

Fig. 49. (a) Top and side views of the thin-layer electrochemical cell with a CaF_2 window. The side view is shown as a cross-section. (b) Arrangement of the electrochemical cell, the photoelastic modulator, and the mirrors in the sample chamber of the FT-IR spectrometer.

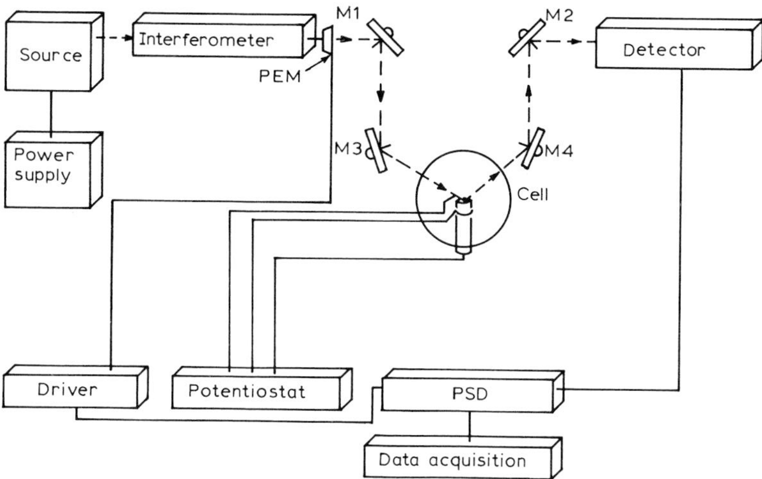

Fig. 50. Block diagram of instrumentation used for FT-IRRAS experiment.

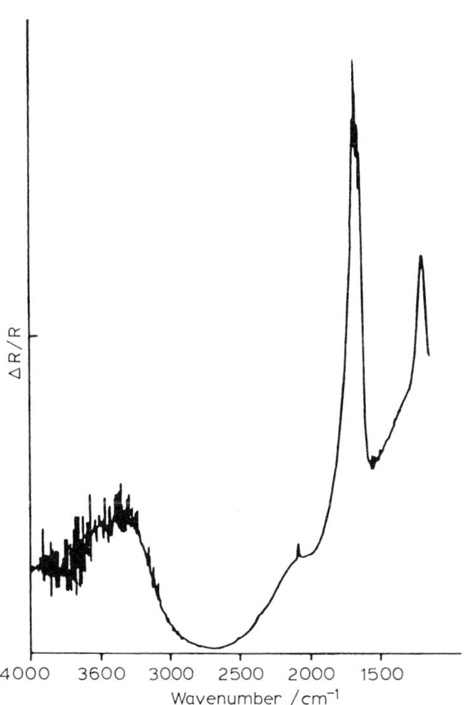

Fig. 51. FT-IRRAS spectrum of the electrode/solution interface for CO adsorption on a smooth platinum electrode in 1 N H_2SO_4 at 0.4 V (NHE).

References pp. 74–77

small organic molecules in aqueous solution, with a minority on aqueous CN^- systems [111, 112]. In both of these cases, the features of interest occur in the "window" region of the IR spectrum and thus may allow "absolute" spectra to be obtained.

The first in-situ PM study was reported by Russell et al. in 1982 [63] and was concerned with the adsorption of CO on Pt. As was discussed above, the first EMIRS paper on this subject [62] had concluded that the dominant poison in the electro-oxidation of MeOH at Pt under aqueous acid conditions was $-C\equiv O$, which exists at high coverage, and another CO species thought at that time to be $\rangle C=O$. The $C\equiv O$ feature appeared as a bipolar band, indicating that its frequency was potential-dependent. Russell and co-workers commented upon the indirect nature of the information derived

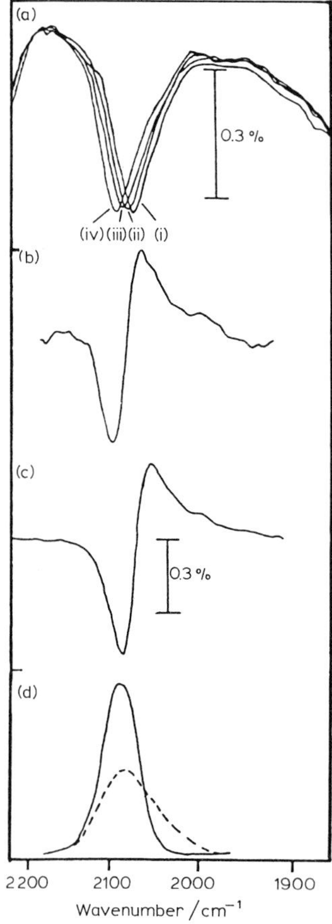

Fig. 52. (a) IRRAS spectra of CO on Pt in 1 M $HClO_4$ saturated with CO. The electrode potential is held constant at (i) 50 mV (NHE), (ii) 250 mV, (iii) 450 mV, and (iv) 650 mV. (b) Difference spectra resulting from subtraction of IRRAS spectra (a) (iii) and (a) (i). (c) EMIRS spectrum of CO-saturated 1 M $HClO_4$ modulated between 50 and 450 mV. (d) A schematic representation of spectra at two potentials which could produce an EMIRS spectrum similar to that shown in (c). The IRRAS spectra shown in (a) rule out this possibility (see text).

from a difference technique such as EMIRS and reported the first in-situ "absolute" spectrum of CO adsorbed on Pt. Figure 52(a) shows IRRAS spectra of a platinum electrode immersed in 1 M $HClO_4$ saturated with CO. This clearly shows the dependence of the CO band frequency on potential. For comparison, the authors included difference spectra resulting from subtraction of IRRAS spectra taken at 50 and 450 mV and the EMIRS spectrum of Pt in CO-saturated 1 M $HClO_4$ modulated between the same two potential limits [Fig. 52(b) and (c)]. The agreement was good, differing only in band intensity and width, effects attributed to the then-limited resolution of the IRRAS technique (30 cm^{-1} at 2100 cm^{-1})*. The authors stated that such a bipolar band might be obtained if the features at the two potential limits were as in Fig. 52(d). However, the IRRAS spectra ruled out this possibility, clearly demonstrating the potential-dependent shift and supporting the interpretation of the EMIRS data. In this paper, the IRRAS spectra show quite distinct curvature, due possibly to the water absorption. This curvature disappears on taking the difference between the spectra at two potentials, illustrating again the lack of surface selectivity of the "so-called" (sic [110]) surface selection rule and the need for this to bear the mantle of being another difference technique. After the report by Kunimatsu and co-workers [94], PM-IRRAS spectra are generally presented as difference spectra, referenced to a base potential, or electrode state, at which no adsorbate is present (or, if the adsorbing species is present at high concentrations, to the electrode in the complete absence of the adsorbing species).

The other studies carried out using the IRRAS technique have concerned the adsorption of CN^- on silver and gold electrodes. Figure 53 shows the change in the FT-IRRAS spectrum of a reflective silver electrode immersed in 0.5 M K_2SO_2/1 M KCN with potential, the spectra being referenced to silver in CN^--free solution (thus guaranteeing monopolar bands). As can be seen from the figure, four bands appear in the chosen spectral region. The features at 2080, 2136, and 2167 cm^{-1} have potential-independent frequencies; whilst the fourth band between those at 2080 and 2136 cm^{-1} has a potential-dependent frequency, the authors took this as a good indication that this band was due to a surface species. The remaining bands were assigned to surface or solution species by the (now necessary) method of probing the surface with first p-polarised light and then s-polarised. The 2080 and 2135 cm^{-1} bands were observed with both polarizations, whilst the potential-dependent 2109 cm^{-1} feature and the band at 2167 cm^{-1} were only observed with p-polarised light and were thus attributed to surface species. Based on the work of Jones and Penneman [115] and Anderson et al. [116], the authors assigned the 2080 cm^{-1} feature to solution CN^{-1} and the 2136 cm^{-1} peak to the $Ag(CN)_2^-$ complexion. The potential-independent (surface) fea-

* N.B. Resolution could be improved at the expense of throughput by using a monochromator slit; however, the PM technique does lead to a high attenuation of the beam and any further loss of beam intensity could be ill-afforded.

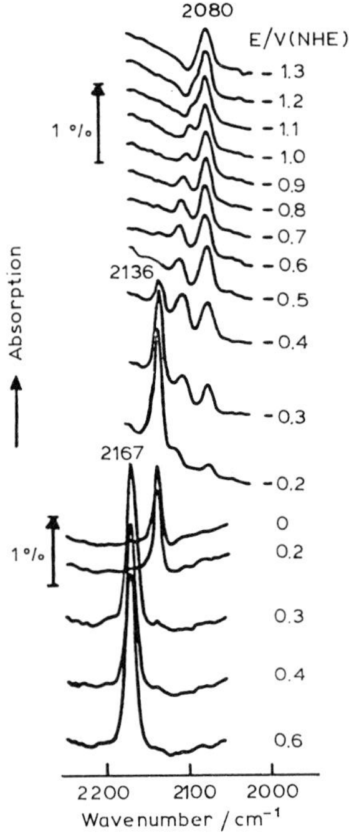

Fig. 53. Change of the FT-IRRAS spectrum with potential of the Ag/0.1 N KCN solution interface. All the spectra are referred to the CN^--free 0.5 M K_2SO_4 solution.

ture at 2167 cm^{-1} was attributed to precipitated AgCN. The potential-dependent surface band was tentatively assigned to adsorbed CN^-, the decrease in its intensity at cathodic potentials being ascribed to partial desorption. The results obtained for the adsorption of CN^- on gold were very similar.

Although the IRRAS technique has assumed the role of a difference method, it has still provided a large amount of information which supplements and supports that derived from the EMIRS studies so that, together with the IR-LSV technique, it helps to provide a much more complete picture of processes occurring at an electrode. It is thus still an important, although experimentally difficult, technique requiring careful interpretation. Any systems investigated by IRRAS involving absorptions in the region of strong solvent absorptions must be investigated using a potential-dependent spectral subtraction. In an area of small, or zero, solvent absorption, i.e. $\approx$ 2000–3000 cm^{-1} for an aqueous system, it can be used as an "absolute" technique so long as the features of interest are strong enough to be seen against the sloping solvent background absorption (see Fig. 51). Any advantage of IRRAS over SNIFTIRS when the former is acting as a difference technique probably lies in the increased sensitivity, once solvent absorptions etc. have been removed, of an a.c. over a d.c. technique.

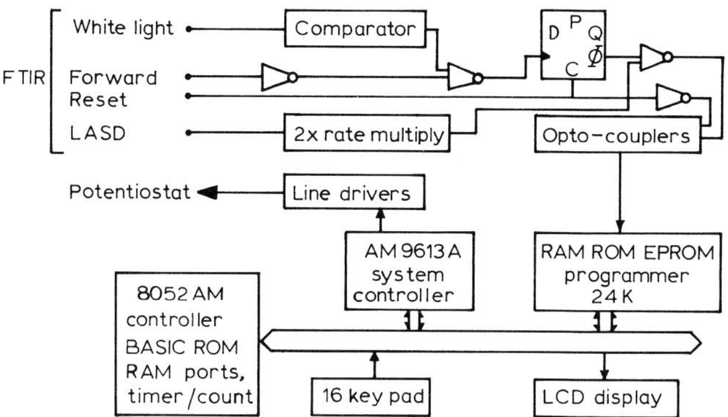

Fig. 54. Block diagram of the timer–sequencer–controller for coupling the electrochemical experiment to the interferometer data collect.

7. Transient techniques

The final section of this chapter is concerned with time-resolved and transient in-situ infrared spectroelectrochemistry. For the former technique, two different approaches have been reported [117, 118] involving both external and internal reflectance. The external reflectance method was reported by Daschbach et al. [117] in 1986 and employs a "standard" SNIFTIRS and EMIRS thin layer cell and optics, with the additional instrumentation shown in Fig. 54. This procedure for performing time-resolved FTIR experiments involves the "time sequence triggering" or "stroboscopic" method [119]. Thus, each point in the interferogram $I(x)$ is acquired over a short period of time (dictated by the digitisation time of the analogue/digital converter) at a specified time after the start of the forward sweep of the interferometer (see Fig. 55). Although the points $I(x)$ in the interferogram are acquired at evenly spaced increments (δ) in terms of the mirror retardation (χ), for a high quality rapid-scanning interferometer, these points have reasonably well-defined times relative to some spatially fixed event during the scan. Thus, these points can be used as a time base for the sample event being monitored. Hence, if the mirror travels 2δ from the point of max retardation ($\chi = -M$) at a fixed velocity V, then the intensity at the detector at this point is $I_{(-M+2\delta)}$. If the external relaxation experiment is triggered by this time base at $\chi = -M$); then, if the time at $-M$ is t_0, at $-\mathrm{M} + 2\delta$ it is $t_0 + (2\delta/V)$ and the intensity at the detector at this retardation or time can be given by $I_{[-\mathrm{M}+2\delta,\, t_0+(2\delta/V)]}$. The second term in the subscript is now the time base of the experiment. If the process is now triggered at $\chi = M + \delta$, then the intensity at the detector at $\chi = -M$ is now $I_{[-M,\, t_0-(\delta/V)]}$. By sequentially triggering the sample event at later times in the scan for successive interferogram files, it is possible to build a set of data consisting of all $I_{[(-\mathrm{M}+n\delta),\, \{t_0 \mp (n\delta/V)\}]}$; $n = 0$ to maximum sample count. The Fourier transform of those files having the same $[t_0 + n(\delta/V)]$ values (i.e. by

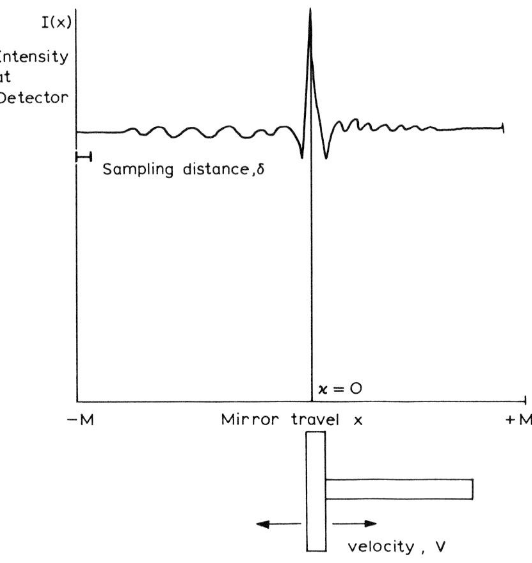

Fig. 55. Schematic representation of an interferogram obtained from an FTIR spectrometer showing the sampling distance and mirror travel. The maximum retardation at the mirror is 1 M.

transforming diagonally across the two-dimensional χ/t domain) yields the spectrum after a particular time after the beginning of the induced (e.g. by a potential step) process (see Fig. 56).

Such a technique is possibly even more severely limited to reversible electrochemical systems than EMIRS. Using this method, the authors obtained time-domain FTIR spectra for the reduction of adsorbed TCNE anion radical at a reflective Pt electrode in MeCN/TBAF, stepped from a base potential of $+0.6$ V to -0.8 V vs. Ag/Ag^+, as shown in Fig. 57. The positive band near 2083 cm^{-1} corresponds to the adsorbed TCNE radical anion and the negative feature near 2065 cm^{-1} to the $C{\equiv}N$ stretch of the solution-soluble dianion. The time resolution was reported to be 1 ms between spectra; with the use of a rapid-scan FTIR, capable of acquiring spectra at almost 100 s^{-1}, the authors postulated temporal resolutions of the order of 10 μs.

The time-resolved studies employing internal reflectance were reported by Yaniger and Vidrine [118] in 1986. The authors took advantage of the rapid scan facility of a modern FTIR instrument to development a relatively simple means of obtaining time-resolved spectra. The spectral acquisition time of such an instrument decreases with increasing resolution, i.e. from 85 scans s^{-1} at 16 cm^{-1} to 39 scans s^{-1} at 4 cm^{-1}. By storing such scans separately (instead of, as in an "invariant" system, co-adding and averaging to improve the S/N ratio), a time-resolved record can be obtained; up to a resolution of $\sim$12 ms.

In the digital time-resolved techniques [120] exemplified by the paper by Daschbach et al. [117] the interferograms of each particular "time snapshot"

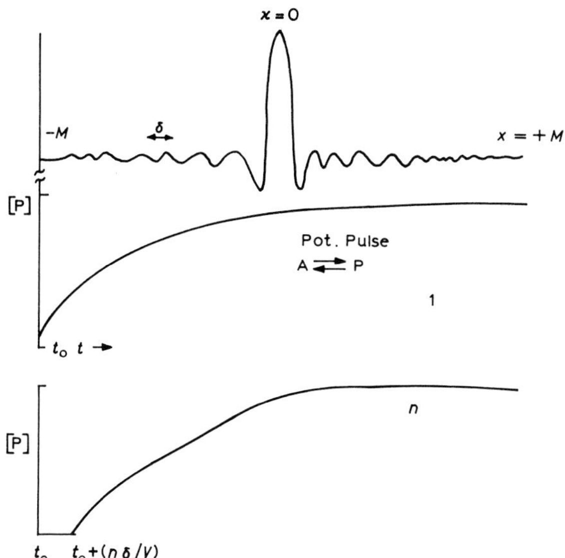

For Run 1,2,3,.... n, interferogram can be written as

Run 1

$$I_{(x)} = I_{(-M,t_o)} + I_{[-M+\delta,t_o+(\delta/V)]} + \cdots\cdots\cdots\cdots I_{[-M+n\delta,t_o+(n\delta/V)]}$$

Run 2

$$I_{(x)} = I_{[-M,t_o-(\delta/V)]} + I_{[-M+\delta,t_o]} + \cdots\cdots\cdots I_{[-M+n\delta,t_o+(\frac{(n-1)\delta}{V})]}$$

Run 3

$$I_{(x)} = I_{[-M,t_o-(2\delta/V)]} + I_{[-M+\delta,t_o-(\delta/V)]} + I_{[-M+2\delta,t_o]} + \cdots\cdots$$

Fig. 56. Fourier transform down diagonal to obtain individual spectra at each time resolution element after $t = t_0$.

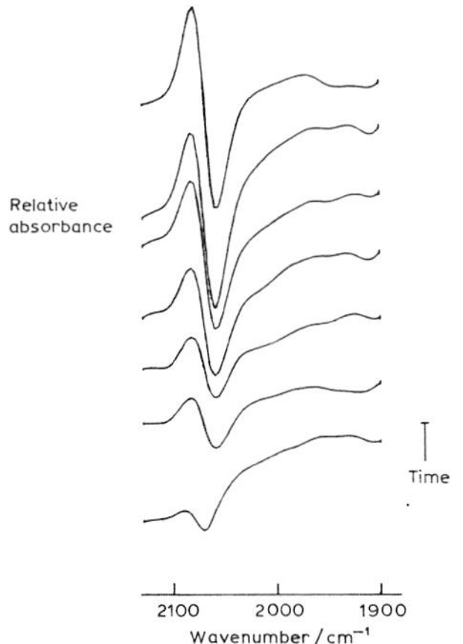

Fig. 57. Time-domain FT-IR spectra for the reduction of adsorbed TCNE anion at a Pt electrode in acetonitrile. The spectra represent sequential acquisitions spaced 0.001 s apart after the electrode potential is stepped. Bottom plot is 0.001 s; top is 0.007 s after step applied.

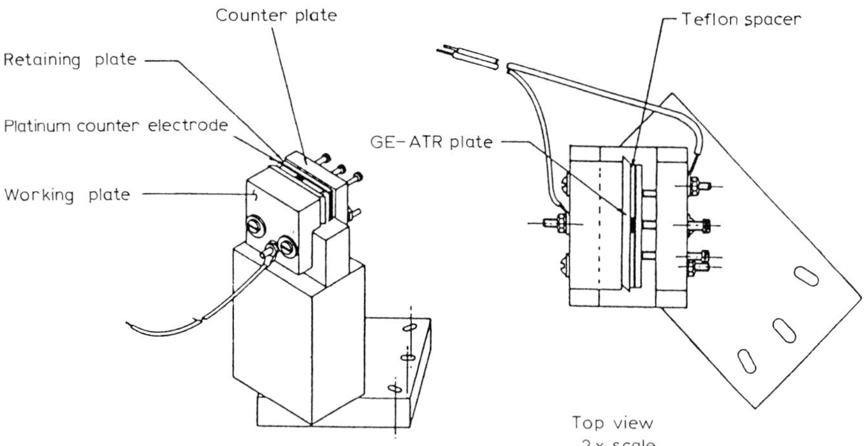

Fig. 58. The ATR spectroelectrochemical cell used in these experiments. The hole in the top of the Teflon spacer/gasket is used for the injection of the solvent/electrolyte and for the insertion of a reference electrode.

taken during a varying process is a composite of points taken from repeated runs of the process. Hence, the major factor determining the S/N ratio and freedom from spectral artefacts is the reversibility of the process under investigation. In contrast, the primary source of noise in a rapid-scan FTIR experiment is instrumental noise, much more favourable and controllable than sample irreversibility. Hence, the technique can be used to study any reaction regardless of its reversibility (cf. EMIRS and SNIFTIRS). As a rapid-scan experiment precludes any of the extensive co-adding and averaging of spectra in a SNIFTIRS experiment, it requires a very sensitive detector as well as a (fast) stable interferometer.

The spectroelectrochemical cell employed by Vidrine and co-workers is shown in Fig. 58. Using this technique, they obtained time-resolved spectra of the galvanostatic reduction of naphthalene to dihydronaphthalene and the naphthalide radical anion (Fig. 59). Each spectrum was the average of four scans collected at 35 scans s^{-1}, i.e. $\sim$115 ms between spectra. The authors also reported the results of studies on the growth and oxidation of polymers. Thus, the electropolymerisation of *N*-methylpyrrole in MeCN/TBAF on an uncoated Ge prism electrode was studied using a growth current density of 1 mA cm^{-2}. The results obtained are shown in Fig. 60(a) and (b); each spectrum was the result of four co-added scans taken at 65 scans s^{-1}, i.e. 60 ms between scans. About 1 s after switching in the galvanostatic control, adsorption and polymerisation of the monomer occurs with new absorbances appearing at 1535, 1450 and 1320 cm^{-1}, the polymer "growing" in its oxidised form. Several overlapping bands in the 1050–1150 cm^{-1} region were assigned by the authors to various fluoroborate species associated with the positively charged polymer, the authors commenting upon the intriguing observation that these species seemed to associate with the oxidised

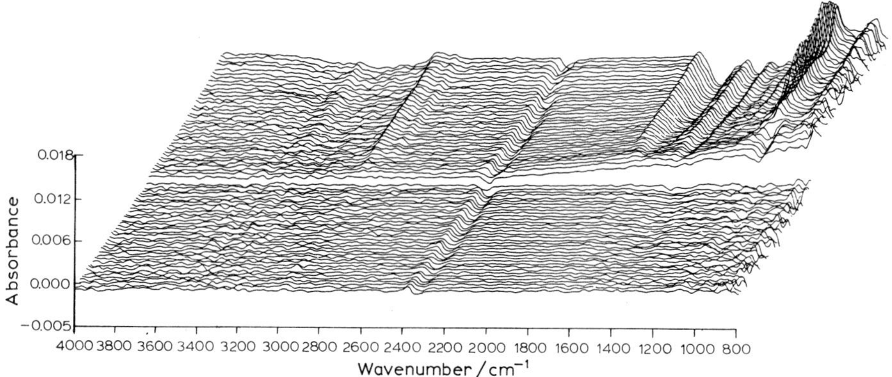

Fig. 59. Rapid-scan time-resolved in-situ FT-IR spectrum of the electroreduction of naphthalene (0.2 M) TBAB/tetrahydrofuran. The downward peaks indicate a decrease in naphthalene concentration near the ATR cathode. The positive peaks indicate the formation of naphthalide radical anion and dihydronaphthalene. The time displacement between successive spectra is 115 ms. Resolution is 10 cm^{-1}.

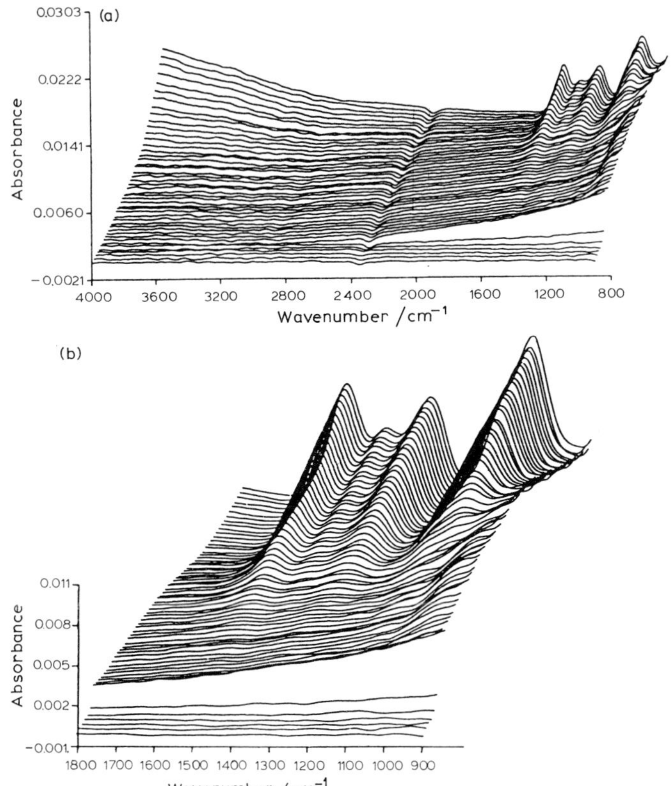

Fig. 60. (a) Rapid-scan time-resolved in-situ FT-IR spectrum of the electropolymerization of *N*-methylpyrrole. The time displacement between successive spectra is 60 ms. Resolution is 17 cm^{-1}. (b) Same run as shown in (a) but covering the spectral region 900–4000 cm^{-1}.

polymer a short time after the initial adsorption and commencement of polymerisation. The broad increase in absorbance above 2000 cm^{-1} was attributed to the characteristic IR absorptions of free charge carriers in conducting polymers [36]. In addition, since these absorptions increase faster than the features due to the oxidised polymer, the authors conclude that the electrical conductivity in a conducting polymer is a three-dimensional phenomenon and requires the presence of bulk polymers.

Using a Pt-coated silicon crystal, the oxidation of neutral polypyrrole was prepared ex-situ by holding the potential of the Pt at 0.8 V vs. SCE for 1 min in MeCN/1 M pyrrole/1 M TBAF electrolyte. The polymer-coated plate was then mounted in the spectroelectrochemical cell and was placed in contact with MeCN/0.5 M TBAF. The polymer was then re-oxidised at a current density of 0.1 mA cm^{-2} and spectra collected at 85 scans s^{-1} and 17 cm^{-1} (12 ms between spectra). In addition to the features due to the oxidised polymer (see Fig. 61), including the electronic absorptions, small negative features are observed due to loss of the neutral form. These features are almost completely obscured by the bands due to the oxidised form. The authors attributed to very large absorptions of the oxidised form to interaction between the vibrations of the polymer and the free charge carriers; thus, the sloshing back and forth of the charge as the chain vibrates leads to large dipole changes during vibration. In addition, the red shift of features due to the neutral form on oxidation was attributed to the progressive weakening of bonds in the π system where the charge resides with increasingly anodic potentials.

Thus, this paper gives tantalising glimpses of the possible applications of in-situ rapid-scan time-resolved FTIR spectroscopy. Unlike the digital time-resolved technique, the rapid-scan method does not require reaction reversibility and may thus have wider application. The technique does have its limitations, however.

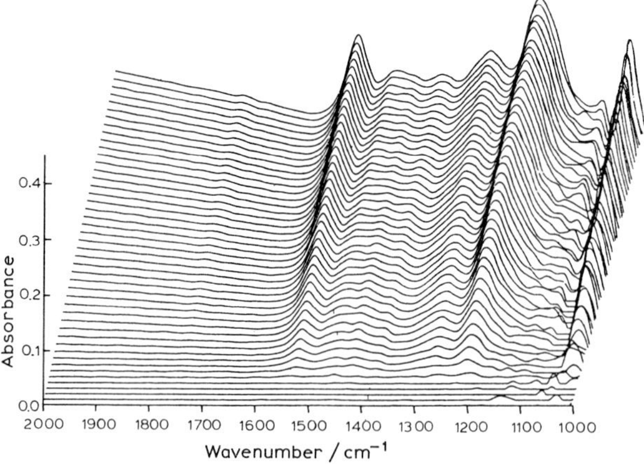

Fig. 61. Rapid-scan time-resolved in-situ FT-IR spectrum of the oxidation (p-doping) of polypyrrole. Time displacement between successive spectra is 12 ms. Resolution is 17 cm^{-1}.

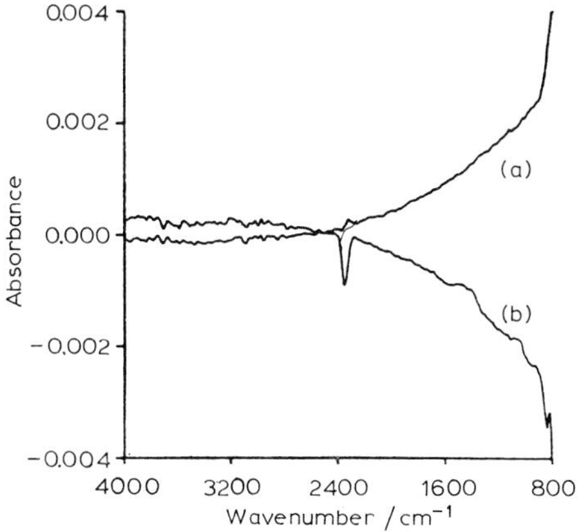

Fig. 62. Change in infrared absorbance of an uncoated germanium ATR plate with applied voltage. (a) +0.5 V, 0.5 M TBAB/acetonirile; (b) −0.5 V, 0.5 M TBAB/tetrahydrofuran.

(a) The temporal resolution does vary inversely with the spectral resolution, higher values of the former being achieved at the expense of the latter.

(b) Due to charge migration to and from the surface of semiconductors at potentials of +0.5 and −0.5 V with respect to the platinum counter electrode, variations in absorbance at low wavenumbers occur (see Fig. 62), hence the "discontinuities" in the various spectra shown above. This phenomenon, as expected, is lessened if platinum-coated prisms are used, at the expense of throughput and thus the S/N ratio. (30 Å of Pt results in an 8-fold decrease in throughput.)

(c) Side reactions may occur at a semiconductor electrode, although this is less of a problem with Si.

(d) Some electrochemical reactions will not proceed on uncoated semiconductor electrodes.

(e) The authors stated that, contrary to general belief, "high interferometer scan speeds are not antagonistic" to the S/N ratio. They quote a "worst case" single scan 100% line taken through a Pt-coated germanium crystal as having noise of $\leqslant 1\%$T. However, the maximum noise allowable in order to observe a monolayer in a SNIFTIRS, etc. experiment is $\leqslant 10^{-2}\%$T ($\Delta R/R \leqslant 10^{-4}$). Hence, this technique is probably limited to "bulk" electrochemical systems, i.e. electropolymerisation, electron-transfer reactions, etc. The digital time-resolved technique, apparently, does not suffer from this limitation.

The final in-situ infrared spectroelectrochemical technique was reported in a preliminary note in 1987 by Roe et al. [121]. The paper reported the development of a new design of a spectroelectrochemical cell with a faster response time to enable the recording of transient IR spectra using the recently developed pulsed IR tunable lasers [122]. A reflection–absorption spectrum was given of CO adsorbed on Pt at +0.2 V (vs. SCE) in 1 M $HClO_4$

References pp. 74–77

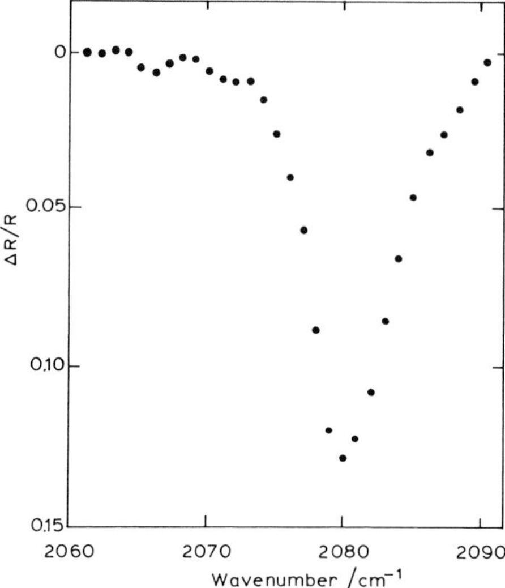

Fig. 63. IRRAS of CO adsorbed on platinum electrode at + 0.2 V (SCE) in 1 M $HClO_4$ without any modulation technique.

(see Fig. 63); each point is the average of 30 laser pulses. No modulation technique was applied (or necessary) as the intensity of the CO feature was so high.

References

1 R.W. Revie, B.G. Baker and J.O'M. Bockris, Surf. Sci., 52 (1975) 664.
2 B.G. Baker, J.O'M. Bockris and R.W. Revie, J. Electrochem. Soc., 122 (1975) 1460.
3 N. Winograd, in J.O'M. Bockris, B.E. Conway, E. Yeager and R. White (Eds.), Comprehensive Treatise of Electrochemistry, Vol. 8, Plenum Press, New York, 1984, p. 445.
4 T.E. Pou, O.J. Murphy, J.O'M. Bockris, V. Young and L.L. Jongson, J. Electrochem. Soc., 131 (1984) 1243.
5 O.J. Murphy, J.O'M. Bockris, T.E. Pole, L.L. Jongson and M.D. Monkowski, J. Electrochem. Soc., 130 (1983) 1792.
6 O.J. Murphy, J.O'M. Bockris, T.E. Pou, D.L. Cocke and G. Sparrow, J. Electrochem. Soc., 129 (1982) 2149.
7 M.R. Philpott, F. Barz, J.G. Gordon, II and M.J. Weaver, J. Electroanal. Chem., 150 (1983) 399.
8 M. Fleischmann, P.J. Hendra and A.J. McQuillan, Chem. Phys. Lett., 26 (1974) 163.
9 M. Fleischmann and I.R. Hill in J.O'M. Bockris, B.E. Conway, E. Yeager and R. White (Eds.), Comprehensive Treatise of Electrochemistry, Vol. 8, Plenum Press, New York, 1984, p. 373.
10 D.L. Jeanmarie and R.P. Van Duyne, J. Electroanal. Chem., 84 (1977) 1.
11 R.P. Van Duyne, J. Phys. (Paris), 38 (1977) C5.
12 A. Campion, Paper presented at Ellipsometry and other Optical Methods for Surface and Thin Film Analysis, Paris, France, June 1983.
13 M. Fleishmann, Paper presented at 34th ISE Meeting, Erlangen, F.R.G., September 1983.
14 W.N. Hansen, Adv. Electrochem. Electrochem. Eng., 9 (1973) 1.

15 J.D.E. McIntyre, in B.O. Seraphin (Ed.), Optical Properties of Solids: New Developments, Elsevier, Amsterdam, 1976, p. 559.

16 E.R. Dobbs, Electromagnetic Waves, Routledge and Kegan Paul, London, 1985.

17 P. Lorrain and D. Corson, Electromagnetic Fields and Waves, Freeman, Reading, 2nd edn., 1970.

18 H.B. Mark, Jr. and B.S. Pons, Anal. Chem., 38 (1966) 119.

19 J.V. Sanders, in J.R. Anderson (Ed.), Chemisorptions and Reactions on Metallic Films, Vol. 1, Academic Press, London, 1971, pp. 1–39.

20 P. Lange, V. Glaw, H. Neff, E. Piltz and J.K. Sass, Vacuum, 33 (1983) 763.

21 (a) J.S. Mattson and C.A. Smith, Anal. Chem., 47 (1975) 1122. (b) A. Prostak, H.B. Mark, Jr. and W.N. Hansen, J. Phys. Chem., 72 (1968) 2576.

22 H. Neff, P. Lange, D.K. Roe and J.K. Sass, J. Electroanal. Chem., 150 (1983) 513.

23 P.R. Griffith and J.A. de Haseth, Fourier Transform Infrared Spectrometry, Wiley-Interscience, New York, 1986.

24 M.A. Habide and J.O'M. Bockris, J. Electroanal. Chem., 180 (1984) 287.

25 D.R. Tallant and D.H. Evans, Anal. Chem., 41 (1969) 835.

26 A.H. Reed and E. Yeager, Electrochim. Acta, 15 (1970) 1345.

27 H. Neugebauer, G. Nauer, N. Brinda-Konopik and G. Gidaly, J. Electroanal. Chem., 122 (1981) 381.

28 A. Tardella and J.-N. Chazalviel, Appl. Phys. Lett., 47 (1985) 334.

29 A. Hatta, Y. Chiba and W. Suetaka, Appl. Surf. Sci., 25 (1986) 327.

30 E.K. Kretschmann, Z. Phys., 241 (1971) 313.

31 D.M. Kolb, in V.M. Agranovich and D.L. Mills (Eds.), Surface Polaritons, North-Holland, Amsterdam, 1982 p. 299.

32 A. Hatta, Y. Chiba and W. Suetaka, Surf. Sci., 158 (1985) 616.

33 K. Nakamoto, Infrared Spectra of Inorganic and Coordination Compounds, Wiley-Interscience, New York, 1970.

34 S.D. Ross, Inorganic Infrared and Raman Spectra, McGraw Hill, London, 1972.

35 H. Neugebauer, G. Nauer, A. Neckel, G. Tourillon, F. Garnier and P. Lang, J. Phys. Chem., 88 (1984) 652.

36 (a) L.W. Shacklette, H. Eckhardt, R.R. Chance, G.G. Miller, D.M. Ivory and R.H. Baughmann, J. Chem. Phys., 73 (1980) 4098. (b) S.I. Yaniger, D.L. Rose, W.P. McKenna and E.M. Eyring, Appl. Spectrosc., 38 (1984) 7.

37 J.F. Rabolt, T.C. Clarke and G.B. Street, J. Chem. Phys., 71 (1979) 4614.

38 (a) F. Abeles, T. Lopez-Rios and A. Tadjeddine, Solid State Commun., 16 (1975) 843. (b) R. Kotz, D.M. Kolb and J.K. Sass, Surf. Sci., 69 (1977) 359. (c) J.G. Gordon and S. Ernst, Surf. Sci., 101 (1980) 499. (d) A. Tadjeddine, D.M. Kolb and R. Kotz, Surf. Sci., 101 (1980) 277.

39 H. Raether, in G. Hass, M.H. Francombe and R.W. Hoffman (Eds.), Physics of Thin Films, Vol. 7, Academic Press, New York, 1977, p. 145.

40 (a) D. Eisenberg and W. Kauzmann, The Structure and Properties of Water, Oxford University Press, London, 1969. (b) K. Kretschmar, J.K. Sass, A.M. Bradshaw and S. Holloway, Surf. Sci., 115 (1982) 183.

41 G.E. Walrafen, J. Chem. Phys., 55 (1971) 768.

42 P. Rhine, D. Williams, G.M. Hale and M.R. Querry, J. Phys. Chem., 78 (1974) 1405.

43 C.D. Russell, J. Electroanal. Chem., 6 (1963) 486.

44 (a) R.W. Murray, W.R. Heineman and G.W. O'Dom, Anal. Chem., 39 (1967) 1666. (b) M. Petek, T.E. Neal and R.W. Murray, Anal. Chem., 43 (1971) 1609.

45 W.R. Heineman, J.N. Burnett and R.W. Murray, Anal. Chem., 40 (1968) 1970.

46 A. Bewick and K. Kunimatsu, Surf. Sci., 101 (1980) 131.

47 T. Davidson, S. Pons, A. Bewick and P.P. Schmidt, J. Electroanal. Chem., 125 (1981) 237.

48 J.S. Clarke, A.T. Kuhn, W.J. Orville-Thomas and M. Stedman, J. Electroanal. Chem., 49 (1974) 199.

49 S. Pons, Ph.D. Thesis, University of Southampton, Gt. Britain, 1979.

50 (a) A. Bewick and J. Robinson, J. Electroanal. Chem., 60 (1975) 163. (b) A. Bewick and J. Robinson, Surf. Sci., 55 (1976) 349. (c) R. Adzic, B.D. Cahan and E. Yeager. J. Chem. Phys., 58 (1973) 1780. (d) A. Bewick, J.M. Mellor and S. Pons, Electrochim. Acta, 25 (1980) 931.
51 A. Bewick, K. Kunimatsu and B.S. Pons, Electrochim. Acta, 25 (1980) 465.
52 E.R. Dobbs, Electromagnetic Waves, Routledge and Kegan Paul, London, 1985, p. 67.
53 A. Bewick, K. Kunimatsu, B.S. Pons and J.W. Russell, J. Electroanal. Chem., 160 (1984) 47.
54 A. Bewick, K. Kunimatsu, J. Robinson and J.W. Russell, J. Electroanal. Chem., 119 (1981) 175.
55 A. Bewick and S. Pons, in R.J.H. Clark and R.E. Hester (Eds.), Advances in Infrared and Raman Spectroscopy, Vol. 12, Wiley-Heyden, London, 1985.
56 J.K. Foley, C. Korzeniewski, J.J. Daschbach and S. Pons, in A.J. Bard (Ed.), Electro-Analytical Chemistry. A Series of Advances, Vol. 14, Dekker, New York, 1986.
57 (a) R.G. Greenler, J. Chem. Phys., 50 (1968) 1963. (b) S.A. Francis and A.H. Ellison, J. Opt. Soc. Am., 49 (1959) 131. (c) J. Pritchard and M.L. Sims, Trans. Faraday Soc., 66 (1970) 427. (d) H.G. Tompkins and R.G. Greenler, Surf. Sci., 28 (1971) 194.
58 R.G. Greenler, J. Chem. Phys., 44 (1966) 310.
59 (a) A. Capon and R. Parsons, J. Electroanal. Chem., 44 (1973) 1. (b) B.D. McNicol, J. Electroanal. Chem., 118 (1981) 71. (c) M.W. Breiter, in J.O'M. Bockris and B.E. Conway (Eds.), Modern Aspects of Electrochemistry, Vol. 10, Plenum Press, New York, p. 161.
60 (a) M.W. Brieter, J. Electroanal. Chem., 14 (1967) 406; 15 (1967) 221. (b) T. Biegler, Aust.J. Chem., 22 (1969) 1583. (c) J. Sobkowski and A. Wieckowski, J. Electroanal. Chem., 34 (1972) 185; 63 (1975) 365. (d) G.C. Allen, P.M. Tucker, A. Capon and R. Parsons, J. Electroanal. Chem., 50 (1974) 335.
61 (a) B. Beden, A. Bewick and C. Lamy, J. Electroanal. Chem., 148 (1983) 147. (b) A. Bewick, K. Kunimatsu, B. Beden and C. Lamy, 32nd ISE Meeting, Dubrovnik, September, 1981, Ext. Abstr. A28, Vol. 4, p. 92.
62 B. Beden, C. Lamy, A. Bewick and K. Kunimatsu, J. Electroanal. Chem., 121 (1981) 343.
63 J.W. Russell, J. Overend, K. Scanlon, M. Severson and A. Bewick, J. Phys. Chem., 86 (1982) 3066.
64 B. Beden, A. Bewick, M. Razaq and J. Weber, J. Electroanal. Chem., 139 (1982) 203.
65 A. Bewick and J.W. Russell, J. Electroanal. Chem., 132 (1982) 329.
66 J. Giner, Electrochim. Acta, 8 (1963) 857.
67 K. Kunimatsu, J. Electroanal. Chem., 140 (1982) 205.
68 K. Kunimatsu, J. Electron Spectrosc. Relat. Phenom., 30 (1983) 215.
69 J.K. Foley and S. Pons, Anal. Chem., 57 (1985) 945A.
70 A. Bewick, J. Electroanal. Chem., 150 (1983) 481.
71 J. Horiuti and T. Toya, in M. Green (Ed.), Solid State Surface Science, Vol. 1, Dekker, New York, 1969.
72 S. Trasatti, J. Electroanal. Chem., 150 (1983) 1.
73 A. Bewick and T. Solomun, J. Electroanal. Chem., 150 (1983) 481.
74 A.J. Bard and L.R. Faulkner, Electrochemical Methods, Wiley, New York, 1980.
75 (a) C. Korzeniewski and S. Pons, J. Vac. Sci. Technol. B, 3 (1985) 1421. (b) C. Korzeniewski and S. Pons, Langmuir, 2 (1986) 468.
76 C. Korzeniewski, R.B. Shirts and S. Pons, J. Phys. Chem., 89 (1985) 2297.
77 K.F. Purcell and J.C. Kotz, Inorganic Chemistry, Holt-Saunders, London, 1977.
78 (a) A.A. Michelson, Philos. Mag., 31 (5) (1891) 256. (b) A.A. Michelson, Light Waves and their Uses, University of Chicago Press, Chicago 1902; reissued in the paperback Phoenix edition, 1962.
79 T. Davidson, S. Pons, A. Bewick and P.P. Schmidt, J. Electroanal. Chem., 125 (1981) 237.
80 K. Chandrasekaran and J. O'M. Bockris, Surf. Sci. 185 (1987) 495.
81 D.S. Corrigan, J.K. Foley, P. Gao, S. Pons and M.J. Weaver, Langmuir, 1 (1985) 616.
82 D.S. Corrigan and M.J. Weaver, J. Phys. Chem., 90 (1986) 5300.
83 D.S. Corrigan, P. Gao, L.-W.H. Leung and M.J. Weaver, Langmuir, 2 (1986) 744.

84 B. Aurian-Blajeni, M.A. Habib, I. Taniguchi and J.O'M. Bockris, J. Electroanal. Chem., 157 (1983) 399.
85 M.A. Habib and J.O'M. Bockris, J. Electrochem. Soc., 130 (1983) 2510.
86 P. Zelenay, M.A. Habib and J.O'M. Bockris, J. Electrochem. Soc., 131 (1984) 2464.
87 M.A. Habib and J.O'M. Bockris, J. Electrochem. Soc., 132 (1985) 108.
88 (a) J.O'M. Bockris, M.A. Habib and J.L. Carbajal, J. Electrochem. Soc., 131 (1984) 3032. (b) J.O'M. Bockris, B.R. Scharifker and J.L. Carbajal, Electrochim. Acta, 32 (1987) 799.
89 S. Pons, J. Electroanal. Chem., 150 (1983) 495.
90 S. Pons, T. Davidson and A. Bewick, J. Electroanal. Chem., 160 (1984) 63.
91 S. Pons and A. Bewick, Langmuir, 1 (1985) 141.
92 J.K. Foley, S. Pons and J.J. Smith, Langmuir, 1 (1985) 697.
93 S. Pons, M. Datta, J.F. McAleer and A.S. Hinman, J. Electroanal. Chem., 160 (1984) 369.
94 H. Seki, K. Kunimatsu and W.G. Golden, Appl. Spectrosc., 39 (1985) 437.
95 S. Pons, T. Davidson and A. Bewick, J. Am. Chem. Soc., 105 (1983) 1802.
96 M. Datta, J.J. Freeman and R.E.W. Janssen, Spectrosc. Lett., 18 (1985) 273.
97 S. Pons, T. Davidson and A. Bewick, J. Electroanal. Chem., 140 (1982) 211.
98 O.A. Petrii and I.G. Khomchenko, J. Electroanal. Chem., 106 (1980) 277.
99 (a) P. Saumagne, Ph.D. Thesis, Université de Bordeaux, 1961. (b) E. Gentric, Ph.D. Thesis, Université de Bretagne Occidentale, 1972. (c) P. Saumagne and M.L. Josien, Bull. Chem. Soc. Fr., (1958) 813.
100 R. Cooney, E.S. Reid, M. Fleischmann and P. Hendra, J. Chem. Soc. Faraday Trans. 2, (1977) 1691.
101 (a) M.P. Soriaga and A.T. Hubbard, J. Am. Chem. Soc., 104 (1982) 3937. (b) M.P. Soriaga, J. White and A.T. Hubbard, J. Phys. Chem., 87 (1983) 3048. (c) M.P. Soriaga and A.T. Hubbard, J. Am. Chem. Soc., 104 (1982) 2742.
102 H. Seki, K. Kunimatsu and and W.G. Golden, Appl. Spectrosc., 39 (1985) 437.
103 P.A. Christensen, A. Hamnett and P.R. Trevellick, J. Electroanal. Chem., 242 (1988) 23.
104 S. Pons and J.J. Smith, in preparation.
105 P.A. Christensen, A. Hamnett and R. Hillmann, J. Electroanal. Chem., 242 (1988) 47.
106 (a) M.S. Katbale and J.P. Devlin, J. Phys. Chem., 83 (1979) 1637. (b) J.C. Moore, D. Smith, Y. Youhne and J.P. Devlin, J. Phys. Chem., 75 (1971) 325. (c) J.J. Hinkel and J.P. Devlin, J. Chem. Phys., 58 (1973) 4750 and references cited therein.
107 C. Amatore and J.M. Savéant, J. Am. Chem. Soc., 103 (1981) 5021.
108 J.P. Suchet, Chemical Physics of Superconductors, Van Nostrand, London, 1965, Chap. 7.
109 W.G. Golden, D.S. Dunn and J. Overend, J. Catal., 71 (1981) 395.
110 W.G. Golden, K. Kunimatsu and H. Seki, J. Phys. Chem., 88 (1984) 1275.
111 K. Kunimatsu, H. Seki and W.G. Golden, Chem. Phys. Lett., 108 (1984) 195.
112 K. Kunimatsu, H. Seki, W.G. Golden, J.G. Gordon, II and M.R. Philpott, Surf. Sci., 158 (1985) 596.
113 K. Kunimatsu, H. Seki, W.G. Golden, J.G. Gordon, II and M.R. Philpott, Langmuir, 2 (1986) 464.
114 K Kunimatsu, J. Electroanal. Chem., 213 (1986) 149.
115 L.H. Jones and R.A. Penneman, J. Chem. Phys., 22 (1954) 965.
116 A.B. Anderson, R. Kotz and E. Yeager, Chem. Phys. Lett., 82 (1981) 130.
117 J. Daschbach, D. Heisler and S. Pons, Appl. Spectrosc., 40 (1986) 489.
118 S.I. Yaniger and D.W. Vidrine, Appl. Spectrosc., 174 (1986) 174.
119 (a) See, for example, J.E. Lasch, D.J. Burchell, T. Masoaks and S.L. Hsu, Appl. Spectrosc., 38 (1984) 351. (b) G. Tourillon and F. Garnier, J. Electroanal. Chem., 135 (1982) 173. (c) E.M.M. Genies and A.A. Syed, Synth. Met., in press.
120 A.W. Mantz, Appl. Opt. 17 (1978) 1347. A.A. Garrison, R.A. Crocombe, G. Masantov and J.A. DeHaseth, Appl. Spectrosc., 34 (1980) 389.
121 D.K. Roe, J.K. Sass, D.S. Bethune and A.C. Luntz, J. Electroanal. Chem., 216 (1987) 293.
122 D.S. Bethune and A.C. Luntz, Appl. Phys. B, 40 (1986) 107.

Chapter 2

Raman Spectroscopic Studies of Species In Situ at Electrode Surfaces

RONALD E. HESTER

1. Introduction

Raman spectroscopy now has become a powerful technique of considerable sensitivity for probing the identities, structures, bonding, and reaction dynamics of molecular species at electrode surfaces. Although the literature on this topic already is extensive, it is the aim of this review to provide an introduction to this still novel spectroscopic technique, to give a glimpse of some of its achievements to date, and to indicate the directions in which further progress is likely to be made in the near future.

The elementary concept and practice of Raman spectroscopy are not particularly new, the effect having been discovered by C.V. Raman in 1926. However, the introduction of laser light sources and other high technology components such as holographic gratings and triple spectrometers to improve stray light refection, intensified multi-channel detectors to enhance the instrumental sensitivity and speed, and tunable ultraviolet and short-pulse lasers to widen the scope of applications, has transformed the technique in recent years. Electronic and rotational spectra can readily be obtained by the Raman method, but it is most widely applied to the measurement and characterization of vibrational modes in the wavenumber range ca. 10–5000 cm^{-1}. Samples in the solid, liquid, and gaseous states are studied with equal facility. Fast and sensitive optical multichannel analyser (OMA) detectors provide a multiplex advantage which may be compared with that of FT infrared instruments but, since there are no moving parts in a multichannel Raman spectrograph, very much faster spectroscopy is done by Raman than by FTIR methods. Later in this article, we shall see Raman spectra obtained from an electroadsorbed transient species in 30 ps.

The old problem of fluorescence interference has largely been eliminated by the use of ultraviolet or far-red excitation, by digital filtering techniques, temporal discrimination with picosecond lasers, or (particularly relevant in the present context) by quenching through adsorption at a solid surface. It is becoming commonplace to use optical microscope objectives both for focusing the excitation laser beam on to a sample and for collection of the Raman light and projection into the spectrometer entrance slit, thus providing excellent spatial resolution (to ca. 1 μm) and highly efficient light collection. Samples in the picogram or femtolitre size range give excellent spectra with the aid of the Raman microscope.

References pp. 103–104

1.1 RESONANCE RAMAN ENHANCEMENT OF SENSITIVITY

The vibrational effect involves inelastic light scattering by a sample with vibrational modes which cause regular periodic oscillations in the sample (usually molecular) polarizability. Elastic light scattering (Rayleigh scattering) invariably is much stronger. Raman spectra therefore are generated by monochromatic light sources (lasers are ideal) and are usually measured on the long-wavelength side (the Stokes side) of the Rayleigh line. Any old wavelength of incident light will do for excitation of Raman spectra. However, if the excitation wavelength does happen to coincide with a strong electronic absorption band of the sample then it is commonly found that a large enhancement of the Raman band intensities occurs. This enhancement, which usually is of several orders of magnitude (up to ca. 10^6 times), is known as the resonance Raman (RR) effect.

It is inappropriate to go into the theory underlying this effect here, but it should be noted that not all Raman-active modes of a sample give enhanced band intensities in RR scattering. The mode selectivity depends on the symmetry of the sample molecule and the nature of its electronic transitions, which are actively involved in the Raman process. Thus, by changing the Raman excitation wavelength from coincidence with one electronic absorption band to another of a different symmetry type, the pattern of vibrational band resonance enhancements may change drastically. This dependence of Raman band intensity on the wavelength of the excitation source is known as the "excitation profile" and may also be used to provide information on the nature of the electronic transitions.

In addition to the vibrational mode selectivity, there is a further selectivity in band intensity enhancements through the RR effect. This further selectivity is associated with the localization of the electronic transition within that part of the sample molecule which is known as the "chromophore". Resonance Raman band enhancements are only possible for vibrational modes associated directly with the chromophoric part of the sample molecule. This feature of RR scattering is at one and the same time limiting and enabling. It clearly limits the information available from RR spectroscopy to the local structure and bonding at the chromophoric site. However, it enables us to probe the structure of such a site with good sensitivity and without interference from spectral features associated with vibrational modes of other, non-chromophoric, parts of the molecule. For the large and complex molecules haemoglobin and cytochrome c, for example, the chromophoric site probed by visible RR excitation happens to be the prosthetic group. Thus, RR spectroscopy of these molecules in the solid state, in free solution, or at an electrode surface, yields structure and bonding information specifically relating to the haem groups. The growing interest in the electrochemistry of biological systems, and in the utilization of enzymes for catalysis of reactions at electrode surfaces, is well served by this aspect of the RR effect. Enzyme active sites or the derived reaction centres of enzyme–substrate complexes are commonly chromophoric (absorbing in the near-UV region), as is the ubiquitous nicotinamide adenine dinucleotide, NAD^+ or NADH, coenzyme. The availability of ultraviolet laser sources has now opened up the possibility of a whole new area of bioelectrochemical systems for study by means of this sharply focused tool of RR spectroscopy.

1.2 SURFACE-ENHANCED RAMAN SPECTROSCOPY (SERS)

For a limited number of metal surfaces, adsorption of a molecular species in a thin (monomolecular layer) film results in a huge increase in the effective vibrational Raman scattering cross-section (again, as with RR scattering, up to ca. 10^6 times) of the adsorbate species. The SERS effect was discovered more than ten years ago for pyridine adsorbed at a silver electrode surface in contact with an aqueous electrolyte [1, 2]. In the intervening period, many hundreds of papers devoted to SERS phenomena have been published, extending the studies to other metals than silver, to non-aqueous as well as aqueous electrolytes, to colloidal dispersions of metals as well as metal electrodes, and even to vacuum-deposited thin film systems under UHV conditions. This review will concentrate on studies of metal–electrolyte interfaces.

Silver was the first, and undoubtedly remains the best, metal for SERS. However, gold and copper electrodes also work very well [3, 4]. For all three metals, the excitation profiles show that red light gives the maximum SERS enhancement and that the effect falls off rapidly in the green-blue region of the spectrum [5, 6]. Nonetheless, green light (particularly argon-ion laser light at 514.5 nm) is commonly used for SERS excitation with silver electrodes due to the fact that the throughput and detector sensitivity of most Raman spectrometers in the green is better than in the red region. However, for gold and copper, red-end excitation is essential for obtaining large SERS enhancements and the kryption-ion laser line at 647.6 nm is often used.

Much weaker SERS effects have been reported for several other metals, including the alkali metals in low-temperature matrices [7], nickel [8], and aluminium [9]. None of these has been used as an electrode material for SERS, however. Other substrates for which weak SERS effects have been claimed are Fe, Pd, Rh, Cd, Hg, In, Sn, TiO_2, and even Pt (but only ca. $7\times$ enhancement).

In spite of the large number of research papers devoted to it, the theory of SERS remains ill-understood. There appears to be general agreement, however, that two major contributing effects need to be considered. These are an electromagnetic (EM) effect and a charge-transfer (CT) effect. Both effects appear to require some degree of surface roughening, although the exact nature of this is the subject of some controversy, as is the relative importance of the two effects in accounting for the observed SERS enhancement factors.

Evidence for an electromagnetic effect comes from several types of experiment which have been summarized helpfully by Chang [10]. The ingenious "spacer" experiments of Murray et al. [11] demonstrated the importance of

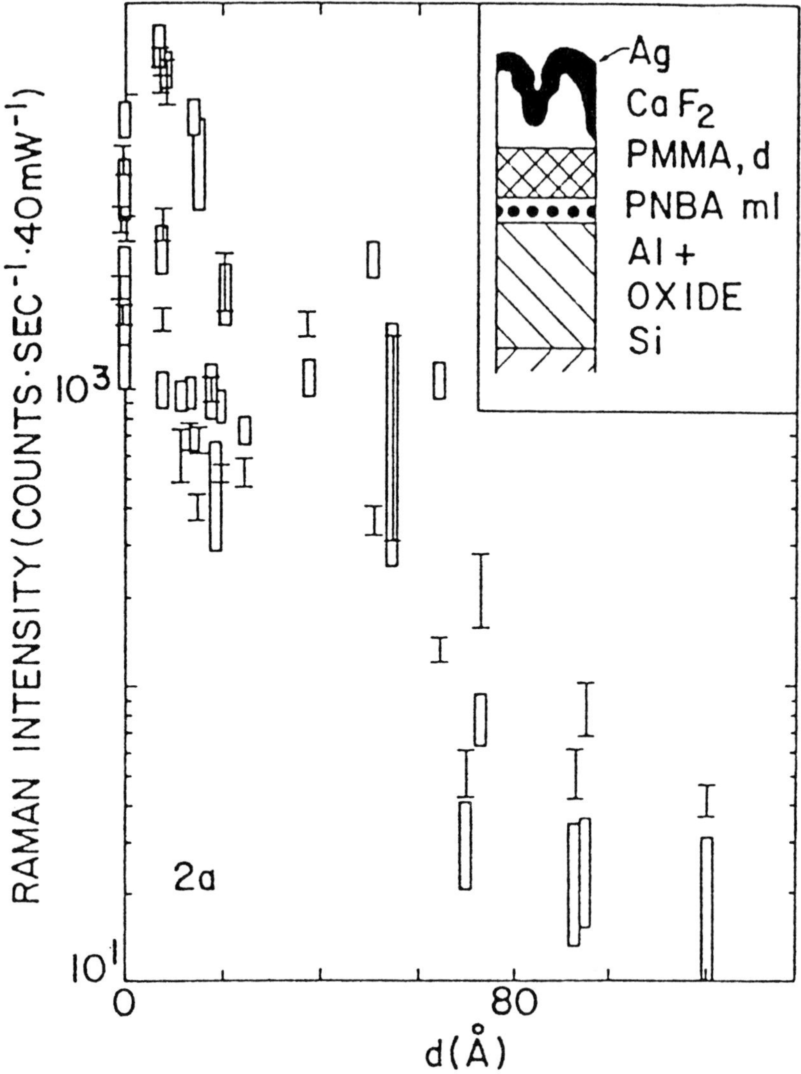

Fig. 1. Intensity of two Raman bands of a *p*-nitrobenzoic acid (PNBA) monolayer, adsorbed in the multilayer structure shown in the inset, vs. a polymethylmethacrylate (PMMA) spacer of thickness *d*. Open symbols, $1100\,cm^{-1}$ band; closed symbols $1597\,cm^{-1}$ band. (Reproduced with permission from ref. 11.)

long-distance enhancement effects which require an EM mechanism for their interpretation. Results from some of these experiments are shown in Fig. 1. These show SERS enhancement effects all the way out beyond 120 Å from the roughened silver surface. Clearly, the EM effect is a property of the metal, in this case silver, and is, as expected, effective for all adsorbates.

It is the CT effect which is adsorbate- as well as metal-dependent. This effect has been associated with the existence of particular "active sites" on

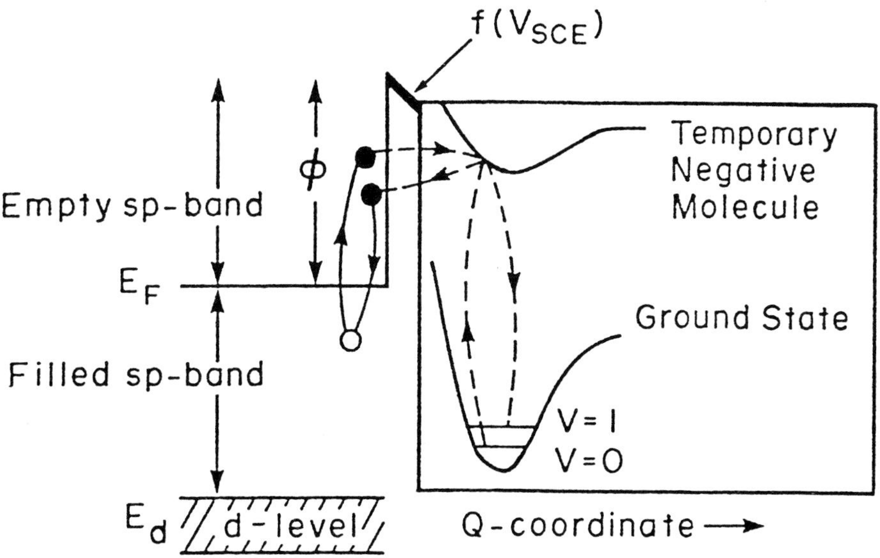

Fig. 2. Schematic representation of the charge-transfer process in SERS. (Reproduced with permission from ref. 12.)

the metal surface, which have variously been attributed to atomic-scale roughness features described as "adatoms" or "cavities". A model [12] for this CT effect is summarized in Fig. 2. Photoexcitation of an electron in the metal results in charge transfer from the metal to the adsorbed molecule, creating a temporary (i.e. short-lived) negative molecule. Reverse transfer of the electron back to the metal leaves the molecule in a vibrationally excited state and the electron then recombines with the hole in the *sp* band. During the time that the electron and hole are separated, they can lose energy of any amount by multiple electron–phonon interactions, etc., and thus their recombination gives rise to emission of the continuum which always accompanies SERS. The resonance condition for such a CT process is established when the Fermi level, E_F, of the metal is separated from the ground or excited state of the adsorbed molecule by an amount corresponding to the energy of the incident or Raman scattered light. It should be borne in mind that the electronic energy levels of the adsorbate may be modified by adsorption and tuned by variations in the applied electric field in the double layer at the electrode–electrolyte interface [10].

One further feature of SERS merits attention at this point. This is the widely recognized fact that strongly fluorescent molecules, whose Raman spectra are obscured by their much stronger and broader fluorescence emission, commonly have their strong fluorescence quenched by adsorption at a SERS-active metal surface. The fluorescence quenching mechanism is presumably closely related to that for the CT effect shown in Fig. 2.

References pp. 103–104

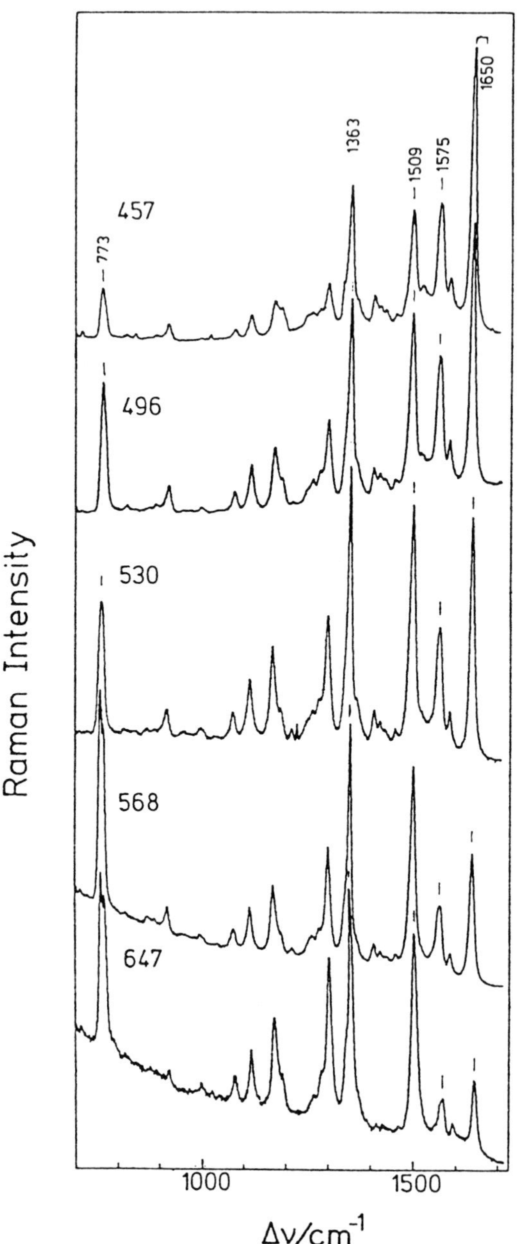

Fig. 3. SERR spectra from 10^{-9} M solutions of the dye R6G excited at a variety of different wavelengths from 457 to 647 nm. (Reproduced with permission from ref. 13.)

1.3 SURFACE-ENHANCED RESONANCE RAMAN SPECTROSCOPY (SERRS)

We have seen that the RR and SER effects can each give rise to enhancement in the effective Raman scattering cross-section for a vibrating molecule of up to ca. 10^6. Clearly, if these two effects can be combined, even more spectacular enhancement of the resulting Raman spectra may result.

It has been amply demonstrated by many published examples that RR and SER enhancement factors are indeed multiplicative, i.e. 10^6 and 10^6 gives 10^{12} times enhancement. The result is an extremely sensitive spectroscopic probe, enabling vibrational spectra of molecules to be measured with good signal-to-noise characteristics from solutions in the sub-nanomolecular concentration range by the simple expedient of adding colloidal silver and exciting the SERR spectra with light of a wavelength which is coincident with an absorption band of the molecule. Sub-monolayer coverages of metal electrode surfaces similarly may be examined by SERRS. The main restriction is, of course, that the excitation profiles of the metal and the adsorbate should match one another.

A good illustration of the great sensitivity which can be achieved by means of SERRS is the set of spectra shown in Fig. 3 for the dye molecule R6G on colloidal silver [13]. As is clearly seen, the quality of these spectra is such that the solutions could be further diluted by at least another order of magnitude, yet already these are from nanomolar solutions.

2. Spectroelectrochemical cells

Many different designs of electrochemical cell suitable for Raman spectroscopic examination of species generated by or adsorbed at the working electrode have been described in the literature. Most of them incorporate the features shown in Fig. 4. This schematic arrangement [14] includes reference and counter electrodes in addition to the working electrode. These are connected to a potentiostat and waveform function generator. This arrangement enables vibrational Raman spectra to be obtained from species in situ at the working electrode surface while the cell is under potentiostat control. Changes in the spectra which occur during the recording of a normal cyclic voltammogram may also be determined with this experimental arrangement.

In the usual arrangement, the working electrode surface is located close to the optical window (glass or fused silica) and the incident laser beam impinges on it at an angle of ca. 60° to the surface normal. The film of electrolyte between the electrode and the optical window is kept as thin as possible in order to minimize absorption of the laser beam and scattered light when the electrolyte is coloured and to minimize bulk solution features in the spectra. However, care is needed when running cyclic voltammograms with such a cell since diffusion in very thin films of electrolyte can be hindered and this may affect the shape and form of the voltammogram.

For Raman spectroelectrochemical studies of electrode materials with

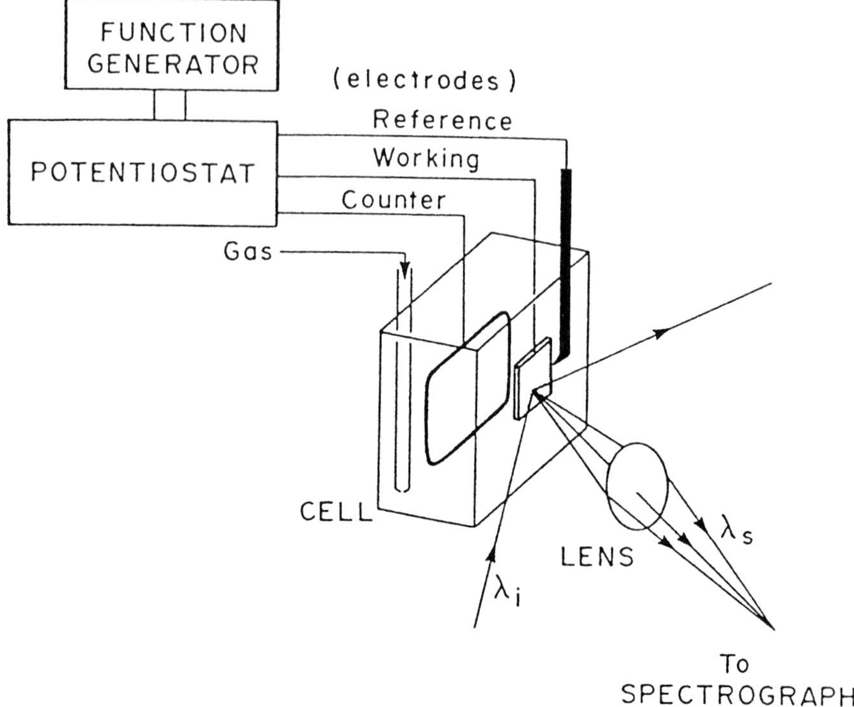

Fig. 4. Schematic arrangement of a three-electrode electrochemical cell suitable for laser Raman spectroscopy. (Reproduced with permission from ref. 14.)

strong optical absorption at the wavelength of the Raman laser, e.g. carbon, a cell with a rotating working electrode has been devized [15]. A cross-sectional drawing of this cell is shown in Fig. 5. The rotating electrode serves to minimize local heating effects at the focus of the excitation laser beam and thus to avoid sample decomposition. An alternative way of achieving this objective is to flow the sample electrolyte solution over the surface

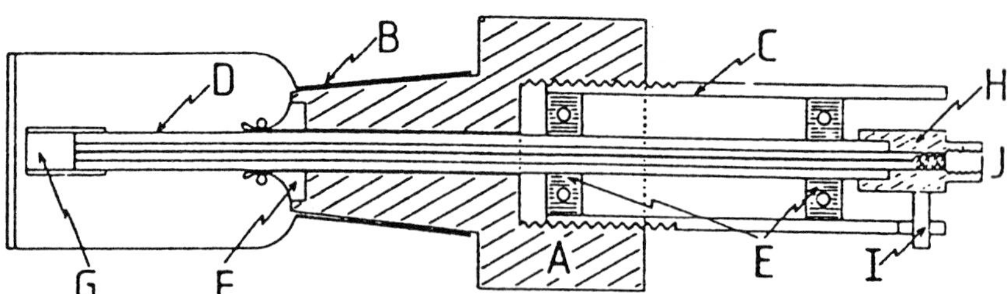

Fig. 5. Rotating electrode unit. A, Nylon block shaped to fit conical joint, B; C, electrode bearing housing (brass); D, rotating electrode mounting shaft (glass); E, ball bearings; F, PTFE seal; G, working electrode; H, brass slip ring; I, carbon brush; J, drive motor connecting point. (Reproduced with permission from ref. 15.)

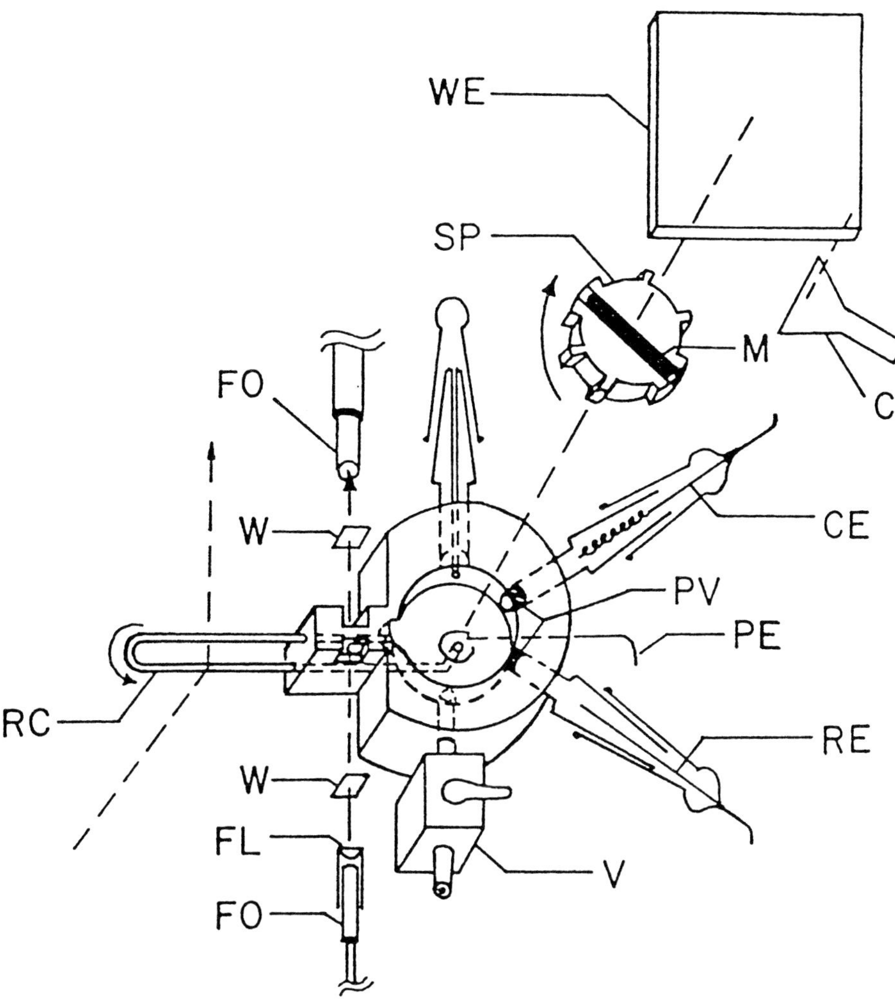

Fig. 6. Spectroelectrochemical circulating flow cell. WE, Working electrode; C, electrical contact to WE; SP, solution propeller; M, magnet to drive SP; CE, counter electrode; PE, potentiometric electrode; RE, reference electrode; PV, porous vycor plug to isolate CE, RE; V, valve; FO, fiber optic for absorption spectra; FL, focusing lens; W, quartz window; RC, Raman capillary for 90° excitation. (Reproduced with permission from ref. 16.)

of the working electrode. Such an arrangement is shown in Fig. 6. This anaerobic circulating flow cell was developed for the study of light-sensitive haem proteins [16].

Finally in this section, a Raman spectroelectrochemical microcell [17] is shown in Fig. 7. Having a total volume of 80 μl, this cell is particularly well suited to work with rare and/or expensive samples, such as biological compounds. Its design enables it to be used with a microscope objective as the laser focusing and Raman collection element. Such an arrangement

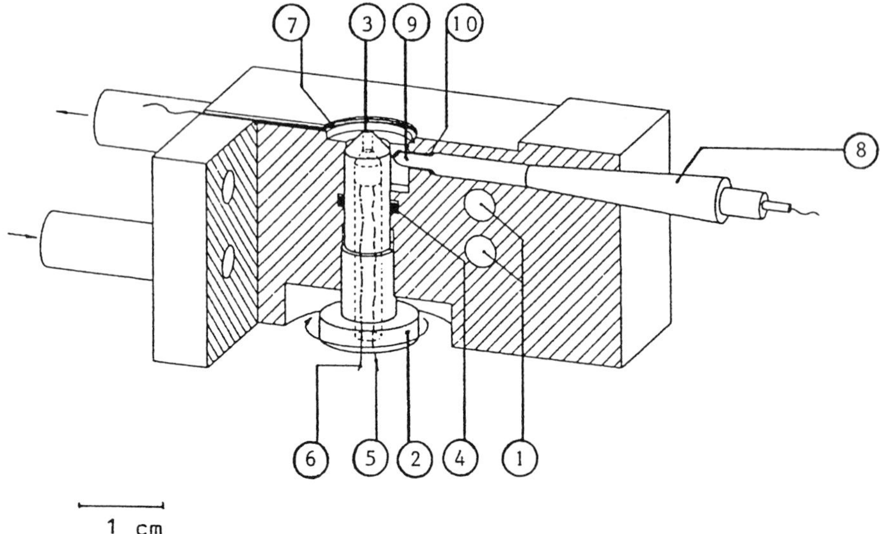

Fig. 7. Raman spectroelectrochemical microcell. 1, Water channels for temperature control; 2, perspex mount; 3, working electrode; 4, rubber O-ring; 5, wire connected to working electrode; 6, thermocouple; 7, counter electrode; 8, reference electrode; 9, capillary; 10, rubber O-ring. (Reproduced with permission from ref. 17.)

produces a very tightly focused illumination of the sample, typically with an illuminated volume of less than 1 μl.

2.1 ELECTRODE SURFACE ACTIVATION FOR SERS

Before an electrochemical cell such as that shown diagrammatically in Fig. 4 can be used for SERS experiments, the working electrode needs to be activated by means of one or more oxidation/reduction cycles (ORC). This activation is evidently associated with the production of microscopic roughness at the electrode surface and is commonly achieved through ORC treatment in one of two ways. Either the electrode is switched from a reducing potential, $-E_1$, to an oxidizing potential, E_2, held at E_2 for a period of time and then switched back again to $-E_1$ (this is called the double potential step method), or it is subjected to a linearly ramped potential from $-E_1$ to E_2, followed by the reverse ramping from E_2 to $-E_1$ (the triangular wave potential sweep). The two procedures are shown [14] in Figs. 8 and 9, respectively. The current vs. time behaviour associated with the double potential step is also shown in Fig. 8, while the current vs. potential (cyclic voltammogram) plot is given for the triangular wave potential sweep in Fig. 9.

Typical cyclic voltammograms for a silver electrode in 0.1 M KCl and in 0.1 M K_2SO_4 supporting electrolytes are shown in Fig. 10. These are produced under ORC conditions typically used for SERS activation. During the oxidation half-cycle, the Ag^0 electrode surface is oxidized to Ag^+, which forms an insoluble layer of AgCl or Ag_2SO_4 on the surface. The amount of

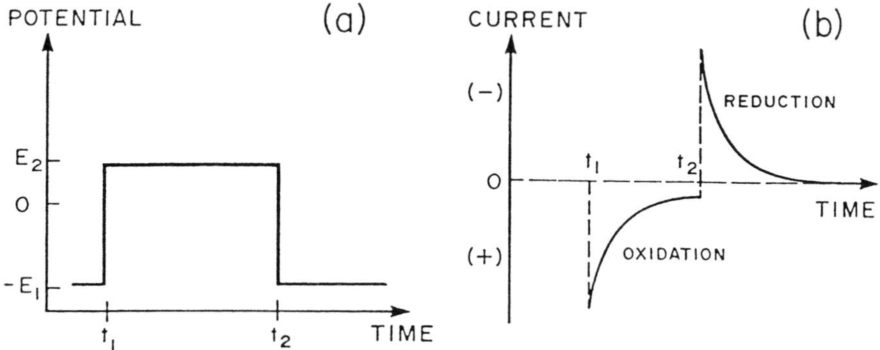

Fig. 8. Double potential step ORC to roughen an electrode surface for SERS activation. (Reproduced with permission from ref. 14.)

Ag^0 oxidized may be determined from the total charge passed during this stage. In the following reduction half-cycle this process is reversed and fresh Ag^0 is deposited at the electrode surface. The detailed topology of the resulting roughened electrode surface is critically influential on the SERS enhancement factors achieved. Larger enhancement factors are commonly observed for surfaces roughened by ORC treatment with concurrent laser illumination of the surface, the photosensitivity of the silver salts evidently leading to favourable modifications in the topology of the reconstructed silver surface.

An illustration of the effect of successive ORC treatments of a silver electrode on the SERS of adsorbed species [6] is shown in Fig. 11. The progressive increase in Raman band intensities for the co-adsorbed NO_3^- and EDFA species through five separate ORCs is clearly seen.

A small number of ORCs is usually sufficient to achieve the maximum enhancement factors for silver electrodes. Indeed, repeated cycling beyond four or five times has been observed to have a detrimental effect on the SERS enhancement. A rather different situation is found for gold, however. It has

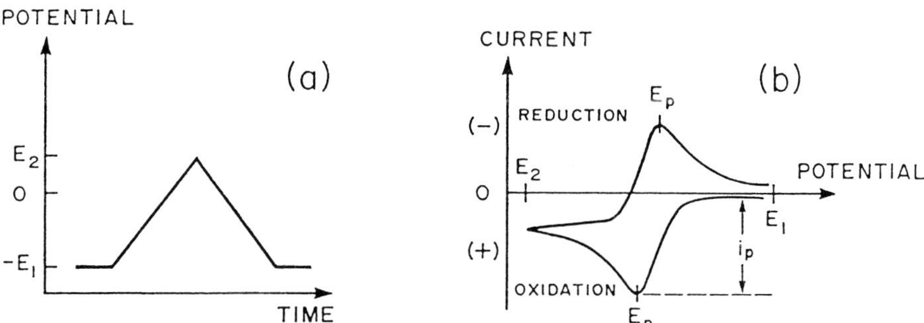

Fig. 9. Triangular wave potential sweep ORC and a typical resultant voltammogram. (Reproduced with permission from ref. 14.)

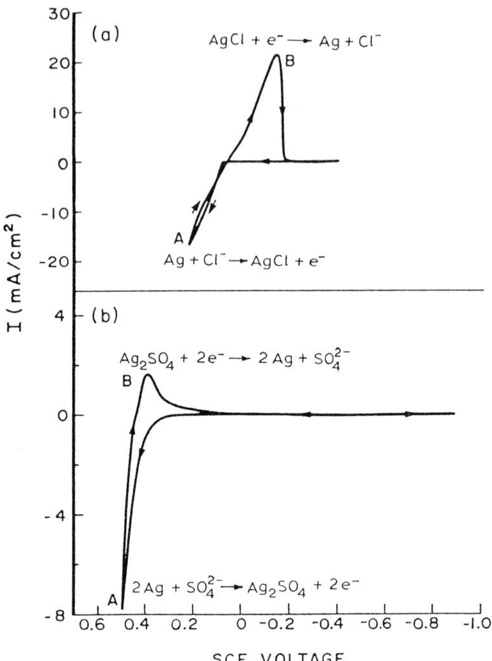

Fig. 10. Experimental forms of the cyclic voltammograms obtained for a silver electrode in an electrolyte of (a) 0.1 KCl and (b) 0.1 M K_2SO_4. 5 mV s^{-1} is the linear potential ramp rate. Voltages are given on the standard calomel electrode (SCE) scale. (Reproduced with permission from ref. 14.)

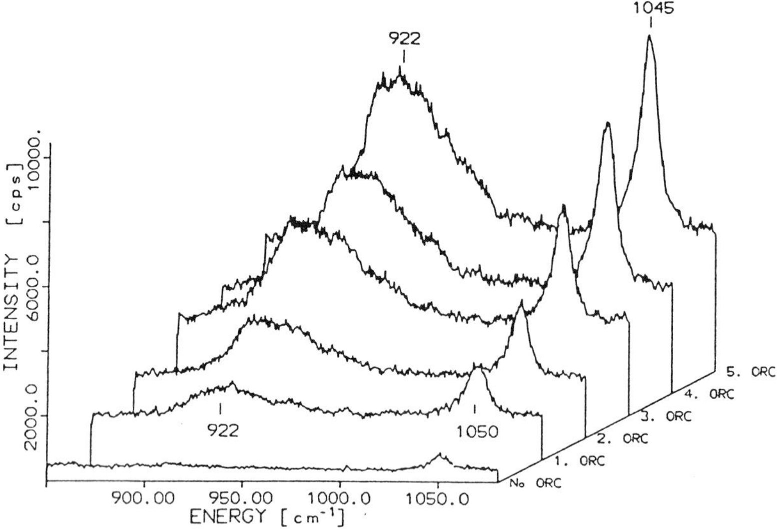

Fig. 11. SER spectra for co-adsorbed NO_3^- and EDTA at a silver electrode showing the effect of five successive ORC treatments. (Reproduced with permission from ref. 6.)

recently been shown that excellent SERS electrodes can be prepared by subjecting gold to large numbers (e.g. 20) of successive ORC treatments, the best results being obtained by using slow anodic–cathodic sweeps of potential rather than steps [18]. Since gold is a more useful electrode material for many purposes than is silver, this discovery of a good method for SERS activation of gold electrodes is particularly important.

One of the more controversial issues in the field of mechanistic studies of the SERS phenomenon is the role played by carbon deposits on the electrode surface. Surface carbon is thought to be formed from CO_2 or carbonate contaminants in the electrolyte or by laser photodecomposition of organic compounds, e.g. pyridine, during the ORC. It has been proposed [19, 20] that intercalation of adsorbed molecules within electrode surface graphitic carbon overlayers results in greatly increased local concentrations of adsorbate, sufficient to account for the observed enhancement factors in SER spectra. It is certainly true that many of the published SER spectra do show the celebrated "cathedral" peaks, which are characteristic of surface carbon in the 1200–1700 cm^{-1} region [21]. However, the consensus of opinion seems to be that, although carbon deposits undoubtedly often (perhaps always) do form under the conditions of a SERS experiment, the weight of evidence points to the carbon behaving as just another adsorbate rather than being the root cause of the SERS enhancement.

2.2 ELECTRODE POTENTIAL DEPENDENCE OF SERS

The electrode potential at which the actual surface charge is zero relative to the electrolyte solution is called the potential of zero charge (PZC). For polycrystalline silver this is -0.94 V (SCE), while for gold and copper it is -0.06 and -0.21 V, respectively [14]. This PZC obviously has a strong influence on the nature of the species attracted to the electrode surface from the electrolyte solution at any given applied electrode potential. Potentials on the positive side of the PZC will attract negative ions while those negative of the PZC will attract positive ions. Neutral species, such as pyridine, may be expected to bind most strongly at potentials in the region of the PZC. The PZC is itself changed by contact adsorption of ions from the electrolyte solution (e.g. Na^+ and K^+) and will therefore vary with the nature of the electrolyte and its concentration.

In order to understand better the SERS effects to be described in the following sections, it is convenient to consider briefly here the generally accepted model of the electrical double layer at an electrode/electrolyte interface. In water, a primary hydration sheath is bound to the electrode surface and to ions with high ionic potentials (charge/size ratio), e.g. Na^+, K^+, and F^-, but not Cs^+, Cl^-, and Br^-. The plane through the centres of the hydrated ions which are outside the electrode hydration sheath but are tightly bound by coulombic forces is called the outer Helmholz plane (OHP). The plane through the centres of ions adsorbed strongly enough at the electrode surface to displace H_2O molecules in the hydration sheath is called the inner Helmholz plane (IHP).

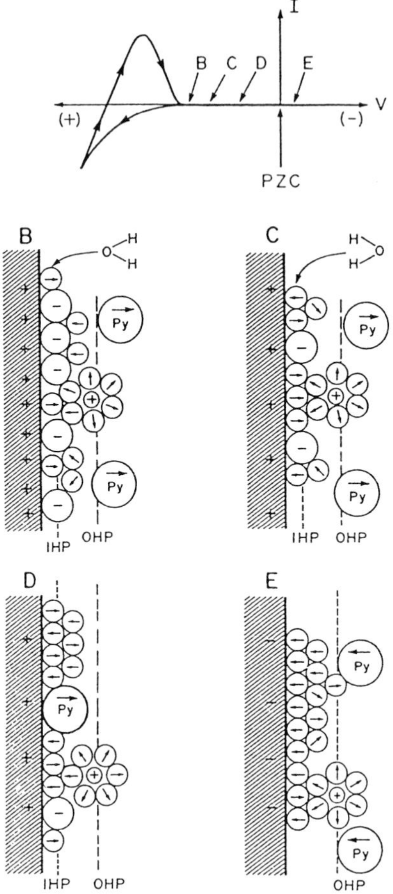

Fig. 12. Cyclic voltammogram and model of the electrical double layer at a silver electrode surface. Arrows indicate the direct-ions of molecular dipoles in the water (smallest circles) and pyridine (largest circles, Py) molecules, the arrow head being the positive end. The cations (solvated) could be Na^+ or K^+, the anions (unsolvated) Cl^- or SO_4^{2-}. IHP and OHP designate the inner and outer Helmholtz planes, respectively, and PZC is the potential of zero charge (see text for further explanations). (Reproduced with permission from ref. 14.)

In terms of this model, Chang and Laube [14] have given a useful description of the changing structure within the electrical double layer as the surface charge of an electrode is changed. Their account is summarized in Fig. 12. This shows a typical cyclic voltammogram used for roughening a silver electrode followed by sweeping the electrode potential up to and through the PZC. Pyridine (Py) molecules are shown moving in to contact the electrode surface as the potential becomes less positive (relative to the PZC) and approaches the PZC value. The fact that this neutral molecule becomes contact bound to the electrode at potentials positive of the PZC is a consequence of its dipolar character and the influence of the electron lone

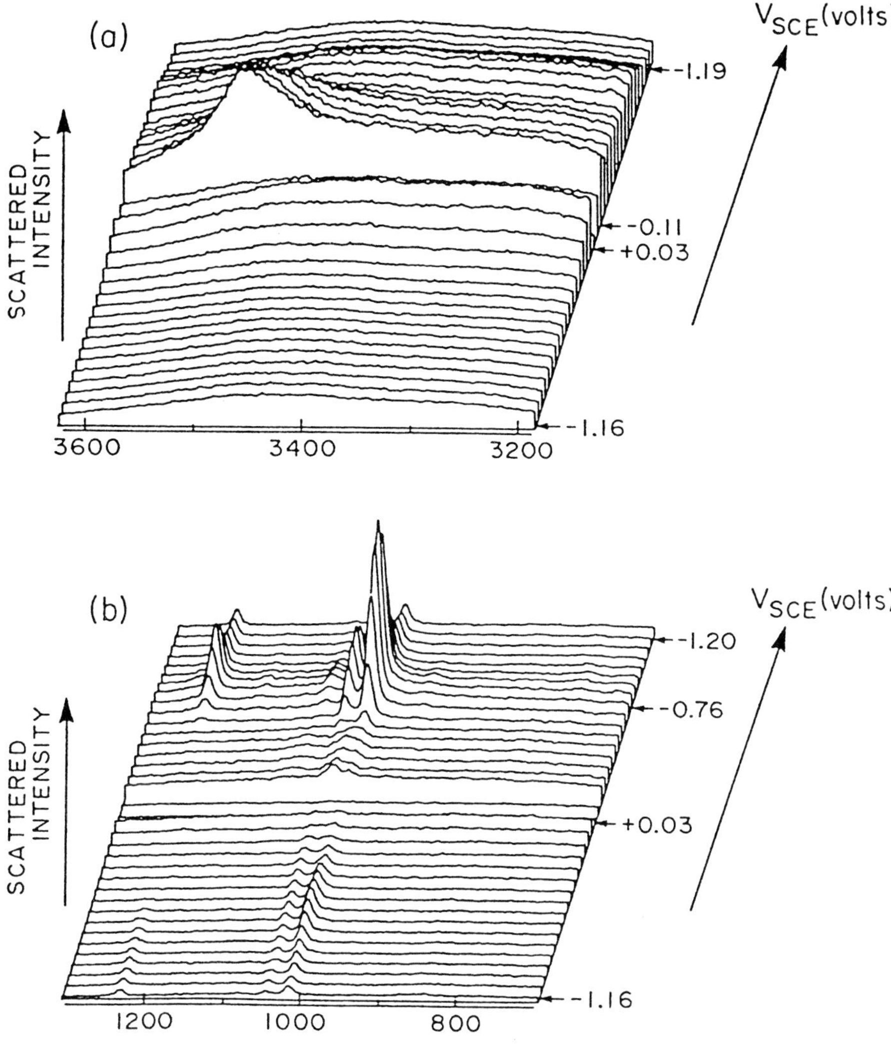

Fig. 13. SER spectra of 0.05 M pyridine in 1 M aqueous KCl recorded as the silver electrode potential was swept through an ORC at $5\,mV\,s^{-1}$. The upper traces (a) show the H_2O band, the lower ones (b) show pyridine bands. (Reproduced with permission from ref. 14.)

pair on the nitrogen atom, which provides a binding mechanism via Lewis base coordination.

Although the foregoing account is greatly simplified in that it takes no account of the vitally important roughness features of the surface, nor of its atomic nature, the presence of defects, the pH of the electrolyte, etc., it nonetheless provides a useful model for understanding the observed potential dependence of SERS effects.

Since colloidal particles of silver and gold also give large SERS effects, it is useful to consider these as disconnected microelectrodes. Controlled variations in the potential at the interface between silver sol particles and electrolyte solution have been achieved by adding the Eu^{3+}/Eu^{2+} redox system and varying the concentration ratio [22]. From measurements of SER spectra of pyridine adsorbed on the silver sol, the potential-dependent effects have been found to be identical with those given by bulk silver electrodes. These experiments also demonstrated that the intensity ratio of the pyridine Raman bands at ca. 1010 and 1040 cm^{-1} may be used as a simple but effective measure of surface potential for colloidal metals.

Pyridine certainly is the most extensively studied compound in the SERS literature. The potential dependence of its vibrational band intensities as a silver electrode is swept through its ORC at a rate of 5 $mV\,s^{-1}$ is shown [14] in Fig. 13. The intensity of the strongest band reaches a maximum at -0.9 V (SCE), in keeping with the model described previously in this section and illustrated in Fig. 12. Also consistent with the model is the fact that the SERS-enhanced H_2O band at ca. 3500 cm^{-1} becomes weaker as H_2O is displaced from the electrode surface by Py. However, there is no increase in the H_2O band intensity when the potential is swept beyond the PZC, as might be expected from the condition illustrated by Fig. 12(E).

Another molecular species much studied by the SERS method is the cyanide ion, CN^-. The real-time, molecule-specific diagnostic value of the SERS data for in situ studies of molecular complexes formed at the electrode /electrolyte interface during cyclic voltammetry is well illustrated by the data shown [14] in Fig. 14. The development of SERS bands during the first ORC for 0.01 M KCN in 0.1 M K_2SO_4 aqueous electrolyte at pH 11 is shown in Fig. 14(a). No SERS bands are seen during the first oxidation half-cycle. On reduction, a band at 2140 cm^{-1} appears first but is soon replaced by a band at 2110 cm^{-1}, which grows in intensity as the potential is swept towards the PZC. The 2140 cm^{-1} band is attributed to the complex ion $Ag(CN)_2^-$, while the 2110 cm^{-1} is characteristic of both $Ag(CN)_3^{2-}$ and $Ag(CN)_4^{3-}$ species. Figure 14(b) shows the potential dependence of the SERS bands during the second ORC. Here, the anodization results first in a decrease of the 2110 cm^{-1} SERS band due to loss of $Ag(CN)_3^{2-}$ and $Ag(CN)_4^{3-}$ as Ag^0 is oxidized to Ag^+. Then, the 2140 cm^{-1} $Ag(CN)_2^-$ band reappears and again disappears as the potential becomes more positive. It is not until the third reduction cycle that a band at 2165 cm^{-1}, characteristic of AgCN, is seen in the SER spectrum. This species appears to require large-scale surface roughness for its SERS enhancement.

3. Chemical species adsorbed on metal electrodes

Since it is claimed that SERS effects are displayed by all adsorbates, it would be fruitless to attempt anything approaching a comprehensive review

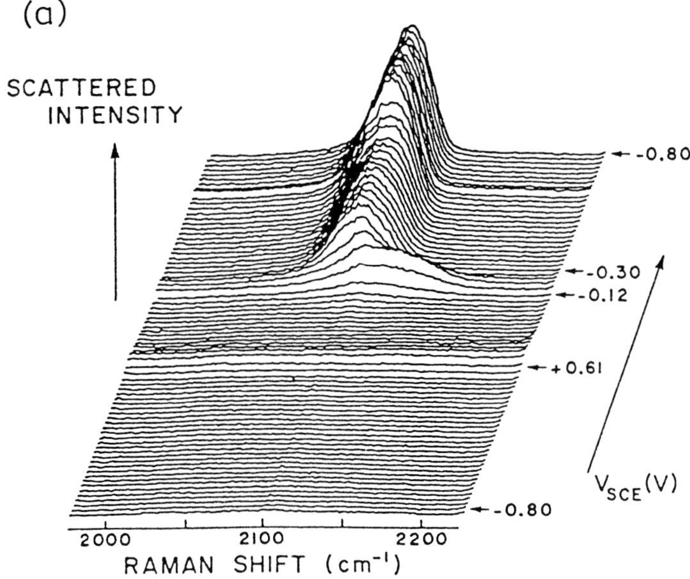

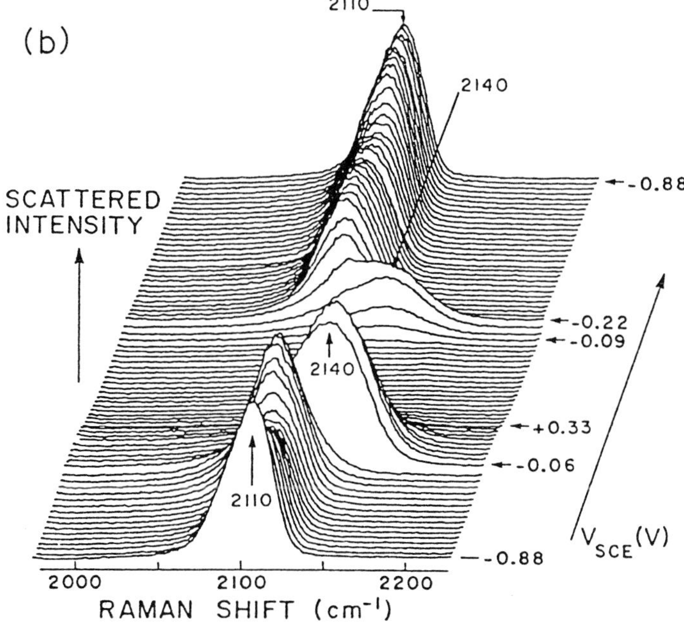

Fig. 14. Development of SERS for Ag–CN complexes during the first (a) and second (b) ORCs for a silver electrode in 0.01 M KCN and 0.1 M K_2SO_4 electrolyte. Sweep rate 50 mV s^{-1}. (Reproduced with permission from ref. 14.)

of SERS studies of chemical species adsorbed at metal electrodes. A few illustrative examples is the most that can sensibly be presented here.

Despite the remarkable success achieved in non-enhanced Raman studies of species at metal surfaces in UHV conditions [23], relatively little attention has been given in recent years to spectroelectrochemical studies of species and reactions at electrodes which do not show enhancement effects. The classical electrode materials Pt and Hg are in this category, although some worthwhile Raman enhancement may be obtained by alloying these metals with Ag or Au.

Gold electrodes have come to look particularly interesting in recent times. Not only are they more inert than silver, but they support a much wider range of potentials in aqueous solution, their derived salts are less photochemically labile, and their SERS-active sites appear to be more stable than those of silver. ORC-roughened gold electrodes remain fully SERS active when transferred from one electrochemical cell to another [18, 24].

In Fig. 15 is shown an SER spectrum [24] of ethene adsorbed at electrochemically roughened gold at +0.20 V (SCE) in ethene-saturated 1 M H_2SO_4. Red laser excitation (Kr ion, 647.1 nm) was used to obtain this spectrum, which shows strong bands centred at 1540 and 1283 cm^{-1}. These have been assigned to C=C stretching and symmetric CH_2 deformation modes, both shifted down from the ethene gas-phase values of 1623 and

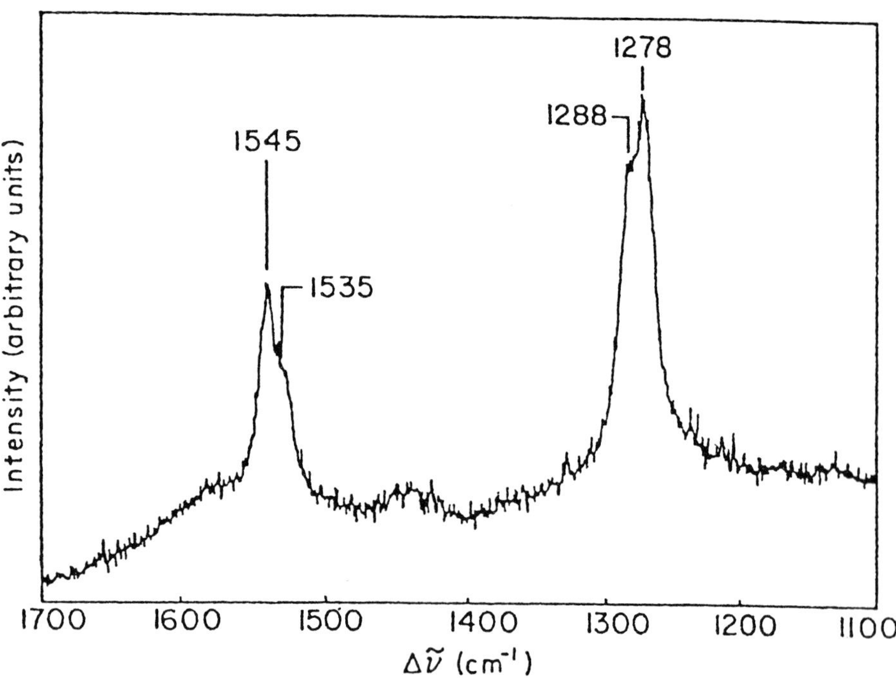

Fig. 15. SER spectrum of ethene adsorbed at a gold electrode. (Reproduced with permission from ref. 24.)

$1342\,cm^{-1}$, respectively. These large shifts are indicative of strong bonding to the gold surface and weakening of the C = C bond. The doublet structures of the two bands are attributed to adsorption at two energetically different surface sites. Gold is a good substrate for the oxidation of ethene and SERS has been used for monitoring the electrocatalytic oxidation processes which give rise to new bands at 990 and $425\,cm^{-1}$, tentatively assigned to a partially oxidized adsorbed intermediate [24]. The SER spectra have also been used to monitor the displacement reactions induced by adding Cl^- or Br^- ions to the solution.

A similar set of results has been reported for benzene and a number of monosubstituted benzenes adsorbed at gold electrodes [25]. Figure 16 shows how the Raman band due to the symmetric ring breathing mode is shifted down on adsorption from 992 to $973\,cm^{-1}$ and is split into a distinct doublet at about 975 and $965\,cm^{-1}$ at electrode potentials more positive than

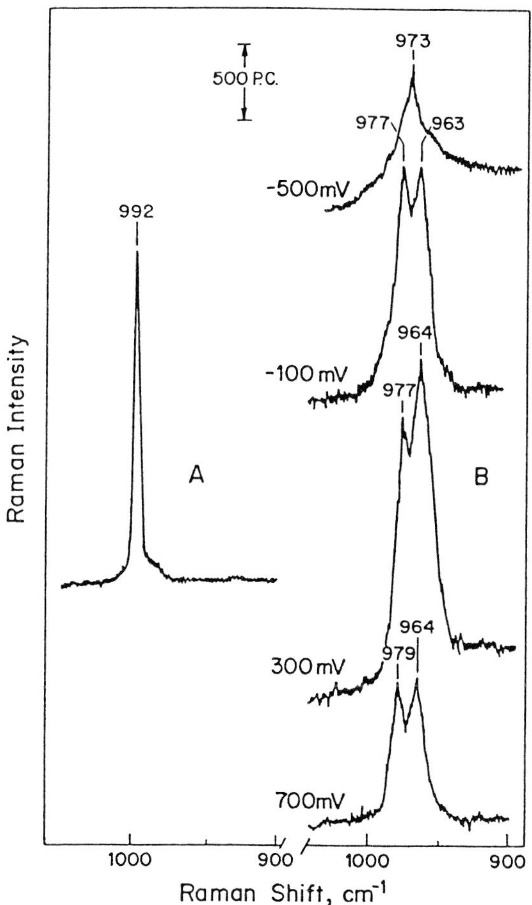

Fig. 16. SER spectra from benzene adsorbed at a gold electrode. A, from liquid benzene; B, from benzene-saturated 0.5 M H_2SO_4. (Reproduced with permission from ref. 25.)

References pp. 103–104

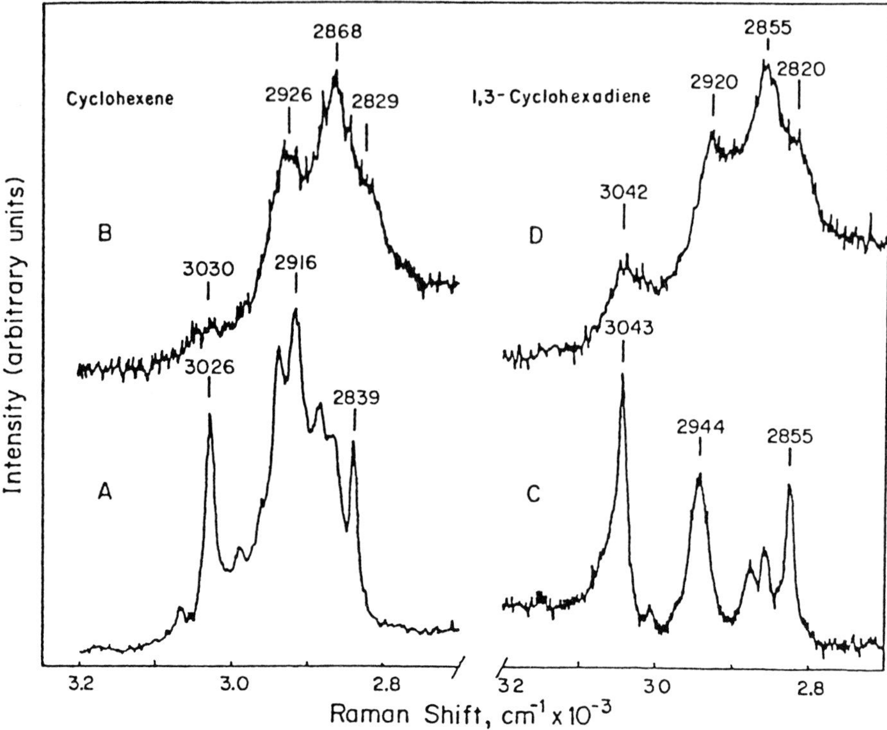

Fig. 17. Normal Raman (A and C) and SER spectra (B and D) of cyclohexene and 1,3-cyclohexadiene. (Reproduced with permission from ref. 26.)

– 300 mV (SCE). These results were obtained from an electrolyte of benzene-saturated 0.5 M H_2SO_4 in which the benzene concentration is ca. 2 mM. The surface coverage, θ, of benzene at the gold electrode when biased at 300 mV was estimated to be ca. 0.2. These results have been interpreted as indicating a flat orientation of the benzene rings on the gold substrate at two different types of surface site. Addition of Br^- ions to the electrolyte caused displacement of the less strongly bound benzene molecules, as indicated by the selective loss of the 975 cm^{-1} band.

The adsorption of various cyclic alkenes and of simple alkynes at gold electrodes has been studied by SERS [26]. As with ethene [24], large shifts of the C = C or C ≡ C bond stretching modes to lower wavenumbers were observed. Differences were also found between the normal Raman (NR) and SER spectra in the wavenumber range characteristic of C–H bond stretching. In Fig. 17, for example, are shown the CH and CH_2 bands for cyclohexene and 1,3-cyclohexadiene obtained by NR and SER methods. It is evident from the NR/SER comparison that a greater SER enhancement effect is experienced in both cases by the paraffinic methylene (CH_2) groups than by the olefinic (CH) groups. For liquid cyclohexene, the ν(CH) is at 3026 cm^{-1} and the $\nu(CH_2)$ gives the lower wavenumber multiplet; for liquid 1,3-cyclohexadiene, the ν(CH) is at 3043 cm^{-1} and the bands below 3000 cm^{-1} are associated with

$\nu(CH_2)$. Since the olefinic C–H bonds are believed to be oriented parallel to the metal surface while the paraffinic C–H bonds are oriented out of the metal surface plane, these NR/SER band intensity effects may be indicative of a surface selection rule for SERS.

The use of a surface selection rule in infrared spectroscopy and in EELS has been widespread and has yielded much valuable information on surface orientation of adsorbed species. The rule is a very simple one: for adsorbates at metal surfaces, only vibrational modes which modulate a component of the molecular dipole perpendicular to the surface are active in these spectroscopies. This rule arises from the fact that metals have high electrical conductivities at vibrational mode frequencies and this results in the parallel (but not the perpendicular) component of the radiation field going to zero at the surface.

There have been a number of reports of analogous surface selection rules for Raman spectra [27–29]. However, for SERS, the situation is complicated by the essential roughness of the metal surface and the mixture of enhancement mechanisms, in addition to the facts that the Raman effect depends upon the molecular polarizability *tensor* and the excitation frequencies are typically high enough to reduce the metal conductivity to levels where finite parallel, as well as perpendicular, electric vectors are established at the surface. An excellent recent review of this subject has been written by Creighton [30]. Suffice it to say here that surface selection rules evidently do exist for Raman spectroscopy but they are more complicated than the rule for infrared and EELS.

3.1 MODIFIED ELECTRODES

The deliberate modification of electrode surfaces by coating with one or more layers of electroactive material has been used for a variety of purposes. Solar energy conversion, electrochromism, corrosion protection, and electrocatalysis are but a few of the applications which are currently of interest. The use of in situ Raman spectroscopic studies can help to determine the structural characteristics of electrode coatings at the molecular level and can provide information on the mechanisms of electrochemical reactions occurring at modified electrode surfaces.

A recent example is the use of Raman spectroscopy to examine a thionine-modified gold electrode [31]. A combination of resonance Raman and SERS effects was used to probe the thionine/leucothionine redox reaction (see Fig. 18) in the immobilized surface coat. Electrodeposition of the coat was shown to result in the linking of adjacent thionine molecules via an amino nitrogen atom in a secondary amine structure, but there was no evidence of chemical bonding between the surface film and the electrode. Analysis of the changes in Raman band intensities as the electrode potential was swept through a range of values to generate the full cyclic voltammogram for the surface coat were used to deduce a mechanism for the movement of charge across the modified-electrode/electrolyte interface. These electrode systems are valu-

References pp. 103–104

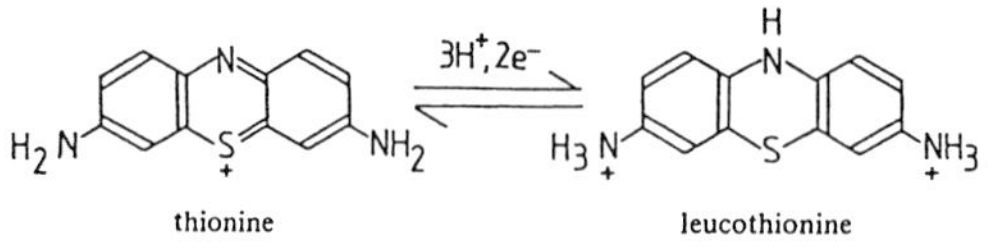

Fig. 18. The thionine/leucothionine redox reactions [31].

able for use in photogalvanic cells where the illuminated electrode must be able to discriminate between the products of the cell photoredox reaction [31].

An application of a different type is the use of a modified electrode for increasing the rate of electron transfer to a solution species which, with an ordinary metal electrode, gives no appreciable voltammetric response due to intrinsically slow kinetics. An interesting example to which SERS has been applied is cytochrome c at a silver [32] or gold [33] electrode, modified by adsorbed 4,4-bipyridine. Major changes of the SER spectra of 4,4-bipyridine have been reported [32, 33] as the electrode potential is changed. These changes could be accounted for in terms of attachment of the bipyridine to the electrode via one nitrogen atom, allowing the molecule to have a vertical orientation to the surface as the neutral species, the radical anion, or the neutral protonated radical species [33].

A novel compacted electrode technique has been described for producing a coating of the insoluble pigment α-copper phthalocyanine on a silver surface [34]. Such films of phthalocyanines and their metal complexes on metal surfaces are useful in a wide range of solid-state devices. Again, a combination of RR and SER spectra has been used to characterize these electrodes and the changes which occur in them when the electrode potential is varied.

4. Non-aqueous systems

It has only recently been discovered that SER spectra can be obtained from systems involving non-aqueous electrolyte solutions [35, 36]. The first report was for pyridine in *N,N*-dimethylformamide solution at a silver electrode [35]. Variations in the relative intensities of the pyridine bands in the 1000 cm^{-1} wavenumber region as a function of electrode potential are shown in Fig. 19. These potential-dependent changes are quite different from those recorded for the corresponding aqueous system and have been interpreted in terms of the solvent effect on the surface morphology of the electrode.

The wider electrochemical potential windows associated with non-aqueous solvents than with water opens the way to a far richer field of reaction studies. A report of SERS of the tris(2,2-bipyridyl)–ruthenium(II) complex ion, $[Ru(bpy)_3]^{2+}$, adsorbed from acetonitrile solution on to a silver electrode [36] has been followed, independently, by a report [37] on an in situ SERS study of the electroreduction reaction to $[Ru(bpy)_3]^{+}$. It had been

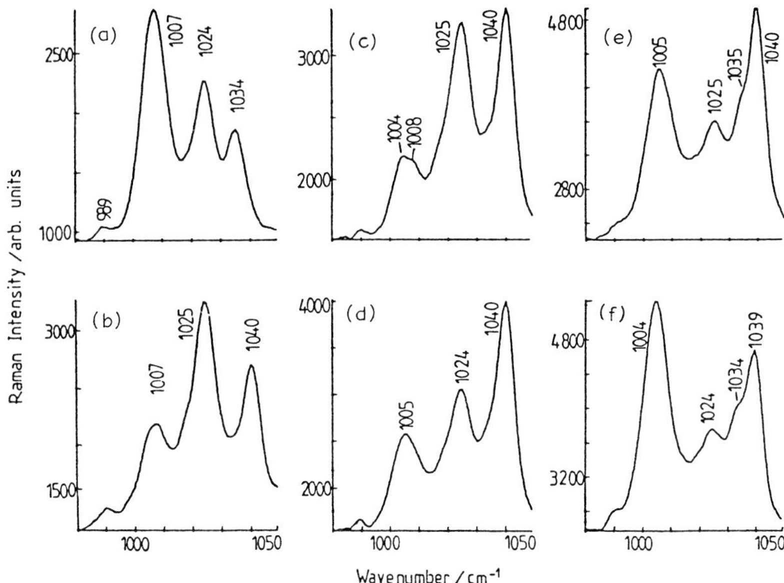

Fig. 19. SER spectra of pyridine in DMF solution at a silver electrode. The spectra show variations due to increasingly negative electrode potentials. (a) 0.0 V, (b) − 0.2 V, (c) − 0.4 V, (d) − 0.6 V, (e) − 0.8 V, (f) − 1.0 V. (Reproduced with permission from ref. 35.)

anticipated that this reduced species would give a vibrational spectrum similar to that of the photoexcited state of $[Ru(bpy)_3]^{2+}$ wherein it has been convincingly demonstrated that the metal-to-ligand charge-transfer (MCLT) process yields a substitutionally asymmetric excited state species [38, 39]. The electroreduced $[Ru(bpy)_3]^{+}$ species itself in free solution had already been shown to be asymmetric [37]. However, the SER spectra were surprising. They showed a quite different pattern of band frequencies, which indicated that adsorption at the metal surface provided an electronic conduction mechanism which resulted in delocalization of the odd electron in the $[Ru(bpy)_3]^{+}$ species over the three bpy ligands.

5. Semiconductor electrodes

There is currently much interest being shown in semiconductor(SC)-based photoelectrochemical (PEC) devices as potential solar energy converters. In developing and improving the properties of PEC devices of this sort, it is important to have methods available for characterizing the surface states of the SC electrodes employed. These surface states are believed to be responsible for many of the significant properties of SC electrodes. In situ Raman spectroscopy provides a convenient and powerful tool for probing the nature of SC electrode surface states and also the changes that occur at SC

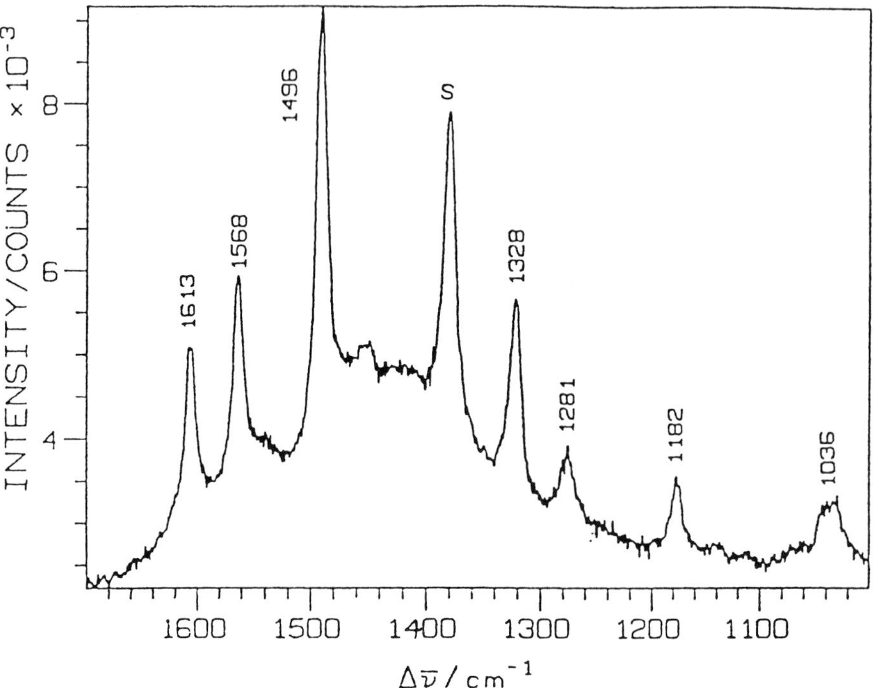

Fig. 20. SERR spectrum of $[Ru(bpy)_3]^{2+}$ at monolayer coverage on an Ag-overlayered *p*-type GaAs semiconductor electrode. S is an acetonitrile solvent band. (Reproduced with permission from ref. 41.)

electrode surfaces during chemical etching or chemical modification by chemisorption or covalent derivatization.

Surface-enhanced resonance Raman (SERR) spectra have been used [40, 41] to study the adsorption of $[Ru(bpy)_3]^{2+}$ and related species at monolayer (or lower) coverage on Si and GaAs SC electrodes. The lack of intrinsic surface-enhancement effects with the GaAs and Si substrates was circumvented by electrodeposition or photochemical deposition of a discontinuous overlayer of Ag on top of the $[Ru(bpy)_3]^{2+}$-coated SC electrodes. An illustration of the excellent quality SERRS data obtained by this overlayer technique is given by Fig. 20 [41]. This spectrum was obtained from a *p*-type GaAs wafer in 10 mM $AgNO_3$/0.1 M tetra-*n*-butylammonium perchlorate (TBAP) in acetonitrile solution. A silver layer was deposited photochemically over one monolayer of the Ru^{2+} complex. From the figure, it is seen that the resulting SERR spectrum of the adsorbed Ru^{2+} complex is comparable in intensity with bands from the acetonitrile solvent (marked S in Fig. 20).

Silver-overlayer SERRS also has been used to study in situ the redox reactions of methyl viologen (MV) adsorbed on to a *p*-type InP SC electrode [42]. These experiments are related to earlier time-resolved resonance Raman spectroscopy (TR3S) work on electron transfer reactions at the surfaces of photoexcited semiconductor colloids (TiO_2 and CdS) involving

MV^{2+} acceptors and SCN^- electron donors in solution [43]. Colloidal semiconductor particles, typically of ca. 10–100 nm diameter, in aqueous sols may be treated as isolated microelectrode systems. Steady-state RRS experiments with c.w. lasers can be used to study phototransients produced at the surfaces of such colloidal semiconductors in flow systems [44], but pulsed laser systems coupled with multichannel detectors are far more versatile. Indeed, a recent TR3S study of methyl viologen reduction on the surface of photoexcited colloidal CdS crystallites has shown important differences in mechanism between reactions occurring on the nanosecond time scale and those observed with picosecond Raman lasers [45]. Thus, it is apparent that Raman spectroscopy may now be used to study very fast interface kinetics as well as providing sensitive information on chemical structure and bonding in molecular species at electrode surfaces.

Acknowledgements

My thanks go to Dr. Hari Virdee for help with the literature survey which preceded this review, to the SERC Rutherford Laboratory for help with pulsed laser experiments, and to the SERC for financial support.

References

1 M. Fleischman, P.J. Hendra and A.J. McQuillan, Chem. Phys. Lett., 26 (1974) 163.
2 D.L. Jeanmaire and R.P. Van Duyne, J. Electroanal. Chem., 84 (1977) 1.
3 M.A. Tadayyoni, P. Gao and M.J. Weaver, J. Electroanal. Chem., 198 (1986) 125.
4 K.A. Bunding, J.E. Gordon, II and H. Seki, J. Electroanal. Chem., 184 (1985) 405.
5 J.A. Creighton, M.G. Albrecht, R.E. Hester and J.A.D. Matthew, Chem. Phys. Lett., 55 (1978) 55.
6 B. Pettinger and H. Wetzel, in R.K. Chang and T.E. Furtak (Eds.), Surface Enhanced Raman Scattering, Plenum Press, New York, 1982, p. 293.
7 M. Moskovits and D.P. DiLella, in R.K. Chang and T.E. Furtak (Eds.), Surface Enhanced Raman Scattering, Plenum Press, New York, 1982, p. 243.
8 W. Krasser and A.J. Renonprez, J. Raman Spectrosc., 11 (1981) 425.
9 P.F. Liao and M.B. Stern, Opt. Lett., 7 (1982) 483.
10 R.K. Chang, Proceedings of the International Conference on Structure and Dynamics of Solid/Electrolyte Interfaces, Berlin, 1986; personal communication.
11 C.A. Murray, D.L. Allara, A.F. Hebard and F.J. Padden, Jr., Surf. Sci., 119 (1982) 449. C.A. Murray, in R.K. Chang and T.E. Furtak (Eds.), Surface Enhanced Raman Scattering, Plenum Press, New York, 1982, p. 203.
12 R.K. Chang, in G.H. Atkinson (Ed.), Time-resolved Vibrational Spectroscopy, Academic Press, New York, 1982.
13 P. Hildebrandt and M. Stockburger, J. Phys. Chem., 88 (1984) 5935.
14 R.K. Chang and B.L. Laube, Crit. Rev. Solid State Mater. Res., 12 (1984) 1.
15 A.J. McQuillan and R.E. Hester, J. Raman Spectrosc., 15 (1984) 15.
16 J.L. Anderson and J.R. Kincaid, Appl. Spectrosc., 32 (1978) 356.
17 C. Otto, A. van Welie, E. de Jong, F.F.M. de Mul, J. Mud and J. Greve, J. Phys. E., 17 (1984) 624.

18 P. Gao, M.L. Patterson, M.A. Tadayyoni and M.J. Weaver, Langmuire, 1 (1985) 173.
19 M.W. Howard, R.P. Cooney and A.J. McQuillan, J. Raman Spectrosc., 9 (1980) 273.
20 R.P. Cooney, M.R. Mahoney and A.J. McQuillan, in R.J.H. Clark and R.E. Hester (Eds.), Advances in Infrared and Raman Spectroscopy, Vol. 9, Wiley, Chichester, 1982, Chap. 4.
21 J.C. Tsang, J.E. Demuth, P.N. Sanda and J.R. Kirtley, Chem. Phys. Lett., 76 (1980) 54.
22 H. Wetzel, H. Gerischer and B. Pettinger, Chem. Phys. Lett., 85 (1982) 187.
23 D.R. Mullins and A. Campion, Chem. Phys. Lett., 110 (1984) 565.
24 M.L. Patterson and M.J. Weaver, J. Phys. Chem., 89 (1985) 1331.
25 P. Gao and M.J. Weaver, J. Phys. Chem., 89 (1985) 5040.
26 M.L. Patterson and M.J. Weaver, J. Phys. Chem., 89 (1985) 5046.
27 C.S. Allen and R.P. Van Duyne, Chem. Phys. Lett., 63 (1979) 455.
28 H. Nichols and R.M. Hexter, J. Chem. Phys., 73 (1980) 965; 75 (1981) 3126.
29 M. Moskovits, J. Chem. Phys., 77 (1982) 4408.
30 J.A. Creighton, in R.J.H. Clark and R.E. Hester (Eds.), Spectroscopy at Surfaces: Advances in Spectroscopy, Vol. 15, Wiley, Chichester, 1988.
31 K. Hutchinson, R.E. Hester, W.J. Albery and A.R. Hillman, J. Chem. Soc. Faraday Trans. 1, 80 (1984) 2053.
32 T.M. Cotton, D. Kaddi and D. Iorga, J. Am. Chem. Soc., 105 (1983) 7462.
33 I. Taniguchi, M. Iseki, H. Yamaguchi and K. Yasukouchi, J. Electroanal. Chem., 186 (1985) 299.
34 A.J. Bovill, A.A. McConnell, J.A. Nimmo and W.E. Smith, Surf. Sci., 158 (1985) 333.
35 K. Hutchinson, A.J. McQuillan and R.E. Hester, Chem. Phys. Lett., 98 (1983) 27.
36 A.M. Stacy and R.P. Van Duyne, Chem. Phys. Lett., 102 (1983) 365.
37 H.R. Virdee and R.E. Hester, J. Phys. Chem., 88 (1984) 451.
38 M. Forster and R.E. Hester, Chem. Phys. Lett., 81 (1981) 42.
39 P.G. Bradley, N. Kress, B.A. Hornberger, R.F. Dallinger and W.H. Woodruff, J. Am. Chem. Soc., 103 (1981) 7441.
40 R.P. Van Duyne and J.P. Haushalter, J. Phys. Chem., 87 (1983) 2999.
41 R.P. Van Duyne, J.P. Haushalter, M. Janik-Czachor and N. Levinger, J. Phys. Chem., 89 (1985) 4055.
42 Q. Feng and T.M. Cotton, J. Phys. Chem., 90 (1986) 983.
43 R. Rosetti, S.M. Beck and L.E. Brus, J. Am. Chem. Soc., 104 (1982) 7322; 106 (1984) 980.
44 K. Metcalfe and R.E. Hester, J. Chem. Soc. Chem. Commun., (1983) 133.
45 R. Rosetti and L.E. Brus, J. Phys. Chem., 90 (1986) 558.

Chapter 3

The Use of Ex-situ UHV Techniques to Study Electrode Surfaces

ROGER PARSONS

1. Introduction

At a metal or semiconductor surface in contact with an electrolyte, the elementary charge-transfer reaction occurs in a region of space of the order of 1 nm thick or less. Hence, in the study of these reactions, the structure and composition of this region, the interphase, is of primary importance. Electrochemists have developed a range of highly sensitive techniques to observe changes in the composition of the interphase which can detect amounts at least as small as 1% of a monolayer in terms of the charge exchanged. They are also capable of giving some information about the nature and strength of bonding of interfacial species by the position of the response on the potential scale, especially when the charge exchange is rapid. However, they provide no general answer to the problem of identifying these species and analysing the nature of their bonding.

For these reasons, electrochemists have turned to spectroscopic techniques in recent years. In principle, the most valuable of these are those which can be used for the electrochemical interface under working conditions and the progress made in the use of IR and Raman vibrational spectroscopies is described in other chapters of this volume. In the last 10 years, these have provided substantial new evidence about the nature of bonding, but they give only limited information about surface structure. X-Ray diffraction and extended X-ray absorption fine structure (EXAFS) are in the early stage of application to the electrode–electrolyte interface and will eventually provide long-range as well as local structural information. Up to the time of writing, much of the spectroscopic work has been done using polycrystalline or otherwise ill-defined electrode surfaces so that the results are a complex average over the heterogeneous surface. This is changing rapidly as the techniques become more widely available and the necessity for well-defined surfaces becomes more clearly recognised.

The alternative route to the solution of the problem of surface structure and composition is the one to be described in this chapter as the ex-situ route. By transferring the electrode from the electrochemical cell into the ultra-high vacuum (UHV), a vast range of techniques developed in the past two decades by surface scientists becomes available, notably those which depend on electron spectroscopy or electron diffraction. The central difficulty then becomes that of verifying that any change in the electrode

References pp. 126–127

during the transfer process is known and allowed for. The ways in which effective transfer may be achieved are discussed here, together with some of the results already obtained. It should be clear from the outset that the in-situ and ex-situ techniques are complementary and that, ideally, both types should be used to solve any given problem.

2. Early work

The use of ex-situ techniques is by no means new. Since the 1920s, electrodeposits have been examined by X-rays or by electron diffraction [1], in the latter case even using single crystal substrates. The electron microscope has also been widely used for the study of deposit morphology [2]. These uses were confined to relatively large-scale structures where the properties approximate to bulk values, i.e. a μm scale. Resolution has improved progressively but the use of UHV and the introduction of low-energy electron diffraction (LEED), X-ray photoelectron spectroscopy (XPS), and Auger electron spectroscopy (AES) in the 1960s represented a major step which allowed the examination of properties at the nm level, i.e. the monolayer level. The first application of XPS to the study of electrodes at monolayer level was made by Winograd and co-workers in 1972 [3], who examined the nature of the oxide film on polycrystalline Pt electrodes. Such films are relatively stable and so may be transferred from the electrochemical cell to the UHV enclosure with comparative ease and this group carried out a series of studies on metal oxides. They were followed by similar experiments using high-area electrodes on which the stable intermediate in methanol oxidation could be detected on Pt [4] as well as a great variety of systems studied by Sherwood [5]. This group also developed a more sophisticated form of transfer, but the pioneer of transfers allowing the use of electrodes well-defined with respect to structure and composition was Hubbard [6] who has pointed out forcefully [7] the problems which arise when attempts are made to interpret surface-sensitive properties using ill-defined surfaces.

Several reviews dealing with various aspects of the use of ex-situ techniques are available [8–12] in addition to those already mentioned [2, 5, 7].

3. Experimental techniques

Three types of transfer between the electrochemical cell and the UHV enclosure have been used. In the first, the electrochemical cell is built on to the stainless steel system and the electrode is transferred directly. In the second, a glove box is used as an intermediate to the transfer which can then be made to a conventional electrochemical cell. In the third type, there is no direct transfer but electrodes prepared under identical conditions are studied in parallel by UHV electrochemical methods.

In the discussion of transfer, it is necessary to consider the ways in which sensitive electrode surfaces may be protected during the transfer and also the question of the nature of the processes occurring when an electrode is put in or taken out of the electrolyte, the "immersion" and "emersion" processes.

3.1 DIRECT TRANSFER SYSTEMS

The first and more successful system of this type is that designed by Hubbard [6, 13] and shown in Fig. 1. This enabled the use of LEED and AES for the characterization of the surface. Strictly speaking, this is not a transfer system because the electrochemical experiments were carried out in the analysis chamber, which was pumped down subsequently. The sample,

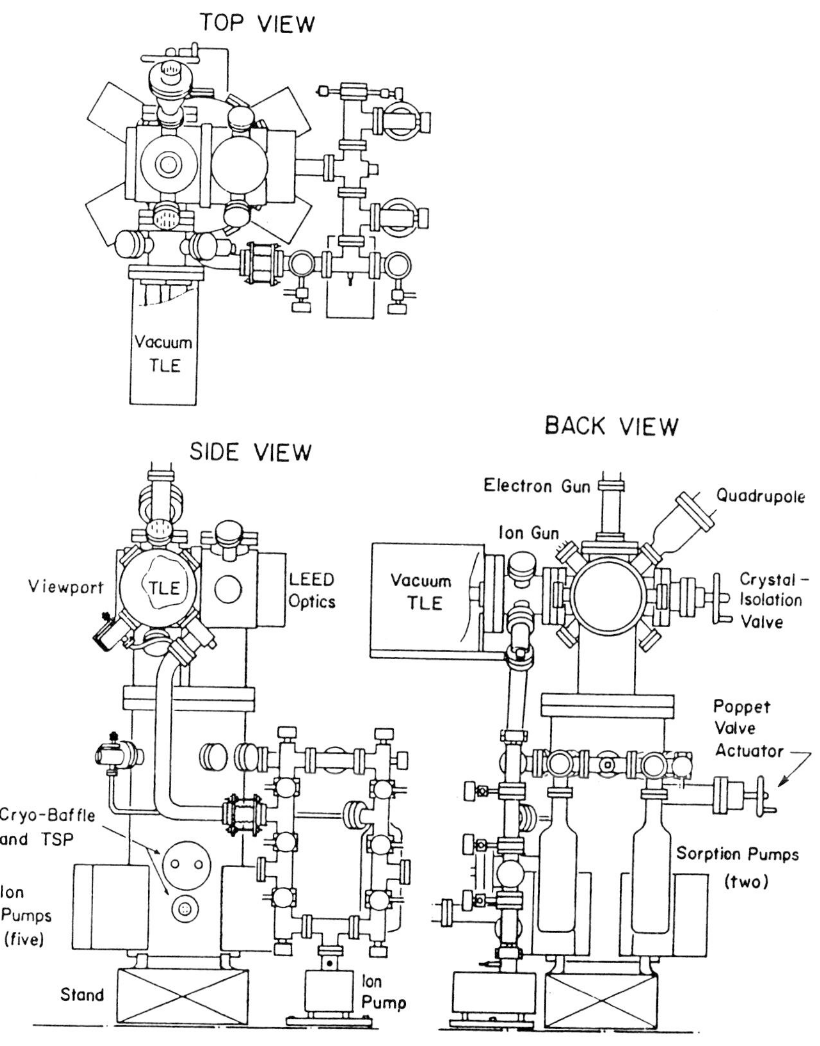

Fig. 1. Sketch of a LEED–thin layer electrochemistry (TLE) system [6].

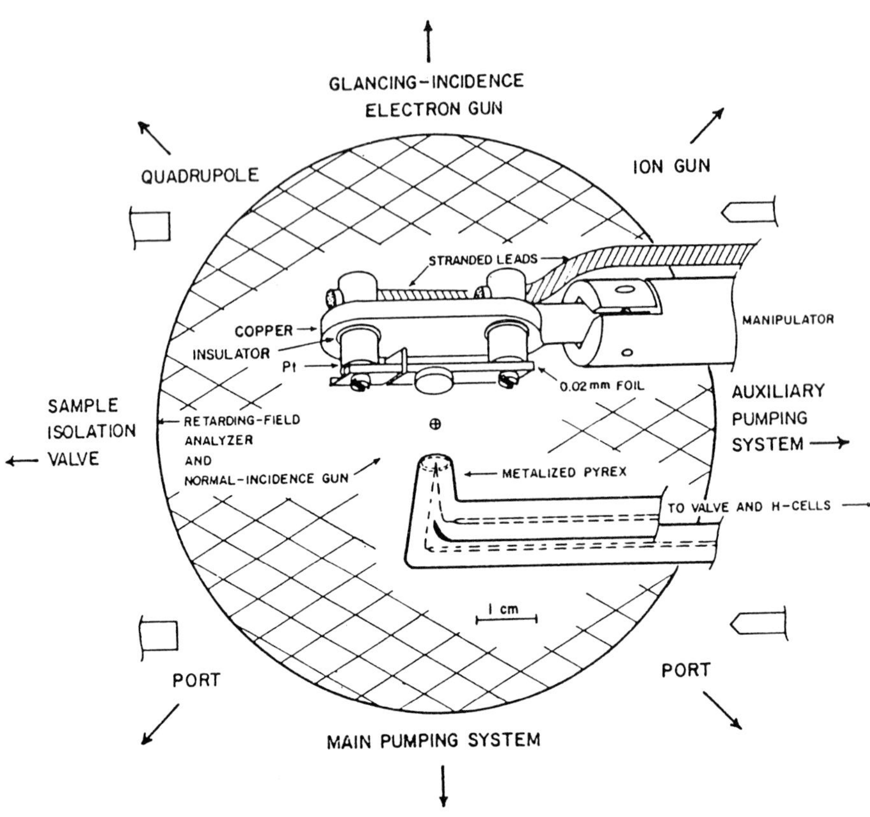

Fig. 2. Sample holder and capillary of the vacuum thin layer cell systems as seen through the view port of the work chamber, with LEED optics in the background [6].

on a manipulator, was positioned in contact with a pair of capillaries formed with a common end as shown in Fig. 2. The sample electrode and the face of this capillary tube formed a thin layer cell. The rest of the outside of the capillary was coated with Pt to prevent accumulation of charge during the vacuum experiments. During electrochemical experiments, the sample and capillary were enclosed in a crystal isolation valve and brought to atmospheric pressure with pure argon. Electrolyte was introduced from microburettes by means of a system of miniature valves and fittings. During the manipulations, the solution came into contact with inert materials only: Teflon, Kel-F, Pt, and glass. After electrochemistry, the chamber pressure could be reduced to 2×10^{-9} Torr in about 5 min.

In a later version [14] shown in Fig. 3, the two capillaries terminated in a cup which could be filled with electrolyte in which the electrode could be immersed. The Pt electrode was then a parallelopiped shaped like the rhombohedral primitive unit cell of the f.c.c. system having all six faces equivalent to (111) orientation or a rectangular solid having all six faces equivalent to (100) orientation. This type of crystal was then used in a transfer

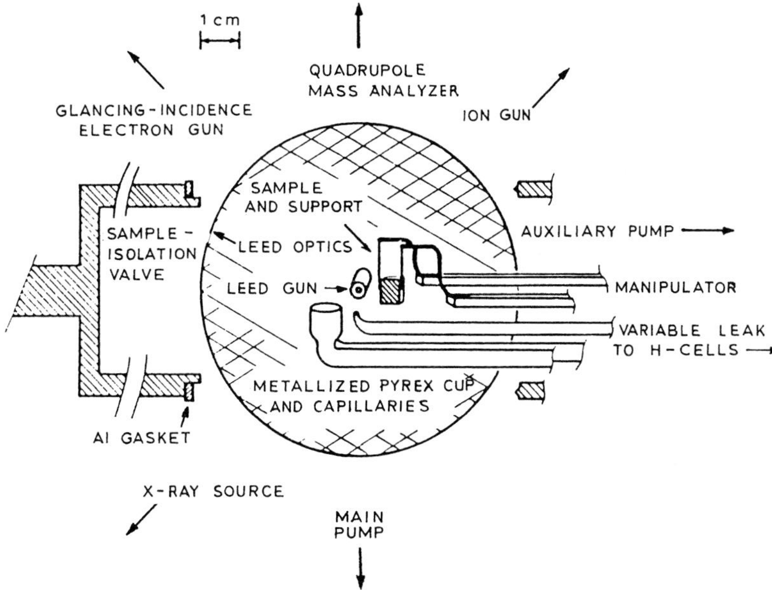

Fig. 3. Single-crystal electrode and electrolysis compartment of the LEED electrochemistry system as seen through the view port of the vacuum chamber [14].

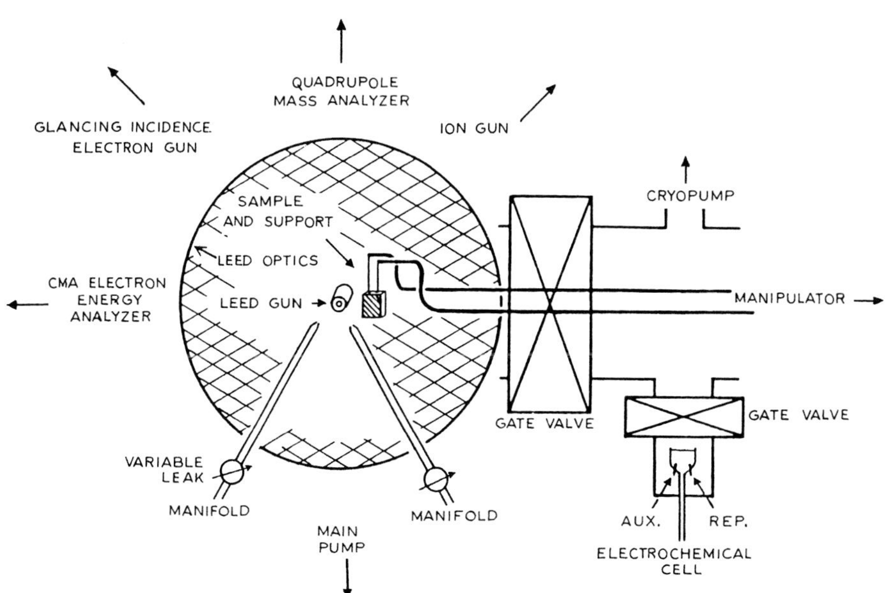

Fig. 4. Schematic of the LEED–electrochemistry instrument [15].

References pp. 126–127

system [15] as shown in Fig. 4, where the electrochemical experiment was separated from the analysis chamber by a gate valve. This arrangement means that the electrochemical cell is totally separated from the analysis chamber and it has been adopted by others attempting this transfer.

An early transfer system in which the cell is connected directly to the UHV chamber was described by Revie et al. [16]. The analysis chamber and the cell, both of glass, were joined by a straight-through UHV valve permitting the sample to be transferred vertically up from the cell using a magnetically operated windlass. This was used for AES analysis of passive films on iron and it does not seem to be readily extendable to use with well-defined electrodes.

A system somewhat similar to that of Hubbard was built by Yeager et al. [17, 18] and is shown schematically in Fig. 5. This uses a flat electrode and a thin layer cell as in Hubbard's first design, Fig. 2, but has two separate chambers for electrochemistry and UHV analysis, the sample holder being transferred between two transfer rods and a manipulator. This allows the use of a carousel on which several electrodes may be mounted. Ross and Wagner also adopted a thin layer configuration for the electrochemical cell [12, 19] with a palladium/hydrogen counter/reference electrode.

A relatively simple system using a cell with a relatively large amount of electrolyte was used by Sherwood and co-workers [20]. As shown in Fig. 6, the electrode was mounted on a long transfer rod and could be moved through seals from the preparation/electrochemical chamber to the analysis chamber. This permitted the use of an almost conventional electrochemical arrangement. Kötz and co-workers [21] adopted a fast insertion lock as shown in Fig. 7 using an "omni-seal" ring between the transfer rod and the

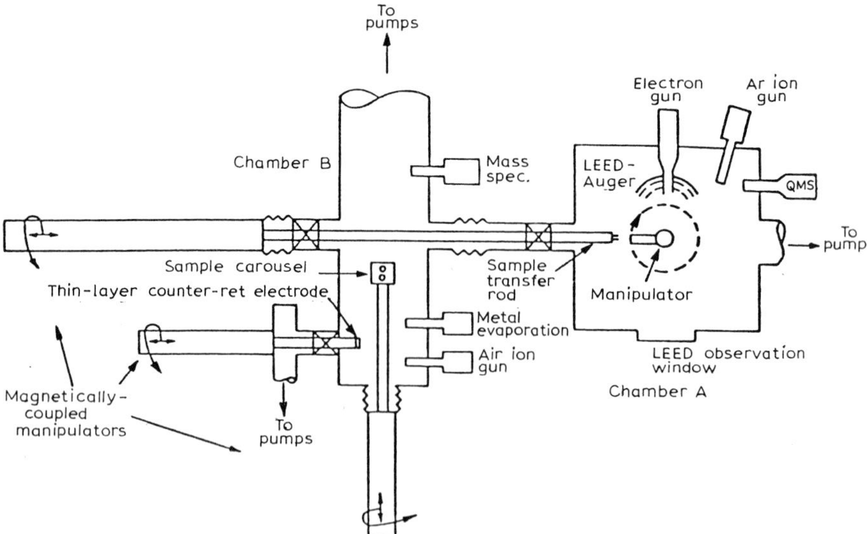

Fig. 5. LEED–AES–thin layer electrochemical cell system with special transfer system [18].

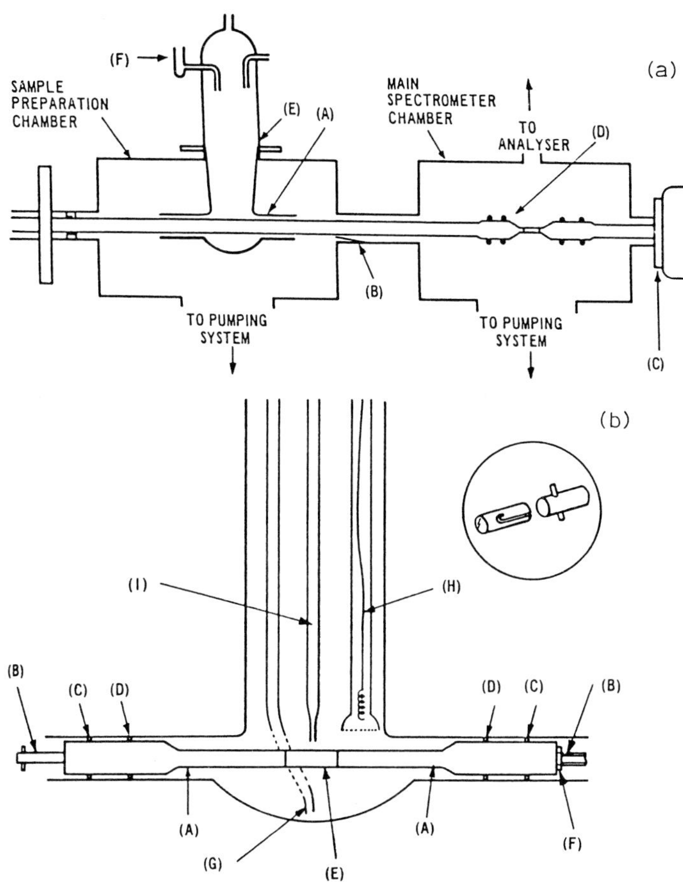

Fig. 6. (a) Schematic diagram of the electrochemical cell attached to the UHV spectrometer. (A), Precision bore glass side arms; (B), flap valve; (C), VG rotatable probe; (D), electrode assembly; (E) glass–metal seal; (F), reference electrode compartment. (b) The electrode mounted on its holder in position in the glass cell. (A), Teflon cylinder; (B), brass studding; (C), Viton "O" rings; (D), Teflon "O" rings; (E), electrode; (F), locking nut; (G), drain tube; (H), Pt counter electrode; (I), Luggin capillary [20].

electrochemical chamber to exclude ambient atmosphere. The more sophisticated system of Strehblow and co-workers [22] resembles the transfer mechanism of Yeager et al. [17] but the electrochemical arrangements use a cell similar to that of Kötz and co-workers. Aberdam et al. [23] used a comparable arrangement which was specifically devised for use with a small hemispherical crystal making a meniscus contact with the electrolyte. This was mounted on a small bridge which could be transferred between the compartments of the UHV system.

Comparison between these different designs is difficult because they have been used for different types of studies. The one type which has been used in several cases is the study of hydrogen and oxygen on platinum, which will be discussed below.

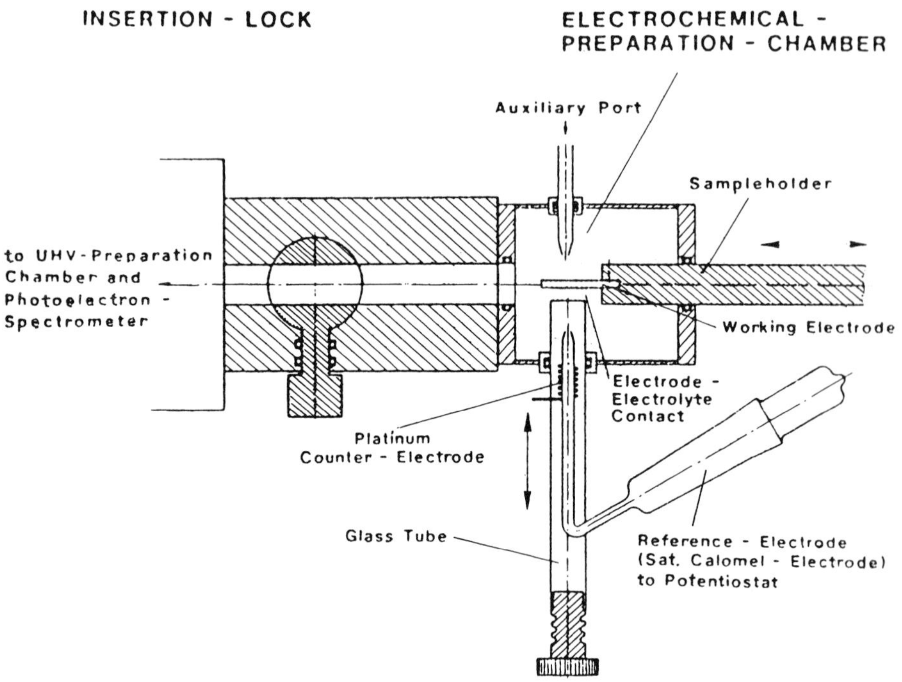

Fig. 7. Schematic diagram of the electrochemical preparation chamber in combination with the fast insertion lock [21].

3.2 TRANSFER USING A GLOVE BOX

This method allows greater flexibility in that the electrode may be remounted during transfer from a conventional UHV support to a conventional electrochemical support, then used in an electrochemical cell of conventional design either within the glove box or outside. Moreover, it requires careful attention to the purity of the gas used in the glove box, the standard required depending on the sensitivity of the samples being used. The early work with this type of transfer was generally with the rather less active electrodes [3, 24–26] but detailed monitoring of each step in the transfer was already used by Hammond and Winograd [27] who designed a special system with miniaturized cells. A large glove box with a specially designed transfer rod was used by Bellier et al. [28] as shown in Fig. 8. The transfer rod could be connected to each of two small evacuable chambers, one attached to the UHV system, the other to the glove box. Thus, the sample could be transferred in either direction in a stainless steel chamber around the transfer rod. The larger intermediate chamber in the glove box served for the transfer of the electrode from the transfer rod to the electrode holder for insertion into the cell, but could also be used for the insertion of electrodes, cells, etc. into the system as it could be pumped down. The glove box was constructed of stainless steel, copper, glass, and teflon. It was not continuously purged

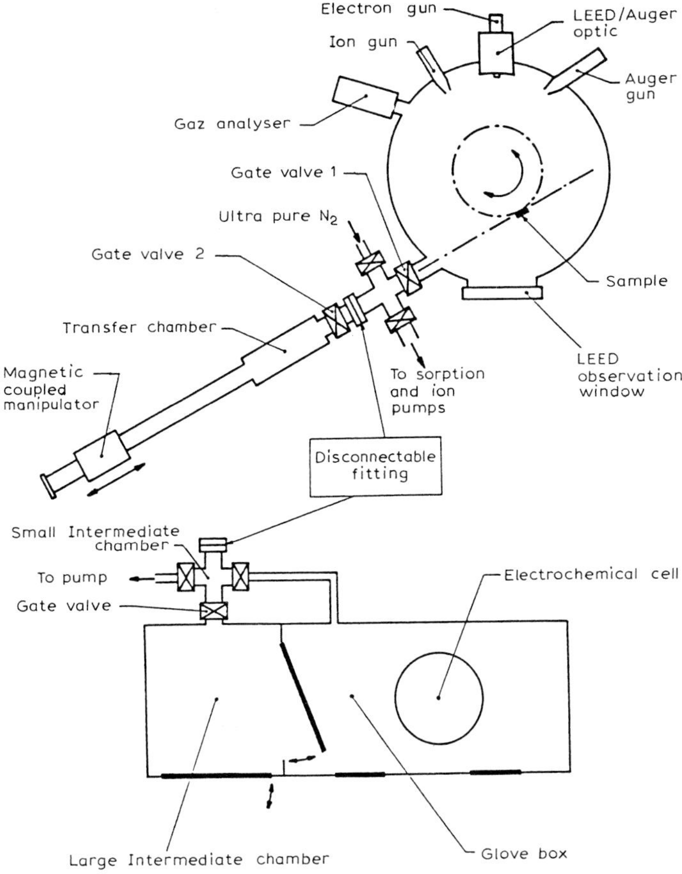

Fig. 8. Schematic diagram of a LEED–AES–electrochemical system using a glove box [28].

but maintained at an overpressure of 5 mbar with N_2 flowing continuously from the glove box to the traps, the atmosphere being renewed about 10 times per hour. Use of a catalyst (BASF R3-11 at 160°C) as one of the traps reduced the oxygen to less than 1 p.p.m. (by weight) as shown by a couloximeter (Chemical Sensor Development). During transfer to the intermediate compartment, the amount of oxygen never rose above 5 p.p.m. Organic material was trapped by activated carbon and particulates by a 0.5 μm filter. The efficacy of the system was demonstrated by transferring Au single crystals prepared and characterized in UHV to an electrochemical cell and back to UHV.

3.3 PARALLEL EXPERIMENTS

This technique avoids the problems of direct transfer by assuming that it is possible to prepare two (or more) electrodes in a way that will result in identical surface states. One may then be characterized in the UHV system while electrochemical experiments are carried out on the other. This method has been used by many in an ad hoc way but the first systematic investigation of clean metal surfaces using this approach seems to be that of Clavilier and Chauvineau [29]. They prepared small ($\sim$2mm diam.) spherical gold electrodes which were treated identically up to insertion into UHV or the electrochemical cell. Surface analysis by AES showed an excellent correlation with the electrochemical behaviour, encouraging the conclusion that the presence or absence of the same impurities was being detected by the two experiments.

The technique was extended to the more active Pt electrodes [30] with similar success and then to single Pt crystals [31]. In the experiments on the latter, it was discovered that preparation of the surfaces outside the UHV could yield surface structures of a quality exceptionally high compared with those habitually used for crystal preparation. The method consists of a high-temperature anneal in a hydrogen + oxygen blowpipe or natural gas + large excess of oxygen blowpipe to a temperature in excess of 1000°C. The hot electrode is quenched with ultra-pure water (which is the product of the blowpipe reaction and so not a foreign component) and the electrode is subsequently protected from the atmosphere by a drop of this water adhering to the surface of interest. This was a flat, ground on the spherical electrode with the desired orientation. Contact with the electrolyte removed this drop without exposing the electrode surface to the atmosphere and the working surface could be made the only part of the electrode in contact with the electrolyte by forming a meniscus and allowing the electrolyte to drain completely from the spherical surface [32, 33] (see Fig. 9). Electrochemical experiments show that the result is a clean surface and this is confirmed by parallel UHV experiments which also show sharp LEED patterns without further treatment of the surface (e.g. ion bombardment) [31]. These experiments show that, if advantage is taken of the particular properties of the electrode material, clean and well-defined surfaces may be obtained by rather simple techniques. It is evident that extension to other materials requires careful thought. Protection of surfaces during transfer is a subject for which the particular properties of a given material, or even a given surface, must be considered. Other examples will be discussed below. Apart from the elimination of any need for masking, the great advantage of this method is that any electrochemical reactions can be monitored from the moment of contact by following the current flowing. The contact is made with the electrode under potentiostatic control. This technique has also been combined with a direct transfer system by Aberdam et al. [23] so that the electrode can be studied after an electrochemical experiment.

Parallel experiments clearly avoid many of the problems encountered in transfer but the final stages of the use of the electrodes in the two environments cannot be identified and it is difficult to cross check the effects of each treatment after the experiments. At the present stage of development, it is

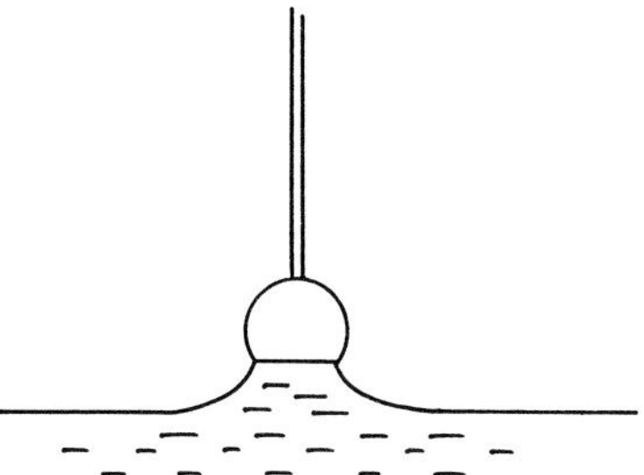

Fig. 9. The dipping electrode technique.

desirable to use a variety of methods to eliminate systematic errors which might occur in one or other.

3.4 TRANSFER OF AN ELECTRODE IN AND OUT OF AN ELECTROLYTE: IMMERSION AND EMERSION

The key problem in ex-situ studies of electrodes is that of identifying and understanding the changes which can occur when the dry electrode is taken from the UHV and put in contact with the electrolyte and when the electrode is removed from the electrolyte, dried, and inserted into the UHV. These processes must be considered in terms of the properties of each system studied and it seems unlikely that there is a single general solution applicable to all systems. The degree of care and control required will depend on the sensitivity of the system.

It would be useful to have a technique which allows the monitoring of the electrode during the whole process of transfer. The only one available at present is that of surface resistivity which is applicable only to a particular form of electrode; a thin film. Experiments of this type were carried out for the immersion and emersion processes by Hansen et al. [34], principally using evaporated Au film on glass supports, although other metals and semiconductors were also examined. It was found that, for hydrophobic electrodes, the bulk solution drained off rapidly as the electrode was emersed but that the double-layer structure remained attached to the solid electrode, i.e. the electrode retained its charge and the potential drop produced by the double layer. The presence of double-layer ions, both specifically and non-specifically adsorbed, was confirmed by XPS measurements. The persistence of the double-layer structure occurred only if the electrode was ideally polarized (blocking). If faradaic processes occurred, leakage of the charge occurred. The persistence of the potential drop was later confirmed by work function measurements [36–38]. This detachment of the electrode with only the double layer adhering depends on the hydrophobic nature of the electrode and it does not occur if the electrode is hydrophilic. However, there is much discussion for gold and for silver as to which class these metals should be put into [39, 40]. Studies of gold electrodes using in-situ reflection IR spectroscopy [41] have shown that organic material can be detected whenever gold electrodes behave hydrophobically.

Emersion of the electrode in transfer systems has relied on either the use of a completely volatile electrolyte (normally aqueous HF [17, 19]) whose residue adhering to the electrode can be pumped off, or a washing procedure in which the electrolyte is replaced by the (volatile) solvent [14]. In either case, it is necessary to consider what processes can occur as potential control is lost when the electrode breaks contact with the solution. The resistance measurements mentioned above suggest that no substantial changes occur provided that no faradaic processes are possible. Traces of oxygen in the ambient gas above the electrolyte can also cause oxidation of surface species and great care is essential to use purified gas when the surface species are susceptible. For example, sub-monolayer deposits of non-noble metallic atoms are readily oxidized and so are observable ex-situ with difficulty [42].

In some cases, the surface can be effectively protected during transfer. The use of a drop of pure water to protect Pt surfaces has been mentioned above. Hubbard and co-workers [43] have shown that iodine adsorption can also be used to protect Pt surfaces. Any type of protection is likely to be specific to a particular substrate and must be studied carefully to ensure that it does not involve modification of the surface structure.

Oxidized surfaces, even when the oxide film is quite thin, seem to be less sensitive to contamination and the transfer problems appear to be less severe. Nevertheless, each system must be examined to verify that contamination or reconstruction does not take place.

4. Some examples of the use of ex-situ techniques in electrochemical problems

4.1 THE NATURE OF OXIDIZED PLATINUM ELECTRODES

It has been known for many years [44, 45] that Pt electrodes subjected to a positive potential form some kind of oxide film which is limited to about one or two monolayers before oxygen evolution occurs. This can be seen clearly in the voltammogram of a Pt electrode in acid solution (see Fig. 10). When the solution and the electrode surface are clean, the charge for the adsorption of this film is exactly equal to that for its desorption, but this charge cannot be ascribed unambiguously to the formation and removal of particular chemical species, using electrochemical measurements alone. An excellent summary of the problems is given by Woods [46].

The first attempt to use ex-situ measurements of the oxidation state of Pt and O in these thin films was made by Kim et al. [3] who used XPS. They were able to study oxide films from the beginning of oxidation about +1.0 V (RHE) up to +2.5 V using simple insertion into the UHV system after washing the electrode. They interpreted their results in terms of three Pt

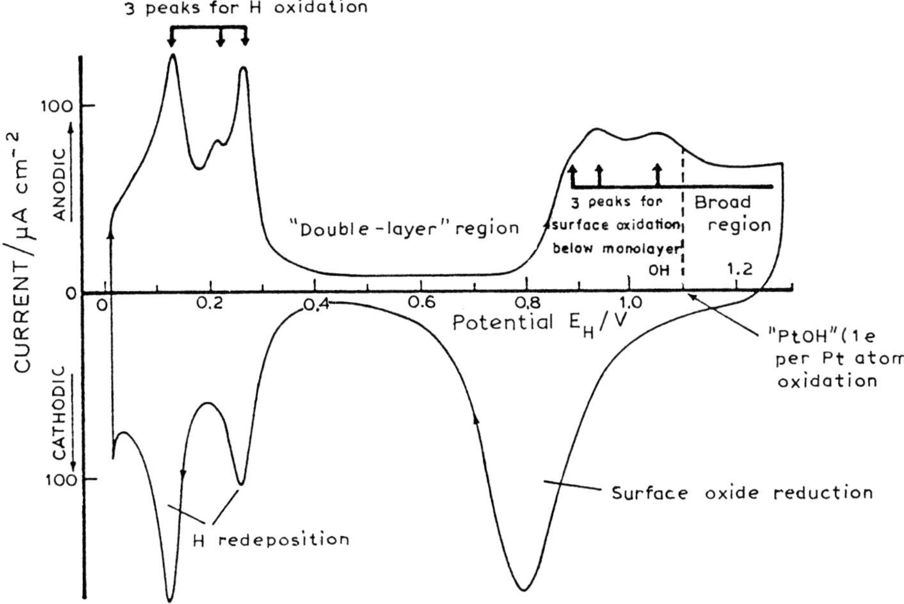

Fig. 10. Potentiodynamic current–potential profile for a polycrystalline Pt electrode in pure aqueous 0.5 M H_2SO_4 at 25°C. Sweep rate 50 mV s^{-1} [57].

states corresponding to PtO_{ads}, PtO, and PtO_2. There was some controversy about the attribution of these states [4, 47] but the thorough study by Hammond and Winograd [27] seems to be conclusive as far as the XPS study of polycrystalline Pt, although the voltammogram (Fig. 11) suggests that some impurities were present even after the initial cycles. They concluded that oxidation to the Pt(II) states occurred under potentiostatic conditions (see Fig. 11) while the Pt(IV) states occurred only with severe and especially galvanostatic oxidation. They also pointed out the problems of changes in the state of the oxide under the X-ray beam or under argon ion bombardment, as well as the participation of anions in the solution.

More recently, a series of experiments on Pt single crystals has been carried out using XPS and electron energy loss spectra (EELS) [48–50] and these have led to the conclusion that the Pt(IV) state is most probably of the form $PtO(OH)_2$. However, the single crystal was strongly perturbed so that it is better to regard these as results for a polycrystalline surface. Further, the oxidation conditions were extremely severe (+3 V vs. Ag/AgCl for 15 h) so that these results are of limited relevance for the normal (early) stages of Pt oxidation. The EELS experiment (Fig. 12 and Table 1) obtained after severe oxidation [50] suggest, however, that the structure is $Pt(OH)_4$ and a similar conclusion was reached by Wagner and Ross [51] using Pt(100) carefully transferred into UHV and studied by AES and TDS.

This type of information has, so far, eluded those using purely electro-

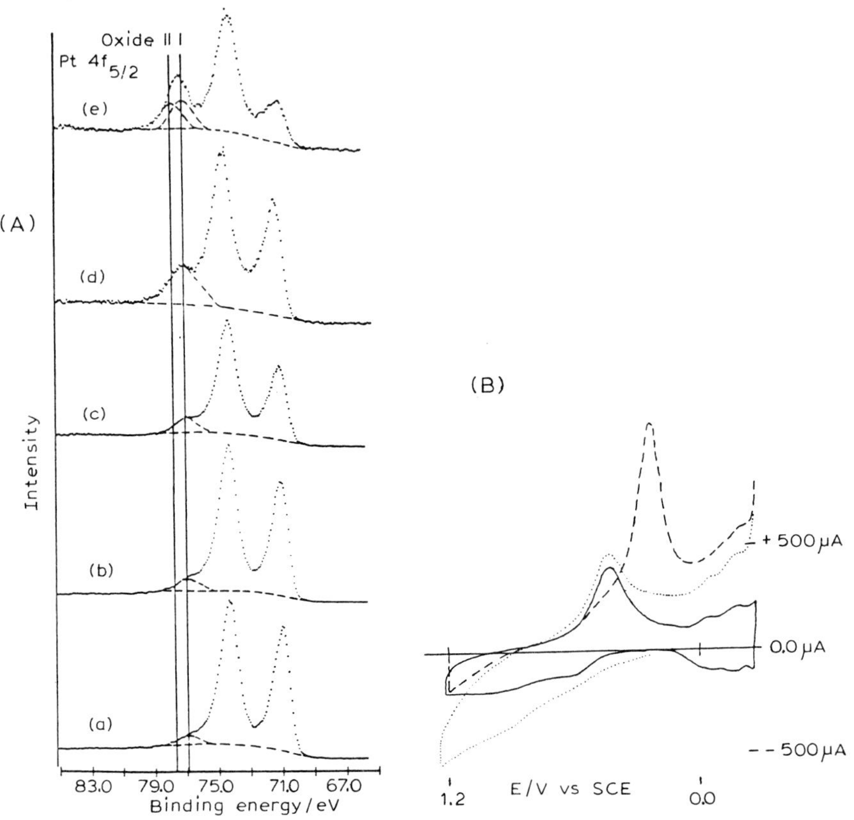

Fig. 11. (A) Pt $4f_{7/2,5/2}$ spectra for potentiostatic oxidation of polycrystalline Pt electrodes in 0.2 M $HClO_4$ (a) 1.6 V for 2 min, (b) 1.8 V for 3 min, (c) 2.0 V for 3 min, (d) 2.2 V for 3 min, (e) 2.2 V for 3.5 min (vs. SCE). (B) (a) Steady-state linear sweep voltammetry of polycrystalline Pt electrode in 0.2 M $HClO_4$ (———); (b) cathodic stripping voltammetry after the electrode was oxidized at + 2.2 V (SCE) for 3 min and analysed by XPS (- - - -); (c) first voltammetric positive sweep immediately following trace (b) (· · · · ·) scan rate 200 mV s^{-1} for all curves [27].

chemical methods or in-situ techniques and consequently it is a valuable extension to knowledge. However, it is clear that the use of a carefully controlled transfer system enables information to be obtained in the monolayer region of greater interest to electrochemists. At the same time, more control over the surface structure could be exercised. More results of this type are to be expected in the near future.

4.2 THE ADSORPTION OF HYDROGEN ON PLATINUM

In contrast to the problem of oxygen adsorption, it has been clearly recognised since the work of Slygin and Frumkin [52] that a monolayer of hydrogen is adsorbed in the atomic form on Pt in the potential range from the equilibrium hydrogen potential to about 0.3 V more positive. They used equilibrium charging curves and later impedance measurements [53] but it

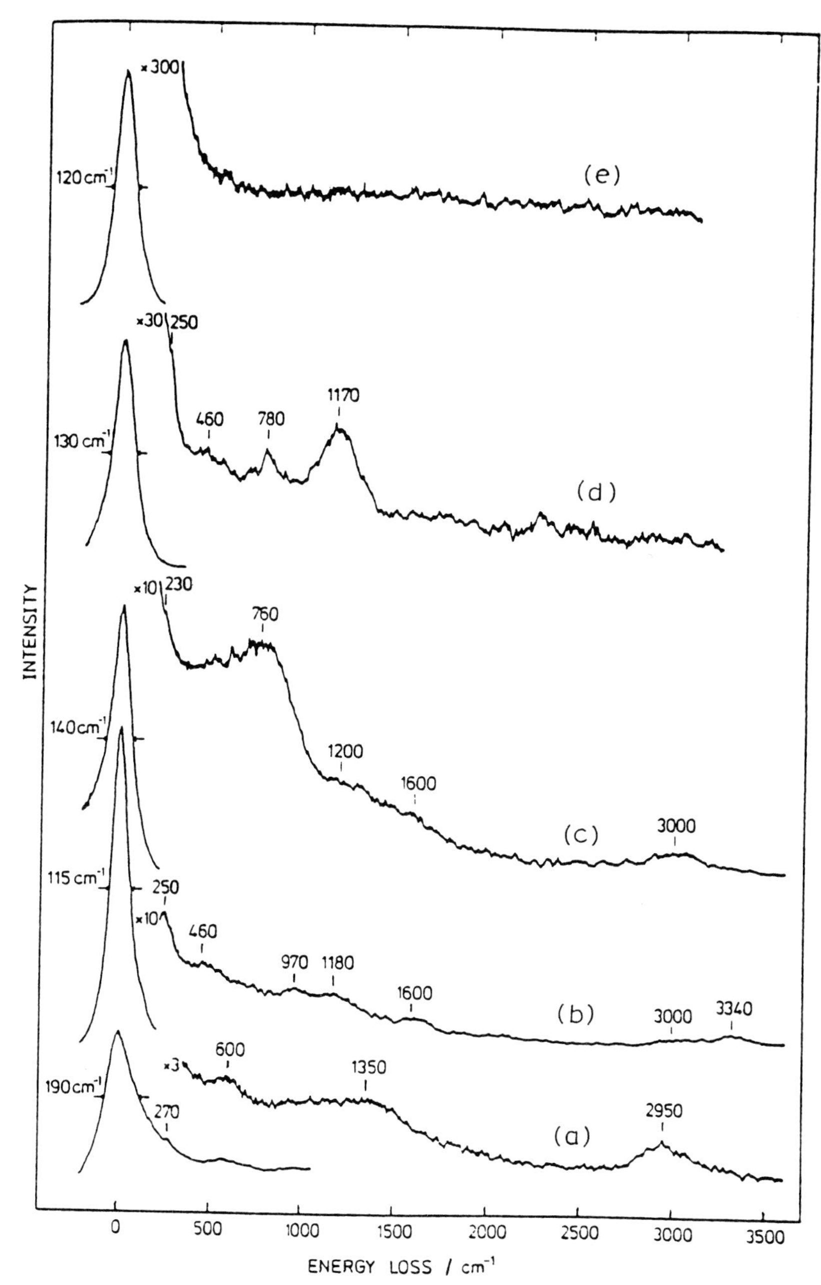

Fig. 12. EEL spectra of a bulk oxidation adlayer on a Pt electrode formed at + 3 V (SHE) in 0.5 M H_2SO_4(aq) (a) at room temperature (290 K), (b) 380 K, (c) 420 K, (d) 700 K, and (e) 900 K. $E_0 = 5.5$ eV; $\theta_1 = \theta_0 = 60°$ [50].

References pp. 126–127

TABLE 1

Energy losses of 5.5 eV electrons on an oxidized platinum electrode surface

Spectra	T (K)	Wavenumber (cm^{-1})	Assignment
(a)	270	600	Pt–OH stretch
		1350	Pt–O–H deformation
		2950	O–H stretch
(b)	380	460	Pt–OH stretch
		970	H_2O frustrated rotation
		1180	Pt–OH deformation
		1600	H_2O scissor deformation
		3000	O–H (hydroxyl) stretch
		3340	O–H (water) stretch
(c)	420	750	Pt oxide (?)
		1200	Pt–OH deformation
		1600	H_2O scissor deformation
		3000	O–H (hydroxyl) stretch
		3340	O–H (water) stretch
(d)	700	460	Pt–O_{ads} stretch
		780	Pt oxide (?)
		1170	Pt oxide (?)
(e)	900		

was not until the impedance measurements of Eucken and Weblus [54] and especially the linear sweep voltammograms of Will and Knorr [55] that two clear states of this adsorption were recognised (cf. Fig. 10). Since then, there has been much discussion about the origins of these different adsorption states. They have been ascribed to different crystallographic sites, to different degrees of charge transfer, and to different types of interaction with co-adsorbed anions (see Woods [46]). It is evident that an effective test of the first hypothesis is to study hydrogen adsorption on single crystal Pt having different orientations. Such an experiment was carried out by Will [56] who sealed single crystals into glass supports so that only a low-index face came in contact with the solution. His results showed that the peak at the more negative potentials (weakly adsorbed hydrogen) was larger for the (111) face while that at the more positive potentials (strongly adsorbed hydrogen) was larger for the (100) face. There was no marked difference between the (110) face and polycrystalline Pt. However, the total amounts of hydrogen adsorbed (obtained by integrating the area under the peaks) was about twice that expected for a smooth surface of the given orientations. These results could be obtained only after a preliminary electrochemical cleaning which involved cycling over 1.5 V for a hundred or so cycles. This suggests that the surfaces were very strongly perturbed and later work has confirmed this.

It is evident that a satisfactory proof of the crystallographic origin of the multiple states of hydrogen adsorption depends on the ability to prepare and

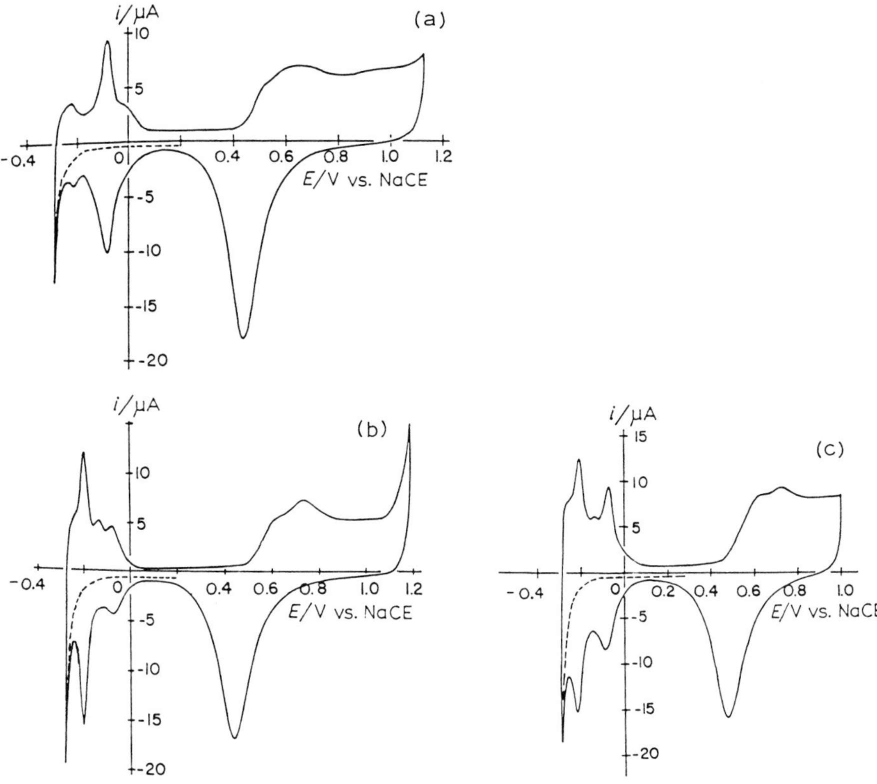

Fig. 13. Cyclic voltammograms for Pt electrodes in 1 M H_2SO_4(aq) (a) Pt(100) rectangular solid single crystal, (b) Pt(111) parallelopiped single crystal, (c) polycrystalline Pt electrode. Sweep rate 10 mV s^{-1}. Temperature 23°C. Electrode area: (a) 0.622, (b) 0.579, (c) 0.800 cm^2 [14].

characterize a well-defined surface and then to transfer it without perturbing it into the electrolyte. This turns out to be exceptionally difficult because Pt surfaces are easily perturbed and are excellent adsorbents for impurities. The first landmark in solving this problem was the remarkable work of Hubbard et al. [15] using the transfer system described above. They studied the (100) and (111) faces, but even these results were obtained only after some decontamination at each stage. After characterization by LEED and AES, the crystal was transferred to the electrochemical cell. A small number of cycles was required to remove the slight contamination and then the results shown in Fig. 13 were obtained. These show clearly that there is a prominent peak for the (111) surface corresponding to the weakly adsorbed H and one for the (100) surface corresponding to the strongly adsorbed hydrogen. Removal of the crystal from the electrochemical cell and transfer to the UHV required a slight cleaning with argon ion bombardment before satisfactory LEED patterns were obtained, indicating that the surface structure remained unchanged. Despite the small degree of uncertainty due to the lack of control during these cleaning processes, there seems little doubt that the electrochemistry was carried out on crystallographically well-defined surfaces, although later work has revealed more subtleties in the behaviour of Pt surfaces.

Subsequent work by other groups [57–59] using transfer systems did not confirm the work of Hubbard's group but it seems likely that this was due to contamination during transfer, since these voltammograms show, for example, incomplete H coverage, imperfect charge balance in the oxygen region, delayed formation of the oxide layer, or time dependence. Further work, particularly by Wagner and Ross [61] has shown that many of these features can be eliminated by improved purification, particularly of the back-fill gas used to raise the pressure around the crystal to atmospheric. In fact, in the period of 5–7 years following the publication of Hubbard's results, there was a controversy about the results obtained for hydrogen adsorption particulary for the (111) face. This resulted on the one hand from the variety of results obtained using UHV transfer systems and on the other from the results obtained by Clavilier et al. [33] using the technique described above in Sect. 3.3. A comparison of these results is shown in Fig. 14. A particular feature of Clavilier's observations is that the voltammogram (e) is obtained immediately after the electrode, pretreated at a high temperature, is contacted with the electrolyte, while (f) results after this electrode is cycled into the oxygen region so that a monolayer of O is formed and removed. Careful study of the LEED patterns of these surfaces [31, 61] had led to the conclusion that (e) corresponds to a surface with a high degree of long-range order, while (f) has lost this order and has a random stepped structure. It is clear that (f) is closely similar to Hubbard's voltammogram (a) and so it must be concluded that the few cleaning cycles used in this work produced the random stepped surface for which this voltammogram is characteristic.

Each low-index face of Pt has comparable subtleties in its behaviour related to the method of preparation and also to the adsorbability of solution components, particularly anions. Pt is probably the extreme example of a surface which must be prepared and transferred with the utmost attention to the details of each step. At the time of writing, there appears to be a reasonable consensus of informed opinion about the voltammetric characteristics of a clean Pt surface of given orientation, although disagreements remain about the origin of some of the observed phenomena.

4.3 THE DEPOSITION OF SILVER ON PLATINUM ELECTRODES PROTECTED BY IODINE

As mentioned above, Hubbard and his co-workers has shown that the adsorption of iodine provides an excellent protective layer for Pt surfaces. This has led them to study the electrochemical properties of this surface and to provide a model of what the combination of ex-situ techniques combined with electrochemistry can provide in terms of detailed information about the structure of the surface layer.

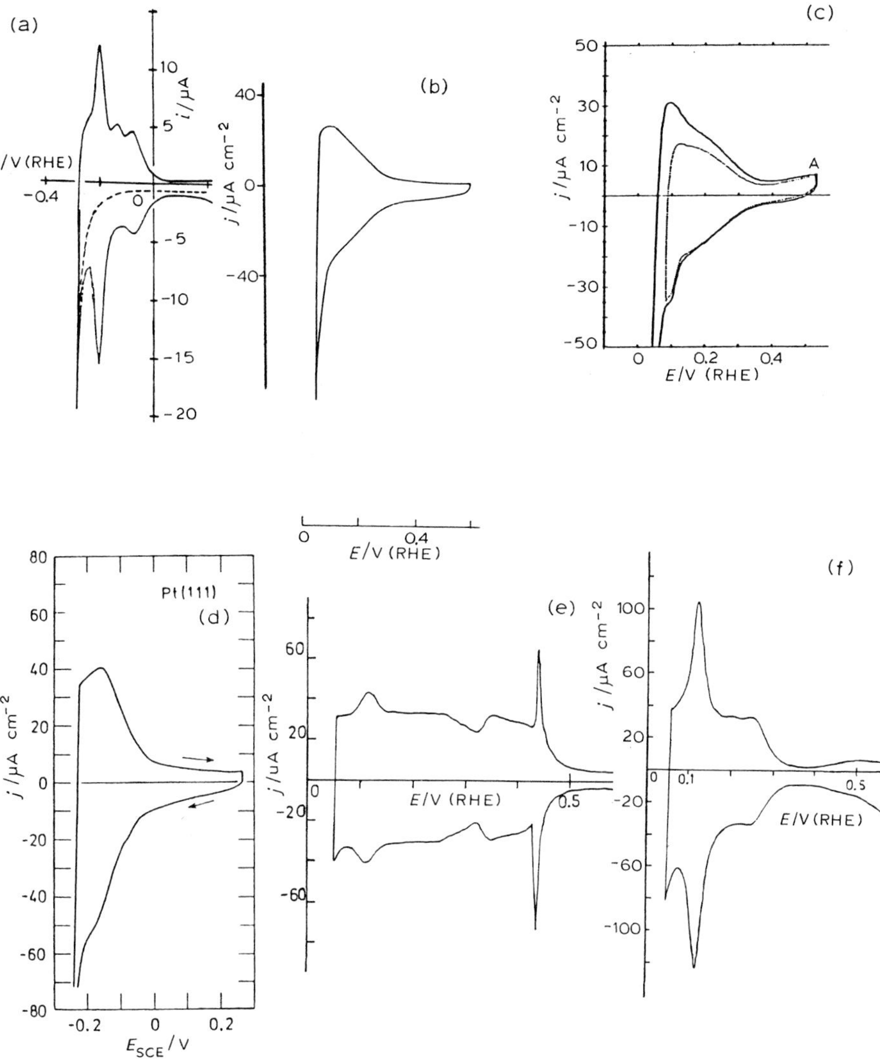

Fig. 14. The hydrogen adsorption region of voltammograms for Pt(111) in aqueous H_2SO_4 (a) Hubbard et al. [14], (b) Adzic et al. [64], (c) Yeager et al. [58], (d) Yamamoto et al. [59], (e) Clavilier et al. [33] without formation of oxide film, (f) Clavilier et al. [33] after formation and removal of oxide film.

In the first place, they found [62] that the adsorption of iodine on a (111) surface disordered by ion bombardment or by electrochemical oxidation would order the Pt substrate so that, after desorption of the iodine, a well-ordered (111) surface was produced. Adsorption of iodine from I_2 vapour under atmospheric pressure yielded the structure shown as (c) in Fig. 15. On heating this, the series of structures indicated in Fig. 15 ensued after which, at 900 K, the iodine evaporated, leaving the well-ordered Pt substrate. (Later work [63] has shown that a similar ordering can be obtained electrochemic-

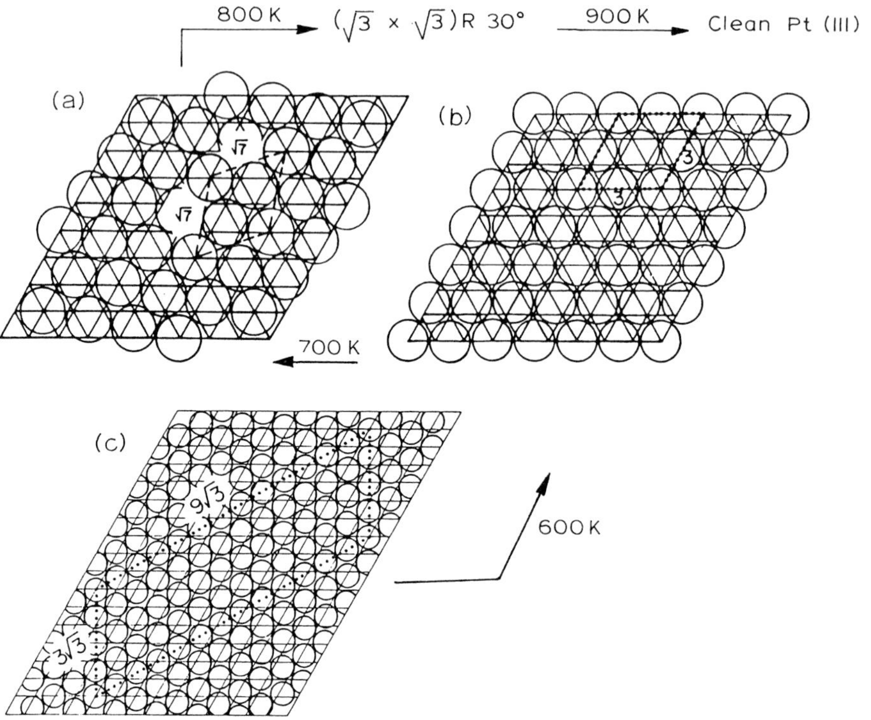

Fig. 15. Surface structures (a) Pt(111) $(\sqrt{7} \times \sqrt{7})$ $R19.1°$–I, $\theta_I = 3/7$, (b) Pt(111) (3×3)–I, $\theta_I = 4/9$, (c) Pt(111) $(3\sqrt{3} \times 9\sqrt{3})$ $R30°$–I, $\theta_I = 52/81$ [62].

ally by replacing the I by CO, which can be oxidized anodically without perturbing the Pt surface.) One interest of this work is that the densest structure Pt(111) $(3\sqrt{3} \times 3\sqrt{3})$ $R30°$–I [Fig. 15(c)] is not obtained by iodine adsorption in the vacuum.

Hubbard and co-workers [62] investigated Ag deposition on the I-protected Pt electrodes and showed that the voltammograms were extremely sensitive to the surface structure, as shown in Fig. 16. Perhaps the most remarkable result is the sensitivity of the Ag voltammogram to very small changes in the I layer; the $\sqrt{7} \times \sqrt{7}$ and 3×3 structures shown in (a) and (b), respectively, in Fig. 15 differ by only about 1% in the total coverage of I, yet there are distinct quantitative changes in the peak heights and widths (Fig. 16). The much less structured voltammogram for the disordered surface is less surprising. A careful study of the AES intensities [15] revealed that the Ag was deposited underneath the I layer. LEED studies at different stages of the Ag deposition enabled the identification of the structure of the layer formed in each peak. Traces of Ag caused an initial $\sqrt{7} \times \sqrt{7}$ superlattice of I to reconstruct to 3×3. Then, in the first peak, the Ag coverage increased to 1 Ag per I atom. In the second peak, this symmetry was retained as the coverage increased to 2 Ag per I atom. Finally, during the third peak, the superlattice reconstructed to $(\sqrt{3} \times \sqrt{3})R30°$ while the coverage of Ag

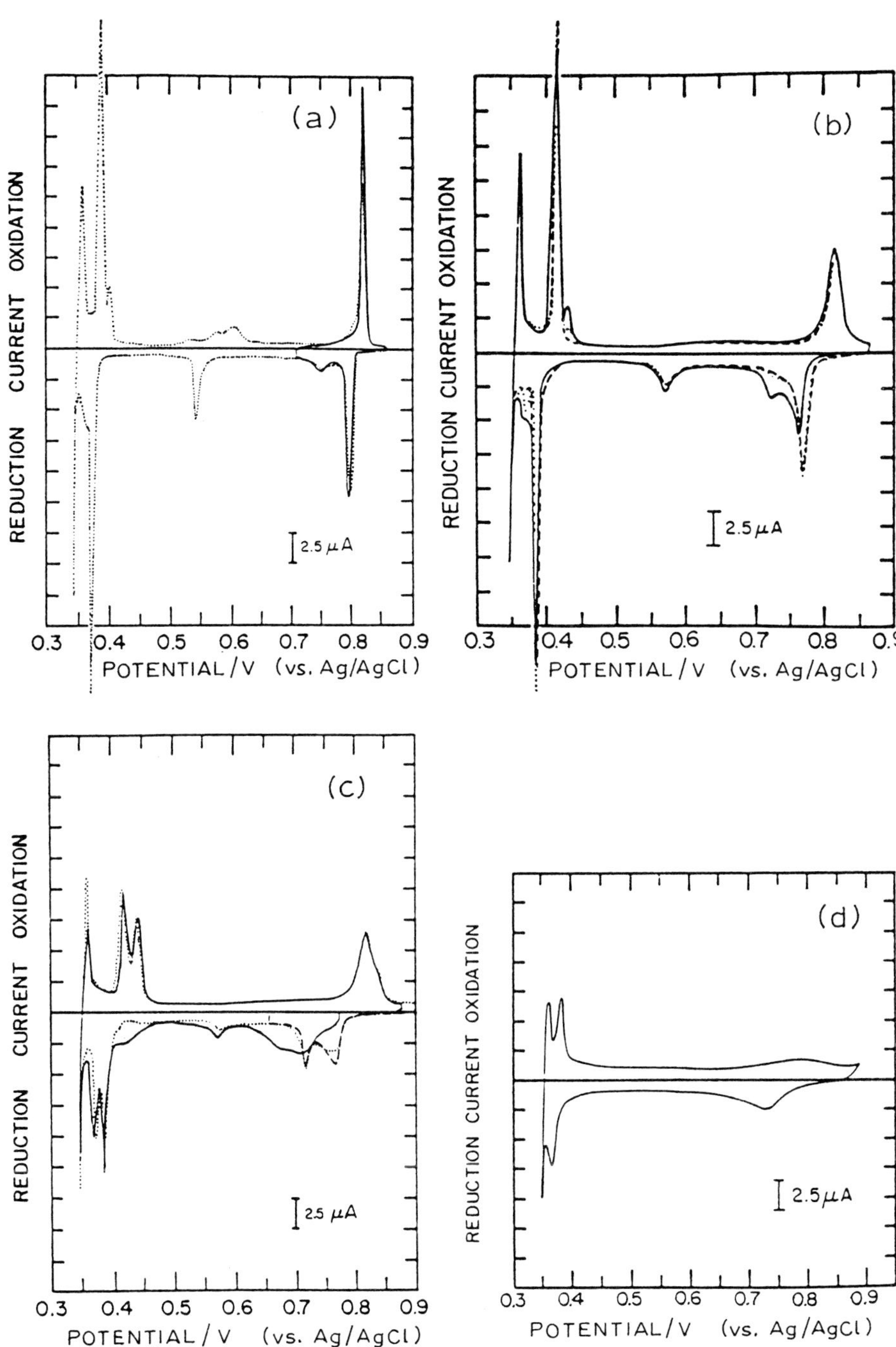

Fig. 16. Cyclic voltammograms for deposition of Ag on to I-ad-lattices on Pt. (a), (b), (c) As specified in Fig. 15, (d) electrochemically disordered Pt(111)–*I*. Scan rate $2\,mV\,s^{-1}$, 1 mM AgCl in 1 M $HClO_4$ starting with structure (a) in (a) (b) and (c) [62]. ——, First; · · · · ·, second; – – –, third sweeps.

indicated the formation of a second layer eventually going over to bulk deposition.

5. Conclusion

The combination of ex-situ UHV techniques with electrochemistry has already resulted in new information about monolayer structures which are based on more convincing evidence than earlier speculations. It is evident that, in the hands of skilled researchers, a great expansion of this knowledge will occur in the next few years. It will be evident from the above discussion that this work needs not only very careful experiments, but also ones that are well-designed with attention to the chemical properties of the system under study.

References

1 See, for example, G.I. Finch, H. Wilman and L. Yang, Discuss. Faraday Soc., 1 (1947) 144.
2 See, for example, H.R. Thirsk and J.A. Harrison, A Guide to the Study of Electrode Kinetics, Academic Press, London, 1972, Chap. 4.
3 K.S. Kim, N. Winograd and R.E. Davis, J. Am. Chem. Soc., 93 (1971) 6296.
4 G.C. Allen, P.M. Tucker, A. Capon and R. Parsons, J. Electroanal. Chem., 50 (1974) 335.
5 P.M.A. Sherwood, Chem. Soc. Rev., (1985) 1.
6 R.M. Ishikawa and A.T. Hubbard, J. Electroanal. Chem., 69 (1976) 317.
7 A.T. Hubbard, Acc. Chem. Res., 13 (1980) 177.
8 B.G. Baker, in J. O'M. Bockris and B.E. Conway (Eds.), Modern Aspects of Electrochemistry, Vol. 10, Plenum Press, New York, 1975.
9 J. Augustynski and L. Balsenc, in B.E. Conway and J.O'M. Bockris (Eds.), Modern Aspects of Electrochemistry, Vol. 13, Plenum Press, New York, 1979.
10 J.S. Hammond and N. Winograd, in R.E. White, J.O'M. Bockris, B.E. Conway and E. Yeager (Eds.), Comprehensive Treatise of Electrochemistry, Vol. 8, Plenum Press, New York, 1984.
11 D.J. Kampe, in R.E. White, J.O'M. Bockris, B.E. Conway and E. Yeager (Eds.), Comprehensive Treatise of Electrochemistry, Vol. 8, Plenum Press, New York, 1984.
12 P.N. Ross and F.T. Wagner, in H. Gerischer (Ed.), Advances in Electrochemistry and Electrochemical Engineering, Vol. 13, Wiley, New York, 1985.
13 A.T. Hubbard, Crit. Rev. Anal. Chem., 3 (1973) 201.
14 A.T. Hubbard, R.M. Ishikawa and J. Katekaru, J. Electroanal. Chem., 86 (1978) 271.
15 A.T. Hubbard, J.L. Stickney, S.D. Rosaco, M.P. Soriaga and S. Song, J. Electroanal. Chem., 150 (1983) 165.
16 R.W. Revie, B.G. Baker and J.O'M. Bockris, J. Electrochem. Soc., 122 (1975) 1460.
17 E. Yeager, A.S. Homa, B.D. Cahan and D. Scherson, J. Vac. Sci. Technol., 20 (1962) 628.
18 A.S. Homa, E. Yeager and B.D. Cahan, J. Electroanal. Chem., 150 (1983) 141.
19 F.T. Wagner and P.N. Ross, J. Electroanal. Chem., 150 (1983) 141.
20 R.O. Ansell, T. Dickinson, A.F. Povey and P.M.A. Sherwood, J. Electroanal. Chem., 98 (1979) 69.
21 H. Neff, W. Foditsch and R. Kötz, J. Electron Spectrosc. Relat. Phenom., 33 (1984) 171.
22 S. Haupt, U. Collisi, H.D. Speckman and H.H. Strehblow, J. Electroanal. Chem., 194 (1985) 179.
23 D. Aberdam, R. Durand, R. Faure and F. El-Omar, Surf. Sci., 171 (1986) 303.

24 T. Dickinson, A.F. Povey and P.M.A. Sherwood, J. Chem. Soc. Faraday Trans. 1, 73 (1977) 327.
25 T.E. Pou, O.J. Murphy, J.O'M. Bockris and L.L. Tongson, J. Electrochem. Soc., 131 (1984) 1243.
26 N.S. McIntyre, S. Sunder, D.W. Shoesmith and F.W. Stanchell, J. Vac. Sci. Technol., 18 (1981) 714.
27 J.S. Hammond and N. Winograd, J. Electroanal. Chem., 78 (1977) 55.
28 J.P. Bellier, J. Lecoeur and A. Rousseau, J. Electroanal. Chem., 200 (1986) 55.
29 J. Clavilier and J.P. Chauvineau, J. Electroanal. Chem., 97 (1979) 199.
30 J. Clavilier and J.P. Chauvineau, J. Electroanal. Chem., 100 (1979) 461.
31 D. Aberdam, C. Corotte, D. Dufayard, R. Durand, R. Faure and G. Guinet, Proc. 4th Int. Conf. Solid Surf., Cannes, 1980, p. 622.
32 D. Dickertman, F.D. Koppitz and J.W. Schultze, Electrochim. Acta, 21 (1976) 967.
33 J. Clavilier, R. Faure, G. Guinet and R. Durand, J. Electroanal. Chem., 107 (1980) 205.
34 W.N. Hansen, C.L. Wang and T.W. Humpherys, J. Electroanal. Chem., 93 (1978) 87.
35 W.N. Hansen, J. Electroanal. Chem., 150 (1983) 133.
36 W.N. Hansen and D.M. Kolb, J. Electroanal. Chem., 100 (1979) 493.
37 G.J. Hansen and W.N. Hansen, J. Electroanal. Chem., 150 (1983) 193.
38 R. Kötz and N. Neff, J. Electroanal. Chem., 215 (1986) 331.
39 T. Smith, J. Colloid Interface Sci., 25 (1980) 51.
40 G. Valette, J. Electroanal. Chem., 134 (1982) 285.
41 O. Hofmann, K. Doblehofer and H. Gerischer, J. Electroanal. Chem., 161 (1984) 337.
42 D. Aberdam, R. Durand, R. Faure and F. El Omer, Surf. Sci., 162 (1985) 782.
43 A. Wieckowski, S.D. Rosaco, B.C. Schardt, J.L. Stickney and A.T. Hubbard, Inorg. Chem., 23 (1984) 565.
44 F.P. Bowden, Proc. R. Soc. London Ser. A, 125 (1929) 446.
45 J.A.V. Butler and G. Armstrong, Proc. R. Soc. London Ser. A, 137 (1929) 504.
46 R. Woods, in A.J. Bard (Ed.), Electroanalytical Chemistry, Vol. 9, Dekker, New York, 1976, p. 1.
47 T. Dickinson, A.F. Povey and P.M.A. Sherwood, J. Chem. Soc. Faraday Trans. 1, 71 (1975) 298.
48 M. Peuckert, F.P. Coenen and H.P. Bonzel, Electrochim. Acta, 29 (1984) 1305.
49 M. Peuckert, Electrochim. Acta, 29 (1984) 1315.
50 M. Peuckert and H. Ibach, Surf. Sci., 136 (1984) 319.
51 F.T. Wagner and P.N. Ross, Appl. Surf. Sci., 24 (1985) 87.
52 A.I. Slygin and A.N. Frumkin, C. R. Acad. Sci. URSS, 2 (1934) 173.
53 A.N. Frumkin, P.I. Dolin and B.V. Ershler, Acta Physicochim. URSS, 13 (1940) 747, 780, 794.
54 A. Eucken and B. Weblus, Z. Elektrochem., 55 (1951) 144.
55 F.G. Will and C.A. Knorr, Z. Elektrochem., 64 (1960) 258.
56 F.G. Will, J. Electrochem. Soc., 112 (1965) 451.
57 H. Angerstein-Kozlowska, B.E. Conway and W.B.A. Sharp, J. Electroanal. Chem., 43 (1973) 9.
58 E. Yeager, W.E. O'Grady, M.Y.C. Woo and P.L. Hagens, J. Electrochem. Soc., 125 (1978) 348.
59 K. Yamamoto, D. Kolb and G. Lempfühl, J. Electroanal. Chem., 96 (1979) 233.
60 P.N. Ross, J. Electrochem. Soc., 126 (1979) 67.
61 F.T. Wagner and P.N. Ross, J. Electroanal. Chem., 150 (1983) 141.
62 A. Wieckowski, B.C. Schardt, S.D. Rosasco, J.L. Stickney and A.T. Hubbard, Surf. Sci., 146 (1984) 115.
63 D. Zuraski, L. Rice, M. Mourani and A. Wieckowski, J. Electroanal. Chem., 230 (1987) 221.
64 R.R. Adzic, W.E. O'Grady and S. Srinivasan, Surf. Sci., 94 (1980) L191.

Chapter 4

New Hydrodynamic Methods

W. JOHN ALBERY, CHRISTOPHER C. JONES and ANDREW R. MOUNT

1. Introduction

There are several advantages in using electrodes with controlled hydrodynamics. First, because the reactant is supplied continuously to the electrode, systems can be studied in the steady state. Secondly, by varying the convection, one can determine quantitatively the role of mass transport in the overall kinetics of the electrode reaction. Thirdly, a double electrode system, such as the ring–disc electrode, can be used in which the downstream electrode measures the flux of reactant, product, or intermediates at the electrode surface. We will consider three hydrodynamic systems. The most popular is the rotating disc system. The other two systems, the wall-jet and tube electrodes, possess the advantage of having no moving parts. This means that these electrodes can be more easily combined with spectroscopic techniques such as ESR. We will start by describing the three different systems.

2. Different hydrodynamic systems

2.1 ROTATING DISC

The velocity pattern for a rotating disc is displayed in Fig. 1 [1]. The disc acts as a pump, sucking solution towards the disc, spinning it round, and flinging it outwards centrifugally. An important feature of this electrode is that it is "uniformly accessible" [2]. The current density does not vary with radial distance but is uniform all over the electrode surface. This feature greatly simplifies the mathematical description of the electrode kinetics. When the convective diffusion equation is solved [1], one finds that there is a diffusion layer of thickness about 10^{-2} cm across which the reactant diffuses to the electrode. There is a temptation to think that, within this layer, the solution is "stagnant". It is certainly true that the velocity component towards the disc (H in Fig. 1) is small because, at low values of z, it varies with z^2. However, the radial component F varies with z at low values of z and hence, inside the "stagnant" diffusion layer, there is considerable radial convection; in fact $v_r/v_z = r/z \sim 10$–100. It is this radial convection

References pp. 147–148

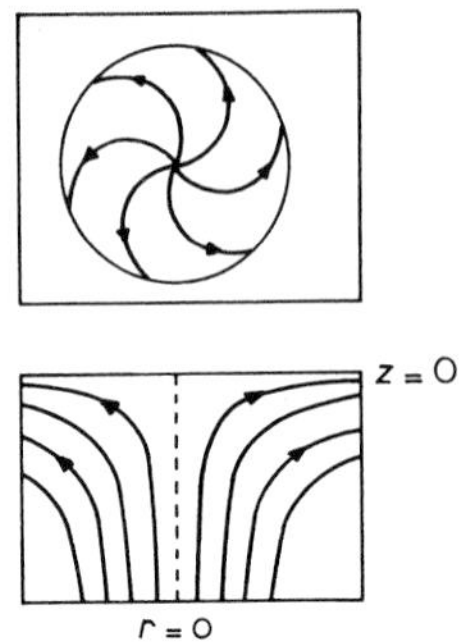

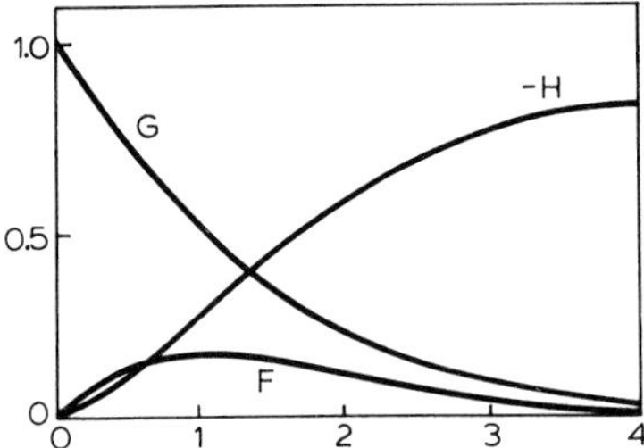

Fig. 1. The velocity pattern for a rotating-disc electrode. The functions F, G, and H describe the radial, angular, and normal components, respectively. The distance normal to the electrode, z, is normalised with the thickness of the hydrodynamic layer, Z_H, where Z_H is given by $\nu/(2\pi W)$ and $\nu/\mathrm{cm^2\,s^{-1}}$ and W/Hz are the kinematic viscosity and rotation speed, respectively.

that carries products and intermediates made on the disc electrode downstream to the concentric ring electrode of a ring–disc combination.

The limiting current, i_L, of the rotating disc electrode is given by the Levich equation [1]

$$i_L = 1.554 nAFD^{2/3}\nu^{-1/6}\ W^{1/2}c_\infty \tag{1}$$

where W/Hz is the rotation speed and $c_\infty/\mathrm{mol\,cm^{-3}}$ is the bulk concentration; the other symbols have their usual meanings.

2.2 WALL-JET

The velocity pattern for the wall-jet system is shown in Fig. 2 [3, 4]. The impinging jet brings the reactant to the centre of the disc electrode; the solution then spreads radially outward. The velocity profile in the radial direction is similar to that for the rotating disc. The component in the normal direction is interesting in that, at large distances away from the electrode, the flow is towards the electrode. Again, this is similar to the rotating disc. But close to the electrode the flow is away from the electrode. This means that the electrode always sees fresh solution that has just passed through the jet. Another advantage of this system is that a relatively high

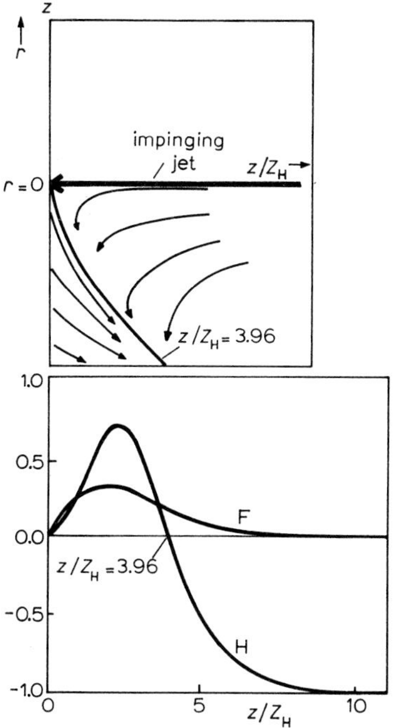

Fig. 2. The velocity pattern for a wall-jet electrode. The functions F and H describe the radial and normal components, respectively. The thickness of the hydrodynamic layer, Z_H, is given by $Z_H = 2.28r^{5/4}a^{1/2}\nu^{3/4}/V_f^{3/4}$ where a/cm is the diameter of the jet and $V_f/\text{cm}^3\,\text{s}^{-1}$ is the volume flow rate. Note that the thickness of the layer varies with $r^{5/4}$.

proportion, 7–10%, of the reactant that passes through the jet can be destroyed on the electrode.

The limiting current for a wall-jet electrode is given by [3, 4]

$$i_L = 1.38nFD^{2/3}\nu^{-5/12}a^{-1/2}r^{3/4}V_f^{3/4}c_\infty \qquad (2)$$

where a is the diameter of the nozzle of the jet and V_f is the volume flow rate. As shown in Fig. 2, the wall-jet electrode is not uniformly accessible. The thicknesses of the hydrodynamic and diffusion layers vary with $r^{5/4}$. We shall see below that this variation can be put to good use in our studies of the kinetics of colloidal deposition.

A further development of the wall-jet system is to insert a packed-bed electrode upstream of the jet to make a packed-bed wall-jet electrode (PBWJE) [5–7]. The packed bed can be used to generate reactant, to generate a fresh electrode surface or, if the bed has immobilised enzyme, to carry out an enzymatic reaction. The packed-bed wall-jet electrode is illustrated in Fig. 3.

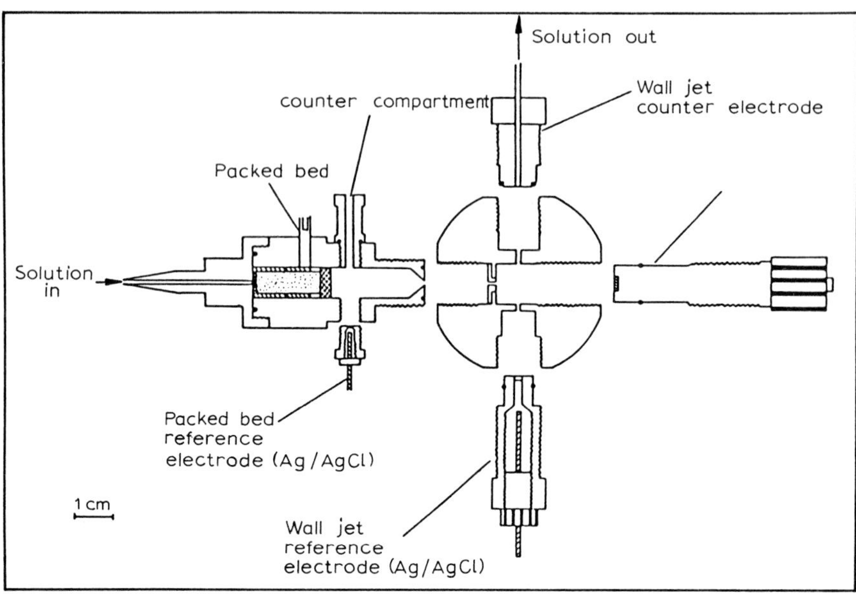

Fig. 3. The packed-bed wall-jet electrode.

2.3 TUBE ELECTRODE

One of the simplest flow systems is the tube electrode with the electrode set into the wall of a tube. The solution flows by under laminar conditions.

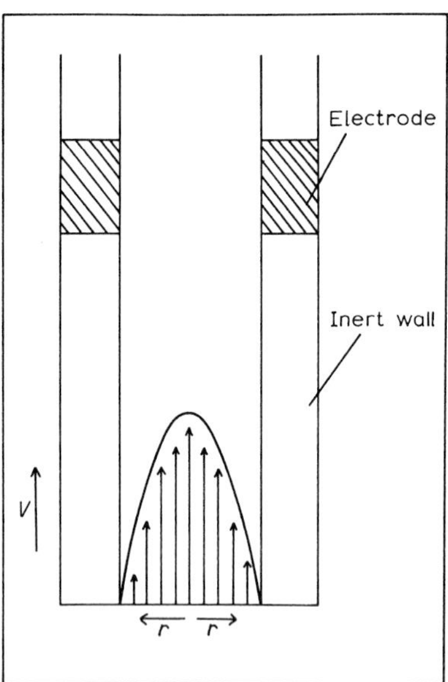

Fig. 4. The channel electrode.

As shown in Fig. 4, the normal parabolic velocity profile is established. The limiting current for this type of electrode is [8]

$$i_L = 5.43nFD^{2/3}l^{2/3}V_f^{1/3}c_\infty \quad (3)$$

where l is the length of the tube electrode. It is interesting that, in this case, the limiting current does not depend on the radius of the tube nor on the kinematic viscosity.

3. Ring–disc electrodes

3.1 THE COLLECTION EFFICIENCY

The use of a second downstream electrode to monitor chemical fluxes at the working electrode is proving to be an important technique for the investigation of electrode mechanisms. This is particularly true for electrodes which have a more complicated structure than a simple metallic surface. Examples are modified electrodes, oxide electrodes, or enzyme electrodes. For these more complex systems, the separate measurement of the fluxes at the electrolyte–electrode interface provides unique and valuable information. Double electrodes can be constructed for all three hydrodynamic systems. A crucial parameter for such a double electrode is the collection efficiency, N, which, in the steady state, relates the flux of material detected as a limiting current on the downstream electrode to the flux of material generated on the upstream electrode. The collection efficiency is a function of the geometry of the electrode and is given for all three systems by [4, 9]

$$N = 1 - \mathrm{F}(\alpha/\beta) + \beta^{2/3}[1 - \mathrm{F}(\alpha)] - (1 + \alpha + \beta)^{2/3}\{1 - \mathrm{F}[(\alpha/\beta)(1 + \alpha + \beta)]\} \quad (4)$$

where

$$\mathrm{F}(\theta) = \frac{3^{1/2}}{4\pi}\ln\left[\frac{(1 + \theta^{1/3})^3}{1 + \theta}\right] + \frac{3}{2\pi}\tan^{-1}\left[\frac{2\theta^{1/3} - 1}{3^{1/2}}\right] + \frac{1}{4}$$

TABLE 1

Gap and ring parameters to be used in eqn. (4) for N_o

	α	β
Rotating ring–disc	$(r_2/r_1)^3 - 1$	$(r_3/r_1)^3 - (r_2/r_1)^3$
Wall-jet ring–disc	$(r_2/r_1)^{9/8} - 1$	$(r_3/r_1)^{9/8} - (r_2/r_1)^{9/8}$
Double channel[a]	$(l_2/l_1) - 1$	$(l_3/l_1) - (l_2/l_1)$

[a] Measuring from the upstream edge of the upstream electrode, the gap lies between l_1 and l_2 and the downstream edge of the collecting electrode is at l_3.

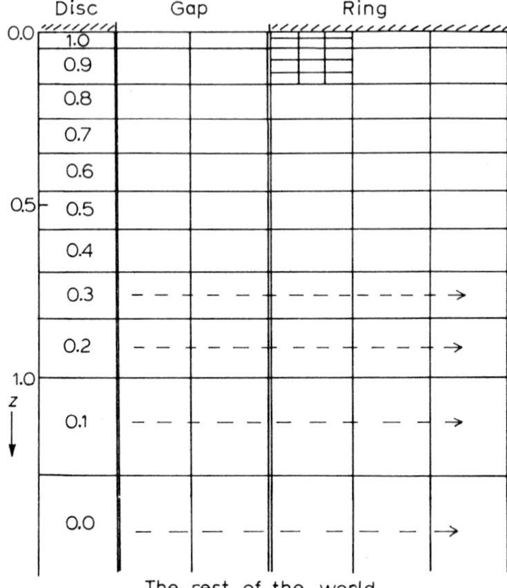

Fig. 5. The grid, distorted by the Hale transformation, used in our simulation programme. A stable species being produced on the disc gives a linear concentration gradient in w. As shown, a finer grid is used on the innermost box on the ring electrode.

Expressions relating α and β to the geometric parameters are collected in Table 1.

3.2 RING–DISC SIMULATION PROGRAMS

Although analytical solutions can be found for many ring–disc systems [10], there are some systems, especially those with more complicated homogeneous kinetics, where the interpretation requires a numerical solution. Bard and Prater [11], following Feldberg [12], developed a simulation program in the early 1970s. More recently, Feldberg and co-workers have improved this program [13]. We have also developed our own simulation program. We have emphasised elsewhere [14] the advantage of using non-Cartesian space. For a rotating ring–disc electrode, a much more efficient program can be produced if one uses the Hale transformation [15] and replaces the Cartesian coordinate normal to the electrode, z, with a coordinate, w, transformed by the Levich function

$$w = \frac{\int_0^x \exp\left(-x^3/3\right) \mathrm{d}x}{\int_0^\infty \exp\left(-x^3/3\right) \mathrm{d}x} \tag{5}$$

where

$$x = z(2D^{-1/3}\nu^{-1/6}W^{1/2}) \tag{6}$$

There are two advantages of the transformation in eqn. (5). First, whereas z and x run from zero to infinity, w runs from zero to unity. Only ten boxes in the direction normal to the electrode are required. The grid is shown in Fig. 5. It can be seen, for instance, that the outermost box covers all space from $x = 1.5$ to $x = \infty$. A stable species in the steady state produces a linear concentration with respect to w; hence, each box is contributing equally to the description of the concentration profile. It is also interesting that, because of the efficient radial convection inside the diffusion layer, the concentration in the four outermost boxes in Fig. 5 does not vary with r. The second advantage of the transformation is that the normal and radial convection terms in the convective diffusion equation collapse into a single term, giving a simpler equation for the simulation. A second feature of our program is the magnified grid that we use for the innermost box of the ring. The concentration profiles are most curved at this point and so one needs a fine grid. However, it is wasteful to use such a fine grid elsewhere. Good results are found on the grid shown in Fig. 5. This grid has many fewer boxes than the grid used by Feldberg and hence our program is very much more efficient. Copies of the program and accompanying notes can be obtained from the author.

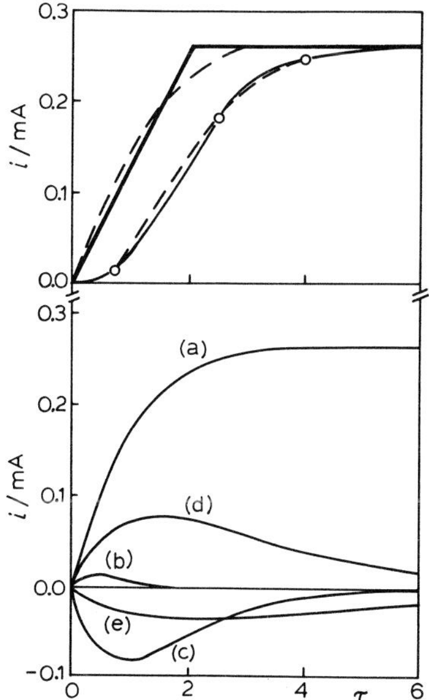

Fig. 6. The procedure for finding the flux at the disc from any ring current transient. The current ramp on the disc produces the ring current transient shown by the solid line. Using the library of functions in the lower diagram, the broken lines display the disc current and ring current transient where the two ring current transients have been fitted at the points shown.

3.3 RING–DISC TRANSIENTS

Ring–disc transients also provide valuable information about the kinetics of complex electrodes. Here, we draw attention to a general procedure we introduced for deriving the flux at the disc electrode from any measured ring current transient [16]. The method works by building up the disc transient by assigning the correct weights to a tanh function and four exponential functions so as to fit the ring transient at four different points. The success of this procedure for a ramp is shown in Fig. 6. Again, we are slightly surprised that this procedure has not been more widely adopted.

3.4 RING–DISC pH TRANSIENTS

We have recently shown that a bismuth oxide ring electrode may be used as a sensitive potentiometric electrode for measuring small local changes in pH at the disc surface [17]. The electrode is particularly sensitive near pH 7 and fluxes as low as 1 pmol cm^{-2} s^{-1} can be measured, providing that stringent precautions are taken to exclude the buffering effects of adventitous CO_2.

We have applied this technique to the study of the proton flux that takes place when a modified electrode, the thionine-coated electrode, is either oxidised or reduced. We were particularly interested in the question as to whether the proton and electron fluxes were in time with one another or not. Typical results for proton and electron fluxes for reduction and oxidation at a number of different values of pH are displayed in Fig. 7. At first sight, we were bewildered by the variety of behaviour. However, we can explain the different transients as follows. In Table 2, we set out the scheme of squares [18, 19] for the thionine/leucothionine system with a number of vital pK_A values. Starting at pH 4 in the oxidation direction ($LH_3^{2+} \rightarrow Th^+ + 2e + 3H^+$), we see that the proton flux is indeed larger than the electron flux and that both fluxes are in time with each other. In the opposite reduction direction, the electron flux is similar but the proton flux is smaller and delayed. The reason for this is that, to start with, protons are used up and the pH crosses the pK_a at 5.5 ($Th^+ + 3H^+ + 2e \rightarrow LH_3^{2+}$). However, for pH > 5.5, the reaction can utilise the H^+ stored in the coat ($Th^+ + 2LH_3^{2+} + 2e \rightarrow 3LH_2^+$). This means that bulk H^+ is not consumed, leading to a smaller H^+ transient. When the electron flux dies away, the pH drifts back to the equilibrium value of 4. As it does so, there is an H^+ flux from the relaxation $LH_2^+ + H^+ \rightarrow LH_3^{2+}$. The explanation of the transients at pH 5 is similar. In the reduction direction, the H^+ flux has almost completely collapsed. In this case, the pH crosses the pK_a boundaries at 8.5 where there will be no H^+ flux ($Th^+ + 2e \rightarrow L^-$). The relaxation flux after the electron flux has died away will also be small since the bulk concentration of H^+ (pH = 5) is so small. At pH 6, the reduction transients are similar to those at pH 5. In the oxidation direction, the pH rapidly crosses the pK_a = 5.5 boundary. Now the coat mops up the H^+, releasing no H^+ to the solution ($3LH_2^+ \rightarrow$

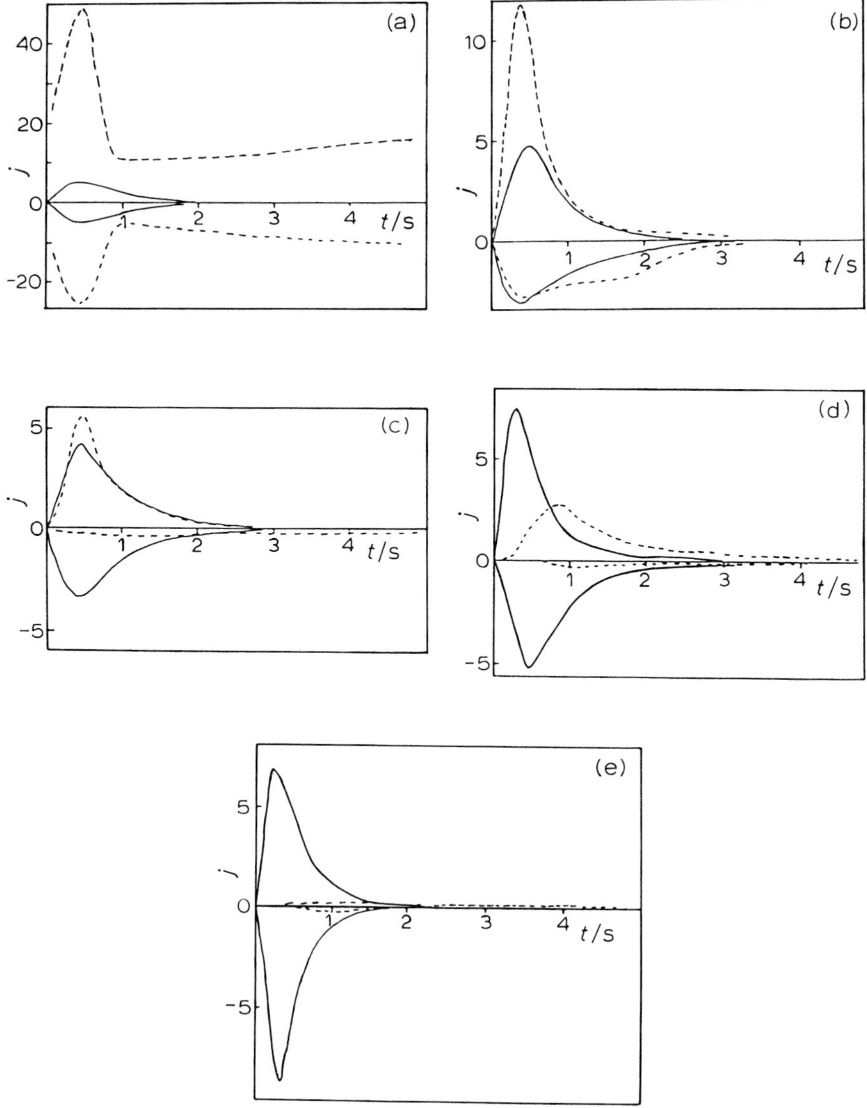

Fig. 7. Electron (solid lines) and proton (broken lines) fluxes for the oxidation and reduction of a thionine-coated electrode at pH values of (a) 3, (b) 4, (c) 5, (d) 6, and (e) 9. The fluxes are measured in nmol $cm^{-2} s^{-1}$.

$Th^+ + 2LH_3^{2+} + 2e$). After a third of the coat has reacted in this way, there is no more LH_2^+ left and so now the H^+ transient starts releasing H^+ to the solution ($LH_3^{2+} \rightarrow Th^+ + 3H^+ + 2e$). This gives a proton flux that is larger than the electron flux, but is in time with it. Finally, at pH 9, we have electron fluxes with negligible proton fluxes. The transients here involve $Th^+ + 2e \rightarrow L^-$ and the reverse reaction, giving no H^+ fluxes. This study

TABLE 2

H_2N — N — S^+ — NH_2

Th^+

H_3N^+ — H N — S — N^+H_3

LH_3^{2+}

			pK_a
Th^+ ⇌	S ⇌	L^-	
⇅	⇅	⇅	8.5
ThH^{2+} ⇌	SH^+ ⇌	LH	
	⇅	⇅	8.5
	SH_2^{2+} ⇌	LH_2^+	
		⇅	5.5
		LH_3^{2+}	

therefore shows how one can use the potentiometric H^+ detector to investigate and understand the changes in protonation of the coats of modified electrodes.

4. Measurement of diffusion coefficient by rotation speed step

A particular example of another transient experiment, which we find most useful, is the determination of diffusion coefficients using the rotation speed step method [20]. The transient response of the limiting current at a rotating disc to a step in the rotation speed say from 5 to 7 Hz is a function of the

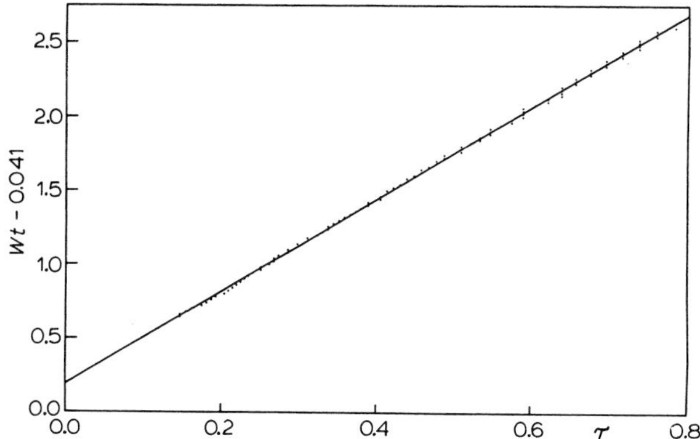

Fig. 8. Typical results for the rotation speed step experiment. The values of τ are found from the current transient using eqn. (8). These are plotted against t according to eqn. (7). A value of D can then be found from the gradient.

normalised, τ, where

$$\tau = t(4.0D^{1/3}\nu^{-1/3}W) \quad (7)$$

Bartlett took the previously published function [20] for the current transient as a function of τ and showed that a good analytical approximation is

$$\tau = -0.2111 + 2.324f - 4.103f^2 + 3.063f^3 - 0.011\sin[4\pi(f - 0.26)] \quad (8)$$

where

$$f = \frac{\Delta i}{\Delta i_\infty}$$

For any value of $\Delta i/\Delta i_\infty$, we can therefore find a value of τ, which can be plotted against t. Typical results for $Fe(CN)_6^{3-}$ are shown in Fig. 8. From the gradient and eqn. (7), we find that $D/\mathrm{cm^2\,M\,s^{-1}} = 6.9$, which can be compared with a value of 6.5 determined from steady-state measurements [21]. The advantage of the rotation speed step technique is that one does not need to know either n or c_∞ to determine D. We are surprised that this simple technique is not used more often by other workers.

5. Electrochemical ESR

This whole volume shows the advantages of combining electrochemistry and spectroscopy. In certain cases, it is valuable to link the electrode and the spectroscopic detection by a controlled hydrodynamic system. Concentration patterns can then be calculated and the spectroscopic signal can be quantitatively interpreted to yield kinetic data for the reaction mechanism. An example of this technique is our work [14, 22, 23] and that of Waller and Compton [24] using channel electrodes in an ESR spectrometer.

References pp. 147–148

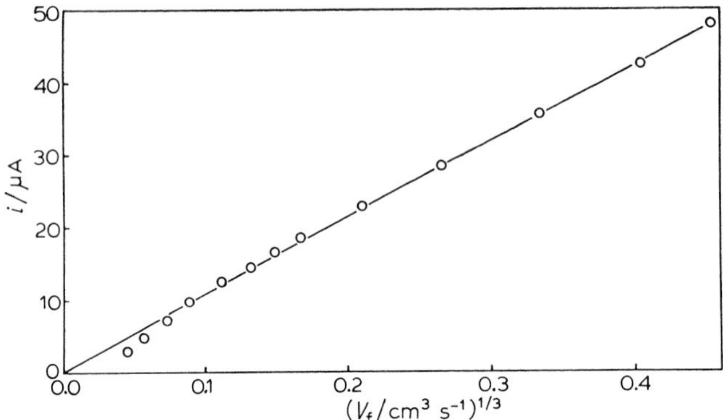

Fig. 9. Variation of the limiting current with flow rate for the reduction of maleimide at a channel electrode, plotted according to eqn. (3).

A recent example from our work concerns a study of the electropolymerisation of maleimide. This system had been studied by Bhadani and Prasad [25] using a twin compartment cell and exhaustive electrolysis. Figure 9 shows the limiting current for the reduction of maleimide in dimethylformamide on a platinum channel electrode placed in the centre of the ESR cavity. As required by the Levich equation [eqn. (3)], the limiting current varies with $V_f^{1/3}$. While this electrochemical reaction is going on, one can also measure the ESR spectrum, obtaining results such as shown in Fig. 10. Good agreement is found between this spectrum and one simulated for the radical anion of maleimide. The variation of the intensity of the ESR signal is shown in Fig. 11. If the radical product of the electrochemical reaction was stable, the signal strength divided by the current should vary with $V_f^{-2/3}$ [22]. The signal decreases with increasing flow rate because the electro-generated radical spends a shorter time in the cavity. The curve in Fig. 11 arises because the radical is unstable and, at lower flow rates, decomposes while still in the cavity. The curve through the points is obtained from our simulation programme, assuming that the radical is decomposing by second-order kinetics. (As discussed above, this simulation programme also uses a distorted space.) Hence from these results, we can deduce

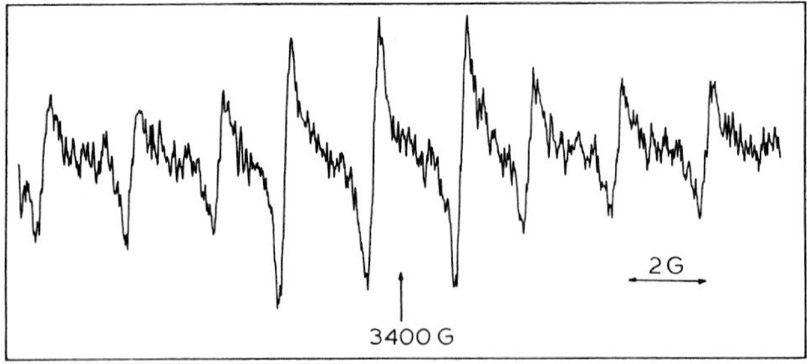

Fig. 10. ESR spectrum of the maleimide radical anion.

that the mechanism for the electropolymerisation is

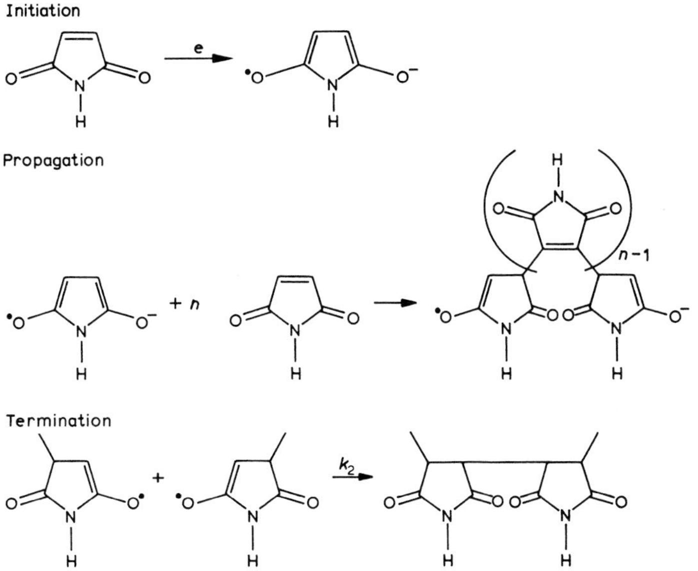

Initiation

Propagation

Termination

where $k_2/\mathrm{dm^3\,mol^{-1}\,s^{-1}} = 41$. This mechanism is the same as that proposed by Bhadani and Prasad. However, we have shown that, in their case where they had *N*-methylpyridinium iodide as the background electrolyte, the actual electroinitiation process is the reduction of the *N*-methylpyridinium ion as opposed to the maleimide. The more detailed knowledge of the nature

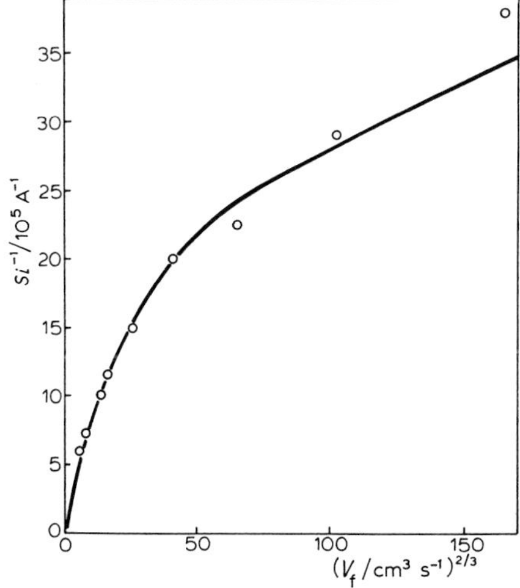

Fig. 11. Variation of the intensity of the ESR signal with flow rate for the maleimide system. The curve is calculated from our simulation programme assuming that the radical species are decomposing by second-order kinetics.

References pp. 147–148

of the intermediates and their lifetimes obtained by electrochemical ESR with controlled hydrodynamics is clear.

6. Wall-jet systems

6.1 PACKED-BED WALL-JET ELECTRODES

One advantage of the wall-jet system is that one can include a packed-bed electrode just upstream of the jet, thus making a packed bed wall-jet electrode (PBWJE) [5–7]. This is a valuable double electrode system in that a packed-bed electrode can achieve complete turnover of a reactant. We have verified that theory and experiment are in good agreement for the collection efficiency by the wall-jet electrode of material generated on the bed [6].

An example of the use of this electrode is our work on developing an amperometric sensor for NO_3^-. Pletcher and Poorabedi showed that NO_3^- can be reduced to NH_3 on a copper electrode [26]. Unfortunately, the electrode tends to poison with time and this makes the sensor unreliable. This problem can be overcome by using a packed-bed electrode of copper to generate a fresh copper wall-jet electrode. The following cycle is used.

Electrode generation

Packed-bed electrode

$$\mathrm{Cu} \longrightarrow \mathrm{Cu}^{2+} + 2\,\mathrm{e}$$

Wall-jet electrode

$$\mathrm{Cu}^{2+} + 2\,\mathrm{e} \longrightarrow \mathrm{Cu}$$

NO_3^- measurement

Wall-jet electrode

$$\mathrm{NO_3^-} + 10\,\mathrm{H}^+ + 8\,\mathrm{e} \longrightarrow \mathrm{NH_4^+} + 3\,\mathrm{H_2O}$$

Electrode stripping

Wall-jet electrode

$$\mathrm{Cu} \longrightarrow \mathrm{Cu}^{2+} + 2\,\mathrm{e}$$

Typical results are shown in Fig. 12. The ability of this system to generate a fresh electrode surface confers on solid electrodes the singular advantage enjoyed for so long by the mercury drop electrode. We expect that this method can be applied with advantage to many other systems.

6.2 COLLOIDAL DEPOSITION

The fact that the wall-jet system has no moving parts means that it is the

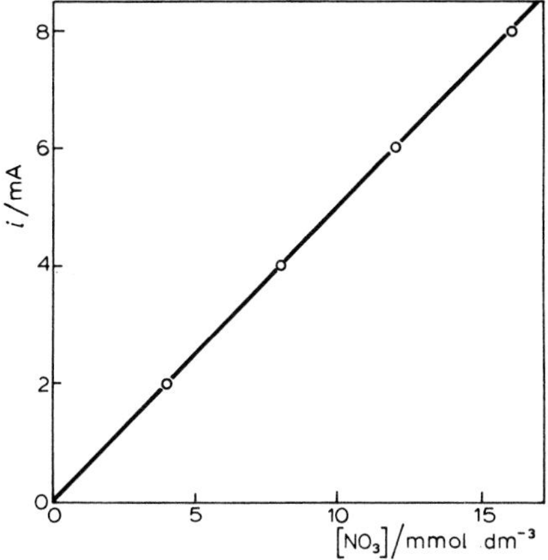

Fig. 12. Typical results for the determination of NO_3^- on the PBWJE.

system of choice when one wishes to carry out in situ investigations of the surface. We have recently constructed the apparatus shown in Fig. 13 [27]. With this apparatus, we can measure the kinetics of the deposition of colloidal particles on to the disc surface. The colloidal particles are brought to the disc surface by wall-jet hydrodynamics. Light from the evanescent wave is scattered by the deposited particles into the microscope where it is measured by a photomultiplier. A typical trace of scattered light intensity with time is shown in Fig. 14. The initial increase on admitting the particles at A is caused by particles in the solution. There is then a steady increase as the particles deposit. At B, the solution is changed to background elec-

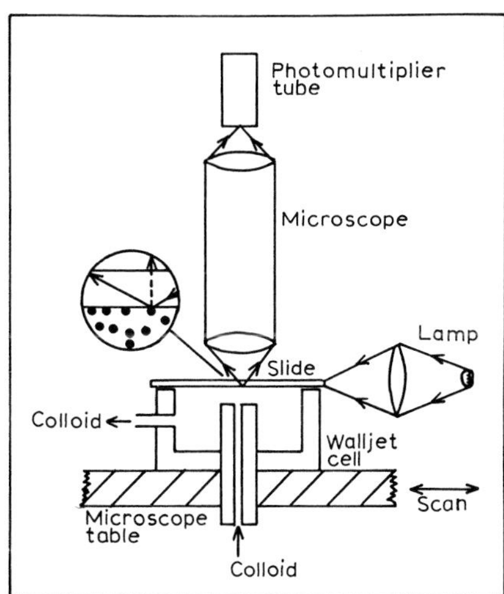

Fig. 13. The wall-jet apparatus for measuring the kinetics of colloidal deposition. The evanescent wave is scattered more strongly by the deposited particles than those in the solution.

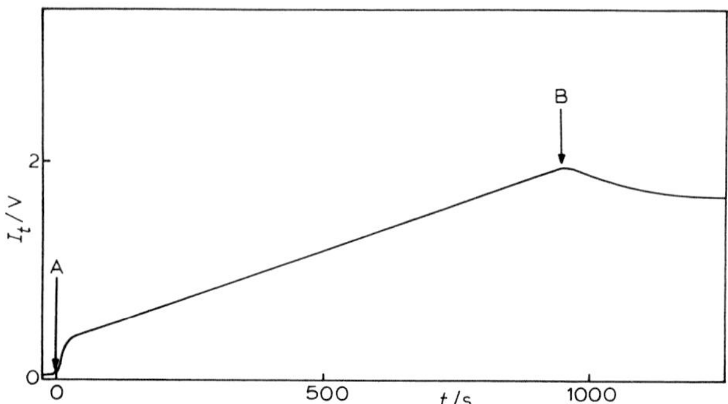

Fig. 14. Typical results for the increase in the intensity of the scattered light as the particles deposit.

trolyte without particles and we lose the contribution to the scattered intensity from the particles in the solution. In this way, we can follow the rate of colloidal deposition in situ.

By moving the microscope table, we can also measure how the rate of deposition varies with the radius. Typical radial scans are shown in Fig. 15. As mentioned above, the wall-jet hydrodynamics do not provide uniform accessibility. In fact, the mass transport-limited flux varies with $r^{-5/4}$ [28]. Hence, if the deposition of the particles is simply controlled by mass transport, then, as shown in Fig. 16, the thickness of the deposit varies with $r^{-5/4}$, giving the "volcano" curve A in Fig. 15. At the other extreme, if surface kinetics are wholly rate-controlling, then the deposit will be uniform. Curve B in Fig. 15 is an intermediate case. It is interesting that the non-uniform accessibility is an advantage in this experiment and allows us to measure the

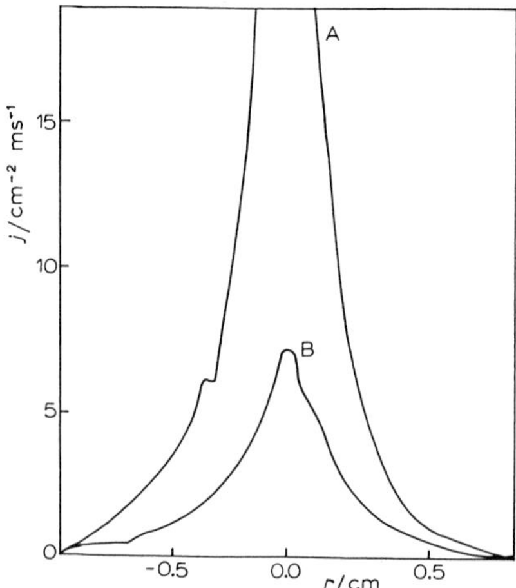

Fig. 15. Typical results for the radial variation of the number of deposited particles. For curve A, the rate-limiting step in the deposition was mass transport, while for curve B, there was mixed control involving both mass transport and surface kinetics.

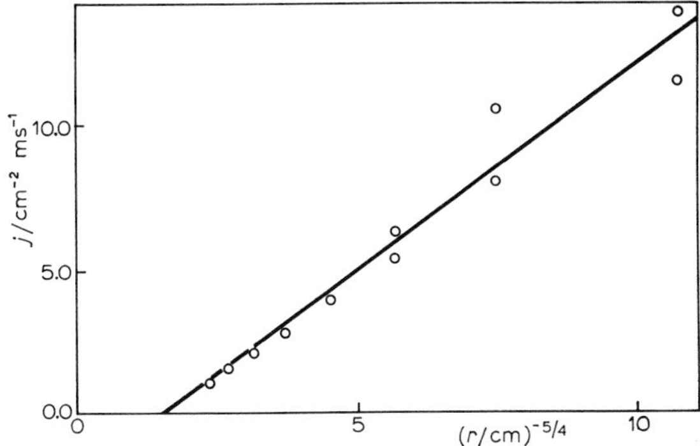

Fig. 16. The data from curve A in Fig. 15 plotted to show that the number of deposited particles varies with $r^{-5/4}$, as required by wall-jet hydrodynamics (see Fig. 2).

effects of mass transport from the radial variation. Furthermore, we need only carry out the experiment at one flow rate.

For those cases where surface kinetics are partially rate-limiting (e.g. curve B in Fig. 15), from the theoretical treatment of the current distribution on a wall-jet electrode [28] we can work out [27] the dimensionless parameter θ, which describes the radial distribution of the deposit, where

$$\theta = 4.5k'r^{5/4}\nu^{5/12}a^{1/2}D^{-2/3}V_f^{-3/4} \tag{9}$$

Figure 17 shows that θ does indeed obey eqn. (9); from the gradient we find that $k' = 7.7 \pm 0.4$ cm M s^{-1}. More recently, we have used a slide coated with tin oxide and have succeeded in measuring how the rate of deposition varies with the potential of the tin oxide electrode [29].

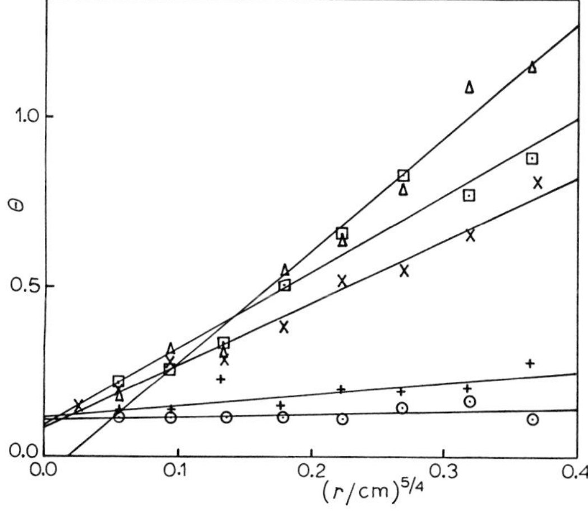

Fig. 17. Results for the dimensionless parameter θ plotted according to eqn. (9) where θ is determined from the radial variation of the deposit. The parameter θ measures the balance between mass transport and surface kinetics.

References pp. 147–148

7. Photoelectrochemistry

The optical rotating disc electrode (ORDE) [30] has proved a useful method for investigating photoelectrochemical systems, especially those where the light is absorbed in the homogeneous solution. The ORDE consists of a transparent tin oxide electrode which is deposited on the end of a quartz rod. Light is shone down the rod and through the electrode into the solution. The advantage of this arrangement is that there is maximum irradiance at the electrode surface itself. This provides the largest possible photocurrents and allows the study of unstable species. We have measured second-order rate constants as large as $10^6\,dm^3\,mol^{-1}\,s^{-1}$ [31]. The rotating-disc hydrodynamics provides the controlled mass transport.

Johnson et al. have pioneered a ring–disc version of this apparatus [32, 33]. Their rotating optical disc–ring electrode (RODRE) has no transparent electrode on the end of the quartz rod. Instead, the disc is surrounded by a concentric ring electrode. As with the ring–disc electrode, the radial convection in the diffusion layer sweeps the photogenerated intermediates and products from the illuminated disc zone to the downstream ring electrode. The advantage of this approach is that the electrode can be made of platinum, giving more reliable electrochemistry than that found on tin oxide. A disadvantage for some systems is that intermediates have to have a lifetime greater than $10^{-2}\,s$ to survive, in detectable quantities, the passage to the ring electrode. We have recently published an analytical theory for the collection at the ring electrode [34].

An interesting feature of these electrodes is that, for stable species, the currents vary as $W^{-1/2}$, as opposed to the more familiar $W^{1/2}$. The reason for

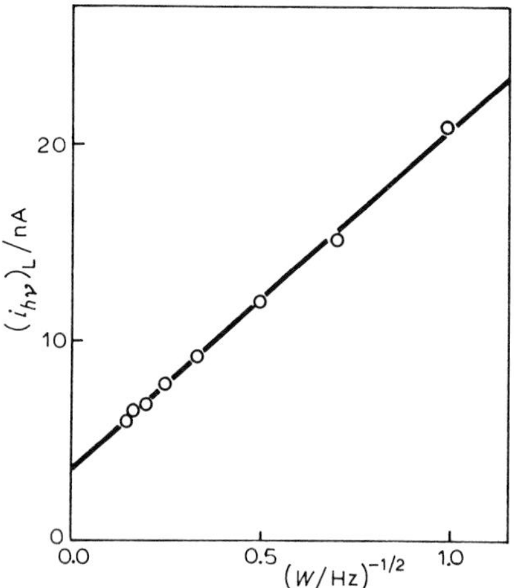

Fig. 18. Photocurrent from the transfer of photogenerated electrons on colloidal CdS particles to the ORDE. Note that the current is plotted against $W^{-1/2}$. The quantum efficiency of the photogeneration can be measured from the gradient.

this behaviour is that only those species generated in the diffusion layer have a good chance of reaching the disc or the ring electrodes. The competition is between diffusion to the electrode surface and loss to the bulk of the solution by convective dilution. Hence, the faster one rotates, the thinner is the diffusion layer, the greater the convective loss, and the smaller the current.

Hitherto, these electrodes have mainly been used to study systems for photogalvanic cells [35]. They are now finding increasing use in the study of photocolloidal systems [36, 37]. Figure 18 shows some typical results on a CdS colloid [37]. The photogenerated electrons on each particle are being transferred to the macroscopic ORDE. The dependance of the limiting currents on $W^{-1/2}$ (Fig. 18) shows that these electrons are stable on the time scale of the experiment (~ 1 s). There are as many as 200 electrons being harvested off each particle. From the size of the current, we can calculate [37] that the overall quantum efficiency of this process is 2×10^{-3}.

This last example shows how the controlled hydrodynamics make it easy to measure quantum efficiencies for these controversial systems.

References

1 V.G. Levich, Physicochemical Hydrodynamics, Prentice Hall, Englewood Cliffs, NJ, 1962, pp. 60–78.
2 W.J. Albery and S. Bruckenstein, J. Electroanal. Chem., 144 (1983) 105.
3 M.B. Glauert, J. Fluid Mech., 1 (1956) 625.
4 W.J. Albery and C.M.A. Brett, J. Electroanal. Chem., 148 (1983) 201.
5 J. Wang and H.D. Dewald, J. Electrochem. Soc., 130 (1983) 1814.
6 W.J. Albery, B.G.D. Haggett, C.P. Jones, M.J. Pritchard and L.R. Svanberg, J. Electroanal. Chem., 188 (1985) 257.
7 W.J. Albery and M.M.P.M. Neto, Port. Electrochim. Acta, 3 (1985) 67.
8 V.G. Levich, Physicochemical Hydrodynamics, Prentice Hall, Englewood Cliffs, NJ, 1962, pp. 112–116.
9 W.J. Albery and S. Bruckenstein, Trans. Faraday Soc., 62 (1965) 1920.
10 W.J. Albery and M.L. Hitchman, Ring–Disc Electrodes, Clarendon Press, Oxford, 1971.
11 A.J. Bard and K.B. Prater, J. Electrochem. Soc., 117 (1970) 207.
12 S.W. Feldberg, Electroanal. Chem., 3 (1969) 199.
13 S.W. Feldberg, M.L. Bowers and F.C. Anson, J. Electroanal. Chem., 215 (1986) 11.
14 W.J. Albery, A.T. Chadwick, B.A. Coles and N.A. Hampson, J. Electroanal. Chem., 75 (1977) 229.
15 J.M. Hale, Batteries, Vol. 2, Pergamon Press, Oxford, 1965, p. 147.
16 W.J. Albery, M.G. Boutelle, P.J. Colby and A.R. Hillman, J. Chem. Soc. Faraday Trans. 1, 78 (1982) 2757.
17 W.J. Albery and E.J. Calvo, J. Chem. Soc. Faraday Trans. 1, 79 (1983) 2583.
18 J. Jacq, Electrochim. Acta, 12 (1967) 1345.
19 W.J. Albery and M.L. Hitchman, Ring–Disc Electrodes, Clarendon Press, Oxford, 1971, pp. 38–72.
20 W.J. Albery, A.R. Hillman and S. Bruckenstein, J. Electroanal. Chem., 100 (1979) 687.
21 M. von Stackleberg, M. Pilgram and W. Toome, Z. Elektrochem., 57 (1953) 342.
22 W.J. Albery, B.A. Coles and A.M. Couper, J. Electroanal. Chem., 65 (1975) 901.
23 W.J. Albery, R.G. Compton and I.S. Kerr, J. Chem. Soc. Perkin Trans. 2, (1981) 825.

24 A.M. Waller and R.G. Compton, This volume, Chap. 7.
25 S.N. Bhadani and J. Prasad, Makromol. Chem., 178 (1977) 187.
26 D. Pletcher and Z. Poorabedi, Electrochim. Acta, 24 (1979) 1253.
27 W.J. Albery, G.R. Kneebone and A.W. Foulds, J. Colloid Interface Sci., 108 (1985) 193.
28 W.J. Albery, J. Electroanal. Chem., 191 (1985) 1.
29 W.J. Albery, R. Fredlein and A.L. Smith, J. Colloid Interface Sci., submitted for publication.
30 W.J. Albery, M.D. Archer and R.G. Edgell, J. Electroanal. Chem., 82 (1977) 199.
31 W.J. Albery, W.R. Bowen, F.S. Fisher and A.D. Turner, J. Electroanal. Chem., 107 (1980) 1.
32 D.C. Johnson and E.W. Resnic, Anal. Chem., 44 (1972) 637.
33 P.R. Gaines, V.E. Peacock and D.C. Johnson, Anal. Chem., 47 (1975) 1393.
34 W.J. Albery, P.N. Bartlett, A.M. Lithgow, J. Riefkohl, L. Romero and F.A. Souto, J. Chem. Soc. Faraday Trans. 1, 81 (1985) 2647.
35 W.J. Albery, Acc. Chem. Res., 15 (1982) 142.
36 W.J. Albery, P.N. Bartlett and J.D. Porter, J. Electrochem. Soc., 131 (1984) 2892.
37 W.J. Albery, P.N. Bartlett and J.D. Porter, J. Electrochem. Soc., 131 (1984) 2896.

Chapter 5

Microelectrodes

J. ROBINSON

1. Introduction

Conventional electrochemical techniques and the associated instrumentation have now been developed to the point where they are often successfully used by non-specialist electrochemists in many areas of chemistry and, indeed, in other scientific disciplines. Also, as evidenced by this book, a wide range of spectroelectrochemical approaches to the study of electrochemical systems is now becoming available. In view of these advances, it is not unreasonable to ask why anyone should wish to work with electrodes of very small dimensions; their construction will inevitably be more difficult than that of more conventional electrodes and it might be expected that the measurement of the very small currents involved will present problems. It is the intention of this chapter to show that, in fact, working with these microelectrodes presents no real difficulties and, more importantly, that microelectrodes have some interesting and useful properties that enable them to be used to investigate systems that are not amenable to study by more conventional approaches, e.g. redox couples in highly resistive media.

When posing the question why use a microelectrode, some of the first answers that might come to mind are that one has only a very small volume of material to study or that one wishes to position the electrode in a very small place. The most obvious example of this type of application is in in vivo studies and these will be discussed briefly in the next section. The remainder of the chapter is concerned with a discussion of the various properties of microelectrodes, such as reduced iR_u drop, that arise as a result of their small dimensions, and the way that these properties can be utilized to good effect. The various methods that have been used for the fabrication of microelectrodes and the electronic equipment required for these studies will also be described.

It is not intended that this chapter should be an extensive review of this rapidly expanding area of electrochemical research but rather, by describing a few examples of the type of work that is possible, that it gives an overview of the subject and will perhaps encourage others to investigate the possibilities of this approach to the study of electrochemical systems.

References p. 172

2. In vivo studies

The first references to the use of microelectrodes appear in the biochemical literature and relate not to microvoltammetric electrodes, as will largely be considered here, but rather to very small ion-sensitive electrodes (particularly pH electrodes) capable of making measurements inside a single biological cell. Biochemists have also used metal microelectrodes to provide direct electrical stimulation of nerves, for example in the design of auditory protheses [1]. Indeed, the two types of electrode have often been used in combination, i.e. an ion-sensitive electrode is used to detect changes caused by the electrical stimulation of a nerve. Applications of both these types of

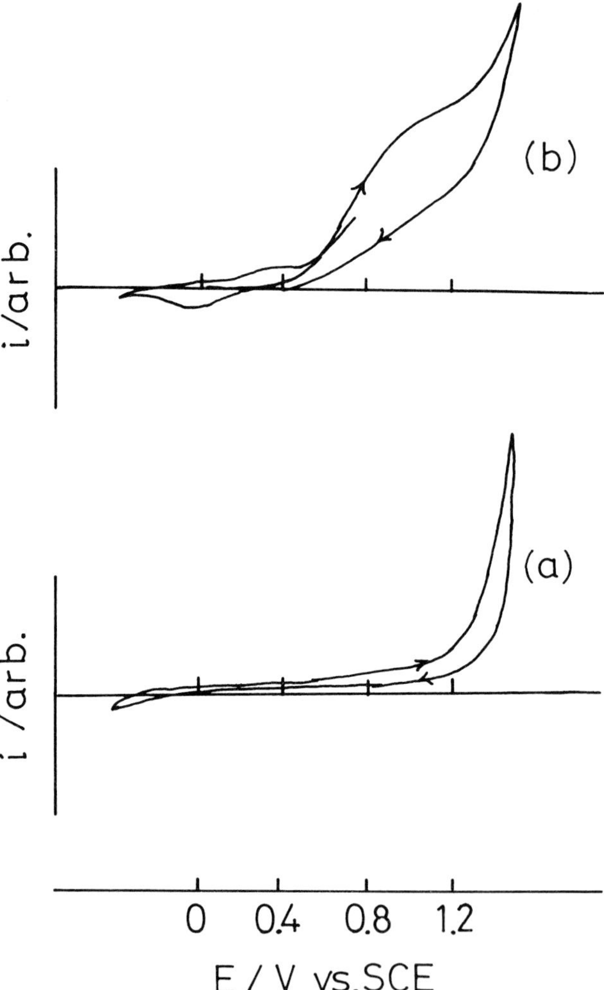

Fig. 1. Cyclic voltammograms obtained with a carbon microelectrode in the cerebro spinal fluid of an anaethetized rat (a) before and (b) after electrical stimulation. (After Wightman et al. [3].)

microelectrode have been thoroughly reviewed by Ferris [2] and will not be considered further here.

It was not until the mid-1970s that microelectrodes were first successfully used to study the voltammetric behaviour of molecules in vivo. This first study by Wightman et al. [3] involved the monitoring of the release of electroactive species into the cerebro spinal fluid of an anaesthetized rat following the electrical stimulation of the *substantia nigra*, a process known to release homovanillic acid, a metabolite of dopamine. A carbon fibre microelectrode was placed in the cerebro spinal fluid and a saturated calomel electrode, and a secondary electrode, were placed nearby. Cyclic voltammograms before and after stimulation were then recorded and, as shown in Fig. 1, an oxidation current at around + 0.8 V, which was attributed to the oxidation of the homovanillic acid, was observed after stimulation.

Wightman and his group have continued with studies of this type and have found that using a variant of normal pulse voltammetry is probably preferable to using cyclic voltammetry because the period at the base potential between pulses helps to prevent film formation on the electrode and thus keeps the electrode clean. In this way, the behaviour of an implanted microelectrode remains reproducible over a long period of time, which is important as it is very difficult to reposition an electrode in exactly the same spot after it has been removed for cleaning.

As an example of this type of study, Fig. 2 shows some back-step corrected normal pulse voltammograms obtained by Wightman's group [4] for three different substances injected into brain tissue. It can be seen that quite well-defined waves are obtained, whereas the respective cyclic voltammograms would be poorly defined.

This area of application of microelectrodes continues to attract attention

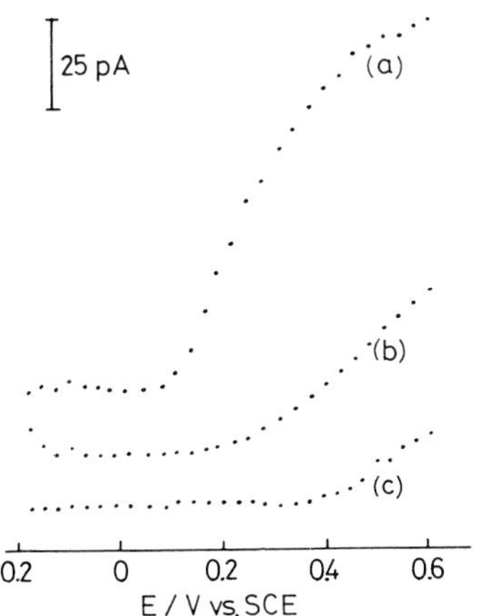

Fig. 2. Back-step corrected normal pulse voltammongrams of substances injected into brain tissue. (a) α-Methyldopamine, (b) ascorbic acid, and (c) dihydroxyphenylacetic acid. The scan rate was 20 m V s^{-1}, the pulse length was 92 ms, and there were 2 s between each step. (After Ewing et al. [4].)

References p. 172

but it is in other areas where the small size of the electrode does not immediately appear to offer advantages that the major growth in the application of microelectrodes has taken place. The remainder of this chapter is concerned with the special properties of microelectrodes that have given rise to this growth.

3. Finite size effects

One of the main characteristics of microelectrodes that makes them so interesting is the unusual mass transport properties that they exhibit. To explain this behaviour, it is simplest to start by considering a simple reduction reaction such as

$$\mathrm{O} + n\mathrm{e}^- \rightleftharpoons \mathrm{R} \tag{1}$$

occurring at a spherical electrode of radius r_s. It can be shown that, for such a system, the current–time profile in response to a potential step to a potential where all O at the electrode is reduced is given by

$$i_s = \frac{nFAD^{1/2}}{(\pi t)^{1/2}} C_0^\infty + 4\pi nFDC_0^\infty r_s \tag{2}$$

A result that will be familiar to all polarographers. This equation clearly has two limiting forms. At short times

$$\underset{t \to 0}{i_s} \simeq \frac{nFAD^{1/2}}{(\pi t)^{1/2}} C_0^\infty \tag{3}$$

i.e. the current is given by the familiar Cottrell equation and is equivalent to that at a planar electrode of the same area. Conversely, at long times

$$\underset{t \to \infty}{i_s} \simeq 4\pi nFDC_0^\infty r_s \tag{4}$$

i.e the current reaches a steady-state value. Thus, it can be seen that, in contrast to semi-infinite planar electrodes, the current at a spherical electrode does not decrease to zero but reaches a steady-state value proportional to the electrode radius. The current density and mass transport coefficient are therefore inversely proportional to the electrode radius. By decreasing the electrode radius, the rate of steady-state mass transport therefore increases and it should be possible to study the kinetics of fast reactions. Unfortunately, in practice, it is difficult to make very small spherical (or hemispherical) electrodes, with the exception of those of Hg [5] or amalgams but, as we will see, finite disc electrodes, in fact, behave in a very similar manner.

Figure 3 shows a representation of the diffusion fields at three electrode geometries. The diffusion profile at the finite disc electrode can be thought of as consisting of two components, one similar to diffusion to a semi-infinite planar electrode and a cylindrical contribution that is often referred to as the edge effect. At short times, or for large electrodes, this edge effect is small and the current is essentially determined by the linear diffusion components. (This approximation is frequently made in practice when working with disc electrodes of conventional size.) At long times, or for small electrodes, these edge effects become significant or, indeed, dominant. Some years ago, the expression

$$i = nFADC_0^\infty \left[\frac{1}{(\pi Dt)^{1/2}} + \frac{b}{\pi^{1/2} r_\mathrm{m}}\right] \tag{5}$$

where b is a time-dependent empirical coefficient, was proposed to describe the current–time transient at a finite disc electrode. Various values of b in the range 1.92–3.21 were proposed [6–10], depending on the time range being studied. The current–time response at a finite disc is therefore complex though, for large electrodes at short times, the first term in eqn. (5) dominates and the familiar Cottrell behaviour is observed. At long times, where the second term in eqn. (5) dominates, it has been shown by many workers [6, 7, 11–14] that the current becomes

$$\underset{t\to\infty}{i} = 4nFDC_0^\infty r_\mathrm{m} \tag{6}$$

which corresponds to a current density of

$$\underset{t\to\infty}{I} = \frac{4nFDC_0^\infty}{\pi r_\mathrm{m}} \tag{7}$$

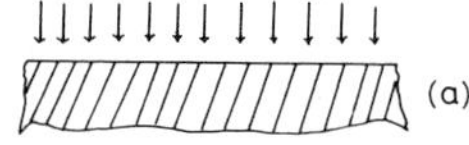

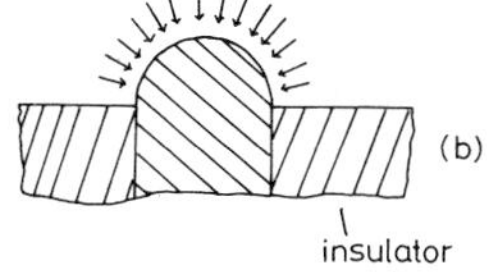

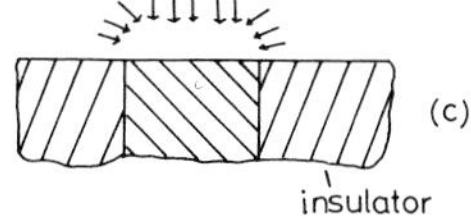

Fig. 3. Representations of the diffusive fields at (a) a semi-infinite planar electrode, (b) a hemispherical electrode, and (c) a finite disc electrode.

References p. 172

Taking a typical value for D of $10^{-5}\,cm^2\,s^{-1}$ and an electrode radius of $10^{-4}\,cm$, this steady-state value will be reached in less than 1 s. Comparing eqns. (6) and (7) with the expression obtained earlier for spherical geometry [eqn. (4)] we can conclude that the steady-state current at a microdisc electrode of radius r_m is identical to that obtained at a hemispherical electrode of raidus $2r_m/\pi$. Alternatively, we can say that the current density at a disc of radius r_m is $4/\pi$ times that at a hemisphere (or sphere) of the same radius. Whilst these formal similarities to the spherical geometry are interesting and, as will be seen later, are also useful in certain circumstances, there is, however, an important distinction which must not be ignored. Whereas at the polarized hemispherical electrode the current density is uniform, the microdisc does not present homogeneous spatial accessibility and the current density decreases from the edge to the centre.

To see the applicability of the foregoing discussion, we can look at some results obtained for the oxidation of ferrocene at a range of platinum microdisc electrodes [15]. Figure 4 shows a steady-state current–voltage curve for this system and, as predicted, we can see that the current reaches a limiting value at large overpotentials. Equation (6) predicts that this limiting current should vary linearly with electrode radius and, as shown in Fig. 5, this is indeed found to be the case, whilst the value of the diffusion coefficient derived from the gradient was in close agreement with that obtained by other methods. Thus, it seems that experimental data are in agreement with the theory presented.

Whilst the currents reported in Figs. 4 and 5 are very small, the current densities, and thus the mass transport coefficients, are in fact very large. For comparison, the current density at a microdisc electrode of radius 4 μm is roughly equal to that at a rotating disc electrode in the same solution rotating at 30 000 rev. min^{-1}. This immediately suggests that it should be possible to study the kinetics of quite fast electrode reactions through the

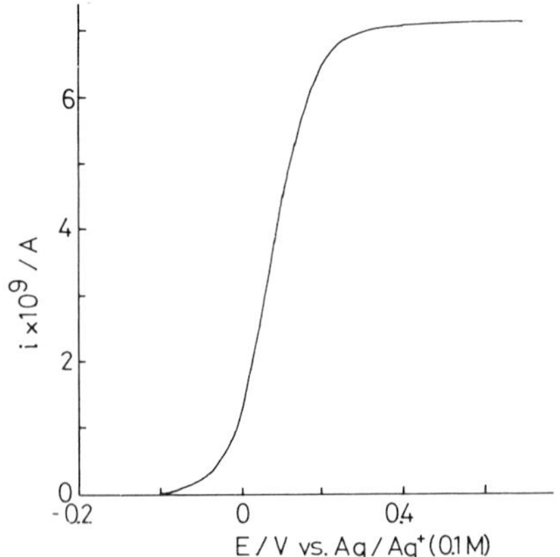

Fig. 4. Voltammogram for the oxidation of ferrocene (3 mM) at a microelectrode (r = 2.5 μm) in acetonitrile/0.1 M tetraethylammonium perchlorate. Sweep rate 2 m V s^{-1}.

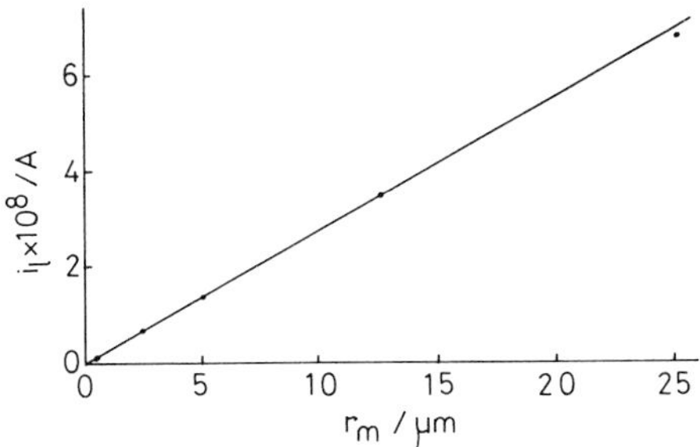

Fig. 5. The variation of limiting current with electrode radius for the oxidation of ferrocene (3 mM) at platinum microelectrodes.

use of steady-state measurements at microelectrodes and, as we will see later, this is indeed the case.

Thus far in this discussion, only small spherical/hemispherical and disc electrodes have been considered. Whilst the disc electrode is perhaps the easiest form of microelectrode to construct, and it is certainly the most widely used to date, other geometries are possible, e.g. microcylinders [16], microbands [17], and microrings [18] and, in practice, the microband and microcylinder in particular have the advantage that they are only small in one dimension and thus the current observed is much larger, and therefore easier to measure, than at a microdisc electrode. The theoretical treatment of these geometries can be difficult but the behaviour is generally found to be similar to that for the disc. Most of the results to be discussed in the remainder of this chapter were obtained with disc electrodes but many of the conclusions would be equally applicable to other geometries. Before proceeding to discuss these results, some of the methods of electrode construction will be described, as will the apparatus required to make microelectrode measurements.

4. Experimental details

4.1 ELECTRODE DESIGN AND CONSTRUCTION

Before considering electrode construction in detail, we ought perhaps consider what exactly we mean by the term microelectrode. It has been used by different people to mean different things (what is micro to an engineer is macro to an electroanalytical chemist) and this has led to the use of the term ultra-microelectrode by some workers. In recent years, the situation has become more clearly defined and now it would probably be accepted that, to be a microelectrode, one of the dimensions should certainly be less than 200 µm whilst the most interesting work has generally been performed on electrodes with dimensions in the range 1–20 µm. We can see immediately from these figures that electrode construction is liable to present some problems.

References p. 172

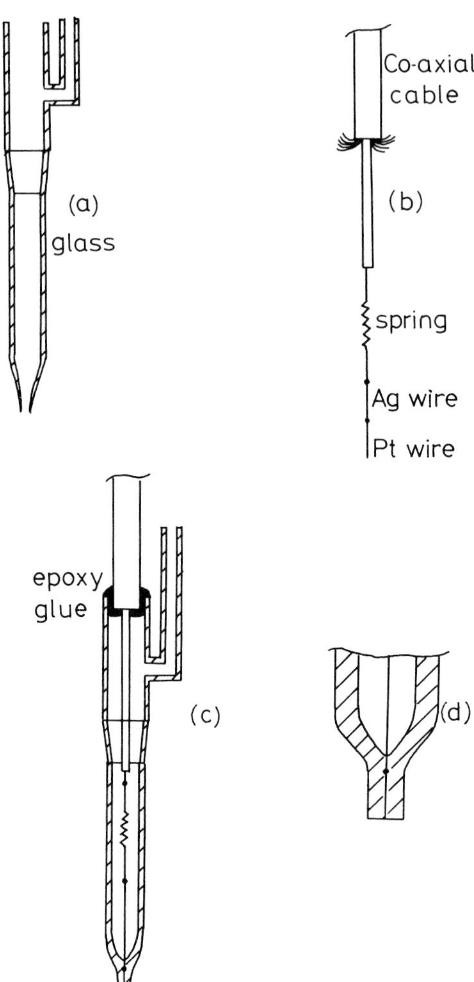

Fig. 6. Construction of platinum microelectrodes.

If we consider first of all disc electrodes, these are clearly going to be most readily made by sealing metal wires into an insulating tube, the most commonly used approach being to seal the wire into a glass tube using an epoxy resin glue. For use in non-aqueous solvents, the glue has to be carefully chosen or, alternatively, the glass tube can be heated under vacuum to form a glass-to-metal seal. This technique has been widely used to prepare electrodes of many types with diameters down to about 5 μm and is the most commonly used technique for preparing carbon fibre electrodes, which have been widely used. For smaller electrodes, greater care has to be taken because of the fragility of the electrode material and Fig. 6 shows the basis of a technique that has been successfully used to make platinum disc electrodes with radii as small as 0.3 μm [15]. In view of the fragility of wires of these dimensions, they are supplied commercially in the form of Wollaston wire, i.e. they are coated in a thick layer of silver. Briefly, the construction of these electrodes can be described as follows. A length of Wollaston wire is first soldered to a length of silver wire which, in turn, is soldered to a small spring attached to the central conductor of a length of low noise co-axial cable as shown in Fig. 6(b). This assembly is then placed inside the glass tube constructed as shown in Fig. 6(a) such that the Wollaston wire extends through the end of the capillary. About 0.5 cm of this wire is then immersed in 50% nitric acid for about 2 h to dissolve the silver. The wire is then slowly withdrawn from the acid back into the capillary and is thoroughly washed with pure water. The electrode is then dried and the co-axial cable is sealed to the glass using silver-loaded epoxy to provide a connection between the glass and the screen of the cable (this helps to reduce noise). Once the epoxy has set, the glass tip is sealed by heating whilst applying a vacuum via the sidearm. Figure 6(c) shows a diagram of the assembled electrode and, as shown in Fig. 6(d), all the bare platinum wire must be sealed in the glass or it will break during polishing. This electrode design has proved to be very robust, exhibiting low noise and also permitting the electrode to be repeatedly repolished without its radius being changed. It also has good chemical resistance to a wide range of aqueous and non-aqueous solutions, including acid cleaning baths.

As we saw earlier, other electrode designs are possible and one that has been reported recently, and which appears to be particularly interesting in view of its ease of manufacture, is the microband electrode [17]. Figure 7 shows a diagram of this design, which essentially consists of a thin film of metal sandwiched between two glass plates (microscope slides) so that only an edge is exposed to the solution. The metal film can either be formed by evaporation on to one of the plates or, alternatively, commercially available films on plastic can be used. In this way, electrodes of very small dimensions can be made if the film thickness is small (in principle, it should be possible to make electrodes only a few tens of Å wide) whilst the large value of the other dimension (up to a few cm) means that currents are not immeasurably small.

For other electrode configurations and for microelectrode arrays, the use of metallo-organic paints has proved effective. For example, a ring can be made by coating a glass rod and then firing the coating before mounting the whole assembly in another tube with epoxy glue.

4.2 ELECTRONIC EQUIPMENT

Let us now turn our attention to the electronic requirements of microelectrode experiments where the main difficulty presented is the measurement of the inherently low current levels (often less than 10^{-10} A). Many measurements have been successfully made using conventional three-electrode techniques with a potentiostatic system that has been modified to increase the gain of the current follower. Potentiostats are, however, inherently

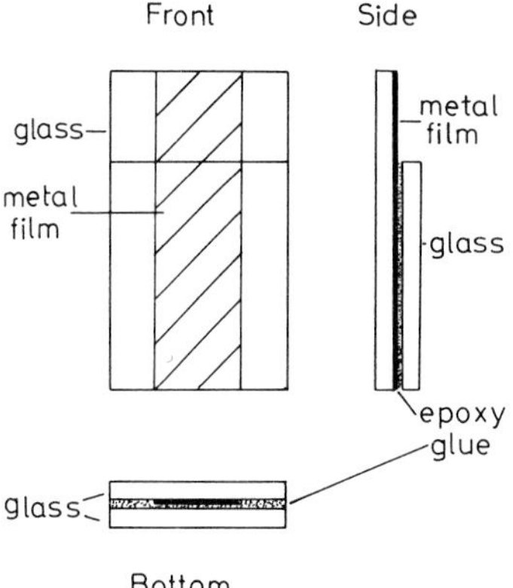

Fig. 7. The construction of microband electrodes.

noisy and when working with certain types of microelectrode, particularly discs, are in fact unnecessary in view of the low currents involved. Instead, a simple two-electrode system is appropriate with the reference electrode serving also as the counter electrode (as in a two-electrode polarographic system). The low level of current flowing in these measurements is insufficient to polarize the secondary/reference electrode significantly and therefore the working electrode potential is well-defined.

As we will see later, iR_u drop is also very small when working with microelectrodes and, in fact, is not greatly affected by the distance between the reference/secondary electrode electrode and the working electrode. This therefore rarely presents a problem with microelectrodes.

Figure 8 is a block diagram of a typical two-electrode configuration for making microelectrode measurements where the function generator and recorder could, of course, be replaced by a microcomputer with appropriate interfaces. To minimize noise, the cell is mounted in a Faraday cage and cables are kept as short as possible. Using a system of this type, noise-free measurements of steady-state currents as small as 10^{-12} A have been made

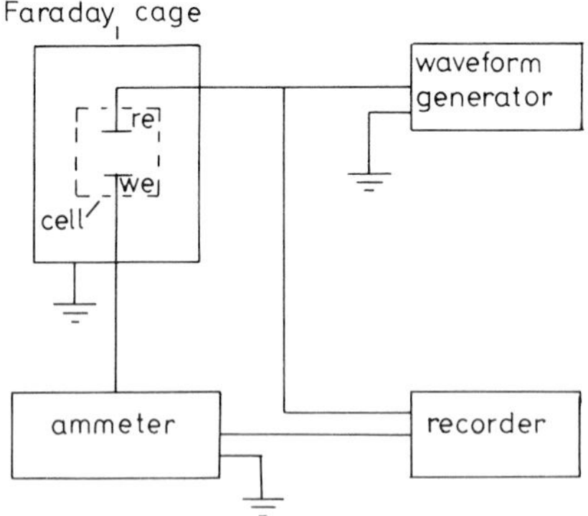

Fig. 8. A block diagram of the apparatus for two-electrode operation.

and, since commercial ammeters capable of measuring currents as small as 10^{-15} A are available, this value could probably be pushed further. Potentiodynamic measurements at these current levels are a little more difficult because the time constant of the ammeter has to be decreased, with a consequent increase in noise, but, with care, they can be made.

Given an appropriate system of potential control and current measurement (or vice versa), microelectrode experiments are essentially identical to those made using conventionally sized electrodes. All aspects of cell design, O_2 removal etc., are therefore normal. Adequate cleaning and solvent/electrolyte purification is, of couse, essential and in fact may be rather more important than with conventional electrodes. The enhanced diffusion to a microelectrode applies also to the impurities, which can therefore build up rapidly near the electrode and, in view of the small electrode size, it can readily be blocked. Having seen how to construct microelectrodes and how to make measurements with them, we will now consider some applications.

5. Application of microelectrodes

5.1 THE STUDY OF KINETICS BY STEADY-STATE MEASUREMENTS

Microelectrodes have now been used to investigate both the kinetics of heterogeneous electron transfer processes and of coupled homogeneous reactions using steady-state and transient methods. The main features of the transient methods are identical to those when using electrodes of conventional size and therefore will not be considered here though, as we will see later, the use of a microelectrode can be beneficial. The essentials of the steady-state method, as has been briefly mentioned earlier, are however unique to these electrodes. By making the electrode very small, the steady-state mass transport rate is increased. If the electrode is small enough, and thus the rate of diffusion high enough, the mass transport rate is no longer rate-determining. This permits the kinetics of the slowest step in the reaction scheme to be investigated. Systems which have been studied in this way include the electron transfer kinetics of the Hg/Hg^{+} reaction [5] and preceding and catalytic homogeneous chemical reactions coupled to electron transfer processes [15]. Here, we will consider as an example of this approach the case of a chemical reaction coupled between two electron transfer steps, the well known ece scheme.

The ece-type reaction scheme is shown in general terms in the equations

$$A \pm n_1 e \rightleftharpoons B \qquad \text{step a} \tag{8}$$

$$B \underset{k_{-1}}{\overset{k_1}{\rightleftharpoons}} C \qquad \text{step b} \tag{9}$$

$$C \pm n_2 e \rightleftharpoons D \qquad \text{step c} \tag{10}$$

$$B + C \underset{k_{-2}}{\overset{k_2}{\rightleftharpoons}} A + D \qquad \text{step d} \tag{11}$$

Amatore and Savéant [19] have identified three limiting cases of this scheme, viz.

(i) ece where step d is unimportant and step b is rate-determining,
(ii) displ where step c is unimportant and step b is rate-determining, and
(iii) disp2 where step c is unimportant and step d is rate-determining.

At one time, all reactions of this type were thought to proceed via the ece route but it was then recognised that the other routes were possible. Whilst the disp2 mechanism can be readily identified because it is second order, it is harder to differentiate between disp 1 and ece processes. Only spectroscopic techniques and double potential steps are capable of this when using conventional electrodes. So let us see how microelectrodes can be used to study this type of reaction.

As we did earlier, we will start by considering the behaviour of a spherical electrode. For an ece reaction with a first-order chemical step, the steady-state diffusion-limited current density is given by [20]

$$I_s = \frac{nFD_A C_A^\infty}{r_s} \left\{ n_1 + \frac{n_2 r_s (k/D_B)^{1/2}}{1 + r_s (k/D_B)^{1/2}} \right\} \tag{12}$$

An alternative way of expressing this is as the apparent number of electrons involved and this is given by

$$n_{APP} = n_1 + \frac{(n_2 r_s (k_1/D_B)^{1/2}}{1 + r_s (k/D_B)^{1/2})} \tag{13}$$

whilst the corresponding expression for the displ scheme is

$$n_{APP} = 2n \left[\frac{(1 + r_s (k_1/D)^{1/2}}{2 + r_s (k_1/D)^{1/2}} \right] \tag{14}$$

From these expressions and for the case where $n_1 = n_2 = 1$, it can be seen that, for either mechanism, the value of n_{APP} will decrease from 2 towards 1 as the electrode gets smaller. A useful similarity to this behaviour is that observed at rotating-disc electrodes; decreasing the electrode radius has the same effect as increasing the rotation rate of a rotating-disc electrode.

The above discussion applies to spherical geometry. Unfortunately, no analytical solution can be found for the finite disc configuration; however, we saw earlier that, for simple reversible electron transfer reactions, the limiting current density at the microdisc electrode was identical to that at a sphere of radius $\pi r_m/4$. Fleischmann et al. [21] have investigated whether this analogy can be extended to systems with coupled chemical reactions. The system chosen for this study was the oxidation of anthracene in very dry acetonitrile at platinum electrodes. This reaction is thought to proceed by the ece mechanism [22]

$-e^-$ ⇌ $\xrightarrow[k]{CH_3CN}$ (H, H, $N{=}C^+{-}CH_3$) ⇌ $-e^-\ -H^+$ ($N{=}C^+{-}CH_3$) (15)

Figure 9 shows a plot of n_{APP} as a function of electrode radius for this system and it can be seen that, as expected, this value decreases as the electrode is made smaller. The solid line is a working curve for a rate constant of $140\,s^{-1}$ for an ece mechanism, assuming the analogy to spherical geometry is valid. It can be seen that the fit is good and the value of k is in close agreement with that obtained by transient techiques. However, a very similar curve, with a similar k value, can be fitted to the data assuming a disp 1 mechanism; thus, it is not possible to differentiate between the two mechanisms in this way. This is largely because of the difficulties involved in fitting experimental data to working curves. Straight lines are much easier to fit.

By making the substitution

$$\Delta n = n_{APP} - n_1 \tag{16}$$

and also incorporating the analogy that the disc electrode behaves like a sphere of radius $\pi r_m/4$, we can derive the following expressions for ece and disp1 schemes, respectively

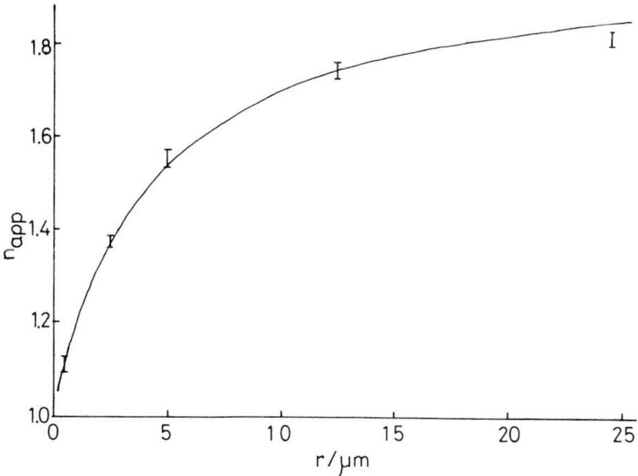

Fig. 9. Determination of the rate constant, k, for the chemical step in the oxidation of anthracene. The solid line is a working curve for a k value of $140\,s^{-1}$; the points are experimental data.

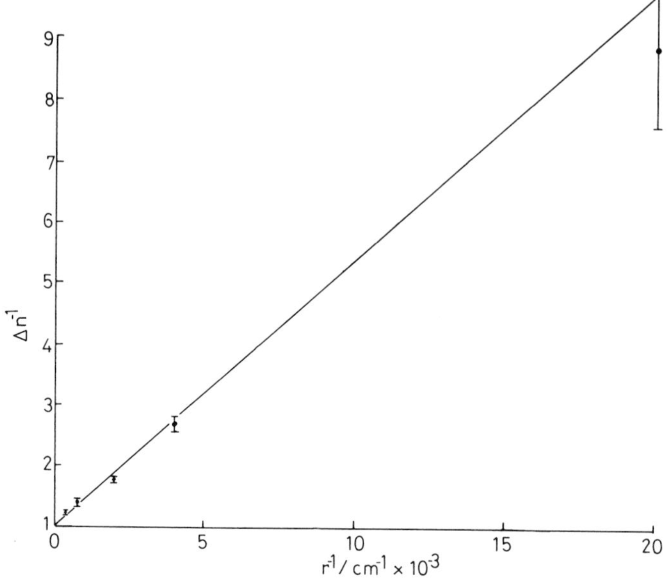

Fig. 10. The data of Fig. 9 replotted as Δn^{-1} vs. r^{-1}.

$$\frac{1}{\Delta n} = \frac{4}{\pi}\left(\frac{D}{n^2 k}\right)^{1/2}\frac{1}{r_{\mathrm{m}}} + \frac{1}{n} \tag{17}$$

$$\frac{1}{\Delta n} = \frac{4}{\pi}\left(\frac{2D}{n^2 k}\right)^{1/2}\frac{1}{r_{\mathrm{m}}} + \frac{1}{n} \tag{18}$$

Thus, for either scheme, a plot of Δn^{-1} vs. r_{m}^{-1} should be a straight line, the only difference being that data treated for an ece scheme will yield a rate constant half that obtained if a disp1 scheme is assumed. Figure 10 shows the data of Fig. 9 treated in this way. Assuming an ece mechanism yields a k value of $190 \pm 50\,\mathrm{s}^{-1}$ whereas assuming a disp1 mechanism, the value would be double this. Comparison with values obtained by transient techniques showed that consistent results were only obtained if reaction did, indeed, follow the ece mechanism. Thus, by using both the steady-state microelectrode technique and transient methods, the ece mechanism for this reaction was confirmed, as was the analogy between spherical and finite disc geometries for this first-order process.

As was pointed out earlier, care must be taken when using the analogy to spherical geometry in view of the non-uniform accessibility of the disc electrode. It has been found to be appropriate in the case of coupled first-order reactions but breaks down with second-order processes [21], as would be expected. In these cases, analysis of the data will require the use of two-dimensional computer simulation methods [15, 23, 24].

It is possible to investigate the kinetics of a wide range of coupled chemical reactions using steady-state methods at microelectrodes, the one exception being following chemical reactions, an exception shared with the rota-

TABLE 1

Values of the rate constant of coupled first-order chemical reactions obtained by various techniques

Technique	Maximum k (s^{-1})
Cyclic voltammetry	1×10^4
Chronopotentiometry	1×10^3
Chronoamperometry	1×10^3
Chromocoulometry	1×10^3
Rotating disc electrodes	1×10^3
A.c. voltammetry	5×10^3
Microelectrodes ($r = 0.5\,\mu$m)	1×10^5

ting disc electrode. In principle, it should be possible to investigate such processes by using two closely spaced microelectrodes, preferably in the disc–ring configuration, and using one electrode to detect the products of the other as with a rotating ring–disc electrode. To date, this does not appear to have been attempted.

From the foregoing discussion, it can be seen that the kinetics of electrochemical systems can be investigated in a very simple manner by making steady-state measurements at microelectrodes with a very minimal amount of equipment. As a comparison with the capabilities of other techniques, Table 1 shows the values of the rate constant of coupled first-order chemical reactions that are accessible to a range of transient and other techniques including steady-state microelectrode measurements. In principle, microelectrodes can be used to study very fast reactions indeed, the only real limitation being the fabrication of a suitably small electrode.

5.2 STUDIES IN RESISTIVE MEDIA

Earlier in this discussion, the low iR_u drop associated with microelectrodes was mentioned. In fact, it can be shown [25] that the total resistance between a microdisc electrode and a large secondary electrode located a long way from it is given by

$$R = \int_{r_m}^{\infty} \frac{dr}{2\pi\kappa r^2} = \frac{1}{2\pi\kappa r_m} \tag{19}$$

Whilst this resistance is undoubtedly large and increases with decreasing electrode radius, the resistance per unit area of electrode actually decreases with the radius of the electrode. This means that, for transient techniques (where the current is proportional to the electrode area), the iR_u drop will decrease as the electrode radius is reduced and, as we will see, this permits experiments not normally possible with larger electrodes. On the other

References p. 172

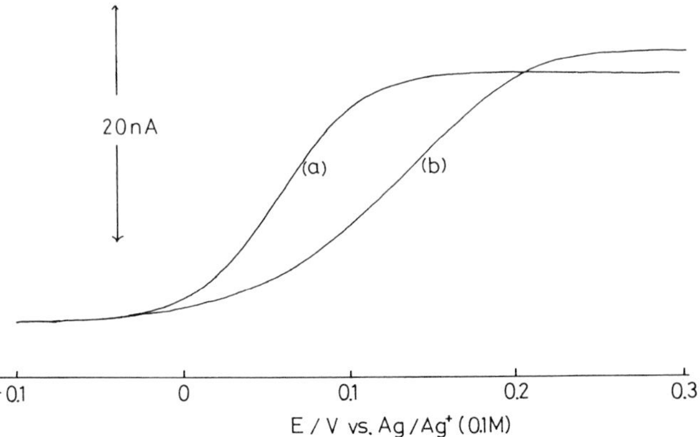

Fig. 11. Voltammograms for the oxidation of 10^{-3} M ferrocene in acetonitrile at 18°C using a 25 μm radius Pt electrode in (a) the absence and (b) the presence of 0.1 M Et_4NClO_4. Scan rate 5 m V s^{-1}.

hand, under steady-state conditions, we have seen [eqn. (4)] that, for a simple reversible electron transfer reaction, the current is proportional to the electrode radius and therefore the iR_u drop can be expected to be independent of the electrode radius. From eqn. (19), we can estimate that, for a 0.5 μm radius electrode in a 10^{-3} M KNO_3 solution, the electrolyte resistance is $2.2 \times 10^5\,\Omega$. This appears to be very large but when it is realized that the Faradaic current flowing in a solution 1 mM in an electroactive species will be of the order of 10^{-12} A, we can see that the iR_u drop will be negligible.

This very low iR_u drop in conventional electrolytic media suggests that it may be possible to undertake studies in media not normally considered amenable to electrochemical study, e.g. in non-polar (or low polarity) solvents containing little or no added electrolyte. Such applications would be

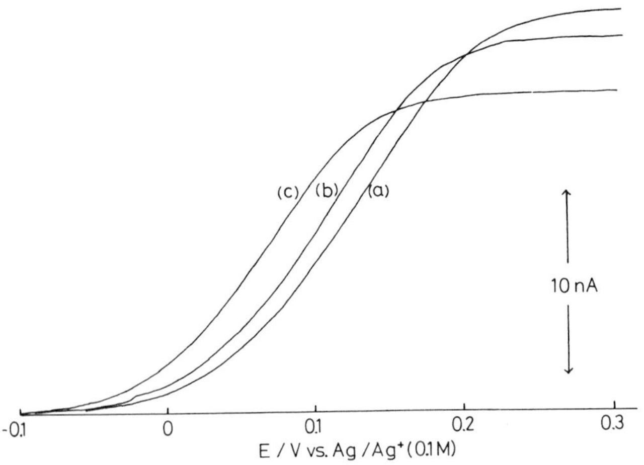

Fig. 12. Voltammograms at a 25 μm radius Pt electrode for a 10^{-3} M ferrocene solution in acetonitrile after the addition of (a) no water, (b) 0.1 M water, and (c) 0.5 M water.

interesting from both a fundamental and a technical point of view. For example, it would be possible to study organic reactions electrochemically in media similar to those used in conventional synthetic organic chemistry, e.g. benzene, whilst a possible technological application of microelectrodes would be as a detector in normal phase liquid chromatography.

Bond et al. [26, 27] have investigated these possibilities by studying the oxidation of ferrocene in acetonitrile under a range of unusual conditions. Figure 11, for example, shows the steady-state voltammogram for this oxidation at a 25 μm radius electrode in the presence and absence of supporting electrolyte. It can be seen that the diffusion-limited current in each case is essentially the same (in fact, the diffusion coefficient in the absence of electrolyte was found to be 3% larger than in its presence) but the voltammogram in the absence of added electrolyte is more drawn out and the half-wave potential is shifted to more positive values. This distortion was thought to be most probably due to the iR_u drop but, despite this, studies based on the limiting currents (such as diffusion coefficient determination at zero ionic strength) are still possible.

In principle, this experiment in the absence of added electrolyte is a one-ion problem, i.e. the generation of a cation in the absence of any anions. In practice, it is probable that attack on the solvent (or water impurities) leads to the generation of H^+ and OH^- ions as well as solvent decomposition products. In reality, therefore, the conditions are more like those conventionally encountered, i.e. it is a three-ion case.

The possibility of water providing charge carriers was tested by the addition of water to a system where no electrolyte had been added. As shown in Fig. 12, this does result in sharper, less drawn out voltammograms, implying that the iR_u drop has been reduced. An alternativeway of improving the voltammetric response was found to be to reduce the electrode size and, as shown in Fig. 13, for a 0.5 μm radius electrode, the behaviour in the absence and presence of added electrolyte is very similar and, in both cases, Nernstian. Thus, contrary to the theory presented earlier, it appears that

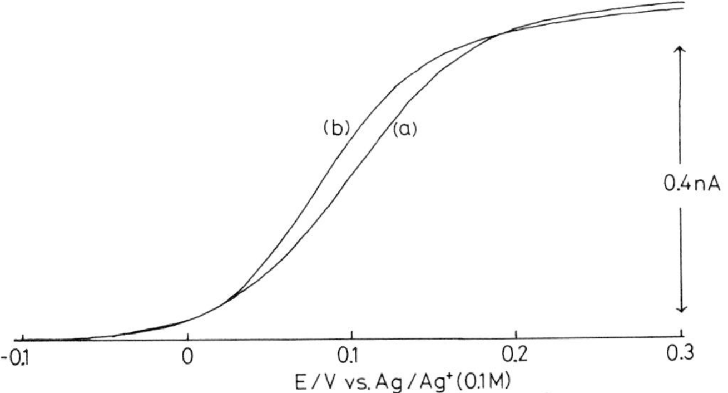

Fig. 13. Voltammograms for a 10^{-3} M ferrocene solution in acetonitrile at a 0.5 μm radius Pt electrode (a) without electrolyte and (b) with 10^{-3} M Et_4NClO_4.

References p. 172

reducing the electrode radius reduces the iR_u drop. The explanation for this is not clear, though it may be due to the non-uniform accessibility of the disc.

In further studies, it was found that, even at 25 μm radius electrodes, the addition of electrolyte had little effect if the concentration of ferrocene was reduced to 10^{-6} M. Under these conditions, voltammograms typical of reversible behaviour were observed, even without added electrolyte. Presumably with the drop in the current of three orders of magnitude, the iR_u drop was no longer significant.

That voltammograms can be recorded under such severe and unusual conditions as this is in itself surprising. What is perhaps even more unexpected is that the voltammograms were found to be essentially Nerstian in shape. This implies that, despite the low concentration of added electrolyte, a recognisable double layer is being formed. Theoretical studies under similar three-ion conditions, though at spherical electrodes [28] have confirmed that this behaviour is indeed to be expected.

From the above results, it can be seen that microelectrodes do fulfil the expectation that, in view of the inherent low iR_u drop associated with their use, they should be useful for making measurements in highly resistive media. As a further example of this type of application, Bond et al. [27] have extended the studies of ferrocene oxidation in acetonitrile to low temperature. Figure 14, for example, shows a series of voltammograms at temperatures down to the freezing point and, in each case, the behaviour was found to be Nernstian. Further, as shown in Fig. 15, good reproducible results could also be obtained in a glass provided a very small electrode (0.5 μm radius) was used, even in the absence of any added electrolyte. These results suggest that the use of microelectrodes at low temperatures may lead to the

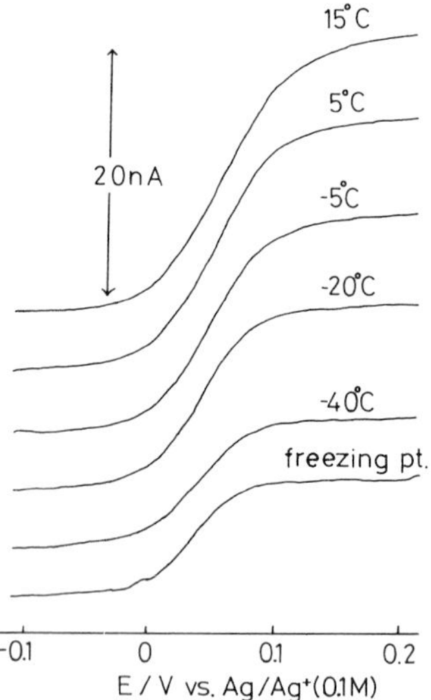

Fig. 14. Steady-state voltammograms for the oxidation of 10^{-3} M ferrocene in acetonitrile (0.05 M Et_4NClO_4) at a 25 μm radius Pt electrode at various temperatures.

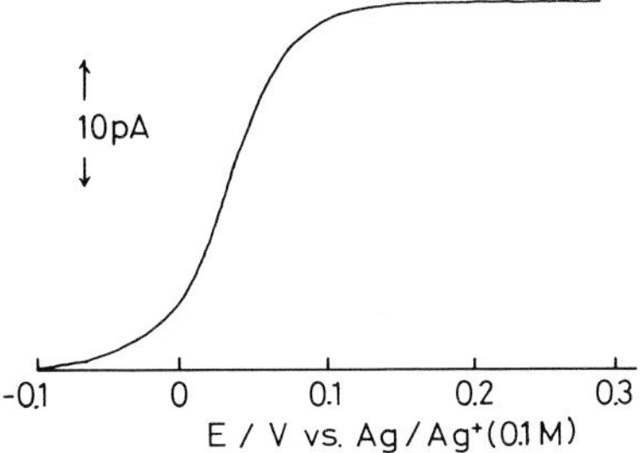

Fig. 15. Steady-state voltammograms for the oxidation of 10^{-3} M ferrocene in acetonitrile glass with no added electrolyte. The electrode was 0.5 μm radius Pt.

simplification of the study of chemical reactions following charge transfers or, indeed, to the elimination of the influence of such coupled reactions.

5.3 TRANSIENT STUDIES OF ELECTRODE KINETICS

We have seen how the reduced iR_u drop associated with the use of microelectrodes can be used to advantage when making steady-state measurements but there are also advantages in using these electrodes for transient methods. We have already seen that the iR_u drop in fast transient studies reduces as the electrode is made smaller but additionally, since the double layer capacity is proportional to the area, charging current problems will also reduce as the electrode is made smaller. This implies that the use of microelectrodes in studies using transient techiques helps both to differentiate against charging currents as well as reducing iR_u drop effects. It should therefore be possible to study the kinetics of faster reactions with microelectrodes as well as reactions in less conducting solutions.

An example of this latter application is that of Pletcher and his coworkers [29] into the deposition of lithium in ether solvents. This system is of particular interest in view of its relevance to lithium battery studies. At conventional electrodes, it is difficult to study transient behaviour because the electrode solution is so resistive but, as shown, in Fig. 16 studies at microelectrodes are very effective. This figure shows a series of current–time transients for the deposition of lithium on 40 μm radius copper disc electrodes and no problems were encountered from either iR_u drop or charging currents. Some might question the relevance of microelectrode studies at very low currents to technological systems such as batteries but it must be remembered that, whilst the currents are small, the current densities involved are, in fact, quite large.

The possibilities for the use of microelectrodes for the study of fast kinetics by transient methods are exemplified by the work of Howell and Wightman [30], which showed that cyclic voltammograms could be obtained at microelectrodes at sweep rates as high as 10^5 V s^{-1} without the need for iR_u drop correction. Figure 17, for example, shows some cyclic voltammograms obtained for the reduction of anthracene whilst Fig. 18 shows how closely

References p. 172

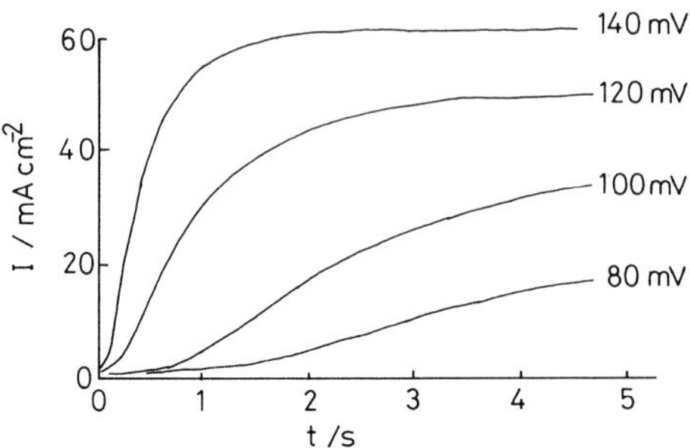

Fig. 16. A series of current–time transients for the deposition of lithium on a copper microelectrode (radius 40 μm) from a solution of $LiAsF_6$ (0.37 mol dm^{-3}) in tetrahydrofuran. The base potential for all pulses was 0.0 V; the potentials stepped to are given on the figure. (After Genders et al. [29].)

the experimentally observed peak current data fit that predicted by theory (as before, the theoretical values were based on an analytical solution for spherical geometry that was adjusted using the analogy between spherical and finite disc geometries). The reduction of anthracene was essentially reversible, even at the highest sweep rates used. At these highest rates, it should be possible to determine standard rate constant values as high as 20 cm s^{-1} and whilst these workers did not investigate a system as fast as this, they did determine the k^0 value for anthraquinone at 1.78 $\pm$ 0.35 cm s^{-1}.

We have seen in this section how microelectrodes can be used in a variety of ways to study both coupled homogeneous chemical reactions and heterogeneous electron transfers. We have also seen one example of a study of metal deposition. The development of microelectrodes has made possible

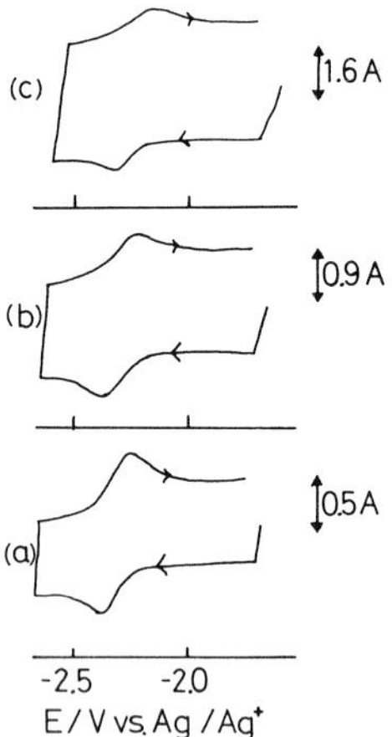

Fig. 17. A series of high sweep rate cyclic voltammograms for the reduction of anthracene at a gold microelectrode (radius 6.5 μm) in acetonitrile containing 0.6 M Et_4NClO_4. The sweep rates were (a) 2 $\times$ 10^4 V s^{-1}, (b) 5 $\times$ 10^4 V s^{-1}, and (c) 10^5 V s^{-1}. (After Howell and Wightman [30].)

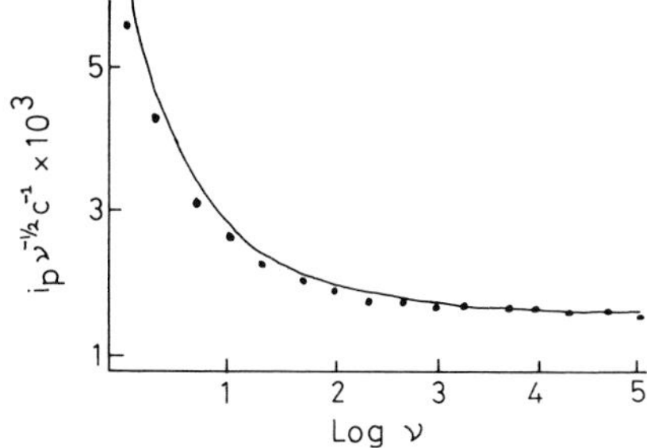

Fig. 18. A plot of the peak current function as a function of logarithm of sweep rate for cyclic voltammograms for the reduction of anthracene at an Au microelectrode (6.5 μm radius.)

a range of new experiments in the area of nucleation and phase growth and these applications will be discussed in the final section of examples.

5.4 NUCLEATION AND PHASE GROWTH

The example of a metal deposition system that we have already seen was, in most senses, conventional with the microelectrode being used largely to reduce the iR_u drop. The small size of microelectrodes does, however, open up the possibilities for new types of experiments.

Phase growth at an electrode proceeds by the formation of nuclei from clusters of sub-critical size and, once formed, these nuclei grow spontaneously. In potentiostatic phase-growth experiments at a conventionally sized electrode, a large number of nuclei, often of different sizes, are present and the current that is recorded is the sum of the currents due to each of these nuclei. Alternatively, in galvanostic measurements, it is an average potential that is measured. If the electrode size is reduced, large fluctuations from these mean values are observed since the number of nuclei present is smaller and the magnitude of the individual transients becomes comparable with the sum. Analysis of these fluctuations can provide information not obtainable from the mean values as has been shown by Bindra et al. [31] for

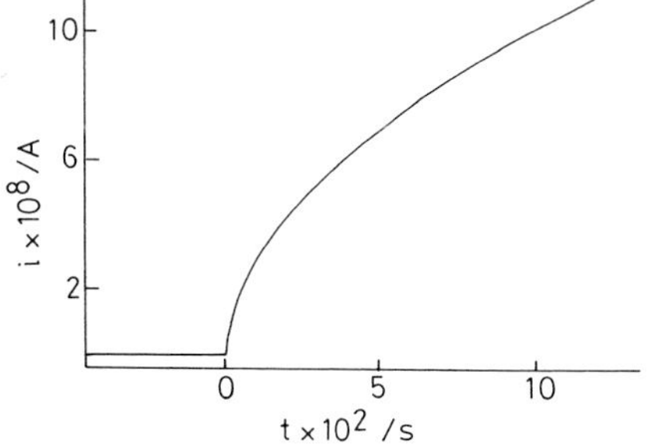

Fig. 19. Potentiostatic transient for the growth of mercury nucleus on a carbon fibre electrode from a 0.1 M solution of $Hg_2(NO_3)_2$ in aqueous KNO_3 at -200 mV. (After Scharifker and Hills [25].)

References p. 172

the nucelation of α-PbO_2 on platinum microelectrodes and for calomel formation on mercury drops. A discussion of the techniques used in these analyses is beyond the scope of this chapter.

As the electrode size is reduced still further, we reach a point where only a single nucleus is able to develop during the period of the experiment and the determination of the nucleation rate is achieved by measuring the induction time for the appearance of the first nucleus and the growth rate of this single nucleus can be determined from the subsequent current–time transient. The nucleation is, of course, a stochastic process and therefore the induction time will vary. The determination of the nucleation rate therefore requires a large number of transients to be studied.

Scharifker and Hills [25] have investigated mercury deposition on carbon fibre electrodes with a view to obtaining nucleation and growth data for a single nucleus and Fig. 19 shows a potentiostatic current–time transient for the growth of such a single Hg nucleus. That it is, indeed, a single nucleus that is growing can be determined from an analysis of the transient. It can be shown that, for the diffusion-controlled growth of a single hemispherical centre, the current is proportional to $t^{1/2}$. As shown in Fig. 20, this was indeed found to be the case for times as short as 1 ms to in excess of 2 min. If more than one nucleus had been growing, then the diffusion fields would have eventually overlapped and at the least there would have been a change of slope in the i vs. $t^{1/2}$ plot and, more probably, a loss of the linear dependence.

An even clearer example of transients arising from the growth of a single nucleus has been presented by Budevski et al. [32]. They have reported data for the two-dimensional nucleation and growth of silver on 100 μm diameter single crystal (100) faces of silver at low overpotential. Figure 21 shows a short section of a current–time transient. The different shapes of the sections (a), (b), and (c) were attributed to the position where the nucleus first formed. Once formed, the nucleus grows (with the current being proportional to its circumference) until it hits the edge of the electrode when

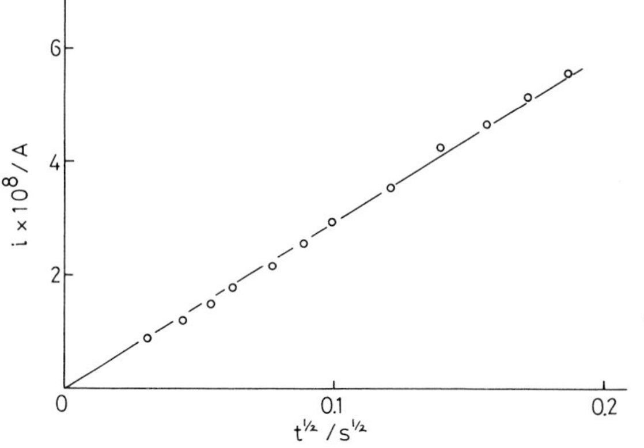

Fig. 20. A plot of current as a function of $t^{1/2}$ for the data of Fig. 19. (After Scharifker and Hills [25].)

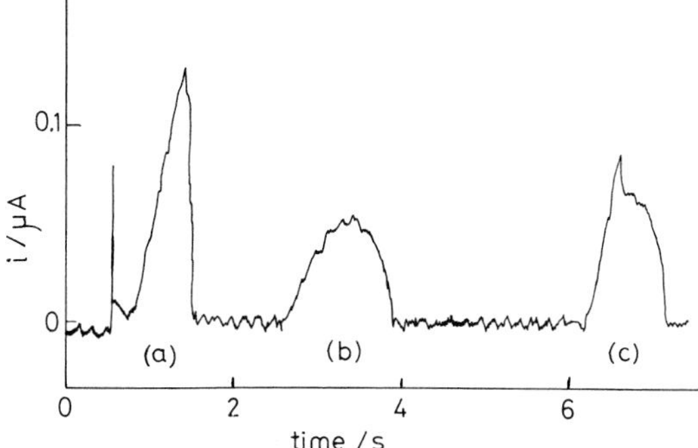

Fig. 21. A section of the current–time transient for the deposition of silver on silver single crystals at an overpotential of 8 mV. (After Budevski et al. [32].)

growth stops. Thus, for (a) the nucleus must have formed near the centre of the electrode since the termination of growth is so abrupt (the nucleus must have hit the edge of the electrode along the length of the perimeter at essentially the same time). For transients (b) and (c), the nucleus was formed midway between the edge and centre and at the edge, respectively. In this study, Budevski et al. went on to obtain kinetic information by a statistical analysis of a large number of transients such as those in Fig. 21.

There have, as yet, been few studies of this type but, in view of their potential, it seems likely that many more will follow in the near future.

6. Closure

It has been the intention of the preceding discussion to give the reader a feeling for the types of experiment that have been performed with microelectrodes and the range of information that can be obtained through their use. In particular, it is hoped that readers will see that microelectrodes have many unique properties and that they will therefore be encouraged to use them in their own investigations. The main factor that is preventing their more widespread adoption is probably the perceived difficulty of construction, though this is really only true of electrodes smaller than 5 μm radius. If one has to pinpoint the one development that will make them more widely accepted, it must therefore be the ready availability of small microelectrodes, either commercially or through the introduction of simpler construction methods. It is undoubtedly true that this could be achieved by the utilization of some of the microfabriation techniques of the semiconductor industry but, unfortunately, the cost of this approach will almost certainly be prohibitive.

List of symbols

A electrode area (cm^2)

References p. 172

C_0^{∞} bulk concentration of species 0 (mol cm^{-3})
F the Faraday (coulombs mol^{-1})
i current (A)
i_s current at a sphere
I current density (A cm^{-2})
I_s current density at a sphere (A cm^{-2})
k first-order rate constant (s^{-1})
n number of electrons
n_{APP} apparent number of electrons
r_s radius of sphere (cm)
r_m radius of disc (cm)
κ specific conductivity (Ω^{-1} cm^{-1})

References

1 R.L. White and H.D. Mercer, IEEE Biomed. Eng., 25 (1978) 494.
2 C.D. Ferris, Introduction to Bioelectrodes, Plenum Press, New York, 1974.
3 R.M. Wightman, E. Strope, P.M. Plotsky and R.N. Adams, Nature (London), 262 (1976) 145.
4 A.G. Ewing, M.A. Dayton and R.M. Wightman, Anal. Chem., 53 (1981) 1842.
5 P. Bindra, A.P. Brown, M. Fleischmann and D. Pletcher, J. Electroanal. Chem., 58 (1975) 31, 39.
6 P.J. Lingane, Anal. Chem., 36 (1964) 1723.
7 Z.G. Soos and P.J. Lingane, J. Phys. Chem., 68 (1964) 3821.
8 C.R. Ito, S. Asakura and K. Nobe, J. Electrochem. Soc., 199 (1972) 698.
9 J.B. Flanagan and L. Marcoux, J. Phys. Chem., 77 (1973) 1051.
10 M.A. Dayton, J.C. Brown, K.J. Stutts and R.M. Wightman, Anal. Chem., 52 (1980) 946.
11 J. Heinze, J. Electroanal. Chem., 124 (1981) 73.
12 M. Kakihana, H. Ikeuchi and G.P. Sato, J. Electroanal. Chem., 117 (1981) 201.
13 K.B. Oldham, J. Electroanal. Chem., 122 (1981) 1.
14 K. Aoki and J. Osteryoung, J. Electroanal. Chem., 122 (1981) 19.
15 M. Fleischmann, F. Lasserre, J. Robinson and D. Swan, J. Electroanal. Chem., 177 (1984) 97.
16 K. Aoki, K. Honda, K. Tokuda and H. Matsuda, J. Electroanal. Chem., 182 (1985) 79.
17 W.R. Kenneth, M.R. Deakin and R.M. Wightman, Anal. Chem., 57 (1985) 1913.
18 M. Fleischmann, S. Bandyopadhyay and S. Pons, J. Phys. Chem., 89 (1985) 5537.
19 C. Amatore and J.M. Savéant, J. Electroanal. Chem., 44 (1973) 169.
20 G.S. Alberts and I. Shain, Anal. Chem., 35 (1963) 1859.
21 M. Fleishmann, F. Lasserre and J. Robinson, J. Electroanal. Chem., 177 (1984) 115.
22 O. Hammerich and V.D. Parker, J. Chem. Soc. Chem. Commun., (1974) 245.
23 D. Shoup and A. Szabo, J. Electroanal. Chem., 140 (1982) 237.
24 B. Speiser and B.S. Pons, Can. J. Chem., 60 (1982) 1352.
25 B. Scharifker and G.J. Hills, J. Electroanal. Chem., 130 (1981) 81.
26 A.M. Bond, M. Fleischmann and J. Robinson, J. Electroanal. Chem., 168 (1984) 299.
27 A.M. Bond, M. Fleischmann and J. Robinson, J. Electroanal. Chem., 180 (1984) 257.
28 A.M. Bond, M. Fleischmann and J. Robinson, J. Electroanal. Chem., 172 (1984) 11.
29 J.D. Genders, W.M. Hedges and D. Pletcher, J. Chem. Soc. Faraday Trans. 1, 80 (1984) 3399.
30 J.O. Howell and R.M. Wightman, Anal. Chem., 56 (1984) 524.
31 P. Bindra, M. Fleischmann, J.W. Oldfield and D. Singleton, Faraday Discuss. Chem. Soc., 56 (1973) 180.
32 E. Budevski, M. Fleischmann, C. Gabrielli and M. Labram, Electrochim. Acta, 28 (1983) 925.

Chapter 6

The Use of Channel Electrodes in the Investigation of Interfacial Reaction Mechanisms

PATRICK R. UNWIN and RICHARD G. COMPTON

"Further fundamental studies are required to delineate those kinds of electrochemical problems that can be most successfully attacked by hydrodynamic steady-state measurements".

W.J. Blaedel
Analytical Chemistry, 39 (1967) 1061.

1. Introduction

Although many of the chapters of this volume testify to the power of spectroelectrochemical methods, it remains the case that the overwhelming majority of mechanistic electrochemical studies are centred around the current–voltage relationship and its dependence upon, for example, concentration, temperature, pH, solvent, and mass transport. The investigation of the kinetics of fast electrode and/or coupled chemical processes demands that mass transport to the electrode is such as to promote competition between, on the one hand, the electrode reaction and, on the other, the transfer of electroactive material away from the electrode before it has chance to react. Whilst this may be achieved in quiescent solution, in principle, by constraining the measurements to short times following a perturbation (of the current or potential) applied to the electrode [1], problems arise when these times are such that double layer charging competes with the Faradaic process, or components of the system under study are highly adsorptive. An alternative to this strategy therefore lies in the addition of convection to the diffusional system. In this way, measurements can be made under steady-state conditions and the rate of mass transport varied by altering the rate of convection.

This chapter is concerned with channel electrodes, which have their origins in the latter concept, forced convection being employed as a variable, by moving the solution over a stationary electrode embedded in the wall of a channel. The aim of the chapter is to show that, of the presently available hydrodynamic systems, the channel electrode (or the closely related tubular electrode) is the most satisfactory in the study of a diverse range of electrochemical and other interfacial problems.

References pp. 290–296

TABLE 1

(a) *The CE mechanism*: the reduction of acetic acid at a platinum electrode.

C $CH_3CO_2H \rightleftharpoons CH_3CO_2^- + H^+$
E $H^+ + e^- \rightarrow \frac{1}{2}H_2$
(CH_3CO_2H and $CH_3CO_2^-$ are electro-inactive at the potentials of interest.)

(b) *The EC mechanism*: the photoreduction of crystal violet at a platinum electrode.

E $[(CH_3)_2NC_6H_4]_3C^+ + e^- \rightleftharpoons [(CH_3)_2NC_6H_4]_3C^{\bullet}$

C $\dot{C}(C_6H_4N(CH_3)_2)_3 \xrightarrow[CH_3CN]{h\nu\ (406\ nm)} H{-}C(C_6H_4N(CH_3)_2)_3$

1.1 MASS TRANSPORT AS A VARIABLE IN THE STUDY OF ELECTRODE PROCESSES

Let us first consider briefly how the use of mass transport as a variable can provide a guide to the reaction mechanism and give quantitative kinetic detail. As an illustration, we consider the behaviour of CE and EC processes (where E signifies electron transfer and C represents a chemical step) at a rotating disc electrode (RDE). This hydrodynamic system has already been discussed by Albery et al. and the reader is referred to Chap. 4 for details. CE and EC processes represent the simplest conceivable electrode reactions involving coupled homogeneous kinetics: mechanistic examples of both types are shown in Table 1. In the discussion which follows, the electron-transfer reaction in the two mechanisms is considered to be a cathodic process; the extension to the anodic case is trivial.

At the RDE, mass transport to the electrode is varied by altering the disc rotation speed (W/Hz). The consequences of this on the current–voltage relationship, for the two reactions, are shown schematically in Fig. 1. If the E step is considered to be electrochemically reversible, then the voltammetric wave is defined by the two parameters $E_{1/2}$ (the half-wave potential) and I_{LIM} (the transport-limited current), as depicted. It is the dependence of these two quantities on the disc rotation speed which allows the deduction of the mechanism, as shown in Fig. 1. For a kinetically uncomplicated reversible electrode reaction, I_{LIM} varies as $W^{1/2}$ [2] and $E_{1/2}$ is independent of W [3]. For a CE process, at fast rotation speeds the limiting current is

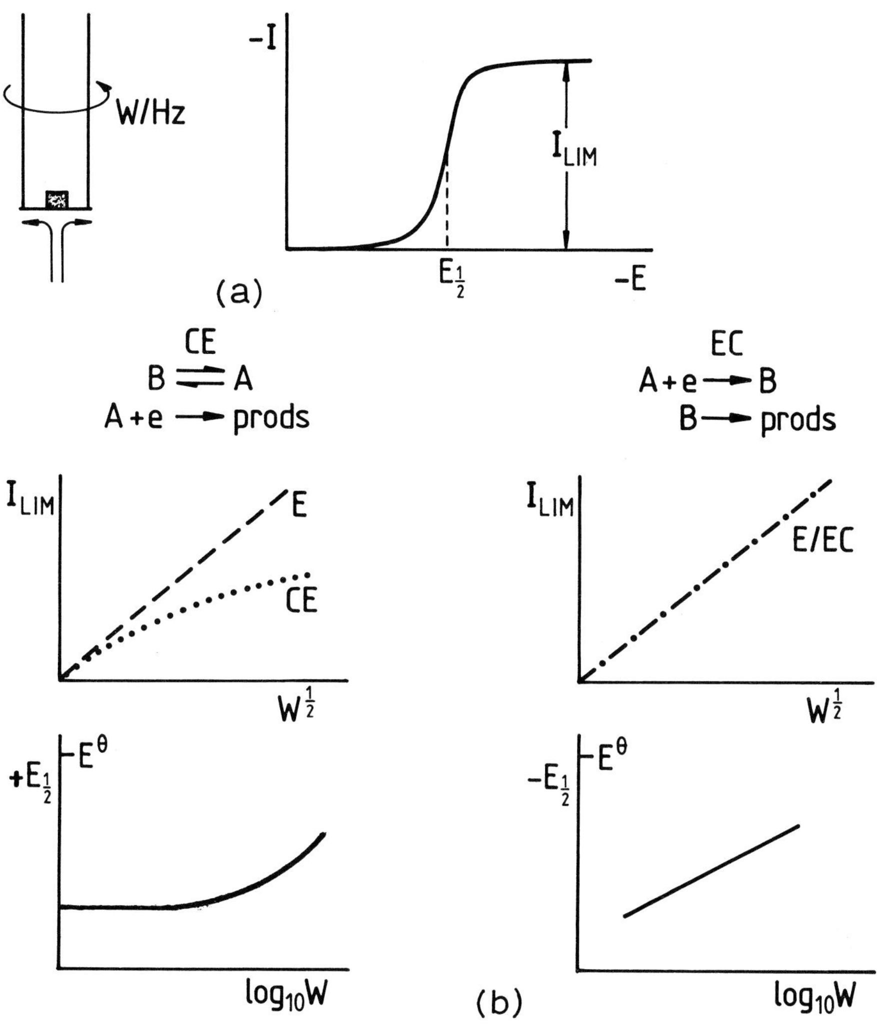

Fig. 1. CE and EC processes at a rotating disc electrode. (a) The current–voltage curve resulting from reverisble electron transfer is characterised by the half-wave potential, $E_{1/2}$, and the mass transport-limited current, I_{LIM}. (b) The effect of coupled homogeneous kinetics (under the reaction layer approximation) on the parameters $E_{1/2}$ and I_{LIM}.

below that expected for the simple E reaction, since the precursor of the electroactive material (CH_3COOH in the example given in Table 1) spends insufficient time in the vicinity of the electrode surface for complete conversion to the electroactive form (H^+). At lower rotation speeds, the "transit time" across the disc surface is increased so as to allow full reduction of the electroactive material and effectively "one-electron" behaviour is observed, as illustrated. Quantitative analysis of the complete I_{LIM}–$W^{1/2}$ behaviour may provide definitive confirmation of a CE mechanism [4–6].

Turning to the EC process, it is evident from Fig. 1 that the chemical step has no influence on I_{LIM}, since it occurs after the electron transfer. The following chemical step manifests itself, however, in the $E_{1/2}$–W behaviour in that the reduction potential is shifted anodically (by about $30\,\mathrm{mV}/\log_{10} W$ at 298 K when C is first order [7, 8] and by about $20\,\mathrm{mV}/\log_{10} W$ at 298 K when C is a second-order process [9]). In contrast, it can be seen that, when the electron-transfer reaction is preceded by a chemical step, an additional free energy barrier is provided to the reduction process and $E_{1/2}$ is shifted cathodically. In general, quantitative analysis of $E_{1/2}$–W data may again give an indication of the electrode reaction mechanism, particularly if the standard electrode potential, E^{-}, of the A/B couple is known under conditions when the kinetic complications are negligible [10].

We have demonstrated how steady-state current–voltage curves measured under conditions of variable mass transport can provide indications of the reaction mechanism. Although we have, by way of example, used the RDE (which is perhaps the most familiar hydrodynamic electrode), the general principle carries over to other electrode types, e.g. the dropping mercury electrode, the channel (or tubular) electrode, the wall-jet electrode, the microelectrode, etc. except that the rotation speed is replaced by the drop lifetime, flow rate (for channel and wall-jet electrodes), or the electrode radius. The question arises as to the hydrodynamic electrode of choice for the study of electrode reaction mechanisms with the protocol suggested above. Some of the criteria by which this question can be approached might reasonably be thought to be the following:

(a) Can the electrode distinguish between similar mechanistic pathways, i.e. probe mechanistic nuances (e.g resolve ECE and DISP1 processes)?

(b) Can the necessary theoretical description of the various mechanisms of interest be carried out with sufficient accuracy so as to permit such distinctions that are feasible according to (a)?

(c) For a given electrode reaction mechanism, can a wide range of rate constants be studied?

Another consideration of practical significance might be the ease with which the electrode and associated flow system (where applicable) can be constructed, particularly since the transit time of material to channel, wall-jet, and microelectrodes is decreased as the electrode is made smaller. This has a direct bearing on condition (c).

The aims of this chapter are to suggest that the channel (or tubular) electrode meets the criteria, set out above, more successfully than the other available hydrodynamic electrodes and to illustrate, by way of practical examples, that the channel electrode and channel electrode methodology can be applied to the study of a diverse range of electrochemical and other interfacial phenomena. By way of introduction, we will firstly consider the channel electrode in the light of the preceding criteria before proceeding with more detailed discussion.

1.2 CHANNEL AND TUBULAR ELECTRODES

In these hydrodynamic systems, solution moves past an electrode which

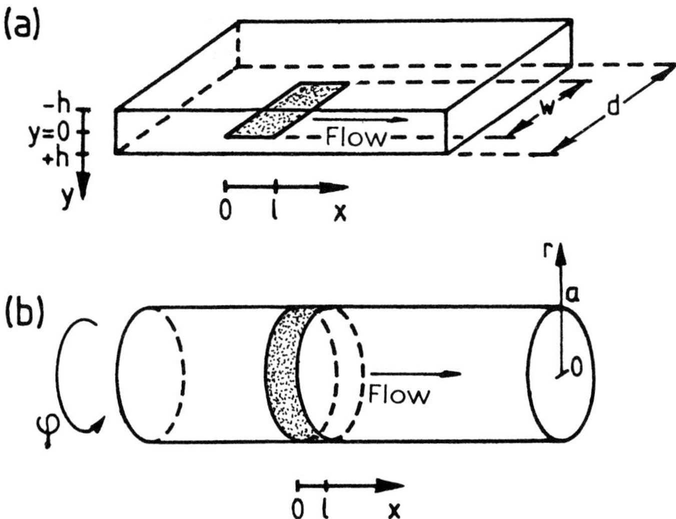

Fig. 2. Co-ordinate system for (a) the channel electrode and (b) the tubular electrode.

is embedded in the wall of a channel or tube, as depicted in Fig. 2. This figure also defines the co-ordinate systems we will adopt. First introduced by Blaedel et al. in 1963 [11] in the tubular form, these electrodes now find widespread and popular application as flow-through electroanalytical detectors of high sensitivity. This aspect of the use of channel electrodes is beyond the scope of this chapter and the reader is referred to refs. 12–18 for further coverage. Reference 12 summarises over 800 papers on flow injection analysis.

Let us compare the channel electrode with other hydrodynamic systems in the context of the criteria suggested above. At this point, however, we eliminate the wall-jet electrode from any further discussion since theories for coupled electrode–solution reactions have yet to appear.

With regard to the first criterion for the optimal hydrodynamic electode, what reasons might we have, a priori, to predict a superior performance from the channel electrode, as compared with, say, the RDE? The most striking difference between the channel electrode and the RDE is that the former is non-uniformly accessible, whereas the latter is uniformly accessible. That is to say that, under transport limited conditions, whereas the current density varies as $x^{-1/3}$ at the channel electrode (or, alternatively the diffusion-layer thickness may be viewed as being $x^{1/3}$-dependent), at the RDE, the current density and diffusion layer are constant over the disc surface. We believe that non-uniformity of current distribution greatly aids in the discrimination between similar electrode reaction mechanisms involving coupled electron transfer and homogeneous kinetics. This may be understood by considering the experiment described earlier in which the transport-limited current is examined as a function of variable mass transport (rotation speed

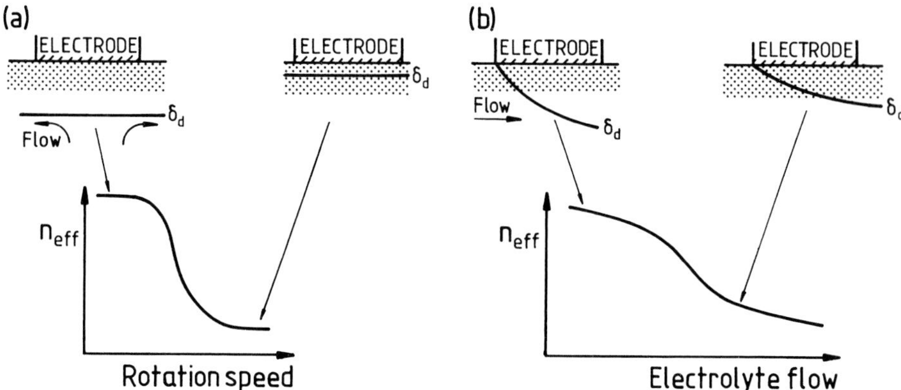

Fig. 3. (a) A rapid transition in the number of electrons transferred (n_{eff}) is observed at the uniformly accessible rotating disc electrode when the diffusion layer becomes sufficiently thin for the intermediate (shown stippled) to survive long enough to cross it (b) A more gradual transition is seen at the channel electrode, where some intermediate escapes (upstream electrode edge) and some is always trapped (downstream electrode edge), due to the shape of the diffusion layer.

of electrode or flow rate), and the corresponding "effective" number of electrons transferred (n_{eff}) measured. The variation of the latter, for complex reactions, effectively depends on the competition between the loss of intermediates through transport into solution or further heterogeneous charge transfer at the electrode surface.

The essential advantage is that, at the non-uniformly accessible channel electrode, the variation in n_{eff} with mass transport occurs more gradually than at a uniformly accessible electrode since, at the upstream edge of the electrode, the steepness of the concentration gradients (thin diffusion layer) promotes the loss of intermediates whereas, further downstream, the intermediates are encouraged to undergo further electron transfer because the shallow concentration gradients (thick diffusion layer) discourage transit of the intermediates away from the electrode surface. In contrast, at the RDE, the loss of intermediates into the bulk solution is controlled by whether the intermediate is sufficiently long-lived to cross the (uniform) diffusion layer, resulting in a sharp transition in the effective number of electrons transferred as the rotation speed is increased. These contrasting effects are illustrated in Fig. 3. In essence, non-uniformly accessible electrodes will be more sensitive in distinguishing between similar mechanisms. To support this statement, we note that non-uniformly accessible microelectrodes can be used to resolve the ECE–DISP1 problem [19], as discussed in detail by Robinson in Chap. 5. Furthermore, in view of the fact that non-uniform accessibility is thought to be a grave disadvantage by some workers [20], since it makes it "much more difficult to calculate kinetic parameters from (current measurements)", it is perhaps revealing that the rotating disc voltammetrician frequently converts his electrode into a rotating ring–disc electrode and performs "collection efficiency" experiments [21] when he wants to study a reaction of any complexity. In other words, he turns his uniformly accessible electrode into a grossly non-uniform one in which reactions driven on the disc are reversed on the ring, the disc-generated products passing over an insulating annulus separating the ring and the disc.

We now turn to the second criterion, in particular bearing in mind the criticism, alluded to above, about the difficulty associated with the theoretical description of processes at non-uniformly accessible electrodes. Again, we will compare and contrast the channel electrode and the RDE. Now the theoretical description of electrode reactions involves, typically, the solution of perhaps several coupled steady-state convective-diffusion equations of the form

$$0 = D_i \nabla^2 c_i - \left(v_x \frac{\partial c_i}{\partial x} + v_y \frac{\partial c_i}{\partial y} + v_z \frac{\partial c_i}{\partial z} \right) \pm \sum_{i,j...} k_{i,j...} c_i^{n_i} c_{j...}^{n_i} \tag{1}$$

where c_i is the concentration of the ith species, D_i is its diffusion coefficient and v_x, v_y, and v_z are the cartesian components of flow. The third term in this equation represents loss or gain of material through chemical reactions involving the species i, j . . . of order n_i, n_j. . . having rate constants $k_{i,j}$. . .

It is evident that the solution of equations such as eqn. (1) requires a knowledge of the terms v_x, v_y, and v_z. It transpires (vide infra) that theoretical treatments of channel electrode problems are generally mathematically simpler than the corresponding analyses for dropping mercury, rotating disc, wall-jet, or microelectrodes. This arises since the solution flowing over the electrode may, to a good approximation, be considered to have a velocity which, close to the electrode surface, increases linearly with distance away from the electrode. Thus, for channel electrodes, v_x, v_y, and v_z are described by the equations

$$v_x \approx 2v_0 \left(1 - \frac{y}{h}\right) \quad \text{for } y \approx \pm h \tag{2}$$

$$v_y = v_z = 0 \tag{3}$$

where v_0 is the solution velocity at the centre of the channel. This greatly simplifies the analytical solution of the relevant convective-diffusion equation(s) as compared with other hydrodynamic electrodes. Close to the rotating disc or the wall-jet electrode, for example, the velocity normal to the electrode surface depends on the square of the distance away from it, a parabolic dependence which generally dictates numerical methods.

Very recently, a further simplification has been introduced by Singh and Dutt [22] in respect of the application of eqn. (1) to channel electrode problems. This involves replacing the concentration gradient in the x direction (i.e. in the direction of flow) by "its average value" [22]. Mathematically, this means that for the electrode process

$$\text{Red} \rightleftharpoons \text{Ox} + \text{e}^-$$

we can write

$$\frac{\partial[\text{Red}]}{\partial x} \approx \frac{[\text{Red}] - [\text{Red}]^{\infty}}{l} \tag{4}$$

and

$$\frac{\partial[\text{O}x]}{\partial x} \approx \frac{[\text{O}x]}{l} \tag{5}$$

where l is the electrode length and $[\text{Red}]^{\infty}$ is the bulk concentration of the electroactive species. ($[\text{Ox}]^{\infty}$ is taken to be zero.) By invoking the approximations as in eqns. (4) and (5), eqn. (1) is reduced to an ordinary differential equation in y. Various time-dependent phenomena have been readily deduced on the basis of the resulting ordinary differential equations since straightforward analytical methods are facilitated [22–25]. Later in this chapter (Sect. 2.3, 3.5 and 4.1), we show that the approximation can be used to attack a variety of complex electrochemical problems at a remarkably successful level when compared with more rigorous theoretical treatments. On this basis, we anticipate that the widespread adoption of the Singh and Dutt method can only further ease the theoretical description of channel electrode problems.

Equations of type (1), as appropriate to the microdisc electrode, are often insoluble analytically. Solutions are only afforded numerically or by making analogy to the uniformly accessible hemispherical micro-electrode (see Chap. 5).

In the consideration of the range of (or fastest) rate constant which can be measured at a particular hydrodynamic electrode, we again refer to our generalised experiment in which we set mass transport in competition with the kinetics of the system in question. The measurement of fast rate constants relies on minimising the transit time of intermediates in the vicinity of the electrode. Since the rotating-disc electrode is uniformly accessible, the ability to minimise the transit time is governed solely by the fastest rotation speeds which can be obtained. The range of measurable rate constants is therefore constrained at the upper limit by the point at which either mass transport to the electrode becomes incalculable (under turbulent conditions) [26] or rotation speeds cannot be realised experimentally. In general, it is the latter condition which places the limit on the fastest measurable rate constants, as the design of high-speed rotators has proved to be complex [27]. Since the microelectrode operates in stationary solution, the rate of mass transport to it is governed solely by the size of the electrode. The limit on the highest measurable rate constant is determined by the smallest electrode that can be constructed.

In contrast to the two systems above, the effective rate of mass transport to the channel electrode is dependent upon both the solution velocity *and* the length of the electrode. We may therefore anticipate that a greater range of rate constants should be measurable at this hydrodynamic system compared with both the rotating-disc and (in principle) the microelectrode.

Having introduced some of the points for discussion in brief, we continue with more detailed argument. An appropriate starting point is a comprehensive description of mass transport in channel electrode systems.

2. Mass transport to the channel electrode

We have seen that the solution of eqn. (1) requires a knowledge of the convective flow within the channel electrode. This fluid flow is described by the Reynolds number [28], given by

$$\mathrm{Re} = \frac{v_0 h}{\nu} \tag{6}$$

where ν is the kinematic viscosity of the solution. Flow is laminar below a certain critical value, $\mathrm{Re}_{\mathrm{crit}} < 2000$ [28–30], and above it flow is turbulent. In virtually all practical channel electrodes, the flow is chosen to be laminar and we confine ourselves to this case alone in this section. Modifications to this description will be introduced for turbulent regimes later (Sect. 6.2). With this proviso, consider a plug flow of solution entering a channel electrode in which "edge effects" can be considered to be negligible ($h \ll d$). The effect of friction at the walls of the channel will be to slow down the flow in their vicinity, as shown in Fig. 4. Eventually, a Poiseuille flow regime is established, downstream of which

$$v_x = v_0\left(1 - \frac{y^2}{h^2}\right) \qquad v_y = v_z = 0 \tag{7}$$

The entry length, l_e, required for this to be set up is given approximately by [31]

$$l_e = 0.1 h \,\mathrm{Re} \tag{8}$$

Given eqns. (1) and (7), the problem of mass transfer may be tackled. The first approach to this was that of Levich [32].

2.1 THE LEVICH EQUATION

Levich [32] solved eqn. (1) for a channel electrode by invoking an approximation originally introduced in 1928 by Lévêque in his theory of heat transfer in pipes [33]. In the present context, this simplification can be written as

$$v_x = v_0\left(1 - \frac{y^2}{h^2}\right) = \frac{v_0(h + y)(h - y)}{h^2} \approx 2v_0\left(1 - \frac{y}{h}\right) \tag{9}$$

which is valid for $y \approx \pm h$, that is in the vicinity of the channel walls (where the electrode is located). Essentially, the Lévêque approximation is to replace the parabolic velocity profile by a linear one near the electrode sur-

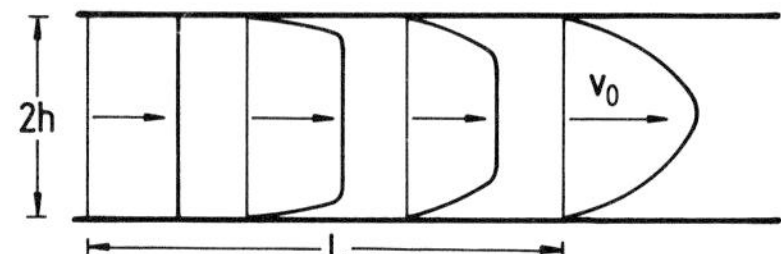

Fig. 4. The establishment of Poiseuille flow in a channel under a laminar regime.

face. This will be satisfactory under conditions in which convection is efficient compared with diffusion, such that concentration changes are confined to being close to the electrode (vide infra).

Assuming both diffusion in the direction of convective flow and diffusional side-edge effects to be negligible, Levich arrived at the equation (under the Lévêque approximation)

$$\frac{\partial c}{\partial t} = D\frac{\partial^2 c}{\partial y^2} - 2v_0\left(1 - \frac{y}{h}\right)\frac{\partial c}{\partial x} \tag{10}$$

which, under steady-state conditions, is

$$D\frac{\partial^2 c}{\partial \sigma^2} = \frac{2v_0\sigma}{h}\frac{\partial c}{\partial x} \tag{11}$$

where $\sigma = h - y$.

The substitution

$$\eta = \left(\frac{v_0}{xh}\right)^{1/3}\sigma \tag{12}$$

in eqn. (11) leads to the second-order differential equation

$$\frac{\mathrm{d}^2 c}{\mathrm{d}\eta^2} + \frac{2\eta^2}{3D}\frac{\mathrm{d}c}{\mathrm{d}\eta} = 0 \tag{13}$$

The missing links between eqns. (11)–(13) are produced in Appendix 1. The solution to the problem of the mass transport-limited current at the channel electrode requires eqn. (13) to be solved subject to the boundary conditions

$$\begin{aligned} h - y &\to \infty \quad c \to c^{\infty} \\ h - y &= 0 \quad c = 0 \end{aligned} \tag{14}$$

The substitution

$$p = \frac{\mathrm{d}c}{\mathrm{d}\eta} \tag{15}$$

in eqn. (13) leads immediately to

$$\frac{1}{p}\frac{\mathrm{d}p}{\mathrm{d}\eta} + \frac{2}{9D}\eta^2 = 0 \tag{16}$$

and so

$$p = \exp\left(-\frac{2\eta^3}{9D}\right) \tag{17}$$

Thus

$$\frac{c}{c^\infty} = \frac{\int_0^\eta \exp(-2\eta^3/9D)\,\mathrm{d}\eta}{\int_0^\infty \exp(-2\eta^3/9D)\,\mathrm{d}\eta} \tag{18}$$

Now, the diffusional flux to the channel electrode is given by

$$J = \int_0^l D\left.\frac{\partial c}{\partial y}\right|_{y=h} \mathrm{d}(x/l) \tag{19}$$

This may be evaluated by expanding the indefinite integral in eqn. (18) as a power series in η and integrating term-by-term. This results in

$$J = \int_0^l \left\{Dc^\infty\left(\frac{v_0}{Dhx}\right)^{1/3}\left[\int_0^\infty \exp\left(-\frac{2\eta^3}{9D}\right)\mathrm{d}\eta\right]^{-1}\right\}\mathrm{d}(x/l) \tag{20}$$

By making the substitution

$$s = \frac{\eta^3}{D} \tag{21}$$

the integral in eqn. (20) can be deduced to be [34]

$$\int_0^\infty \exp\left(-\frac{2\eta^3}{9D}\right)\mathrm{d}\eta = D^{1/3}\int_0^\infty s^{-2/3}\exp\left(-\frac{2s}{9}\right)\mathrm{d}s = \frac{D^{1/3}\Gamma(\frac{1}{3})}{3(2/9)^{1/3}} \tag{22}$$

Thus, the final expression for J is

$$J = \int_0^l 0.67c^\infty D^{2/3}\left(\frac{v_0}{hx}\right)^{1/3}\mathrm{d}(x/l) \tag{23}$$

If this is expressed in terms of a diffusion layer of thickness δ_d, then

$$J = \frac{Dc^\infty}{\delta_\mathrm{d}} \tag{24}$$

and so

$$\delta_\mathrm{d} = \frac{1}{0.67}\left(\frac{Dhl}{v_0}\right)^{1/3} \tag{25}$$

Hence the total mass-transport limited current to the electrode may be calculated to be

$$I_\mathrm{LIM} = nFwlJ \tag{26}$$

$$= \quad 0.925nFc^{\infty}\left(\frac{V_f D^2 l^2}{h^2 d}\right)^{1/3} w \tag{27}$$

where F is Faraday's constant, n is the number of electrons transferred, and $V_f(=4v_0hd/3)$ is the volume flow rate ($cm^3\,s^{-1}$) of the solution.

2.2 NUMERICAL METHODS

Finite difference methods have been used both to test the assumptions made in the derivation of eqn. (27) under the Lévêque approximation [35] and to solve electrochemical diffusion-kinetic problems with the full parabolic profile [36–38]. The suitability of the various finite difference methods commonly encountered has been thoroughly investigated by Anderson and Moldoveanu [37], who concluded that the backward implicit (BI) method is to be preferred to either the simple explicit method [39] or the Crank–Nicholson implicit method [40].

The basis of any of these methods is illustrated in Fig. 5, the (x, y) plane being covered with a two-dimensional finite difference net with increments Δx and Δy in the variables x and y. The electrode is assumed to be of sufficient width so that it can be considered to be uniform in the z-direction. The extension to three-dimensions, if required, would however present no conceptual or computational problems to what follows.

The net is such that the difference between grid points in the x-direction is lK^{-1} and in the y-direction $2hJ^{-1}$. The co-ordinates of the grid points (see Fig. 5) are then

$$x_k \quad = \quad \left(\frac{k}{K}\right) * l \quad k \quad = \quad 0, 1, 2 \ldots K \tag{28}$$

$$(h - y)_j \quad = \quad \left(\frac{j}{J}\right) * 2h \quad j \quad = \quad 0, 1, 2 \ldots J \tag{29}$$

The BI method, in particular, involves replacing the derivatives in eqn. (1)

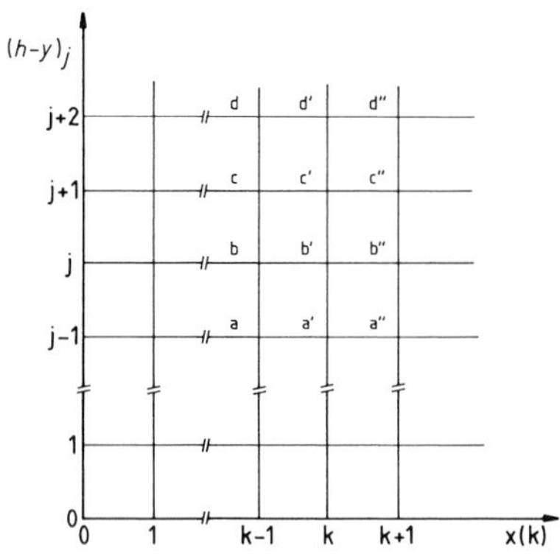

Fig. 5. Two-dimensional (x–y plane) finite difference net (from ref. 37).

by the expressions

$$\left(\frac{\partial c}{\partial x}\right)_{j,k} = (c_{j,k+1} - c_{j,k})\left(\frac{K}{l}\right) \tag{30}$$

and

$$\left(\frac{\partial^2 c}{\partial y^2}\right)_{j,k} = (c_{j-1,k+1} - 2c_{j,k+1} + c_{j+1,k+1}) * \left(\frac{J}{2h}\right)^2 \tag{31}$$

Moldoveanu and Anderson have shown [36] that this approach can be utilised to generate steady-state concentration profiles $c_{j,k}(y_j, x_k)$ from the equation

$$\frac{\partial c}{\partial t} = 0 = D\frac{\partial^2 c}{\partial y^2} - v_0\left(1 - \frac{y^2}{h^2}\right)\frac{\partial c}{\partial x} \tag{32}$$

This therefore represents a more rigorous theoretical treatment than that of Levich in that the full (parabolic) expression for v_x is retained.

Moldoveanu and Anderson found the BI method attractive compared with alternative numerical methods on a number of grounds. Firstly, the BI method is more economical in computer time since it operates via vector calculations based on the two-dimensional grid representing the (x, y) space, rather than the matrix calculations employed in alternative procedures. The vectors describe concentrations in the y-direction for different values of x. The calculation proceeds "downstream", starting from the vector defining the boundary conditions specified for upstream of the electrode, each vector enabling the calculation of the next. Secondly, the entire computation is independent of the value at (0, 0), i.e the concentration at the upstream edge of the electrode. This would otherwise be problematic since one would have to decide whether (0, 0) was part of the electrode or part of the upstream wall. In the calculation of the transport-limited current for example, the concentration at (0, 0) would have to be given a value of either zero (corresponding to complete conversion of the electroactive material arriving at the electrode) or that corresponding to the bulk value (with no flux through the wall). Alternative numerical strategies may give answers dependent on this choice, and so may be considered to be less than satisfactory in that respect [37].

Since Anderson's method is used to solve eqn. (32) without any preconceptions concerning the magnitude of the diffusion layer, it obviously results in an expression for the limiting current valid over a wider range of diffusion-layer thicknesses than permitted under the constraints of the Levich analysis. The results are shown in Fig. 6, which points out the region ($\mathrm{d}Dl/V_f h \ll 1$) in which the Levich analysis is valid. At the other extreme, for particularly low flow rates or long electrodes, the diffusion layer becomes comparable in size with the channel height resulting in exhaustive electrolysis of the material flowing over the electrode. In this case, the limiting

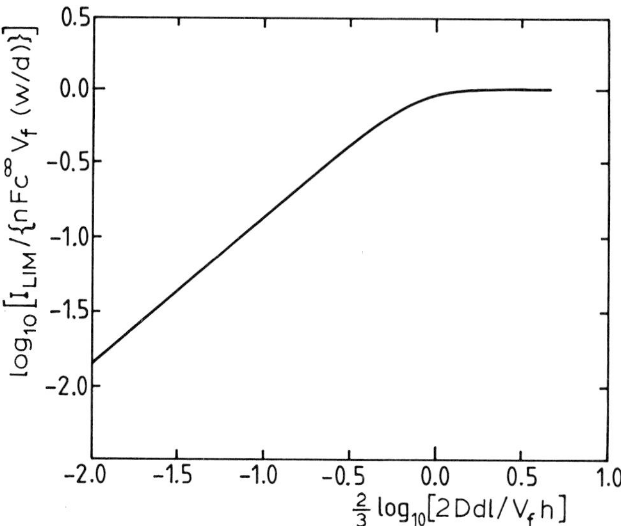

Fig. 6. Diffusion-limited current–flow rate dependence, as calculated by Anderson [37]. At the left-hand side of the diagram, the Levich equation [eqn. (27)] applies, whilst under the conditions denoted by the right-hand side of the figure, the electrode behaves as a thin-layer cell and eqn. (33) is applicable.

current becomes proportional to the first power of the flow rate

$$I_{\mathrm{LIM}} = nFV_f c^{\infty} \tag{33}$$

In practice, cells in which the Levich equation holds are easy to design (see Sect. 5) and there are numerous reported examples of cells for which eqn. (27) has been experimentally verified [41]. These are generally used in the mechanistic investigation of electrode reactions. (By contrast, the "thin-layer" limit is often more appropriate in electroanalytical work, for example in cells used as chromatographic detectors where optimal sensitivity is desired [42]).

The other assumptions of the Levich equation were investigated by Flanagan and Marcoux, using the simple explicit method [35], who concluded, in respect of axial diffusion, that for electrode geometries and flow rates for which the other Levich assumptions hold, axial diffusion has no observable effect. The deviations from the Levich equation observed by a number of workers [43–47] at very low flow rates, can be attributed either to natural convection becoming significant compared with forced convection or to so-called "edge effects" [48]. The latter phenomenon arises when lateral diffusion is no longer negligible compared with vertical diffusion, i.e. for short electrodes and/or slow flow rates, and manifests itself in larger mass transport-limited currents than would be predicted by the Levich equation. The problem of edge effects at channel electrodess has been approached analytically by Aoki et al. [49] using a method which complements the numerical simulations of Flanagan and Marcoux [35]. It was shown that the upstream edge effect was the dominant one and that, in general, the downstream effect could be neglected. Mathematically, the upstream and downstream edge effects were found to be described by the equation

$$I(p) = I(0)\left[1 + 0.5644p^{3/4} - 0.2457p + 1.393p^{1/2} \sum_{n=0}^{\infty} \alpha_n^{-3/2} \exp\{-(\alpha_n/p)^{3/4}\}\right] \quad (34)$$

where

$$p = \left(\frac{Dh}{2v_0 l^2}\right)^{2/3} \quad (35)$$

$-\alpha_n$ is the nth zero of the Airy function, i.e. $\mathrm{Ai}(-\alpha_n) = 0$ [50], and $I(0)$ is the mass transport-limited current without edge effects, as given by eqn. (27). In eqn. (34), the upstream edge effect is represented by the second and third terms, whilst the last term corresponds to the downstream effect. This equation thus provides a basis by which the edge effect for a particular electrode and solution flow rate may be assessed. The relationship between the current and the parameter p is shown in Fig. 7.

The problem of edge effects at channel electrodes has also been considered by Cope and Tallman [51], but under the condition of inviscid flow; that is, the solution was assumed to have a constant linear velocity, V. Thus, the convective-diffusion equation [as opposed to eqn. (11)]

$$2hV\frac{\partial c}{\partial x} = D\frac{\partial^2 c}{\partial y^2} + D\frac{\partial^2 c}{\partial x^2} \quad (36)$$

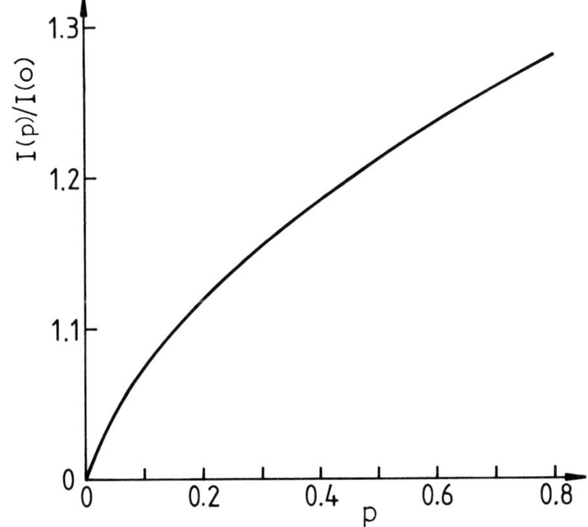

Fig. 7. The relationship between $I(p)/I(0)$ and p, calculated from eqn. (34), derived by Aoki et al. [49].

was solved where the 2nd term on the right-hand side corresponds to the effect of lateral diffusion. Since inviscid flow was assumed, the results obtained give only "qualitative information about behaviour at a real electrode" [51].

Whilst the cells/electrodes used in mechanistic work are designed so that edge effects may be neglected, it is evident that enhanced sensitivity will result by maximising the edge effect. To this end, micro and micro-array channel electrodes have been developed [52–60].

2.3 THE SINGH AND DUTT APPROXIMATION

This approximation, as indicated previously (Sect. 1.2), involves replacing the concentration gradients in the convective term by "average" values as defined by eqns. (4) and (5). The effect of this approximation on the calculation of the diffusion-limited currents at a channel electrode can be illustrated as follows. The concentration gradient in eqn. (11) is replaced by

$$\frac{\partial c}{\partial x} = \frac{c - c^{\infty}}{l} \tag{37}$$

so that

$$D\frac{\partial^2 (c - c^{\infty})}{\partial \sigma^2} = \frac{2v_0 \sigma (c - c^{\infty})}{hl} \tag{38}$$

or

$$\frac{\partial^2 (c - c^{\infty})}{\partial \xi^2} = \xi (c - c^{\infty}) \tag{39}$$

where

$$\xi = \left(\frac{2v_0}{hDl}\right)^{1/3} (h - y) \tag{40}$$

and it has been assumed that $(lD/h^2 v_0) \ll 1$. The solution to eqn. (39) is readily shown to be

$$c - c^{\infty} = \frac{\mathrm{Ai}(\xi) c^{\infty}}{\mathrm{Ai}(0)} \tag{41}$$

where $\mathrm{Ai}(\xi)$ is the Airy function [50].

The mass transport-limited current is then calculated as

$$I_{\mathrm{LIM}} = 0.834 nFc^{\infty} \left(\frac{V_f D^2 l^2}{h^2 d}\right)^{1/3} W \tag{42}$$

Notice that, except for the numerical factor of 0.834 as compared with 0.925, this equation is identical to that derived by Levich [eqn. (27)]. Thus, in this application, the approximation entails an error of around 10%.

Given this level of success in the description of the steady-state transport-

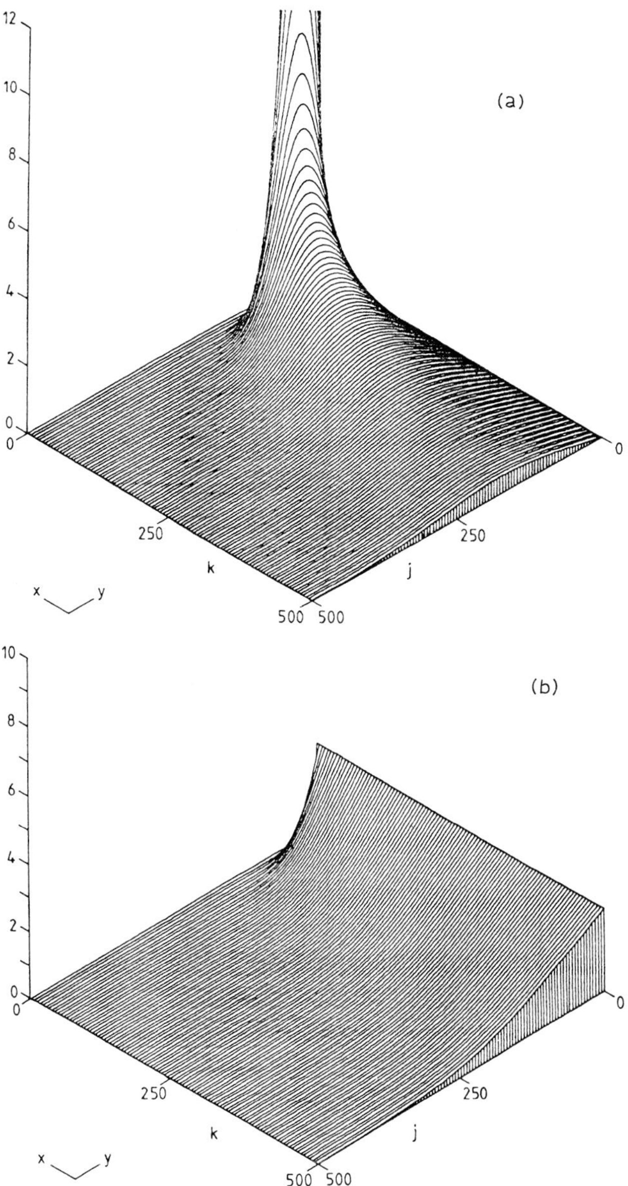

Fig. 8. **Three-dimensional plots** comparing the quantities (a) $(g_{j,k+1} - g_{j,k})(K/l)$, proportional $\partial[\text{Red}]/\partial x$, and (b) $(g_{j,k} - 1)/l$, proportional to $([\text{Red}] - [\text{Red}]^{\infty})/l$, over the cartesian (x, y) space of a channel electrode. The quantities have been multiplied by (-1) for ease of viewing.

limited current resulting from simple electron transfer and that demonstrated by Singh and Dutt themselves for certain linear sweep and cyclic voltammetric problems [22–25] leads one to search for the physical basis to the approximation embodied in eqns. (4) and (5). An appropriate line of enquiry is to generate the quantities $\partial[\text{Red}]/\partial x$ and $([\text{Red}] - [\text{Red}]^{\infty})/l$ as a function of x and y using the BI method when the transport-limited current for a simple electron transfer is being passed. The results, shown in Fig. 8, relate to a channel electrode and an electroactive species, Red, characterisd by the parameters: $2h = 0.04$ cm, $l = 0.4$ cm, $D = 10^{-5}\,\text{cm}^2\,\text{s}^{-1}$, $w = 0.6$ cm, and $v_0 = 0.0625\,\text{cm s}^{-1}$. Concentrations have been normalized, viz. $g_{j,k} = c_{j,k}/c^{\infty}$. It is evident that there is no great correlation in the shape of the two surfaces. Disagreement is most extreme at the upstream edge of the electrode (around $x = 0, y = 0$), where eqn. (4) predicts a zero concentration gradient whereas, in reality, the current density tends to infinity at this point. Figure 8 might be thought to indicate the lack of a sound physical basis for the Singh and Dutt approximation, at least under steady-state conditions.

To examine whether the reasonable agreement with the Levich equation is fortuitous and perhaps limited to a few cases, we have, elsewhere, examined a diverse range of channel electrode problems for which rigorous theory (at the Lévêque level) is available using the approximation [61]. For the chronoamperometric response to a potential step [62], linear sweep and cyclic voltammetry [22–25] of reversible, quasi-reversible, and irreversible systems, ECE [63] and DISP1 [64] mechanisms, agreement with rigorous theory is found to be, at worst, 10%, and is often very much better. We shall discuss the reasons for agreement in these cases specifically in Sects. 3 and 4. For the moment, let us consider why the approximation works in general (to the 10% level) and identify under what conditions it will hold. To this end, let us define an "average concentration", $\bar{c}(y)$, corresponding to the appropriate y co-ordinate

$$\bar{c}(y) = \int_0^l \frac{c(y)}{l}\,\mathrm{d}x \tag{43}$$

Then, integrating eqn. (10) with respect to x

$$\int_0^l \frac{\partial c}{\partial t}\,\mathrm{d}x = \int_0^l D\frac{\partial^2 c}{\partial y^2}\,\mathrm{d}x - v_x\int_0^l \frac{\partial c}{\partial x}\,\mathrm{d}x \tag{44}$$

gives

$$\frac{\partial}{\partial t}\left\{\int_0^l c\,\mathrm{d}x\right\} = D\frac{\partial^2}{\partial y^2}\left\{\int_0^l c\,\mathrm{d}x\right\} - v_x\int_0^l \frac{\partial c}{\partial x}\,\mathrm{d}x \tag{45}$$

or

$$\frac{\partial \bar{c}}{\partial t} = D\frac{\partial^2 \bar{c}}{\partial y^2} - \frac{v_x}{l}\int_0^l \frac{\partial c}{\partial x}\,\mathrm{d}x \tag{46}$$

and so

$$\frac{\partial \bar{c}}{\partial t} = D\frac{\partial^2 \bar{c}}{\partial y^2} - \frac{v_x}{l}\{c(l, y, t) - c(0, y, t)\} \tag{47}$$

This equation reduces to the Singh and Dutt form of the convective-diffusion equation if

$$c(l) \quad = \quad \bar{c}(y) \tag{48}$$

is a good approximation.

What happens upon the addition of kinetic terms to eqn. (10)? Considering first-order kinetics, we obtain the general equation

$$\frac{\partial c}{\partial t} \quad = \quad D\,\frac{\partial^2 c}{\partial y^2} \; - \; v_x\,\frac{\partial c}{\partial x} \; \pm \; k_a a \; \pm \; k_b b \; \pm \; k_c c \tag{49}$$

from which it is readily deduced that

$$\frac{\partial \bar{c}}{\partial t} \quad = \quad D\,\frac{\partial^2 \bar{c}}{\partial y^2} \; - \; \frac{v_x}{l}\,\{c(l,\,y,\,t) \; - \; c(0,\,y,\,t)\} \; \pm \; k_a \bar{a} \; \pm \; k_b \bar{b} \; \pm \; k_c \bar{c} \tag{50}$$

That is to say, no additional approximations are involved. Thus, the Singh and Dutt approximation is expected to work best for both time-dependent and steady-state behaviour with or without the presence of first-order kinetics, when the concentrations near the electrode are not too different from the concentration gradient averaged over the length of the electrode. Whilst this requirement might be considered to be contradictory to the essential non-uniformity of accessibility to the electrode, we shall see in Sect. 4.1 that this is not the case for steady-state problems.

The extent to which eqn. (47) works may be investigated by using the BI method to generate the quantities $g(j,\,K)$ and $\bar{g}(j)$ [equivalent to the l–y dependence of c and $\bar{c}(y)$] as a function of j [i.e. across the short dimension of the channel, see eqn. (29)]. The results of this process are shown in Fig. 9. Whilst the agreement is reasonable, it is certainly not quantitative, the discrepancy between the two curves arising, of course, as a consequence of the non-uniformity of the electrode. However, notice that, as the electrode surface is approached ($j \rightarrow 0$), it is apparent that

$$g(j,\,K) \;\rightarrow\; \bar{g}(j) \tag{51}$$

and

$$\frac{\partial g}{\partial y} \;\rightarrow\; \frac{\bar{g}}{y} \tag{52}$$

The same was found to be true for the species involved in DISP1 or ECE processes [61]. Figure 10 compares the quantities $g(j,\,K)$ and $\bar{g}(j)$ as a function of j for the intermediate formed in the (first-order) chemical step (rate constant k) of an ECE process (see Sect. 4.1), with normalised rate constant $K = 1.0$, where

$$K \quad = \quad k\left(\frac{h^2 l^2}{4 v_0^2 D}\right)^{1/3} \tag{53}$$

Figure 11 shows the same variation, but for the initial electroactive species

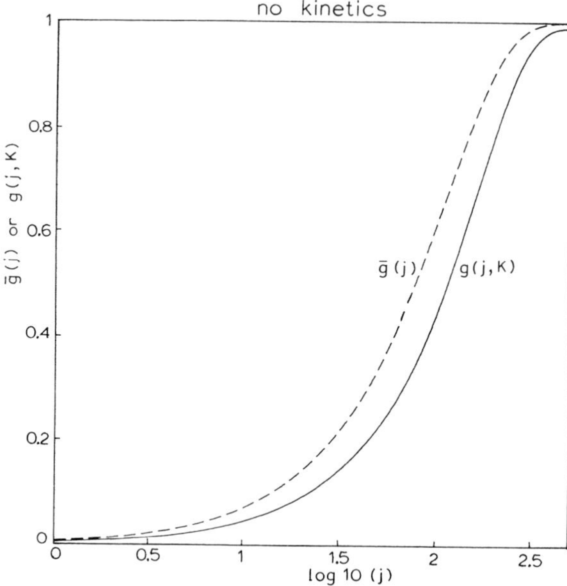

Fig. 9. The variation of the quantities g(j, K) and ḡ(j) with j for the species Red undergoing a simple kinetically uncomplicated electrode reaction Red $\rightleftharpoons$ Ox + e^-. The parameters used in the generation of the data are as in Fig. 8 (see text).

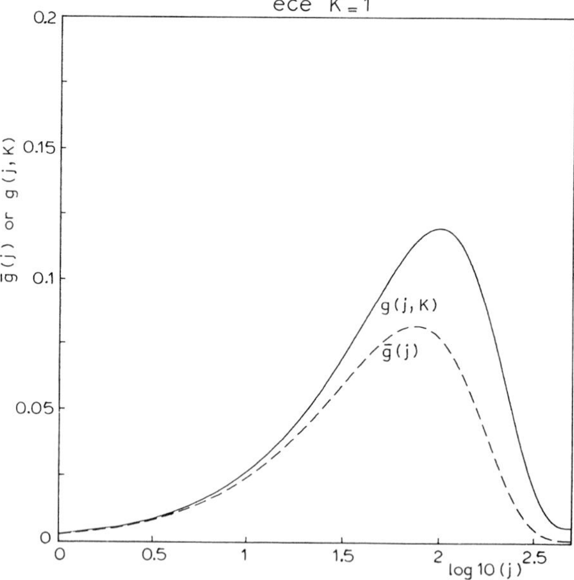

Fig. 10. The variation of the quantities g(j, K) and ḡ(j) with j for the product of the chemical step formed within an ECE pathway with K = 1.

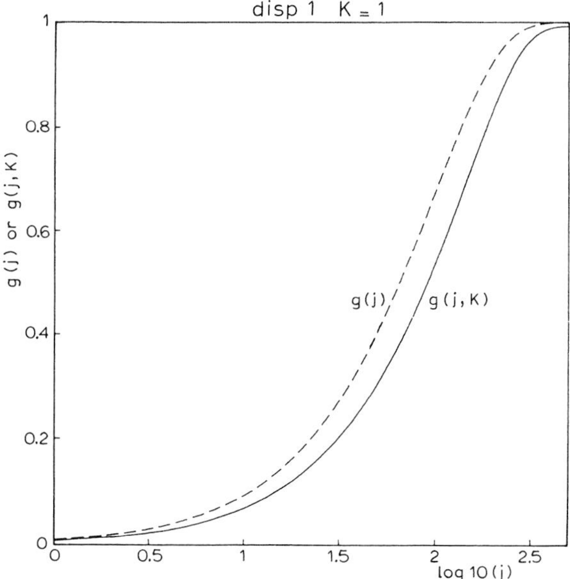

Fig. 11. The variation of the quantities g(j, K) and $\bar{g}(j)$ with j for an electroactive species undergoing reduction via the DISP1 pathway with $K = 1$.

involved in a DISP1 process (see Sect. 4.1 for definition), characterised by K = 1.0. (The parameters used to generate Figs. 10 and 11 were: $2h = 0.04\,\text{cm}$, $l = 0.4\,\text{cm}$, $D = 10^{-5}\,\text{cm}^2\,\text{s}^{-1}$, $w = 0.6\,\text{cm}$, and $v_0 = 0.0625\,\text{cm}\,\text{s}^{-1}$). Equations (51) and (52) were seen to be reasonable approximations as $j \to 0$ for a wide variety of rate constants describing the kinetic step within both ECE and DISP1 mechanisms [61]. Therefore, provided that the interest is in calculating variables which depend on concentrations (or their gradients) in the near vicinity of the electrode, the Singh and Dutt approximation is valid. Conversely, in applications where the complete concentration profile might be important, such as in spectroelectrochemistry, less satisfactory employment might be anticipated.

3. Voltammetry and related experiments at channel electrodes

One of the advantages of hydrodynamic electrodes over those employed in stationary solution is that both steady-state and time-dependent measurements can be made. Whilst, for the majority of applications, channel electrodes are operated under steady-state conditions, switching to the "time-dependent" mode is often useful. Firstly, the sensitivity is enhanced (this is of particular benefit in analytical applications) and secondly, the concentration of species adsorbed or deposited on the electrode can be controlled. In systems in which the electrode is prone to fouling, for example, the time required to make the measurement can therefore be reduced. This section discusses voltammetry at the channel electrode both under steady-state and in various time-dependent configurations.

3.1 STEADY-STATE VOLTAMMETRY

In the course of their pioneering studies on tubular electrodes, Blaedel and Klatt analysed the form of the current–voltage curves for reversible [65], quasi-reversible [66], and irreversible [66] processes. The theoretical treatment was at the level of the Levich approximation, proceeding exactly as described in Sect. 2.1, with the suitable modification of the boundary conditions appropriate to the case in question. The boundary conditions relevant to these three cases are delineated in Table 2.

For a reversible process, Blaedel and Klatt deduced the current–voltage relationship as

$$E = E^{\ominus} - \frac{RT}{nF}\ln\left(\frac{D_{\mathrm{Ox}}}{D_{\mathrm{Red}}}\right)^{2/3} + \frac{RT}{nF}\ln\left(\frac{I_{\mathrm{Lim}}}{I} - 1\right) \tag{54}$$

where $E^{\ominus}$ is the formal electrode potential of the couple in question and I is the current.

For the quasi-reversible case, where the forward and back rate constants are given by

$$k_f = k_s \exp\left[(-\alpha n_a F/RT)(E - E^{\ominus})\right] \tag{55}$$

$$k_b = k_s \exp\left[\{(1 - \alpha) n_a F/RT\}(E - E^{\ominus})\right] \tag{56}$$

where α is the transfer coefficient, n_a the number of electrons preceding (and including) the rate-determining step of the electrode reaction, and k_s the standard formal rate constant, the resulting current is conveniently written as

$$\frac{I}{I_{\mathrm{rev}}} = 1 - 2u + 2u^2\ln(1 - u^{-1}) \tag{57}$$

where

$$u = \frac{0.67 D_{\mathrm{Red}}^{2/3}(v_0/hl)^{1/3}}{k_b + (D_{\mathrm{Red}}/D_{0x})^{2/3}k_f} \tag{58}$$

and I_{rev} is the current that would flow at the same electrode if the charge transfer was reversible. Evidently, I/I_{rev} increases monotonically from zero to unity as the term $[k_b + (D_{\mathrm{Red}}/D_{0x})^{2/3}k_f]$ becomes increasingly large. The current–voltage relationship may be determined by consideration of eqns. (55)–(58), but the transcendental nature of eqns. (57) and (58) dictates the use of a computational method.

Extension of the quasi-reversible to the irreversible case is facilitated by simply neglecting the reverse reaction. The half-wave potential for a catho-

TABLE 2

Boundary conditions for reversible, quasi-reversible, and irreversible electrode reactions at channel electrodes

Reaction	Boundary $h - y \to \infty$	$h - y = 0$	
Reversible $Ox + n\,e^- \rightleftharpoons Red$			$\frac{[Ox]^0}{[Red]^0} = \exp\left\{\frac{nF}{RT}(E - E^{\ominus})\right\}$
Quasi-reversible $Ox + n\,e^- \underset{k_b}{\overset{k_f}{\rightleftharpoons}} Red$	$[Ox] \to [Ox]^\infty$; $[Red] \to [Red]^\infty$	$D_{Ox}\frac{\partial[Ox]}{\partial y} + D_{Red}\frac{\partial[Red]}{\partial y} = 0$	$D_{Ox}\frac{\partial[Ox]}{\partial y} = k_f[Ox]^0 - k_b[Red]^0$
Irreversible $Ox + n\,e^- \xrightarrow{k_f} Red$			$D_{Ox}\frac{\partial[Ox]}{\partial y} = k_f[Ox]^0$

Note: D_{Ox}, D_{Red} and $[Ox]^0$, $[Red]^0$ are, respectively, the diffusion coefficients and surface concentrations of Ox and Red.
The rate constants of the forward and back reactions in the quasi-reversible and irreversible cases are k_f and k_b, given by eqns. (55) and (56).

dic process is then given by

$$E_{1/2,\mathrm{irrev}} = E^{\ominus} + \frac{RT}{\alpha n_a F} \ln \left[\frac{k_s}{1.1 D_{0x}^{2/3} (v_0/hl)^{1/3}} \right] \tag{59}$$

The electrode geometry-dependence of the half-wave potential is in total contrast to the reversible case

$$E_{1/2,\mathrm{rev}} = E^{\ominus} + \frac{RT}{nF} \ln \left(\frac{D_{\mathrm{Red}}}{D_{0x}} \right)^{2/3} \tag{60}$$

Thus, reported values of $E_{1/2,\mathrm{irrev}}$ should be accompanied by data on flow rates, and electrode and channel dimensions.

For $0.1 \leq I/I_{\mathrm{LIM}} \leq 0.9$, numerical calculations [66] showed that the shape of irreversible waves is accurately represented by

$$E = E_{1/2,\mathrm{irrev}} + \frac{0.618}{\alpha n_a} \log_{10} \left[\frac{I_{\mathrm{LIM}}}{I} - 1 \right] \tag{61}$$

3.2 LINEAR SWEEP AND CYCLIC VOLTAMMETRY

With reference to reversible electrode reactions, we have tackled the question of how slow the rate of change in potential has to be for the resulting voltammogram to be indistinguishable from that under true steady-state conditions [67]. It was concluded that, in order to obtain both the correct half-wave potential (to within 1 mV) and Tafel slope, the scan rate, v_s, must be such that

$$\sigma = \left(\frac{nFv_s}{RT} \right) \left(\frac{l^2 h^2}{4 v_0^2 D} \right)^{1/3} < 0.1 \tag{62}$$

with the half-wave potential dependence on σ, for $\sigma < 1.0$, being deduced as

$$E_{1/2,\mathrm{rev}} = E^{\ominus} + \left(\frac{0.550RT}{nF} \right) \sigma \tag{63}$$

Equation (63) was confirmed experimentally, as shown in Fig. 12, by measurements on the well-characterized one-electron reversible reduction of the dye fluorescein in the presence of 0.1 M aqueous NaOH.

$^{-}$O ... O (F) $\underset{-e^-}{\overset{+e^-}{\rightleftharpoons}}$ $^{-}$O ... O^{-} ($S^{\bullet}$)

F — R — R — $S^{\bullet}$

$-R = $ $C_6H_4CO_2^-$

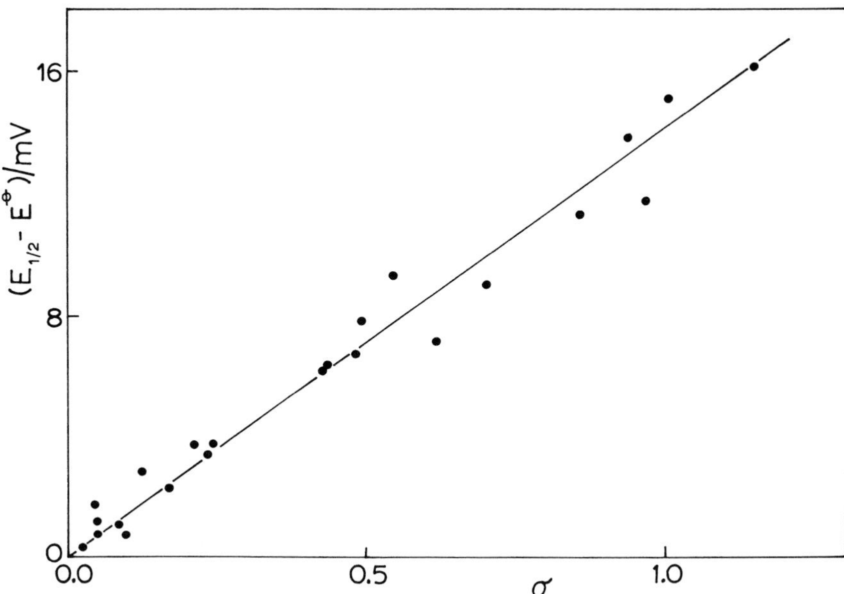

Fig. 12. The variation of experimentally determined half-wave potentials with the parameter σ. The straight line corresponds to the theoretically predicted behaviour [eqn. (63)]. The data are from ref. 67.

Aoki et al. [62] have recently extended the analysis of this problem, at the level of the Levich approximation, to include both the half-wave (or half-peak) potential and peak (or maximum) current (in terms of I_p/I_{LIM}) dependences on σ for $\sigma > 1.0$. The variations of these two quantities with σ are shown in Fig. 13 and 14. Good agreement between the analysis of Compton and Unwin [67] and Aoki et al. [62] is displayed (Fig. 13) for $\sigma < 1.0$.

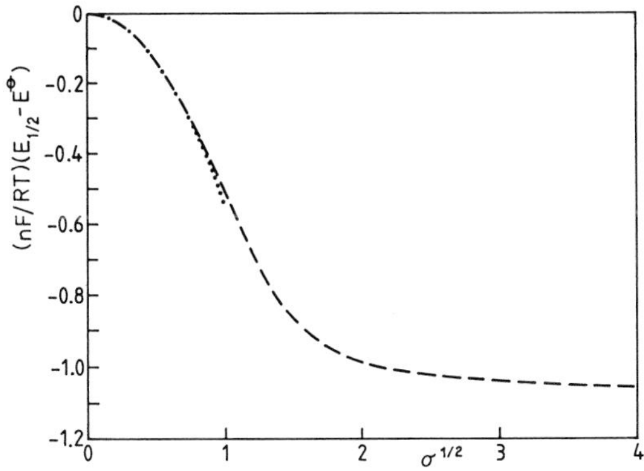

Fig. 13. The variation of half-wave (half-peak) potential with $\sigma^{1/2}$. ····, As determined by Compton and Unwin [67] for $\sigma < 1.0$ [eqn. (63)]; – – –, calculated by Aoki et al. [62] with no restriction on σ.

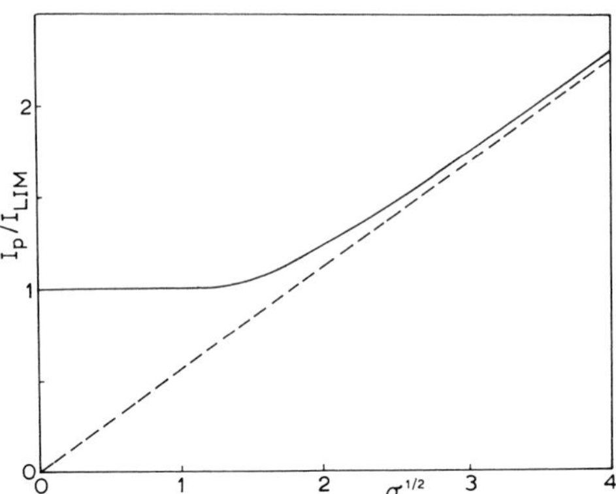

Fig. 14. The relationship between the peak (or maximum) current and $\sigma^{1/2}$. The broken line represents the behaviour in the absence of convection. Taken from ref. 62.

Singh and Dutt, using the approximation described in Sect. 2.3, have theoretically predicted, and experimentally verified, the behaviour when very fast scans are applied in both the linear sweep and cyclic voltammetric modes, for reversible [22, 23], quasi-reversible [25], and irreversible [24] electrode kinetics. Very attractive agreement with experiment is typically found, of which Fig. 15 is representative.

3.3 ALTERNATING CURRENT VOLTAMMETRY

Alternating current impedance measurements enjoy appreciable application in the study of a diverse range of electrochemical processes including

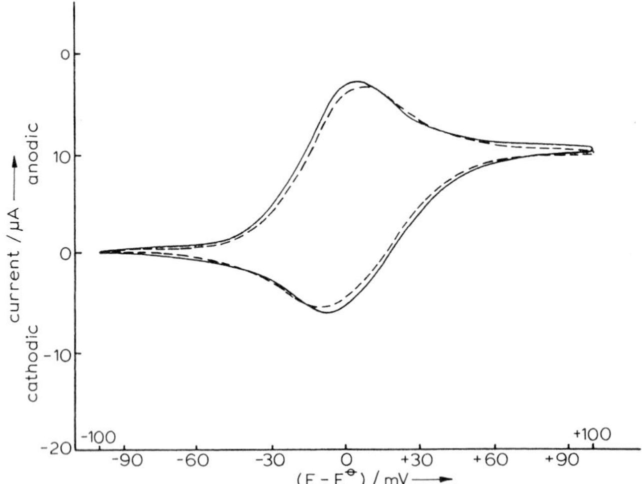

Fig. 15. Experimental (– – –) and calculated (——) current–potential curves for the oxidation of o-dianisidine at a tubular electrode, with $v_0 = 3.98\,\text{cm s}^{-1}$, $l = 1.0\,\text{cm}$, $a = 0.10\,\text{cm}$, and $v_s = 16.7\,\text{mV s}^{-1}$. Taken from ref. 24.

adsorption, charge transfer kinetics, and corrosion [68]. A theory describing the alternating current response to a modulation of the electrode potential, $E(t)$, about a value $E_{\text{d.c.}}$ at a frequency f(rad s^{-1}), i.e.

$$E(t) = E_{\text{d.c.}} + \Delta E \cos(ft) \tag{64}$$

has been developed for the channel electrode under the conditions of the Levich approximation [69]. Provided that the amplitude of the a.c. perturbation to the applied potential, ΔE, is such that $\Delta E \ll RT/nF$, the expression for the a.c. component of the current, $I_{\text{a.c.}}$, was found to have the form

$$I_{\text{a.c.}} = nFwl^{2/3}D^{2/3}\left(\frac{2v_0 d}{hf'}\right)^{1/3} \frac{c^{\infty} F \Delta E}{4\cosh^2(\theta_{\text{d.c.}}/2)RT} A\cos(f'\tau + \phi) \tag{65}$$

where A and ϕ are the amplitude and phase angle between current and voltage and

$$\theta_{\text{d.c.}} = \left(\frac{nF}{RT}\right)(E_{\text{d.c.}} - E^{\ominus}) \tag{66}$$

$$f' = f\left(\frac{l^2h^2}{4v_0^2D}\right)^{1/3} \tag{67}$$

$$\tau = \left(\frac{4v_0^2D}{l^2h^2}\right)^{1/3} t \tag{68}$$

Figure 16 shows the theoretically predicted influence of the normalised frequency, f', on the phase angle between current and voltage, ϕ, along with experimental verification [70] using the ferri–ferrocyanide reversible couple

$$\text{Fe(CN)}_6^{3-} \underset{-\text{e}}{\overset{+\text{e}}{\rightleftharpoons}} \text{Fe(CN)}_6^{4-}$$

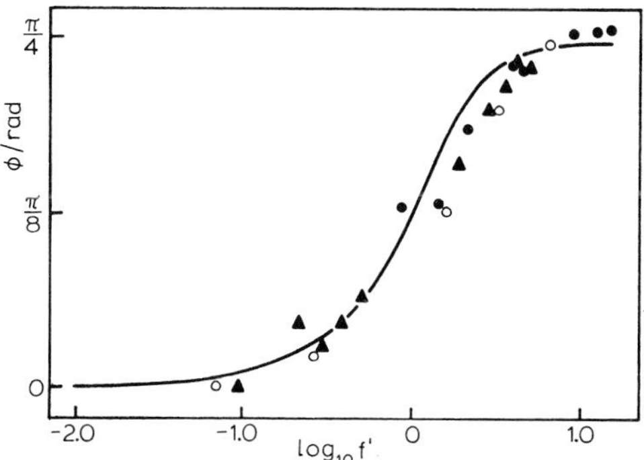

Fig. 16. The influence of f' on the phase angle ϕ between current and voltage. ——, Theoretical prediction [69] and experimental verification [70] at flow rates of ▲, 1.0×10^{-1}; ○, 4.8×10^{-2}; and ●, 1.1×10^{-2} $cm^3\ s^{-1}$.

Evidently, for sufficiently large values of f' ($f' > 10$), that is at high frequencies and/or low flow rates, the resulting a.c. current is phase-shifted through $\pi/4$ as predicted by the ordinary Warburg impedance expression applicable to quiescent solutions [71]. At lower frequencies ($f' < 10$), the effect of convection is to cause the phase shift to fall below this value. An implication of the above is that channel electrodes are ideal hydrodynamic electrodes with which a.c. impedance measurements can be made. The cell can be designed specifically with eqn. (67) in mind so that the condition $f' > 10$ holds over a wide range of frequencies, at convenient flow rates (see Sect. 5), thus retaining the usual advantages of hydrodynamic voltammetry (sensitivity and reproducibility) whilst convective effects can be neglected in the analysis of a.c. data.

3.4 CHRONOPOTENTIOMETRY

The measurement of the change in potential with time upon application of a constant current to an electrode in a solution containing electroactive material is a powerful technique for studying electrode kinetics [72]. Results are given in terms of the transition time, which is the time taken for the concentration of the electroactive species to fall to zero at the electrode surface. Whilst this technique is generally confined to stationary solutions, translating it to the channel electrode offers advantages in that convective replenishment of electroactive material at the electrode surface increases accuracy by suppressing irregular natural convection and lengthening the transition time.

Theory for chronopotentiometry of reversible channel electrode reactions under the Levich approximation has been presented by Aoki and Matsuda [73]. The transition time in flowing solution, τ_c, was found to be related to that in stationary solution, τ_0, via the approximations

$$\frac{\tau_c}{\tau_0} = 1 + 0.2928\left(\frac{I_{\mathrm{LIM}}}{I}\right)^3 + 0.1763\left(\frac{I_{\mathrm{LIM}}}{I}\right)^6 + 0.1273\left(\frac{I_{\mathrm{LIM}}}{I}\right)^9 \ldots$$

$$\text{for } 0 < \frac{I_{\mathrm{LIM}}}{I} < 0.75 \tag{69}$$

and

$$\frac{\tau_c}{\tau_0} = \left(\frac{I_{\mathrm{LIM}}}{I}\right)^{-2}\left[-0.5731\ln\left(1 - \frac{I_{\mathrm{LIM}}}{I}\right) - 0.1326\right]$$

$$\text{for } 0.70 < \frac{I_{\mathrm{LIM}}}{I} \leqslant 1 \tag{70}$$

3.5 CHRONOAMPEROMETRY

The application of a step in the electrode potential, from a region in which all species present are electro-inactive, to one in which the redox process of interest is diffusion-controlled, results in a current transient. In quiescent solution, this is described by the Cottrell equation [74].

$$I = \frac{nFAD^{1/2}c^{\infty}}{(\pi t)^{1/2}} \tag{71}$$

where A is the area of the electrode. Such measurements are occasionally useful at channel electrodes: firstly, the diffusion coefficient of the electroactive species can be evaluated without any prior knowledge of its concentration [75] (in contrast to steady-state measurements); secondly, switching to the transient mode can circumvent (to some extent) problems which may arise, due to film formation or adsorption, under steady-state conditions [76].

The most sophisticated treatment of this problem to date is that due to Aoki et al. [62], at the Levich approximation level, for reversible electron transfer. Figure 17 shows the resulting dimensionless chronoamperomogram, in terms of the normalised time, τ. The curve is identical to that obtained by Flanagan and Marcoux using the simple explicit numerical method [35]. At short times ($\tau < 0.24$), the behaviour is Cottrellian and the equation

$$\frac{I(\tau)}{I(\tau \to \infty)} = 2\Gamma(\tfrac{1}{3})(\pi\tau)^{-1/3}3^{-4/3} \tag{72}$$

describes the current–time behaviour to within 5%. This approximation is illustrated by the broken line in Fig. 17.

An alternative method is to solve eqn. (10) under Levich conditions, but invoking the Singh and Dutt approximation [61]. The boundary conditions relevant to the problem for the reaction

$$\text{Ox} + n\,\text{e}^- \rightleftharpoons \text{Red}$$

are shown in Table 3.

Defining α_n as the roots of the equation $\text{Ai}(\alpha_n) = 0$ [50], the chronoamperometric behaviour is readily shown to be

$$\frac{I(\tau)}{I(\tau \to \infty)} = 1 + \left[\frac{\text{Ai}(0)}{\text{Ai}'(0)} \sum_{n=1} \frac{\exp(\alpha_n \tau)}{\alpha_n}\right] \tag{73}$$

where

$$\text{Ai}(0) = 0.355028; \quad \text{Ai}'(0) = -0.258819 \tag{74}$$

The form of this equation is compared with that of Aoki et al. in fig. 18. Notice that, as the steady-state is approached, there is excellent agreement between the two methods but at short times ($\tau < 0.4$) the method utilising the Singh and Dutt approximation overestimates the quantity $I(\tau)/I(\tau \to \infty)$ by about 10%. Can we rationalise this? Implicit in the analysis of the problem via the Singh and Dutt approximation is that the current ratio $I(\tau)/I(\tau \to \infty)$ is defined with reference to "Singh and Dutt" steady-state currents, which have been shown to be lower than the corresponding cur-

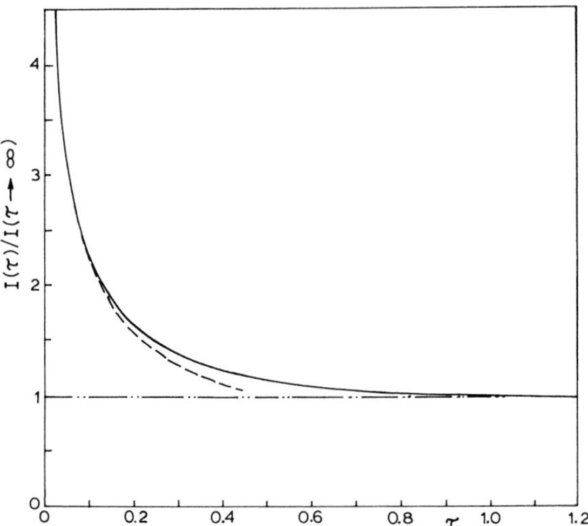

Fig. 17. Theoretical chronoamperometric response for a simple electron transfer process at a channel electrode to a potential step, from a region in which no current flows to one at which the current is mass transport-limited (under steady-state conditions) (——). Also shown is the Cottrellian behaviour described by eqn. (72) (– – –). Taken from ref. 62.

rents calculated under the Levich approximation by about 10% (see Sect. 2.3). What happens if we therefore define the calculated values of $I(\tau)$ with respect to "Levich" $I(\tau \to \infty)$? Figure 19 compares the outcome of this alternative definition with the rigorous treatment by Aoki. As expected, there is now excellent agreement between the two analyses for $\tau < 0.4$ but, as the steady-state is approached, the current ratio tends towards 0.902, the ratio of numerical factors when comparing the treatments of the steady-state problem via the Singh and Dutt and Levich approximations.

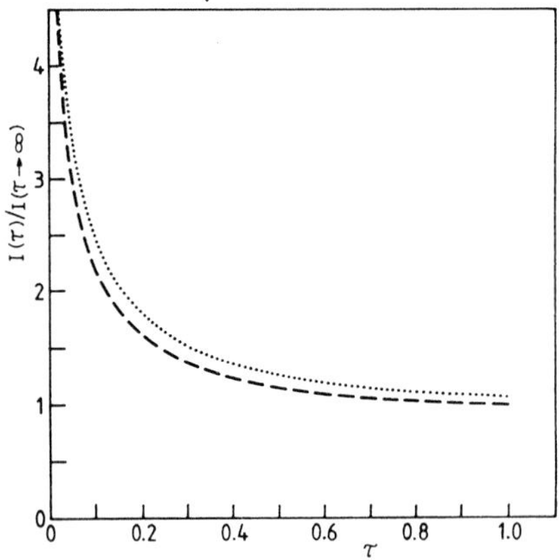

Fig. 18. Theoretical chronoamperomograms calculated using the Singh and Dutt approximation (····) [eqn. (73)] and compared with the rigorous theory of Aoki et al. [62] (– – –).

TABLE 3

Boundary conditions for the reaction Ox + $n\,e^- \rightleftharpoons$ Red in the region of the electrode upon application of a potential step

Boundaries		Conditions
t	$h - y$	
0	≤ 0	[Ox] = $[\text{Ox}]^\infty$
>0	$\rightarrow \infty$	[Ox] $\rightarrow$ $[\text{Ox}]^\infty$
>0	0	[Ox] = 0

The analysis of chronoamperometric data at hydrodynamic electrodes often involves recording the time, t_n, at which the current $I(t_n)$ is some fraction above the steady-state value:

$$I(t_n) \;=\; nI(t \;\rightarrow\; \infty) \tag{75}$$

In general, $n = 2$ [75]. What, therefore, are the implications of deconvoluting t_2 data using the "Singh and Dutt" chronoamperometric working curve shown in Fig. 19? It is apparent from this figure that the "Singh and Dutt" working curve may be used to analyse chronoamperometric behaviour down to $t_{1.2}$ with no loss in accuracy compared with analysis via the rigorous theory.

As to why the Singh and Dutt approximation works to such a high degree of accuracy (when definition is with respect to "Levich" steady-state cur-

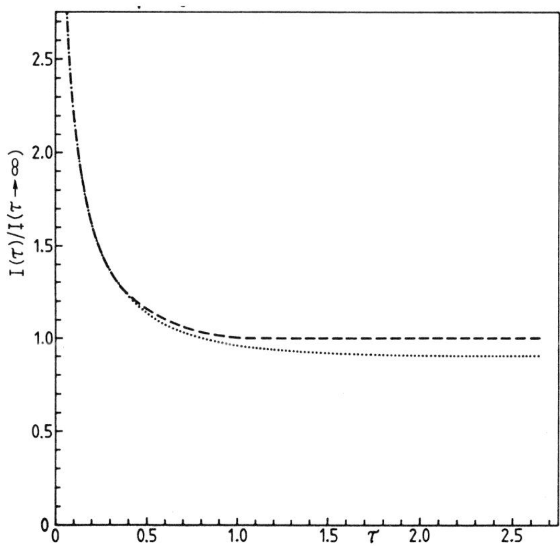

Fig. 19. Theoretical chronoamperomograms. ····, Calculated via the Singh and Dutt approximation, but referenced to "Levich" steady-state values; – – –, according to the theory of Aoki et al. [62].

rents) is answered by considering the current density at the electrode for these times. Aoki et al. [62] have calculated the current distribution at a channel electrode for the problem of chronoamperometry. The results are shown in the 3-D plot in Fig. 20. Whilst it is only at the shortest time that the electrode behaves as one with ideal uniform accessibility, true steady-state concentration profiles are not set up until $\tau \rightarrow 1.0$. The data in Fig. 20, along with those in Fig. 19, would appear to indicate that, for $\tau < 0.4$, most of the electrode is more accurately approximated as being uniformly accessible (implicit in the Singh and Dutt approximation) than under a steady-state regime.

3.6 ANODIC STRIPPING VOLTAMMETRY (ASV)

"Anodic stripping voltammetry is the most sensitive technique available for the determination of some metals in water samples" [77]. Trace metal ions, preconcentrated on to the electrode surface by reduction, are analysed by taking the electrode positive after a selected deposition time. The charge passed in the resulting current (stripping) peak gives a measure of the

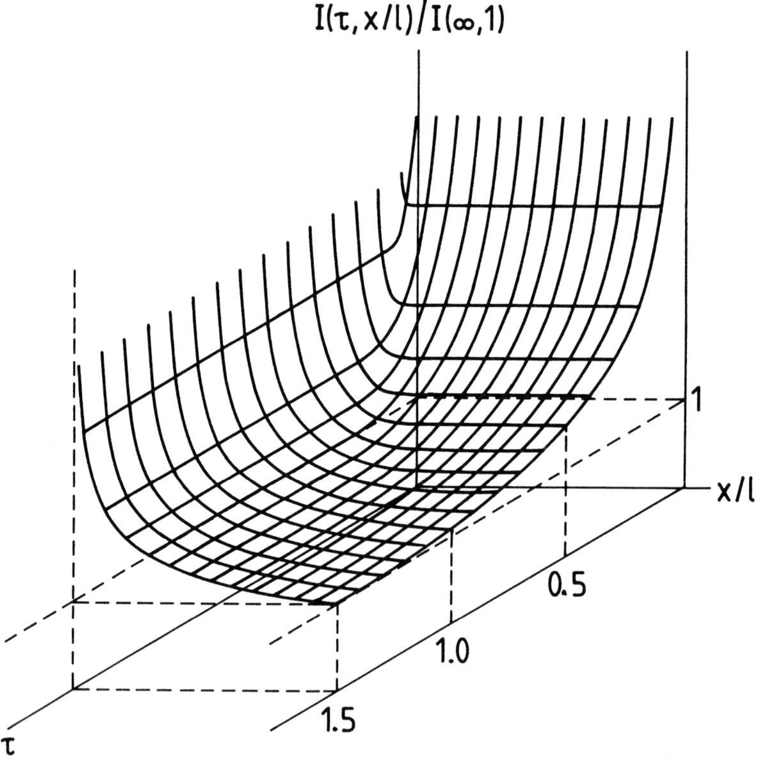

Fig. 20. Current distribution along the channel electrode as a function of the normalised time (following a potential step).

concentration of the metal in the original solution. The combination of ASV and a flow-through electrode [78–80] is thought to be more sensitive since larger currents flow during the plating process than in stationary solutions. The sensitivity is estimated to be nanomolar compared with the micromolar level that can be achieved by conventional channel electrode voltammetry. By using a pair of electrodes in series, the upstream electrode can serve to remove oxygen and other interferents before analysis downstream [77].

Johnson et al. [81] have developed a pulsed anodic technique at platinum flow-through electrodes and have applied it successfully to the detection of a number of classes of molecules such as alcohols and carbohydrates, amines, and sulphur compounds. The method has also been extended to the detection of analytes at potentials where they are not electroactive. This procedure has been used to detect Cl^- and CN^- [82, 83].

The technique employs a triple-step potential waveform, typically as illustrated schematically in Fig. 21(a), although others are possible [81]. The potential E_1 is such that the platinum electrode surface is rapidly oxidized, any contaminants being desorbed. The electrode is then stepped to a potential E_2, which is large and negative enough to reduce the oxidized surface and allow adsorption of the analyte. Stepping the potential to E_3 oxidises the species on the electrode surface. The analyte is detected in the chronoamperometric response of the electrode to this last step. Figure 21(b) illustrates schematically the chronoamperometric behaviour in the absence of adsor-

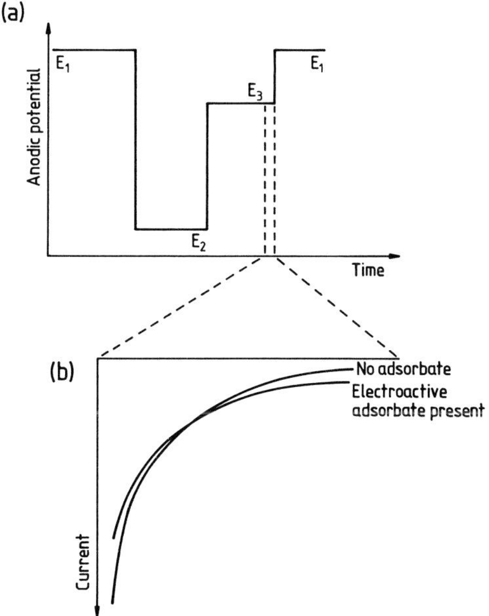

Fig. 21. Principles behind the pulsed anodic detection technique of Johnson et al. [81]. (a) Potential-step form. (b) Current–time behaviour of the electrode, at the potential E_3, with and without electroactive adsorbate.

bed analyte compared with that when the electroactive analyte is present. Evidently, the effect of adsorbed electroactive material is to block surface sites on the electrode, thereby suppressing the current due to the oxidation of the electrode at short times. At longer times, the analyte is desorbed, the (once-occupied) surface sites are oxidised, and a larger current flows.

The key to the characterization of a particular analyte is thus the extent to which it modifies the chronoamperometric response of the platinum electrode to an anodic potential step.

3.7 PULSED-FLOW VOLTAMMETRY (PFV)

For analysis, steady-state voltammetry at the channel electrode is limited to concentrations at or above the micromolar level. Whilst ASV provides an increase in sensitivity via pre-concentration of the analyte on the electrode, PFV achieves sub-micromolar concentrations through current measurements at two different flow rates. Subtracting the two signals gives a result free from the (transport-independent) surface processes which dominate the current at low concentrations. At the rotating-disc electrode, surface processes can be masked by superimposing a sinusoidal modulation on the rotation speed, and picking out the transport-controlled process with a lock-in amplifier [84]. Clearly, the superimposition of a sinusoidal modulation on the flow rate through a channel electrode is more difficult to realise, although work in this area has been reported [85–87]. Alternative, more convenient, arrangements involve stopped-flow [88] or pulsed-flow [89, 90] techniques. An inherent problem in PFV is the response time of 15–30 s in typical detectors, although some degree of improvement has resulted by employing very thin channel electrodes [90].

4. Electrode kinetics and coupled homogeneous reactions

In this section, we will provide a detailed analysis of the ECE mechanism (and its nuances) so as to examine the channel electrode, by way of practical example, under the "criteria for hydrodynamic electrodes" set out in the introduction. A brief survey of the theoretical analyses of other mechanisms will follow.

4.1 ECE AND RELATED MECHANISMS

The ECE and DISP mechanisms, which describe overall two-electron processes, are expressed by

Electrode

$$A \pm e^- \rightleftharpoons B \qquad \text{(i)}$$

Solution

$$B \xrightarrow{k_1} C \qquad \text{(ii)}$$

Electrode

$$\mathrm{C} \pm \mathrm{e}^- \rightleftharpoons \mathrm{F} \tag{iii}$$

Solution

$$\mathrm{B} + \mathrm{C} \xrightarrow{k_2} \mathrm{A} + \mathrm{F} \tag{iv}$$

Savéant and co-workers [91–93] have identified three limiting cases arising from steps (i)–(iv); ECE, DISP1, and DISP2. The first two steps are common to all the mechanisms, the difference between them being the fate of the intermediate, C. The second electron transfer may occur either at the electrode via reaction (iii) (ECE mechanism) or via disproportionation in the bulk solution through reaction (iv) (DISP mechanism). Two DISP mechanisms are possible; step (ii) rate determining (DISP1) or step (iv) rate determining (DISP2).

First consider the theory for an ECE process. In terms of the normalized variables χ, ξ, and K where

$$\chi = \frac{x}{l} \tag{76}$$

the relevant convective-diffusion equations under the Lévêque approximation, assuming equal diffusion coefficients for A, B, and C, are

$$\frac{\partial^2[\mathrm{A}]}{\partial \xi^2} = \xi \frac{\partial[\mathrm{A}]}{\partial \chi} \tag{77}$$

$$\frac{\partial^2[\mathrm{B}]}{\partial \xi^2} - K[\mathrm{B}] = \xi \frac{\partial[\mathrm{B}]}{\partial \chi} \tag{78}$$

$$\frac{\partial^2[\mathrm{C}]}{\partial \xi^2} + K[\mathrm{B}] = \xi \frac{\partial[\mathrm{C}]}{\partial \chi} \tag{79}$$

The boundary conditions appropriate to the problem when the electrode potential is such that the flux of A to the electrode is diffusion-limited, and all C arriving there is destroyed, are defined in Table 4.

We will show below how this problem is solved when the Singh and Dutt

TABLE 4

Boundary conditions for the ECE problem at the channel electrode [eqns. (77)–(79)]

Boundaries		Conditions		
$0 < \chi < 1;$	$\xi = 0$	$[\mathrm{A}] = 0;$	$\frac{\partial[\mathrm{B}]}{\partial \xi} = -\frac{\partial[\mathrm{A}]}{\partial \xi};$	$[\mathrm{C}] = 0$
$\chi = 0;$ $\chi > 0;$	$\xi > 0$ $\xi \to \infty$	$[\mathrm{A}] \to [\mathrm{A}]^\infty;$	$[\mathrm{B}] \to 0;$	$[\mathrm{C}] \to 0$

approximation is invoked. The method of analysis may be compared with more rigorous theory, at the level of the Levich approximation, by referring to ref. 63. Applying the Singh and Dutt approximation with respect to the species A, B, and c gives

$$\frac{\partial[\mathrm{A}]}{\partial\chi} \simeq [\mathrm{A}]^{\infty} - [\mathrm{A}] \quad (80)$$

$$\frac{\partial[\mathrm{B}]}{\partial\chi} \simeq [\mathrm{B}] \quad (81)$$

$$\frac{\partial[\mathrm{C}]}{\partial\chi} \simeq [\mathrm{C}] \quad (82)$$

so that eqns. (77)–(79) are modified to

$$0 = \frac{\partial^2([\mathrm{A}] - [\mathrm{A}]^{\infty})}{\partial\xi^2} \quad (83)$$

$$0 = \frac{\partial^2[\mathrm{B}]}{\partial\xi^2} - \xi[\mathrm{B}] - K[\mathrm{B}] \quad (84)$$

$$0 = \frac{\partial^2[\mathrm{C}]}{\partial\xi^2} - \xi[\mathrm{C}] + K[\mathrm{B}] \quad (85)$$

Equation (84) is readily solved to give

$$[\mathrm{B}] = [\mathrm{A}]^{\infty}\frac{\mathrm{Ai}'(0)\mathrm{Ai}(K+\xi)}{\mathrm{Ai}(0)\mathrm{Ai}'(K)} \quad (86)$$

By adding together eqns. (84) and (85), solving the resulting differential equation, and applying the boundary conditions (Table 4), it can be shown that

$$[\mathrm{B}] + [\mathrm{C}] = [\mathrm{A}]^{\infty}\frac{\mathrm{Ai}'(0)\mathrm{Ai}(K)\mathrm{Ai}(\xi)}{\mathrm{Ai}(0)\mathrm{Ai}'(K)\mathrm{Ai}(0)} \quad (87)$$

From eqns. (86) and (87), we deduce that

$$\left.\frac{\partial[\mathrm{C}]}{\partial\xi}\right|_{\xi=0} = [\mathrm{A}]^{\infty}\frac{\mathrm{Ai}'(0)}{\mathrm{Ai}(0)}\left[\frac{\mathrm{Ai}'(0)\mathrm{Ai}(K)}{\mathrm{Ai}(0)\mathrm{Ai}'(K)} - 1\right] \quad (88)$$

and the effective number of electrons transferred in the electrode reaction can be written as

$$n_{\mathrm{eff}} = 2 - \frac{\mathrm{Ai}'(0)\mathrm{Ai}(K)}{\mathrm{Ai}(0)\mathrm{Ai}'(K)} \quad (89)$$

The form of the n_{eff}–K relationship deduced via the above analysis is shown in Fig. 22, along with the behaviour predicted when eqns. (77)–(79) are solved at the Levich approximation level. Evidently there is excellent agreement between the two approaches, particularly for $K > 1.0$. Given that the

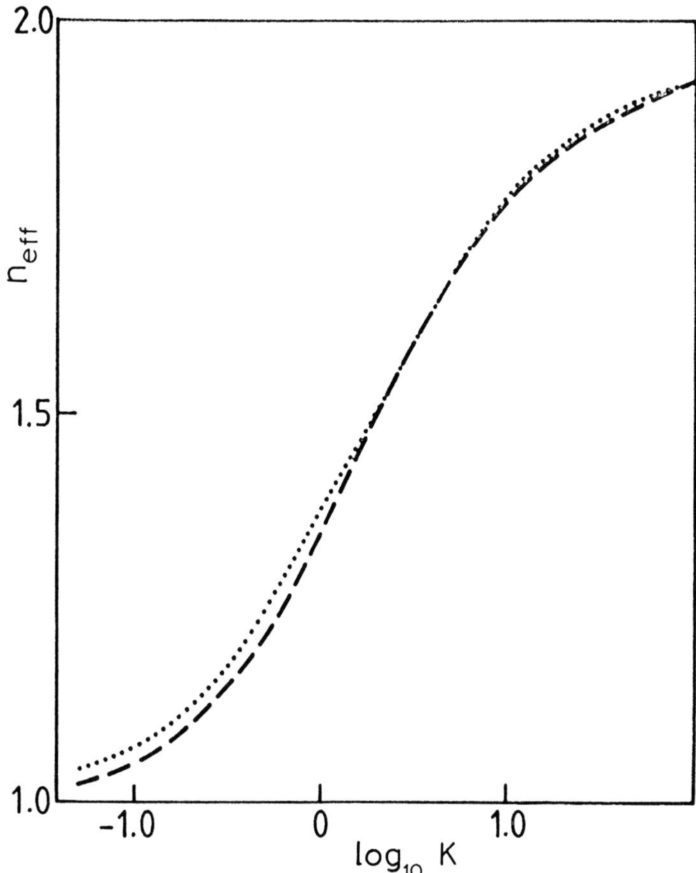

Fig. 22. The variation of n_{eff} with K for an ECE process calculated using the Singh and Dutt approximation (····) and via the "Levich" approach as described in ref. 63 (– – –).

"Levich" approach requires Laplace transformation of the partial differential equations (77)–(79) with subsequent (difficult) inversions into real space, the merits of the Singh and Dutt approximation are apparent.

We next consider the behaviour for DISP1 reactions at channel electrodes. The normalised steady-state convective-diffusion equations for this case, under the Levich approximation, are

$$\frac{\partial^2[A]}{\partial\xi^2} = \xi\frac{\partial[A]}{\partial\chi} - K[B] \tag{90}$$

$$\frac{\partial^2[B]}{\partial\xi^2} = \xi\frac{\partial[B]}{\partial\chi} + 2K[B] \tag{91}$$

$$\frac{\partial^2[F]}{\partial\xi^2} = \xi\frac{\partial[F]}{\partial\chi} - K[B] \tag{92}$$

Again, the species A, B, and F are assumed to have equal diffusion coefficients. The reader is referred to ref. 64 for coverage of the solution to the problem under the Levich approximation. Here, we seek a solution by invoking the Singh and Dutt approximation under which the modified convective-diffusion equations may be written as

$$\frac{\partial^2[\mathrm{A}]}{\partial\xi^2} = \xi[\mathrm{A}] - K[\mathrm{B}] \tag{93}$$

$$\frac{\partial^2[\mathrm{B}]}{\partial\xi^2} = \xi[\mathrm{B}] + 2K[\mathrm{B}] \tag{94}$$

$$\frac{\partial^2[\mathrm{F}]}{\partial\xi^2} = \xi[\mathrm{F}] - K[\mathrm{B}] \tag{95}$$

The boundary conditions appropriate to this problem are defined in Table 5. Equation (94) is readily solved

$$[\mathrm{B}] = \frac{\mathrm{Ai}(2K+\xi)}{\mathrm{Ai}'(2K)} \cdot \left.\frac{\partial[\mathrm{B}]}{\partial\xi}\right|_{\xi=0} \tag{96}$$

It may also be deduced that

$$[\mathrm{A}] - [\mathrm{F}] - [\mathrm{A}]^{\infty} = \frac{\mathrm{Ai}(\xi)}{\mathrm{Ai}'(0)} \cdot \left.\frac{\partial[\mathrm{A}]}{\partial\xi}\right|_{\xi=0} \tag{97}$$

Equations (96) and (97), together with the boundary conditions, can then be used to show that

$$-\left.\frac{\partial[\mathrm{A}]}{\partial\xi}\right|_{\xi=0} = 2[\mathrm{A}]^{\infty}\frac{\mathrm{Ai}'(0)}{\mathrm{Ai}(0)}\left[1 + \frac{\mathrm{Ai}'(0)\mathrm{Ai}(2K)}{\mathrm{Ai}(0)\mathrm{Ai}'(2K)}\right] \tag{98}$$

from which the effective number of electrons transferred in a DISP1 process is deduced to be

$$n_{\mathrm{eff}} = 2\left[1 + \frac{\mathrm{Ai}'(0)\mathrm{Ai}(2K)}{\mathrm{Ai}(0)\mathrm{Ai}'(2K)}\right]^{-1} \tag{99}$$

This function is shown in Fig. 23 along with the working curve previously obtained [64] under the "Levich" analysis. As with the ECE case, there is

TABLE 5

Boundary conditions for the DISP1 problem at the channel electrode [eqns. (90)–(92)]

Boundaries		Conditions		
$0 < \chi < 1$;	$\xi = 0$	$[\mathrm{A}] = 0$;	$\frac{\partial[\mathrm{A}]}{\partial\xi} = -\frac{\partial[\mathrm{B}]}{\partial\xi}$;	$\frac{\partial[\mathrm{F}]}{\partial\xi} = 0$
$\chi = 0$; $\chi > 0$;	$\xi > 0$ $\xi \to \infty$	$[\mathrm{A}] = [\mathrm{A}]^{\infty}$;	$[\mathrm{B}] = 0$;	$[\mathrm{F}] = 0$

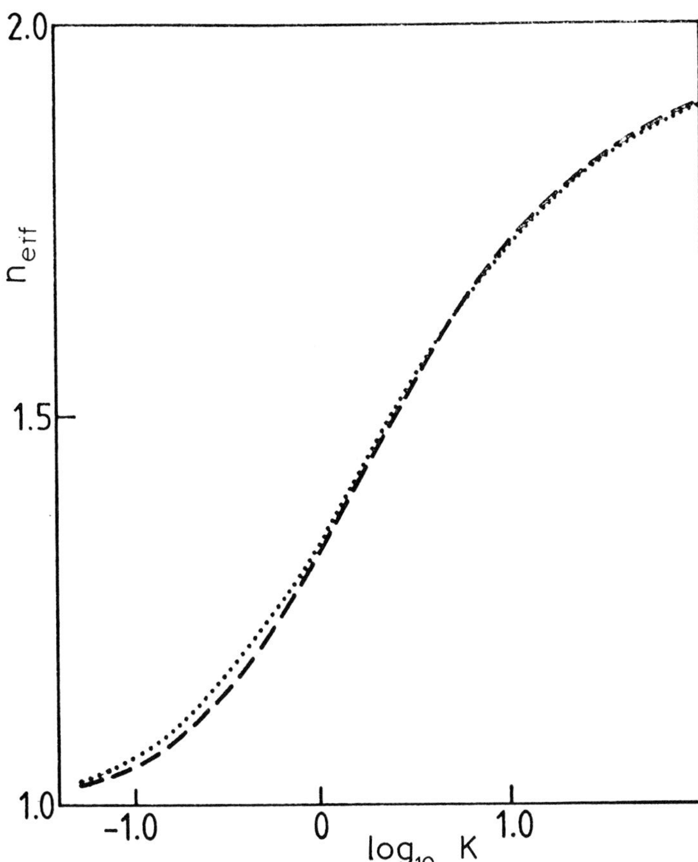

Fig. 23. The variation of n_{eff} with K for a DISP1 process, calculated (a) using the Singh and Dutt approximation (····) and (b) via the "Levich" approach as described in ref. 64 (– – –).

excellent agreement between the two methods, again vindicating the use of the Singh and Dutt approximation. Furthermore, the two "Singh and Dutt" working curves are accurate enough to allow discrimination between mechanisms. The question arises as to why the approximation is successful in this application. Obviously, the working curves resulting from the two theoretical approaches would be expected to be exact agreement in the limit of pure one- or pure two-electron transfer, since n_{eff} values are determined with reference to the kinetic-free steady-state limiting current behaviour calculated using the relevant theoretical method. However, the results of the two approaches coincide for $K \geq 2$ [and for $K < 2$ the error between the two sets of data (in Figs. 22 and 23) is no greater than 10%, for the reasons given in Sect. 2.3]. As to why the "Singh and Dutt" method works to within 1% (compared with rigorous analytical theory) when $K \geq 2$ is explained by considering Figs. 9–11, 24 and 25, which compare the quantities $g(j, K)$ and $\bar{g}(j)$, defined in Sect. 2.3, for the following cases under diffusion-controlled

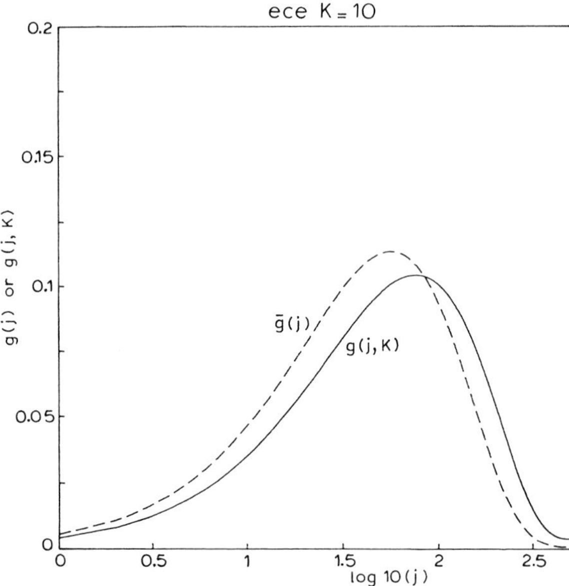

Fig. 24. The variation of the quantities g(j, K) and ḡ(j) with j for the species C formed within an ECE pathway with K = 10. The parameters used in the generation of the data are, again, as in Fig. 8.

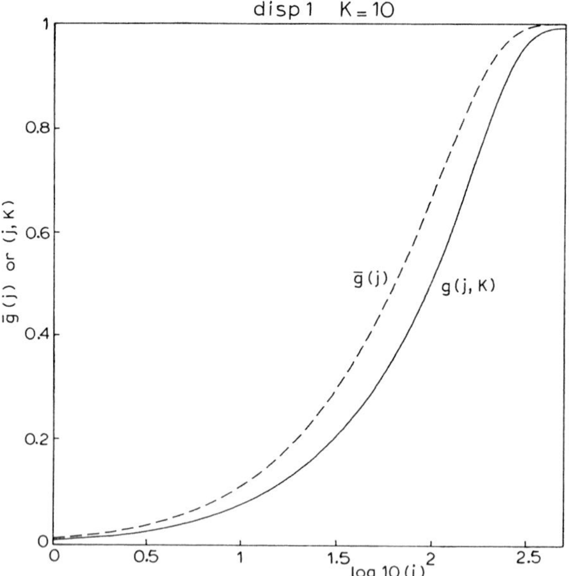

Fig. 25. The variation of the quantities g(j, K) and ḡ(j) with j for the species A undergoing reduction via the DISP1 pathway with K = 10. The generation of the data is as for Fig. 8.

electron-transfer

(i) an electroactive species undergoing heterogeneous electron transfer, uncomplicated by homogeneous chemical reactions (Fig. 9);

(ii) species C in an ECE process with $K = 1$ (Fig. 10) and $K = 10$ (Fig. 24);

(iii) species A in a DISP1 process with $K = 1$ (Fig. 11) and $K = 10$ (Fig. 25).

As mentioned previously (Sect. 2.3), interest is confined to the concentration profile close to the electrode surface. Notice that with this constraint, the difference between the concentration profiles calculated at the "Levich" and "Singh and Dutt" levels is about the same for both the species involved in DISP1 and ECE processes with $K = 10$ (Figs. 24 and 25) and the corresponding "reference" one-electron concentration profiles (Fig. 9). It would thus appear that the success of the Singh and Dutt approximation in this kinetic domain is due to self-compensating errors in the concentration gradients (at the electrode surface) calculated via the Singh and Dutt method for ECE/DISP1 and kinetic-free processes.

When the operating homogeneous kinetics are slower ($K < 2$), the Singh and Dutt approximation actually leads to a more accurate representation of the true concentration gradient at the downstream edge of the electrode (Figs. 10 and 11). However, since n_{eff} values are calculated with reference to kinetic-free steady-state currents which themselves are in error by 10%, the Singh and Dutt working curve deviates from that calculated using the Levich approach.

The preceding derivations of the n_{eff}–flow rate behaviour for ECE and DISP1 processes serve to illustrate that the theoretical description of electrode reaction mechanisms is readily achieved, fulfilling the second criterion of Sect. 1.

What of the other criteria? In the Introduction, it was suggested that steady-state voltammetry at the channel electrode could discriminate between similar mechanisms (such as ECE and DISP1 reactions) as a consequence of the electrodes' non-uniform accessibility. This leads to n_{eff} varying between 1 and 2 electrons over a wider convective (flow rate) range than at the uniformly accessible rotating-disc electrode, thus providing a larger "hydrodynamic window" in which to probe the electrode reaction mechanism. Figure 26 emphasises this point with respect to an ECE process. Figure 27 shows that it is easier to discriminate between the two mechanisms at a channel electrode, rather than a rotating-disc electrode, since a greater difference in the n_{eff}–convection behaviour, between the two mechanisms, emerges at the channel electrode.

Channel electrode steady-state voltammetry has been used to show that the reduction of fluorescein to leuco-fluorescein, in aqueous solution at around pH 9.5–10.0, proceeds via a DISP1 mechanism [64]

Electrode

$$\mathrm{F} \underset{-\mathrm{e}^-}{\overset{+\mathrm{e}^-}{\rightleftharpoons}} \mathrm{S}^{\cdot}$$

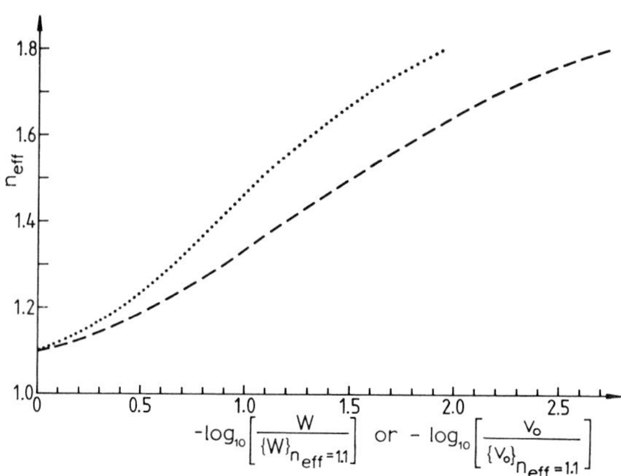

Fig. 26. Change in n_{eff} for an ECE process as a function of the change in rotation speed or solution velocity at the rotating disc (····) and channel (– – –) electrodes, respectively. The larger "hydrodynamic window" provided by the channel electrode is apparent.

Solution

$$\mathrm{S}^{\cdot} + \mathrm{H}^{+} \xrightarrow{k} \mathrm{SH}^{\cdot +}$$

$$\mathrm{SH}^{\cdot} + \mathrm{S}^{\cdot} \rightleftharpoons \mathrm{F} + \mathrm{L}$$

where L is

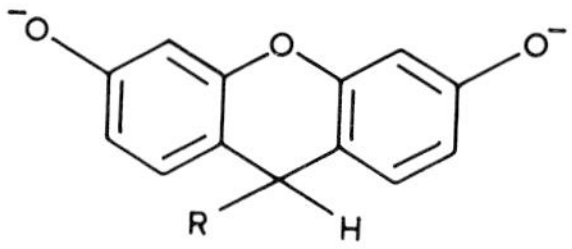

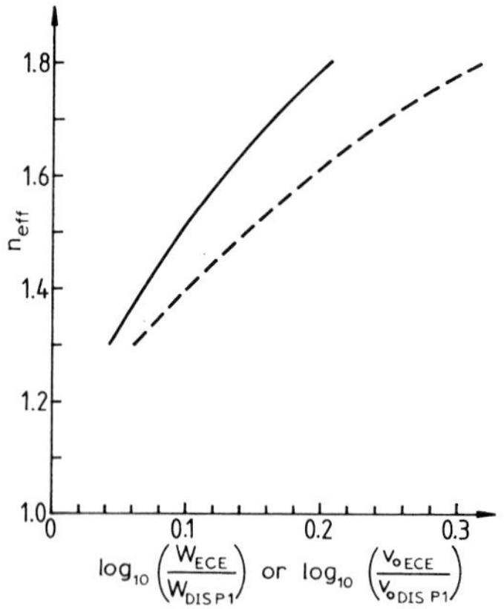

Fig. 27. The difference in rotation speed (——) or flow rate (– – –) for ECE and DISP1 mechanisms, as a function of n_{eff}.

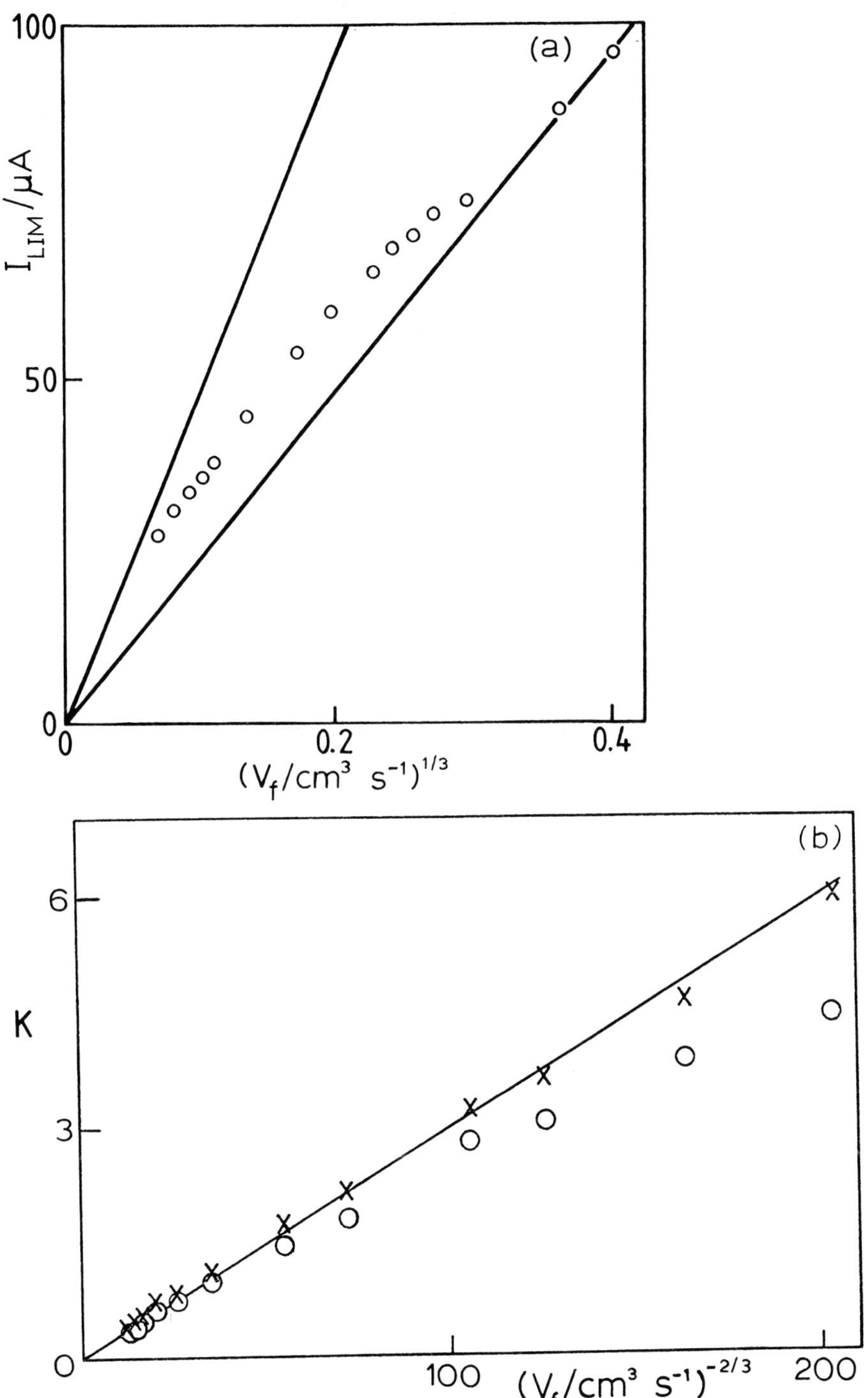

Fig. 28. (a) The transport-limited current–flow rate behaviour for fluorescein reduction. The solid lines represent the calculated one- and two-electron behaviour. (b) Analysis of current–flow rate data assuming either an ECE (○) or a DISP1(×) mechanism.

The experimental current–flow rate behaviour is presented in Fig. 28(a). Notice that the characteristic transition between a two- and a one-electron process with increasing flow rate, expected for an ECE or DISP1 process, is evident. When converted to K values via the working curves in Figs. 22 and 23 and plotted as a function of $V_f^{-2/3}$ [see eqn. (53)], the data in Fig. 28(a) results in a straight line on the basis of a DISP1 mechanism, whilst the assumption of an ECE mechanism produces a curve, as shown in Fig. 28(b). In this way, the channel electrode can be used to discriminate between ECE and DISP1 processes.

The final criterion with which hydrodynamic electrodes might be judged concerns the range of rate constants which might be studied for a given electrode reaction mechanism. In principle, both the range of (and the fastest) rate constant measurable under steady-state conditions at the channel electrode ought to be larger than that at other hydrodynamic electrodes, since the effective solution velocity over the electrode is controlled by two variables, i.e. the solution velocity itself and the length of the electrode (rather than just the one variable as in rotating-disc or microelectrode voltammetry).

As an illustration of this point, let us calculate the fastest rate constant we might measure at a channel electrode with reference to the chemical step in an ECE process. The maximum solution velocity which can be employed under laminar conditions is governed by eqn. (6), with the restrictions previously outlined (Sect. 2). For a conventional channel electrode, with $h = 0.1\,\mathrm{mm}$, and a solution with a kinematic viscosity of $0.01\,\mathrm{cm^2\,s^{-1}}$ (e.g. H_2O at 25°C) mean solution velocities up to $0.667\,\mathrm{m\,s^{-1}}$ may be used. Under these conditions, we can then calculate the minimum electrode length [via eqn. (35)] so that edge effects are negligible ($\leq 1\%$). Aoki et al. [49] have shown that an edge effect of 1% corresponds to $p = 0.0013$. Hence, assuming a typical diffusion coefficient of $10^{-5}\,\mathrm{cm^2\,s^{-1}}$ and with the other parameters as above, an electrode length of $3.37 \times 10^{-3}\,\mathrm{cm}$ can be used without any serious consequence in terms of edge effects.

If 1.9 is deemed to be the maximum n_{eff} that can be measured accurately at hydrodynamic electrodes then, for an ECE process, normalised rate constants, K, up to 54 are attainable [Fig. (22)]. Hence, via eqn. (53), the maximum first-order rate constant we might hope to measure, with the parameter constraints outlined above, is $3.9 \times 10^4\,\mathrm{s^{-1}}$. Measurement of a homogeneous rate constant of the same magnitude for an ECE process under exactly the same conditions at the rotating-disc electrode would require a rotation speed of $1.2 \times 10^4\,\mathrm{rad\,s^{-1}}$ [94]. We are not aware of any rotating-disc electrode systems which can rotate at such speeds [95]. In practice, the maximum rotation speed employed is $10^3\,\mathrm{rad\,s^{-1}}$ [96]. Microdisc electrodes of radius $1.4\,\mu\mathrm{m}$ would be necessary if rate constants of this magnitude were to be measured via steady-state voltammetry in stationary solution [19]. Whilst electrodes of this size can be constructed [97], we are again approaching the point at which this becomes difficult [98].

The above examples serve to illustrate the suitability of the channel electrode to the measurement of fast homogeneous chemical reactions. As a further point of interest, note that we confined our channel electrode example, above, to a conventional cell (of channel half-height 0.1 mm). Of course, there is nothing to prevent the diminution of this parameter. This would have the effect of increasing the maximum solution velocity attainable under a laminar regime [eqn. (6)]. A "knock-on" effect is that the electrode length (under conditions of negligible edge effect) could then be reduced. It should then be apparent that the 2/3 and −2/3 dependence of measurable rate constant on flow rate and electrode length, respectively, ensures that there is a dramatic effect on the maximum rate constant which can be measured via this technique.

The consequence of the above discussion in terms of electrode design and construction presents no problems. The pumping systems necessary for the attainment of fast flow rates have been designed [99] and short electrodes are easily fabricated (see Sect. 5.1).

4.2 EC REACTIONS

This reaction sequence, as mentioned in the Introduction, is expressed by

E

$$A + ne^- \rightleftharpoons B \tag{v}$$

C

$$B \xrightarrow{k} \text{products} \tag{vi}$$

The case where a first-order chemical reaction follows reversible electron transfer has been considered using the Levich analysis [100]. Results, in

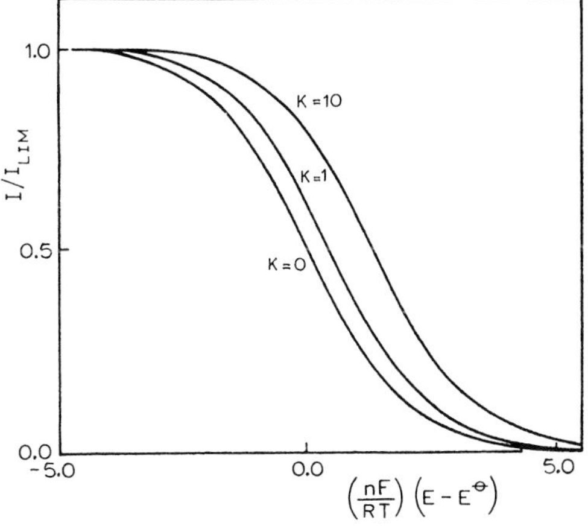

Fig. 29. Voltammetric waves for an EC reduction process at a channel electrode calculated at the "Levich" approximation level for $K = 0$, 1, and 10.

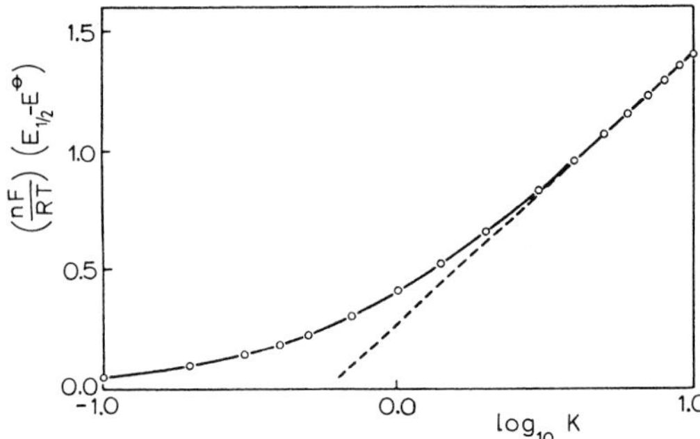

Fig. 30. Variation of $E_{1/2}$ with $\log_{10}K$ for an EC cathodic process. The broken line represents eqn. (100), which is valid for large K.

terms of the normalised rate constant K [defined by eqn. (53)], are shown in Fig. 29. As predicted in the Introduction, the influence of the chemical step is to shift the reduction half-wave potential anodically. For fast following kinetics ($K > 4$), this behaviour is represented by

$$\frac{nF}{RT}(E_{1/2} - E^{\ominus}) = 0.27 + 0.50 \ln K \tag{100}$$

where $E^{\ominus}$ is the standard electrode potential which, in the limit of small k, becomes the half-wave potential. The $E_{1/2}$–K behaviour is illustrated graphically in Fig. 30. Notice that, if the electrode geometry can be "tuned" so that K lies between 0.1 and 4, values of the rate constant, k, can be deduced by fitting experimental data to the working curve in Fig. 30. For $K > 4$, k can only be evaluated, via eqn. (100), if there is an independent knowledge of $E^{\ominus}$.

4.3 CE REACTIONS

This mechanism was introduced earlier in the chapter and is defined by

C

$$\mathrm{Y} \underset{k_{-1}}{\overset{k_1}{\rightleftharpoons}} \mathrm{A} \tag{vii}$$

$$K_{\text{eqm}} = \frac{k_{-1}}{k_1} = \frac{[\mathrm{Y}]^{\infty}}{[\mathrm{A}]^{\infty}}$$

E

$$\mathrm{A} + n\mathrm{e}^{-} \longrightarrow \mathrm{B} \tag{viii}$$

The effect of the preceding chemical reaction is to reduce the concentration of the electroactive species A (to less than expected for a simple E reaction). The observed transport-limited current, I_{k}, is therefore less than the diffusion-limited current, I_{d}, which would be observed if $K_{\text{eqm}} = 0$ (the A/Y equilibrium lying entirely in favour of A).

At the channel electrode, this problem has been treated by Matsuda, initially employing the reaction-layer approximation [101], but subsequently the full coupled convective-diffusion equations were solved [102] (at the level of the Lévêque approximation). It was shown that, to within 1%, the equation

$$\frac{I_k}{I_d} = \frac{1.290\Lambda_{CE}}{\tanh(1.290\Lambda_{CE}K_{eqm}) + 1.290\Lambda_{CE}} \quad (101)$$

held where the dimensionless parameter Λ_{CE} is given by

$$\Lambda_{CE} = \left(\frac{hl}{v_0}\right)^{1/3} [k_1(1 + K_{eqm})]^{1/2} K_{eqm}^{-1} D^{-1/6} \quad (102)$$

4.4 EC′ REACTIONS

This scheme is represented by

Electrode

$$A + ne^- \rightleftharpoons B \quad (ix)$$

Solution

$$B + Z \xrightarrow{k} A \quad (x)$$

Under conditions where the conversion of B to A in solution is pseudo-1st-order, approximate solutions, valid for either extremely large or small k, were obtained by Klatt and Blaedel [103]. A more complete treatment, valid for all values of k, was given by Matsuda and co-workers [104]. The analysis showed that the catalytic reaction caused the diffusion-limited current in the presence of the catalytic cycle, I_{cat}, to be increased over the current that would be observed in the absence of the cycle ($k = 0$), I_d, by an amount depending on the parameter

$$\Lambda_{cat} = \left(\frac{k}{D^{1/3}}\right)^{1/2} \left(\frac{hl}{2v_0}\right)^{1/3} \quad (103)$$

It was found that, for $\Lambda_{cat} \leqslant 1.25$

$$\frac{I_{cat}}{I_d} = 1 + 0.55274\Lambda_{cat}^2 - 0.06969\Lambda_{cat}^4 + 0.01217\Lambda_{cat}^6 \quad (104)$$

while for $\Lambda_{cat} > 1.25$

$$\frac{I_{cat}}{I_d} = 1.23832\Lambda_{cat} + 0.30958\Lambda_{cat}^{-2} \quad (105)$$

The theory was verified experimentally using the Fe^{2+}/H_2O_2 system and was shown to be satisfactory over a wide range of electrolyte flow rates.

5. Design and fabrication of channel electrode cells and the associated flow system

5.1 CHANNEL ELECTRODE CELLS

We first consider the construction of cells designed to obey the Levich equation [eqn. (27)]. This dictates that the flow conforms to a laminar regime, which firstly implies that obstructions and rough edges have been eliminated from the path of the solution, and secondly places an upper limit on the solution velocity in accordance with eqn. (6) such that the Reynolds number $\leq 2 \times 10^3$.

Channel electrode cells are generally designed to be demountable, comprising a channel unit and a coverplate which bears the working electrode. Two typical designs are shown in Fig. 31. Figure 31(a) shows a cell which is constructed in perspex, the channel unit and coverplate being mated through the application of mechanical pressure [70, 104, 105], a thin silicone rubber layer between the channel unit and the coverplate acting as a seal; the cell Fig. in 31(b) is assembled by cementing the (silica) channel unit and coverplate together. The cell depicted in Fig. 31(b) has found application in in-situ electrochemical ESR studies [106].

The working electrode, consisting of a thin metal foil ($< 1\,\mu$m thick) is usually glued to the coverplace. We have found that, for most of the common electrode materials (Pt, Au, Cu, Ag), rubber-based glues are optimal for this if aqueous electrolyte solutions are used. Photocatalysed acrylic adhesive is suitable for attaching most metal foils to silica [31]. Alternatively, the metal foil can be cast in an epoxy resin coverplate. This ensures that, after polishing, there are no ridges between the edges of the electrode and the rest of the coverplate. This last method of fabrication is especially appropriate when very short electrodes (< 0.5 mm) are to be employed. Electrical contact to the electrode is made either via a hole in the back of the coverplate [Fig. 31(b)] or by extending the foil beyond the edge of the plate, as shown in Fig. 31(a).

Once the cell is assembled, the electrode should be located sufficiently downstream so that the entry length requirement for Poiseuille flow [eqn. (8)] is fulfilled. It is also essential that the electrode is positioned away from the edges of the channel (i.e. $w < d$) since, in this region, v_x deviates from the expression given in eqn. (7). Such edge effects are only significant over distances $\sim h$. Electrodes constructed as in Fig. 31(a) are thus insulated from the solution in the vicinity of the channel edge, for example, by "blocking out" the appropriate area of foil with polystyrene solution.

The associated counter and reference electrodes can either be included in the cell body or plumbed into the flow system separately. The cell designed by Meyer et al. [105] and used by Aoki et al. [104] incorporates both a platinum counter electrode and a silver pseudo-reference electrode within the channel unit. The cell shown in Fig. 31(a) contains a silver–silver chloride reference electrode in one of the ducts at the end of the channel. For a.c. impedance measurements, where ohmic drop must be minimised, this is

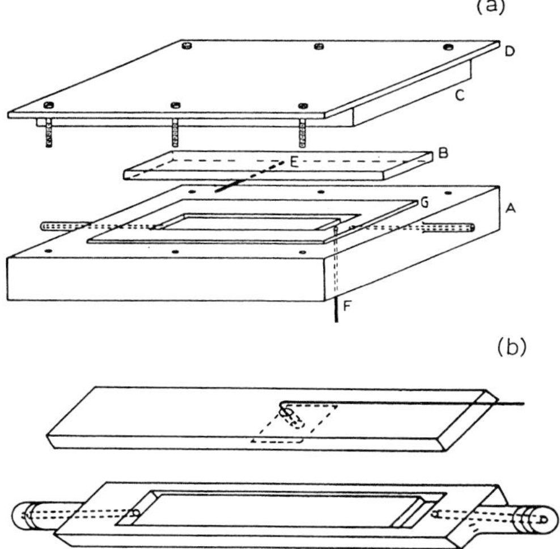

Fig. 31. (a) Perspex channel electrode cell. A, Channel unit; B, cover plate; C, rubber block; D, metal plate; E, working electrode; F, reference electrode; G, silicone rubber gasket. From ref. 70. (b) Silica channel electrode cell (unassembled) showing the cover plate with electrode and lead-out wire and the channel unit.

replaced by a silver foil located on the coverplate about 0.5 mm upstream of the working electrode [70]. The counter electrode used in conjunction with this cell consists of a mesh gauze of platinum wire coiled into a spiral and enclosed in a glass tube downstream of the cell. Counter electrode products are therefore flushed away, and cannot interfere with the channel electrode reaction.

The tubular electrode (the close relative of the channel electrode) can be either integral or demountable [11, 107–110]. Integral tubular electrodes are readily construced [11, 107] by pushing a machined annular electrode over a forming rod, either together with connection tubes which are then cemented to the electrode, or by casting epoxy resin round the forming rod, removing the former and polishing to remove any ridges in the tube [108]. Electrodes of this type, whilst readily fabricated, suffer from the disadvantage that the electrode surface is not visible and so its smoothness and cleanliness can only be ascertained indirectly. Accordingly, demountable tubular electrodes have been designed [109, 110] as shown in Fig. 32. These may, however, be more prone to (local) turbulance, particularly where the individual sections are bolted together. Newly designed and constructed flow cells should be tested against the Levich equation [eqn. (27)] over a wide range of flow rates using a "standard" electroactive couple (e.g. ferri-ferrocyanide).

One of the advantages of flow-through electrodes over other types of hydrodynamic electrodes is that they involve no moving parts. They are therefore ideal for use in conjunction with both spectroscopic and microscopic techniques. Whilst both channel and tubular electrodes can be used in

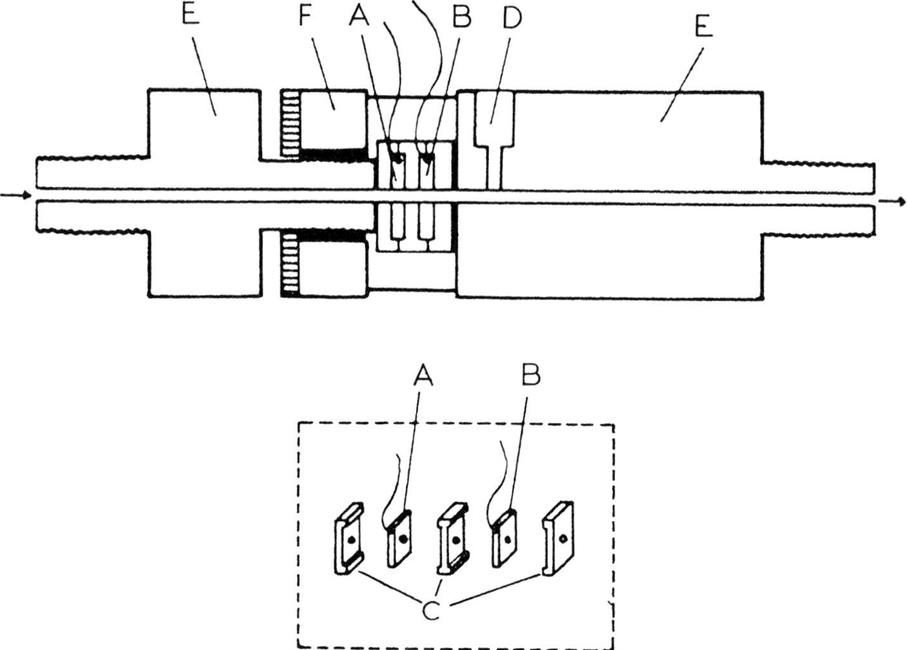

Fig. 32. Demountable tubular electrode design. A, Working electrode 1; B, working electrode 2; C, teflon spacers; D, reference electrode; E, teflon cell body; F, brass thread. From ref. 109.

conjunction with ESR spectroscopy, techniques such as light microscopy can only be employed in-situ with the channel electrode configuration. This is facilitated by incorporating a "window" in the channel unit in the region of the electrode [111–113].

5.2 FLOW SYSTEMS

Flow systems can be designed to have the solution either recirculating or flowing to waste once it has passed through the channel electrode cell. Matsuda and co-workers use a recirculating flow system [104] shown schematically in Fig. 33. The pipes and valves are constructed in glass and the connections between them are made with silicone rubber tubing. The thermostatted solution reservoir has a capacity of about 3 l. Solution is propelled through the system with a pump, the flow velocity being controlled by regulating the valves V1 and V2 and the pump's motor voltage. In this way, solution volume flow rates up to $2.7\,\mathrm{cm^3\,s^{-1}}$ [104] can be obtained.

Whilst pumping is the preferred method if very large flow rates are required ($> 5 \times 10^{-1}\,\mathrm{cm^3\,s^{-1}}$) the resulting flow pulsation can be a problem. This can be damped to some extent by incorporating a length of thin tubing and a hollow partially filled ball between the pump and the cell [3]. Regular flow can be achieved using a gravity feed system, shown schematically in Fig. 34. It consists of a glass reservoir and several metres of PTFE tubing,

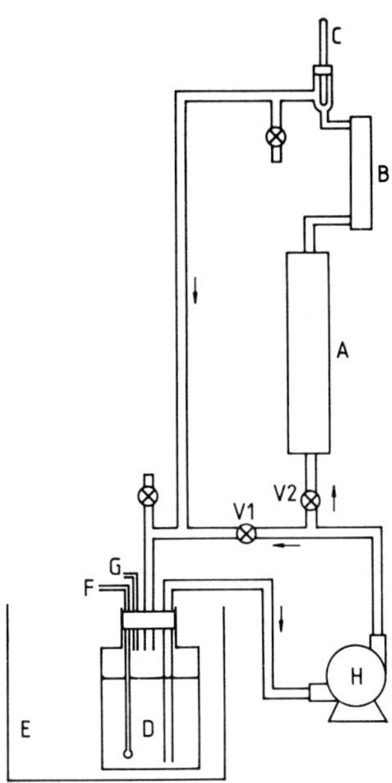

Fig. 33. Schematic representation of pumped-recirculating flow system employed by Matsuda and co-workers (from ref. 104). A, Channel electrode cell; B, flow meter; C, thermometer; D, solution reservoir; E, thermostat; F, degas inlet; G, degas outlet; H, pump.

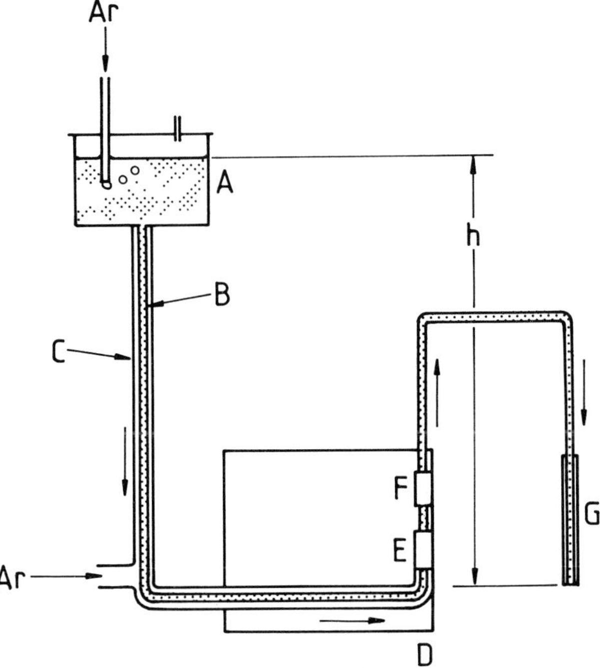

Fig. 34. Schematic representation of a gravity-fed flow system. A, Solution reservoir; B, PTFE tubing; C, PVC argon jacket; D, thermostat; E, channel electrode cell (including reference electrode); F, counter electrode; G, calibrated capillary.

typically of 1–2 mm bore. This is jacketed with an argon-purged tube to prevent the ingress of oxygen. Connections between the PTFE tubing, the cell, and the counter electrode are made using silicone rubber tubing.

Deoxygenated electrolyte is gravity fed from the reservoir via one of several calibrated glass capillaries, each providing a different flow range, the total range, in general, being 10^{-4} to $5 \times 10^{-1}\,\mathrm{cm^3\,s^{-1}}$. The adjustment of the rate of flow within each range is achieved by varying the height, h, between the reservoir and the tip of the capillary. Notice that, in this system, solution runs to waste once it has passed through the channel electrode cell. The temperature of the cell and the solution is kept constant by enclosing the cell and about 1 m of the preceding tubing in a perspex air-thermostatted box. The cell is mounted on one wall of the box to ensure that the direction of solution through it is upwards, thus preventing the formation of bubbles.

5.3 APPARATUS FOR STUDIES IN TURBULENT FLOW REGIMES

Vielstich and co-workers [99, 114, 115] have suggested that fast electrode reaction mechanisms can be studied more advantageously by switching to a turbulent flow regime, as discussed in Sect. 6.2. One of the experimental systems by which this is achieved is illustrated in Fig. 35. The electrode cell is tubular with a diameter of 0.22 cm and is an integral part of the reservoir, as shown. The working electrode is a thin metal ring, 0.1 mm in length. A second ring located further downstream (by about 0.05 mm from the working electrode) serves as a reference electrode. Solution is pumped out of the reservoir, circulated through a series of tubes, and then back in through the tubular electrode, the maximum achievable volume flow rate being $66\,\mathrm{cm^3\,s^{-1}}$. Heat generated in the system is removed with a water cooling jacket.

That the flow is turbulent can be tested by consideration of the mass transport limited current–flow rate behaviour in this regime, which is predicted theoretically to be [99, 114–116]

$$I_{\mathrm{LIM}} = \frac{0.276 nFAD^{2/3}\nu^{1/3}\mathrm{Re}^{7/12}c^{\infty}}{l^{1/3}(2a)^{2/3}} \tag{106}$$

Hence a plot of I_{LIM} versus $\mathrm{Re}^{7/12}$ should be linear and the slope should accurately predict the diffusion coefficient of the "test species". For the experimental system described above, Vielstich and co-workers obtained good agreement with eqn. (106), provided that $l \geq 100\,\mu\mathrm{m}$. A lower rate of mass transport than predicted to shorter electrodes was attributed to surface roughness at the electrode and the wall of the tube.

6. Extension of channel electrode methodology: practical examples

In this section, we show how channel electrode methodology is readily extended to facilitate the investigation of a diverse range of interfacial

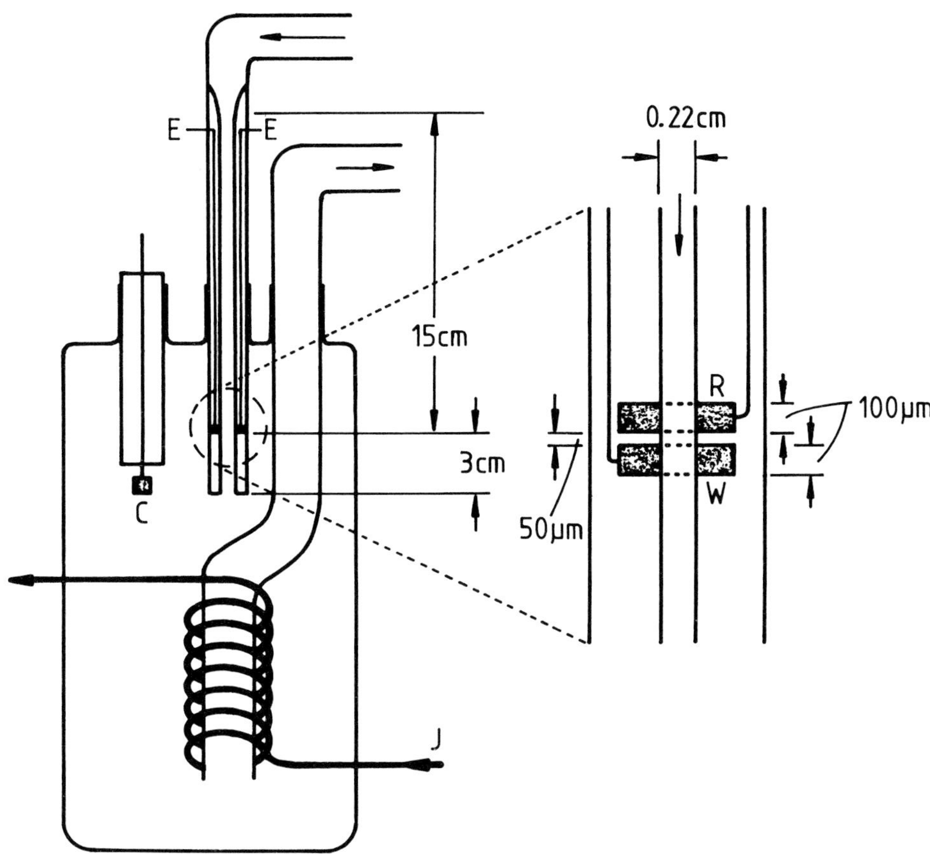

Fig. 35. Turbulent tubular electrode set-up employed by Vielstich and co-workers [115]. The counter, reference, and working electrodes are denoted by C, R, and W. J represents the cooling jacket and E labels the electrical contact to R and W.

processes. With reference to studies on ionic solid dissolution, the anodic dissolution of iron, pitting corrosion, along with electrode reaction mechanisms in general, the material in the preceding sections is developed to encompass turbulent as well as laminar flow conditions and the use of double channel electrode systems.

6.1 THE DOUBLE CHANNEL ELECTRODE

As with the rotating disc [117], the channel electrode may be used in a double electrode configuration, as illustrated in Fig. 36. This figure also defines the electrode geometry. Introduced by Gerischer et al. in 1965 [118], it comprises two neighbouring rectangular electrodes located flush with one of the surfaces of the rectangular channel. Tubular analogues have also been constructed [119, 120]. In this sub-section, we first discuss the application of the double channel electrode in the study of electrode reaction

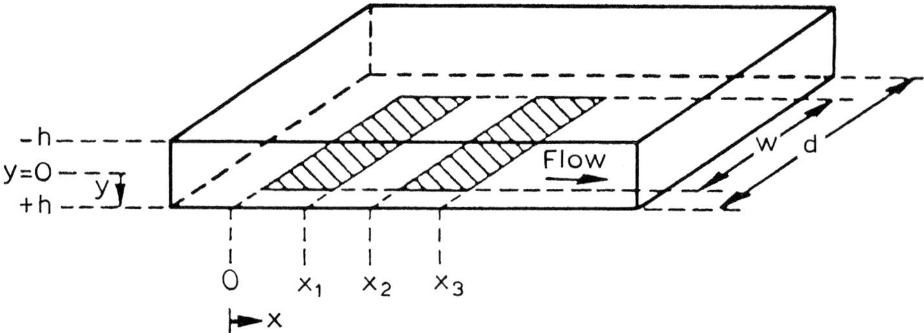

Fig. 36. Double channel electrode geometry.

mechanisms in which homogeneous kinetics operate. Subsequently, we consider their use in the investigation of the mechanism of anodic metal dissolution, in particular in the determination of whether the adsorption of intermediates plays a role in this process.

6.1.1 The double channel electrode in the investigation of electrode reaction mechanisms

In this section, the discussion centres on the application of double channel electrodes in the study of electrode reaction mechanisms under conditions of laminar flow. (The modifications necessary to what follows when turbulent flow operates can be found in Sect. 6.2.) When employed in this way, the upstream (generator) electrode produces the species of interest, which is then detected on the downstream electrode. This procedure is illustrated schematically in Fig. 37. In general, the detector electrode is held at a potential at which the destruction of the species produced upstream is diffusion-controlled. Kinetic and mechanistic information about the electrogenerated species is then available from "collection efficiency", N, measurements, given by

$$N = \frac{\text{current at downstream electrode}}{\text{current at upstream electrode}} = \frac{I_{\text{det}}}{I_{\text{gen}}} \tag{107}$$

The theory for the case where the electrogenerated intermediate is stable has been treated by Matsuda [121], Braun and co-workers [118, 122], and Cook [123]. These theoretical treatments differ in that Cook assumes a uniform current distribution on the upstream electrode (applicable to an irreversible voltammetric wave when little current flows) whereas Matsuda and Braun and co-workers consider a uniform surface concentration on the upstream electrode, which applies to all points on a reversible voltammetric wave, and near or at the diffusion-limited current for an irreversible process. The Matsuda/Braun treatment leads to the expression

$$N = 1 + \lambda^{2/3}[1 - \mathrm{F}(\theta)] - (1 + \theta + \lambda)^{2/3} \times [1 - \mathrm{F}\{(\theta/\lambda)(1 + \theta + \lambda)\}] - \mathrm{F}(\theta/\lambda) \tag{108}$$

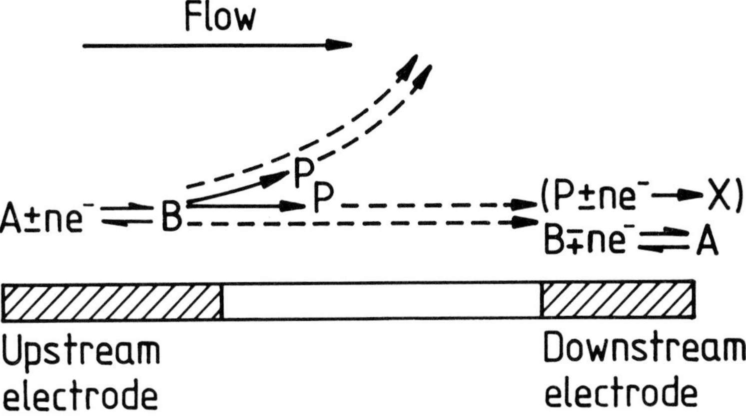

Fig. 37. Principle behind double electrode operation under steady-state conditions. A proportion of the species B, electrogenerated upstream, can be lost before detection downstream via diffusion or reaction in solution. The collection efficiency, N, provides information about these two processes. Where the product of the homogeneous reaction, P, is electroactive, it too can be detected downstream.

for the collection efficiency where

$$\theta = \frac{x_2 - x_1}{x_1} \tag{109}$$

and

$$\lambda = \frac{x_3 - x_2}{x_1} \tag{110}$$

The function $F(\theta)$ has been tabulated by Albery and Bruckenstein [124] and is given by

$$F(\theta) = \frac{3^{1/2}}{4\pi} \ln \frac{(1 + \theta^{1/3})^3}{(1 + \theta)} + \frac{3}{2\pi} \tan^{-1}\left(\frac{2\theta^{1/3} - 1}{3^{1/2}}\right) + \frac{1}{4} \tag{111}$$

Collection efficiencies for typical electrode geometries are readily calculated from eqn. (108). Table 6 (after Brett and Oliveira Brett [3]), gives results for a range of electrode geometries. The Cook theory leads to the expression

$$N = 1 - (1 + \theta + \lambda)\,F\left(\frac{1 + \theta}{\lambda}\right) + (\theta + \lambda)\,F\left(\frac{\theta}{\lambda}\right) + \frac{3^{3/2}\lambda^{2/3}}{2\pi}\,[(1 + \theta)^{1/3} - \lambda^{1/3}] \tag{112}$$

for the collection efficiency. Note that both treatments of the problem predict N to be dependent on the electrode geometry only. This has been experimentally verified on the ferri–ferrocyanide couple at gold double electrodes [125].

In general, the Cook theory leads to slightly higher values for N than

TABLE 6

Steady-state collection efficiencies as a function of double channel electrode geometry when the product of the generator electrode reaction is kinetically stable on the channel electrode timescale

x_3/x_2	x_2/x_1						
	1.05	1.10	1.15	1.20	1.30	1.40	1.50
1.10	0.1368	0.1297	0.1247	0.1209	0.1154	0.1115	0.1085
1.20	0.1991	0.1907	0.1846	0.1798	0.1726	0.1675	0.1635
1.30	0.2433	0.2346	0.2280	0.2228	0.2148	0.2090	0.2045
1.40	0.2799	0.2691	0.2623	0.2569	0.2485	0.2423	0.2374
1.50	0.3063	0.2975	0.2907	0.2851	0.2765	0.2700	0.2650
1.60	0.3304	0.3216	0.3148	0.3092	0.3004	0.2939	0.2887
1.70	0.3512	0.3425	0.3357	0.3301	0.3213	0.3147	0.3094
1.80	0.3696	0.3610	0.3542	0.3486	0.3398	0.3331	0.3278
1.90	0.3859	0.3774	0.3707	0.3651	0.3563	0.3496	0.3443
2.00	0.4006	0.3922	0.3855	0.3800	0.3713	0.3646	0.3592

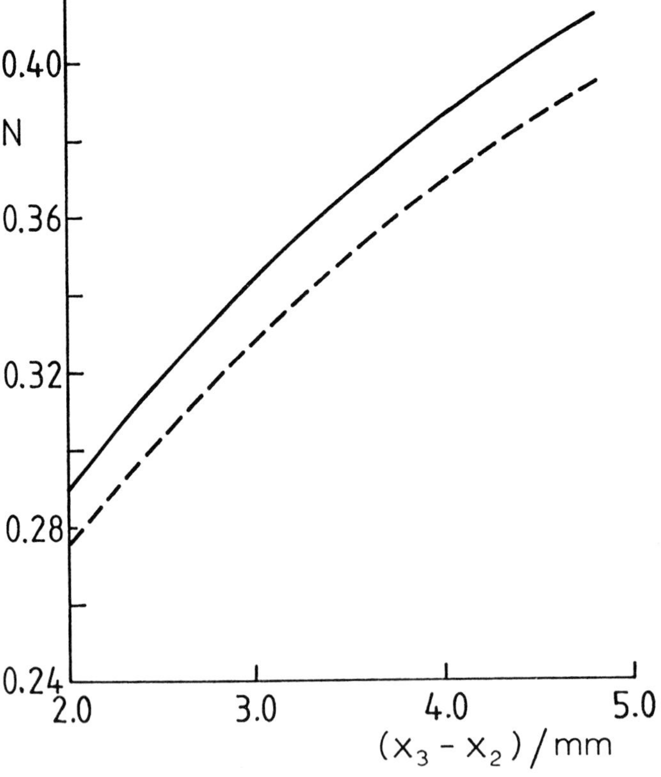

Fig. 38. The collection efficiency as a function of the length of the downstream electrode, according to Cook (——) and Matsuda (- - -), for a channel electrode for which $x_1 = 3$ mm and $(x_2 - x_1) = 1$ mm.

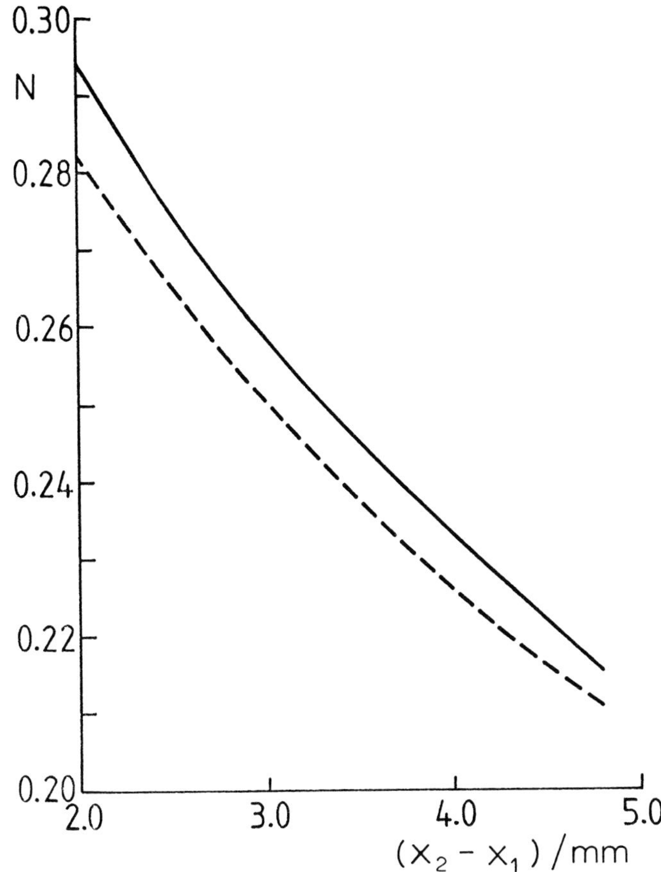

Fig. 39. The collection efficiency as a function of the gap thickness, according to Cook (——) and Matsuda (– – –), for a channel electrode with $x_1 = 3\,\text{mm}$ and $(x_3 - x_2) = 3\,\text{mm}$.

those predicted via the treatment of Matsuda and Braun, as illustrated in Figs. 38 and 39. This difference arises since, under the constant surface concentration approach, more of the current on the generator electrode is located towards the upstream end (as a consequence of the non-uniform accessibility). Hence the electrogenerated material has a longer average transit time to the detector electrode and thus a greater chance to escape into the bulk of the solution.

Implict in the above discussion is that, whereas for a reversible electrode reaction a plot of I_{det} vs. I_{gen} will be linear as the voltammetric wave on the upstream electrode is scanned, the behaviour for an irreversible process will result in a slight curve as N changes from the value predicted by the Cook theory to the (slightly lower) value expressed by the Matsuda–Braun theory.

The collection efficiency theory has been extended to cover the case where the electrogenerated intermediate decays via first-order kinetics for both EC and EC′ mechanisms (see Sects. 4.2 and 4.4 for detailed schemes). An early

TABLE 7

Mechanisms involving following chemical reactions treated by Matsuda and co-workers [126] for the double channel electrode

Mechanism	Upstream electrode	Solution	Downstream electrode	Current density on upstream electrode
$E_{rev}C$	$A \pm n_1e^- \rightleftharpoons B$	$B \longrightarrow$ prod	$B \pm n_2e^- \longrightarrow Y$	$I(x)$ = const. $x^{-1/3}$
$E_{rev}C'$	$A \pm n_1e^- \rightleftharpoons B$	$B \longrightarrow A$ (A regenerated, returns to upstream electrode)	$B \pm n_2e^- \longrightarrow Y$	$I(x)$ = const. $g(\Lambda^3 x/x_1)$
$E_{irrev}C$	$A \pm n_1e^- \longrightarrow B$	$B \longrightarrow$ prod	$B \pm n_2e^- \longrightarrow Y$	$I(x)$ = const.

approximate treatment by Braun and co-workers [118, 122] was subsequently improved upon by Matsuda and co-workers [126] who solved the problem numerically for the three cases outlined in Table 7. Also shown are the relevant representations of the current density distribution on the upstream electrode for each case.

The function $g(t)$ employed in the description of the current density on the upstream electrode for the $E_{rev}C'$ process is given by

$$g(t) = \frac{3^{1/3}t^{-1/3}}{\Gamma(\frac{1}{3})}\left[1 + \Gamma(\tfrac{2}{3})\sum_{j=1}^{\infty} q_j \frac{t^{2j/3}}{\Gamma\left(\frac{2j}{3} + \frac{2}{3}\right)}\right] \tag{113}$$

where the coefficients q_j are represented by

$$q_1 = \chi_1 - \omega_1 \tag{114}$$

$$q_m = \chi_m - \omega_m - \sum_{k=1}^{m-1} q_k\omega_{m-k} \quad m = 2, 3\ldots \tag{115}$$

with

$$\omega_1 = \frac{3^{1/3}\Gamma(\frac{2}{3})}{\Gamma(\frac{1}{3})} \tag{116}$$

$$\omega_2 = 0 \tag{117}$$

$$\omega_3 = \tfrac{1}{6} \tag{118}$$

$$\omega_m = \frac{\omega_{m-3}}{m(m-1)} \quad m = 4, 5\ldots \tag{119}$$

and

$$\chi_m = \frac{(m+1)\omega_{m+1}}{\omega_1} \quad m = 1, 2\ldots \tag{120}$$

The results, expressed in terms of the collection efficiency in the presence of first-order kinetics, N_k, as a function of the parameter Λ [given by eqn. (103)] for various electrode geometries, can be consulted in ref. 126 for each of the mechanisms. Matsuda et al. also derived simple approximate equations for the N_k–Λ relationship, which can be written as

$$\frac{N_k}{N_0} = \exp\{-\alpha\Lambda^2 + \beta\Lambda^4 - \gamma\Lambda^6\} \tag{121}$$

and

$$\frac{N_k}{N_0} = \exp\{-\alpha' + \beta'\Lambda^2 - \gamma'\Lambda^4\} \tag{122}$$

The ranges of N_k/N_0, θ, and λ in which eqns. (121) and (122) hold to within a few percent are delineated in Table 8. The coefficients α, β, γ, α', β', and γ' depend upon the electrode geometry and are given by summations of the type

$$\alpha = \sum_{m=0}^{3} \sum_{n=0}^{3} \alpha_{mn} \theta^m \lambda^n \tag{123}$$

where α represents α, β, γ, α', β', or γ'. The values of the coefficients for each of the mechanisms are tabulated in Tables 9 and 10.

That double channel electrodes are ideally suited to the study of electrode reaction mechanisms involving following chemical reactions is illustrated by reference to studies on the electrochemical oxidation of 4-amino-*N*, *N*,-dimethylaniline (ADMA) in basic solution at a platinum electrode [125]. This reaction is thought to proceed via the scheme

Electrode

$$H_2N{-}C_6H_4{-}N(CH_3)_2 \;-2\,e^- \rightleftharpoons\; HN{=}C_6H_4{=}N^+(CH_3)_2 \tag{xi}$$

Solution

$$HN{=}C_6H_4{=}N^+(CH_3)_2 \xrightarrow[OH^-]{k} HN{=}C_6H_4{=}O + NH(CH_3)_2 \tag{xii}$$

The oxidation step is diffusion-controlled at about $+0.4$ V versus the saturated calomel electrode (SCE). On setting the generator electrode to this potential and scanning the detector electrode in the range $+0.6$ to -0.7 V, two reduction waves were apparent. The first reduction wave, around 0.0 V versus SCE, was assigned to the reduction of the oxidized ADMA, whilst the

TABLE 8

Regions of applicability of eqns. (121) and (122) in terms of N_k/N_0, θ, and λ

Mechanism	$\frac{N_k}{N_0} = \exp\{-\alpha\Lambda^2 + \beta\Lambda^4 - \gamma\Lambda^6\}$			Accuracy (%)	$\frac{N_k}{N_0} = \exp\{-\alpha' + \beta'\Lambda^2 - \gamma'\Lambda^4\}$			Accuracy (%)
	N_k/N_0	θ	λ		N_k/N_0	θ	λ	
$E_{rev}C$	1.0–0.08 1.0–0.02	0.0–1.0 0.2–1.0	0.25–2.0 0.25–2.0	2 3	0.1–0.002	0.05–0.7	0.7–3.0	2
$E_{rev}C'$	1.0–0.08	0.0–1.0	0.25–2.0	3	0.1–0.002	0.05–0.7	0.7–3.0	2
$E_{irrev}C$	1.0–0.08	0.0–1.0	0.25–2.0	3	0.1–0.002	0.05–0.7	0.7–3.0	2

TABLE 9

The values of the coefficients used to calculate α, β, γ, α', β', γ' [via eqn. (123)] for the $E_{rev}C$ mechanism

m n	$\alpha_{m,n}$	$\beta_{m,n}$	$\gamma_{m,n}$	$\alpha'_{m,n}$	$\beta'_{m,n}$	$\gamma'_{m,n}$
0 0	1.0191E0	2.0224E − 1	2.3139E − 2	−9.9994E − 2	−8.6814E − 1	−6.1625E − 2
0 1	4.7604E − 1	−3.4945E − 2	−1.4887E − 2	6.0931E − 1	−1.8720E − 1	−6.3848E − 3
0 2	−1.9917E − 1	5.7600E − 2	1.2265E − 2	−2.3869E − 1	9.4952E − 2	2.3591E − 3
0 3	4.0072E − 2	−1.1940E − 2	−1.8591E − 3	3.3748E − 2	−1.6921E − 2	−4.8508E − 4
1 0	2.6344E0	−2.2514E − 1	−2.9141E − 2	1.6482E0	−2.1414E0	1.4544E − 1
1 1	−1.6944E0	2.1858E − 1	5.5107E − 2	−2.3220E0	1.2392E0	−1.8966E − 2
1 2	1.1264E0	−1.9006E − 1	−2.5154E − 2	6.6628E − 1	−1.2994E0	−5.9076E − 2
1 3	−2.3247E − 1	3.8814E − 2	1.3668E − 3	−5.3692E − 2	2.8266E − 1	1.6780E − 2
2 0	−2.3079E0	6.7679E − 1	6.2330E − 2	−4.6772E0	2.3822E0	−4.4010E − 1
2 1	2.9096E0	−4.5329E − 1	−1.8852E − 2	5.8737E0	−5.0082E0	−1.8386E − 2
2 2	−1.8598E0	2.9534E − 1	−4.6552E − 2	−2.2224E0	4.1354E0	1.5123E − 1
2 3	3.4469E − 1	−3.4239E − 2	2.9681E − 2	2.5628E − 1	−8.6295E − 1	−4.2241E − 2
3 0	1.1972E0	−3.5639E − 1	−2.2546E − 2	4.1372E0	−1.6415E0	3.3496E − 1
3 1	−1.6474E0	2.5321E − 1	−1.6230E − 2	−5.2472E0	4.3979E0	−1.5051E − 2
3 2	1.0080E0	−1.3319E − 1	5.8548E − 2	2.3546E0	−3.3752E0	−7.8071E − 2
3 3	−1.7168E − 1	2.0130E − 3	−2.8266E − 2	−3.2398E − 1	6.9319E − 1	2.4822E − 2

TABLE 10

The values of the coefficients used to calculate α, β, γ, α', β', γ' for the E_{rev} C′ and E_{irrev} C cases

m n	$\alpha_{m,n}$	$\beta_{m,n}$	$\gamma_{m,n}$	$\alpha'_{m,n}$	$\beta'_{m,n}$	$\gamma'_{m,n}$
0 0	8.9953E − 1	2.0833E − 1	2.7627E − 2	− 4.2700E − 1	− 8.8241E − 1	− 7.5533E − 2
0 1	5.8226E − 1	2.2754E − 2	1.8513E − 2	1.8085E0	4.2964E − 1	9.5692E − 2
0 2	− 2.7649E − 1	6.2753E − 2	− 1.2062E − 2	− 1.1029E0	− 3.6093E − 1	− 7.1734E − 2
0 3	6.0111E − 2	− 3.1880E − 2	2.3084E − 3	2.0463E − 1	7.4217E − 2	1.4200E − 2
1 0	2.7982E0	4.8814E − 2	6.5984E − 2	5.4209E0	− 7.0979E − 1	4.6030E − 1
1 1	− 2.3925E0	− 3.0285E − 1	− 3.0181E − 1	− 1.8539E1	− 7.0335E0	− 1.4525E0
1 2	2.1791E0	− 4.0137E − 3	3.2668E − 1	1.2946E1	5.1297E0	1.0161E0
1 3	− 6.3045E − 1	1.0113E − 1	− 1.0165E − 1	− 2.5214E0	− 1.0235E0	− 1.9853E − 1
2 0	− 2.3579E0	1.5484E − 1	− 1.0567E − 1	− 1.0541E1	1.1715E0	− 1.0200E0
2 1	4.2897E0	6.2286E − 1	6.8875E − 1	4.5240E1	1.3961E1	3.5963E0
2 2	− 4.4093E0	− 5.3587E − 1	− 1.0127E0	− 3.4104E1	− 1.1953E1	− 2.7013E0
2 3	1.3686E0	1.2120E − 1	4.0746E − 1	6.8595E0	2.5287E0	5.4285E − 1
3 0	1.1499E0	− 3.3409E − 1	− 1.0004E − 1	4.9643E0	− 3.0551E0	5.1014E − 1
3 1	− 2.3873E0	5.9335E − 1	1.8496E − 1	− 3.1034E1	− 7.0294E0	− 2.4697E0
3 2	2.5480E0	− 4.3625E − 1	1.0100E − 1	2.5182E1	7.6457E0	1.9958E0
3 3	− 8.0934E − 1	1.3555E − 1	− 9.1527E − 2	− 5.2129E0	− 1.7356E0	− 4.1234E − 1

second wave (at around $-0.2\,V$) was attributed to the electro-reduction of the quinone monoimine formed in the chemical step. Hence the *total* cathodic current at the detector electrode is a measure of the proportion of the species generated on the upstream electrode reaching the downstream electrode. The value of the collection efficiency thus calculated should therefore be that expected of a stable intermediate and independent of flow rate for a particular electrode geometry. The results, shown in Fig. 40 in terms of the mean solution velocity, $\bar{U}\,[=(2/3)v_0]$, (obtained by Aoki and Matsuda [125]) show that this is indeed the case. Furthermore, the value of the collection efficiency calculated on this basis was found to be in excellent agreement with eqn. (108) for a number of electrode geometries.

Also shown in Fig. 40 is the I_{LIM}–$\bar{U}^{1/3}$ behaviour for the first reduction wave alone (at the detector electrode). Extrapolation of this data to zero current produces a positive intercept on the x ($\bar{U}^{1/3}$) axis, indicating that the electrode reaction is influenced by following chemical kinetics. The data in Fig. 40 can be used to calculate N_k as a function of $\bar{U}$, which in turn can be converted to the corresponding values of the kinetic parameter Λ via eqns. (121) and (122). It follows from eqn. (103) that a plot of Λ against $\bar{U}^{-1/3}$ allows the 1st-order rate constant for the chemical step to be elucidated. Figure 41 shows the Λ–$\bar{U}^{-1/3}$ behaviour for the reduction of the oxidized ADMA at the detector electrode at a series of basic conditions. As predicted by eqn. (103), a linear relationship between Λ and $\bar{U}^{-1/3}$ holds and intercepts the origin. The data in Fig. 41 allowed Aoki and Matsuda [125] to evaluate the 2nd-order rate constant, k_2, for the deamination of oxidized ADMA

$$k_2 = \frac{k}{[OH^-]} \tag{124}$$

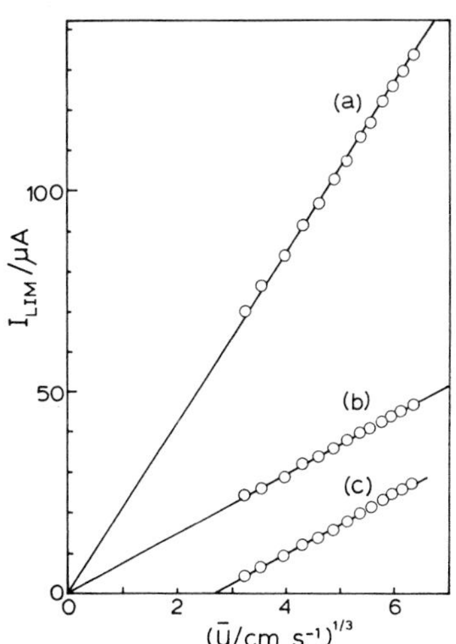

Fig. 40. Transport-limited current–flow rate behaviour for the ADMA (EC) system at (a) the upstream electrode (due to the oxidation of ADMA), (b) the downstream electrode (total cathodic current), and (c) the downstream electrode (reduction of oxidized ADMA). Electrode geometry is defined by $x_1 = 0.105$, $x_2 = 0.129$, $x_3 = 0.237$ cm. From ref. 125.

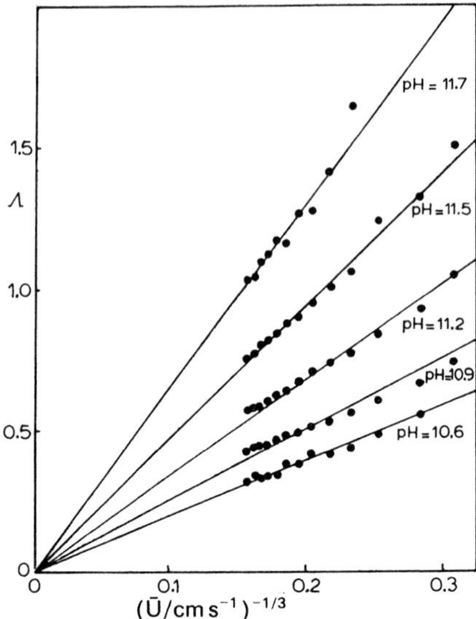

Fig. 41. Λ–$\bar{U}^{-1/3}$ relationship for the reduction of oxidized ADMA at the downstream electrode in a double electrode cell with geometry as given in the legend to Fig. 40. From ref. 125.

as $1.8 \times 10^4\,M^{-1}\,s^{-1}$, in good agreement with the value obtained via spectrophotometric methods [125] and illustrating the success of the double channel electrode technique in this application.

6.1.2 The double channel electrode in the investigation of anodic metal dissolution

In its simplest form, the anodic dissolution of a metal, M, may be represented by

$$M \rightarrow M^{n+} + n\,e^- \qquad \text{(xiii)}$$

However, this superficially simple process often comprises a complex series of steps, several of which may involve the adsorption of intermediates [127]. Although it has long been recognised that double electrode systems, such as the rotating ring–disc electrode, are uniquely well-suited tools with which such processes may be studied [128], it was not until comparatively recently that double channel electrodes were put to this task [129–131]. In essence, the technique (as introduced by Bruckenstein and Napp [128]) involves stepping either the potential or the current at the upstream electrode, thereby switching the anodic dissolution process on and measuring the transient current at the detector electrode set at a potential at which a product of the reaction at the upstream electrode is oxidized or reduced at a mass-transport controlled rate. Figure 42 illustrates the basis of the technique for the case where the upstream electrode is stepped galvanostatically. It is apparent that, when the anodic dissolution process involves adsorption

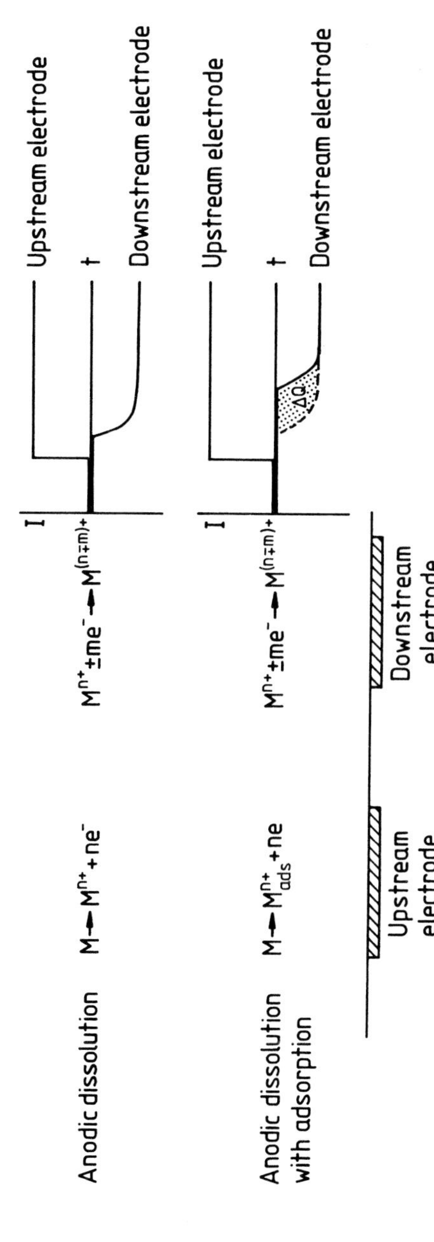

Fig. 42. Investigation of anodic metal dissolution by application of a galvanostatic step to the upstream electrode. When the dissolution process is complicated by adsorption of the intermediate, the current transient at the detector electrode has a time lag longer than that without adsorption.

of the product (metal ions) of the dissolution process

$$M \rightarrow M^{n+}_{ads} + n\,e^- \qquad \text{(xiv)}$$

$$M^{n+}_{ads} \rightarrow M^{n+}_{aq} \qquad \text{(xv)}$$

the delay time in the transient current at the detector electrode is longer than for immediate loss of the product to solution

$$M \rightarrow M^{n+}_{aq} + n\,e^- \qquad \text{(xvi)}$$

The extent to which the surface of the upstream electrode is covered by the adsorbed species can be evaluated by comparing the current–time curve at the detector electrode with that predicted under the condition where adsorption does not take place. The area between the two curves represents the total charge, ΔQ, of the adsorbed intermediate. Hence, the number of ions, N_i, adsorbed on the metal surface is

$$N_i = \frac{\Delta Q L}{nF} \qquad (125)$$

where L is the Avogadro Number and n is the number of unit charge on the adsorbed ion.

The procedure for the determination of the adsorption of intermediates in the dissolution mechanism, as outlined above, relies on a knowledge of the current–time behaviour at the detector electrode for the case where the upstream generation of the electroactive material is uncomplicated by adsorption processes. This problem was solved recently by Aoki et al. [132], whose analysis, at the level of the Levich approximation, led to the following expressions for the collection efficiency at the normalized time τ [given by eqn. (68)], $N(\tau)$.

Large τ

$$N(\tau) = N(\tau \rightarrow \infty) - \frac{3^{4/3}\Gamma(\frac{2}{3})}{2\Gamma(\frac{1}{3})} \sum_{n=1}^{\infty} \sum_{j=0}^{\infty} \frac{q_j(-1)^j(\alpha_n')^{-3/2}}{\Gamma((2j+2)/3)}$$

$$\times \int_0^{\lambda} \chi_1^{(2j-1)/3} \times \{\mathrm{H}(\tfrac{3}{2} - j, \tau, (1 + \theta + \lambda - \chi_1)(\alpha_n')^{-3/2}$$

$$- \mathrm{H}(\tfrac{3}{2} - j, \tau, (\theta + \lambda - \chi_1)(\alpha_n')^{-3/2}\}\,\mathrm{d}\chi_1 - N' \qquad (126)$$

where $N(\tau \rightarrow \infty)$ is given by eqn. (108) and

$$N' \approx \left(\frac{3^{5/3}}{7.150\Gamma(\frac{2}{3})}\right) \int_0^{\lambda} \{(1 + \lambda + \theta - \chi_1)^{1/3} - (\theta + \lambda - \chi_1)^{1/3}\}$$

$$\times\ \mathrm{H}\left(\frac{1}{2}, \tau, \frac{\chi_1}{3.575}\right) \mathrm{d}\chi_1 \qquad (127)$$

The terms α_n' are the solutions of the equation $\mathrm{Ai}'(-\alpha_n') = 0$ [50]

$$\chi_1 = \frac{x}{x_1} \tag{128}$$

q_j are coefficients given by

$$q_0 = 1 \tag{129}$$

and eqns. (114)–(120) for $j = 1 \ldots$

H$(3/2 - j, \tau, z)$ is readily evaluated via the recurrence formula

$$\begin{aligned} \partial \mathrm{H}(\tfrac{3}{2} - j, \tau, z) &= \mathrm{H}(\tfrac{1}{2} - j, \tau, z) \\ &= 2\left[\frac{(-1)^{-j}}{3\pi}\right] \sum_{n=j}^{\infty} \tau^{n-j} \cos\left[\frac{(2n+1)\pi}{6}\right] \\ &\quad \times \frac{\Gamma(2n+1)/3}{n - j!\, z^{(2n+1)/3}} \end{aligned} \tag{130}$$

Small τ

$$\begin{aligned} N(\tau) = \frac{1}{3^{1/2}\pi} \int_0^{\lambda} \Bigg\{ & E_1(\varrho_1) - E_2(\varrho_2) + 0.167\left[\frac{E_2(\varrho_1)}{\varrho_1} - \frac{E_2(\varrho_2)}{\varrho_2}\right] \\ & + 0.0245\left[\frac{E_3(\varrho_1)}{\varrho_1^2} - \frac{E_3(\varrho_2)}{\varrho_2^2}\right]\Bigg\} \mathrm{d}\chi_1 \end{aligned} \tag{131}$$

where

$$\varrho_1 = \frac{3(\theta + \lambda - \chi_1)^2}{\tau^3} \tag{132}$$

$$\varrho_2 = \frac{3(1 + \theta + \lambda - \chi_1)^2}{\tau^3} \tag{133}$$

and the functions E_1, E_2, and E_3 are exponential integrals [133], given by

$$E_n(\varrho) = \int_0^{\infty} \exp\{-\varrho t\} t^{-n} \mathrm{d}t \quad n = 1, 2, 3 \tag{134}$$

Equations (126) and (131) were evaluated numerically; full details are given in ref. 132. The results, plotted in terms of $N(\tau)/N(\tau \to \infty)$ against τ for $\lambda = 1$ and $\theta = 0.2, 0.5, 1.0$, and 1.5 are shown in Fig. 43. This figure indicates the range of τ in which eqns. (126) and (131) are applicable for particular electrode geometries. For comparison, Aoki et al. [132] also solved the same problem using the digital simulation (simple explicit) method of Flanagan and Marcoux [35]. The results of this procedure are also illustrated in Fig. 43. Values of $N(\tau)/N(\tau \to \infty)$ can be estimated to within 4.5% for $0.1 \leq \theta \leq 1.0$ and $0.5 \leq \lambda \leq 1.5$ via the approximate equation

$$\frac{N(\tau)}{N(\tau \to \infty)} = \frac{1 + \tanh(\Xi)}{2} \tag{135}$$

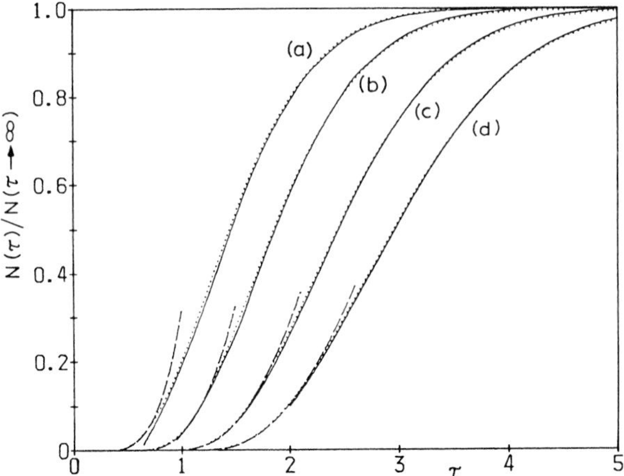

Fig. 43. Collection efficiency–normalized time curves calculated by Aoki et al. [132] ——, via eqn. (126); – – –, via eqn. (132); and · · · ·, via digital simulation for (a) $\theta = 0.2$, (b) $\theta = 0.5$, (c) $\theta = 1.0$, and (d) $\theta = 1.5$. In each case $\lambda = 1.0$. The values of $N(\tau \rightarrow \infty)$ for the four electrode geometries are (a) 0.375, (b) 0.316, (c) 0.258, and (d) 0.222.

where

$$\begin{aligned}\Xi = {} & [1.44 - 0.386\theta - \{2.53\theta + 0.38\lambda + 0.9\}^{-2}]\tau \\ & + \{(0.386\theta - 1.44)(2.53\theta + 0.38\lambda + 0.9)^{2/3} \\ & + 1.5(2.53\theta + 0.38\lambda + 0.9)^{-4/3}\} - (2\tau)^{-1}\end{aligned} \tag{136}$$

Tsuru et al. [129–131] applied the double channel electrode transient technique to the investigation of the anodic dissolution of iron in sulphate and chloride aqueous media, in the pH range 1–3. The upstream electrode served as the generator electrode at which the anodic dissolution reaction occurred. Overall this involves

$$\text{Upstream electrode}: \quad Fe \rightarrow Fe^{2+} + 2e^- \tag{xvii}$$

The upstream electrode reaction was monitored at either a gold or glassy carbon detector electrode, which oxidized the incoming ferrous ions to ferric

TABLE 11

The Heusler and Bockris mechanisms for the anodic dissolution of iron

Mechanism	Steps involved	Rate-determining step
Heusler	(xix), (xxi), (xxii), and (xxiii)	(xxii)
Bockris	(xix), (xx), and (xxiii)	(xx)

ions at a transport-controlled rate

$$\text{Downstream electrode}: Fe^{2+} \rightarrow Fe^{3+} + e^- \qquad \text{(xviii)}$$

In sulphate solution, the anodic dissolution of iron is thought to proceed via one of two pathways: the Bockris (non-catalysed) mechanism [134] or the Heusler (catalysed) mechanism [135]. These two processes may be represented by the following scheme, Table 11 identifying the steps important to each mechanism.

$$Fe + H_2O \underset{k_{-1}}{\overset{k_1}{\rightleftharpoons}} [FeOH]_{ads} + e^- + H^+ \qquad \text{(xix)}$$

$$[FeOH]_{ads} \xrightarrow{k_2} FeOH^+ + e^- \qquad \text{(xx)}$$

$$[FeOH]_{ads} + Fe \underset{k_{-3}}{\overset{k_3}{\rightleftharpoons}} Fe[FeOH] \qquad \text{(xxi)}$$

$$Fe[FeOH] + OH^- \xrightarrow{k_4} FeOH^+ + [FeOH]_{ads} + 2e^- \qquad \text{(xxii)}$$

$$FeOH^+ + H^+ \underset{k_{-5}}{\overset{k_5}{\rightleftharpoons}} Fe^{2+}_{aq} + H_2O \qquad \text{(xxiii)}$$

The pathway which the dissolution process follows is governed, to a large extent, by the surface activity of the iron [136–140]. Iron of high purity, with very low surface activity, undergoes anodic "non-catalysed" dissolution [136], whilst highly active material, containing a large density of imperfections, is observed to follow the catalysed mechanistic pathway [136, 137]. For both mechanisms, mass transport has no effect on the kinetics at the potentials of interest [141, 142]. Whilst the two processes can be distinguished by conventional electrochemical methods (for example by measurement of the steady-state anodic Tafel slope and the electrochemical reaction order with respect to pH) as illustrated by Table 12, these criteria do not, in themselves, provide direct evidence for the adsorption of intermediates in the reaction. That adsorbed intermediates are involved in the mechanism can be proved

TABLE 12

Criteria for the discrimination between the Bockris [134] and Heusler [135] mechanisms for the anodic dissolution of iron under conditions for which mass transfer has no effect on the kinetics [141, 142]

Mechanism	Anodic Tafel slope $\frac{\partial E}{\partial \log i} \Big/ \text{mV}$	$\frac{\partial \log i}{\partial \text{pH}}$
Bockris	39.4	1
Heusler	29.6	2

directly by using the double channel electrode technique described (vide infra).

In their studies, Tsuru et al. [129] worked with iron of purity $> 99.99\%$, which was cold rolled and annealed at 1223 K in a hydrogen atmosphere for several hours. Preparation of the iron electrode in this way ensures that a low dislocation density is achieved. As expected, the current–potential characteristics measured as a function of pH [Tafel slope of 44 ± 4 mV decade^{-1} and $(\partial \log i/\partial \mathrm{pH}) = 1.1$] revealed the dissolution mechanism to proceed via the Bockris (non-catalysed) route. That this was the mechanistic pathway was firmly established by measuring the quantity of adsorbed intermediates as a function of potential for $1 < \mathrm{pH} < 3$ using the double channel electrode technique. Typical experimental chronoamperomograms, obtained by Tsuru et al. [129] at a series of flow rates, are shown in Fig. 44. Notice that the quantity of adsorbed species (deduced from the area between the experimental and theoretical "no adsorption" curves) is independent of solution flow rate. The surface coverage, θ_s, is readily evaluated from chronoamperomograms, such as those in Fig. 44, by applying eqn. (125). Knowledge of the current–potential behaviour for the dissolution process then allows the θ_s–potential relationship to be deduced. Figure 45 illustrates the

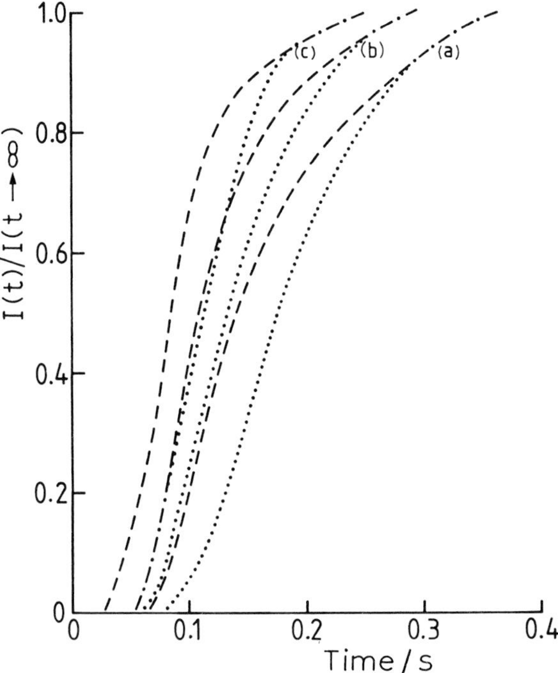

Fig. 44. Current transients at the downstream electrode when a current step is applied to the upstream iron electrode. ····, Experimental curves; – – –, corresponding theoretical curves for the no adsorption cases calculated from the theory of Aoki et al. [132]. The data, redrawn from ref. 129, relates to mean solution velocities of (a) 10, (b) 30, and (c) 50 cm s^{-1}.

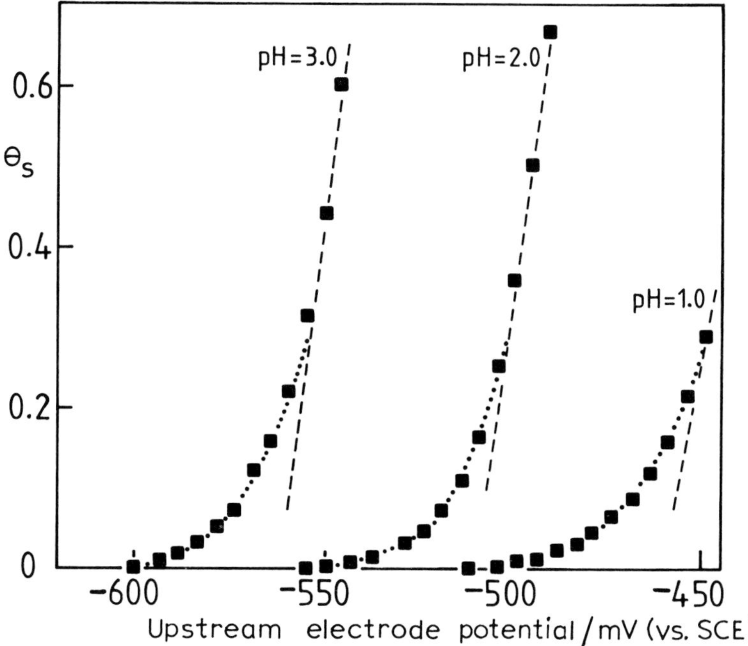

Fig. 45. Coverage of surface by the adsorbed intermediate as a function of potential in the anodic dissolution of iron. Adsorption follows a Langmuir isotherm at low coverages (····) and a Tempkin isotherm at higher coverages. The data are as calculated by Tsuru et al. [129].

outcome of this process at pH 1, 2, and 3. The data would appear to corroborate the view that, for the Bockris mechanism, the adsorption isotherm is Langmuirian at low coverages [139, 143], i.e.

$$k_1(1 - \theta_s)[\mathrm{OH}^-]\exp\{\alpha_1 FE/RT\} = k_{-1}\theta_s \exp\{-(1 - \alpha_1)FE/RT\} \tag{137}$$

and for $\theta_s \ll 1$

$$\alpha_s = k_1[\mathrm{OH}^-]\exp\{FE/RT\} \tag{138}$$

At higher coverages, Lorenz and co-workers [139, 140] postulated that adsorption would follow a Tempkin isotherm

$$k_1(1 - \theta_s)[\mathrm{OH}^-]\exp\{\alpha_1 FE/RT\}\exp\{-\beta\theta_s f\}$$
$$= k_{-1}\theta_s \exp\{-(1 - \alpha_1)FE/RT\}\exp\{(1 - \beta)f\theta_s\} \tag{139}$$

where α_1 is the transfer coefficient for the electrode reaction in step (xix), β is the symmetry factor for adsorption process, and f is the coverage-dependent rate of change of the Gibb's free energy of adsorption

$$f = \frac{1}{RT}\left(\frac{\partial \Delta G^0_{\theta_s}}{\partial \theta_s}\right) \tag{140}$$

As $\theta_s \rightarrow 1$

$$\exp(f\theta_s) = k_1[\mathrm{OH}^-]\exp\{FE/RT\} \tag{141}$$

A linear relationship between θ_s and the potential is therefore predicted. That this is the case is again borne out in the data in Fig. 45.

The above description serves to illustrate the power of the double electrode technique in the investigation of this type of process; it provides quantitative data on the coverage of surfaces by the intermediate, thereby allowing the nature of the operating adsorption isotherm to be deduced.

It should finally be pointed out that Tsuru et al. [129] have applied the same technique to the study of anodic iron dissolution in chloride (0.05–3.0 M) media, again in the pH range 1–3. Their results lead them to postulate the mechanism

$$\mathrm{Fe} + \mathrm{OH}^- + \mathrm{Cl}^- \rightleftharpoons [\mathrm{FeClOH}^-]_{\mathrm{ads}} + \mathrm{e}^- \tag{xxiv}$$

$$[\mathrm{FeClOH}^-]_{\mathrm{ads}} \rightarrow \mathrm{FeOH}^+ + \mathrm{Cl}^- + \mathrm{e}^- \tag{xxv}$$

$$\mathrm{FeOH}^+ + \mathrm{H}^+ \rightleftharpoons \mathrm{Fe}^{2+} + \mathrm{H_2O} \tag{xxvi}$$

The adsorption of the intermediate $[\mathrm{FeClOH}^-]$ again follows Langmuir (at low coverages) and Tempkin (at high coverages) isotherms.

6.2 THE STUDY OF ELECTRODE REACTION MECHANISMS IN A TURBULENT FLOW REGIME AT A MICRO-TUBULAR ELECTRODE

A recurring theme throughout this chapter has been the need for high rates of mass transfer to electrodes when fast electrochemical reactions are to be studied. Hitherto, we have considered this problem in the context of laminar flow. Whilst it is the case that the majority of electrode reaction mechanisms at the channel electrode have been investigated under such conditions [41], an alternative strategy by which high rates of mass transfer may be attained lies in switching over to a turbulent regime [114–116, 120, 144–148]. In this section, we describe some of the recent developments in this area with particular reference to the work of Vielstich and co-workers [99, 114, 115, 120, 145, 146]. Since virtually all of the reported studies concern the tubular electrode, we will confine our description to this system.

6.2.1 Mass transport to channel and tubular electrodes under a turbulent flow regime

The description of mass transport to channel and tubular electrodes given in Sect. 2 was restricted to laminar conditions. Once Re $>$ 2000, the pattern of flow is no longer smooth and steady: fluctuating, irregular (eddying) motions become superimposed on the main stream. Consequently, a complete theoretical description of mass transport, under such a regime, is impossible [149] and, as a result, empirical methods are introduced. In particular, a simplified representation of turbulence is afforded by considering the motion in terms of a "time-averaged" velocity, $\bar{v}$, and a velocity of fluctuation, v', such that [115, 149]

$$\begin{aligned} v_x &= \bar{v}_x + v'_x \\ v_y &= \bar{v}_y + v'_y \\ v_z &= \bar{v}_z + v'_z \end{aligned} \tag{142}$$

Then, provided that the time-averages are considered over an adequate length of time, t_T, that they become time-independent [149]

$$\bar{v} = \frac{1}{t_T} \int_t^{t+t_T} v \, dt \tag{143}$$

the time-averaged fluctuations may be neglected

$$\overline{v'_x} = \overline{v'_y} = \overline{v'_z} = 0 \tag{144}$$

That is to say that we work in terms of a "fully developed" turbulent flow profile, the form of which has been "influenced" by the fluctuations v'_x, v'_y, and v'_z.

The fully developed turbulent velocity profile within the tubular electrode cell can satisfactorily be represented by the empirical equation [150, 151]

$$\frac{v_x}{v_0} = \left(\frac{a - r}{a}\right)^{1/n} \tag{145}$$

where a and r are defined in Fig. 2 and, as for the laminar case, v_x represents the velocity in the x direction at some value of r. The velocity at the centre of the tube is again denoted by v_0. Provided that $4 \times 10^3 < \text{Re} < 10^5$, the velocity distribution is described by eqn. (145) with $n = 7$. The profile is then said to follow the "1/7-th power velocity distribution law" [150]. This velocity profile is compared with the laminar profile [given by eqn. (7)] in Fig. 46. It is apparent that greater rates of mass transfer (within the vicinity of the tube walls) can be achieved by employing turbulent rather than laminar flow. At higher Reynolds numbers ($\text{Re} > 10^5$), n takes on higher values, the velocity profile becoming "fuller". The lead-in length required to establish fully developed turbulent flow is much shorter than under laminar conditions [eqn. (8)], a length of 40–80 tube diameters typically being needed [152, 153].

Von Kárman [153] has identified three zones within the turbulent boundary layer at tubes.

(a) The laminar sublayer, close to the wall of the tube, in which flow may be considered to be laminar as a consequence of the turbulent eddies being damped out. In this region, v_x varies linearly with distance from the wall $(a - r)$ such that [115]

$$v_x = \frac{(a - r)v^{*2}}{\nu} \tag{146}$$

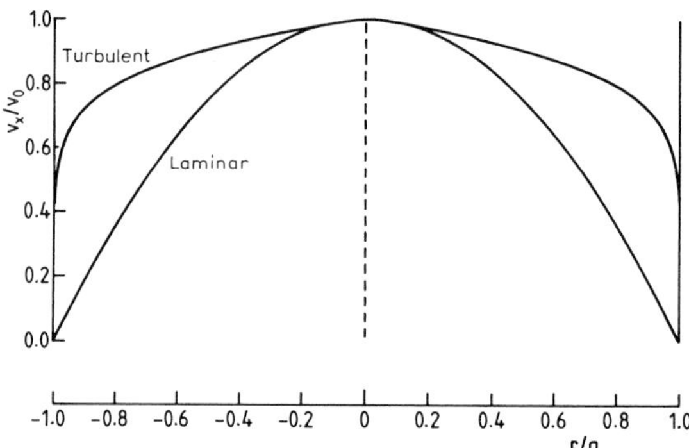

Fig. 46. Comparison of laminar and turbulent (1/7th power) velocity distributions in a tubular electrode cell.

where v^* is the friction velocity [154] given by the equation

$$v^* = \left(\frac{\sigma_0}{\varrho}\right)^{1/2} \tag{147}$$

with σ_0 representing the sheer stress (kg m^{-1} s^{-2}) at the wall of the tube in the laminar boundary sublayer and ϱ the density of the solvent.

(b) A transition region in which flow changes from laminar to turbulent [155].

(c) A turbulent core in which flow may be assigned to follow the 1/7-th power velocity distribution law

$$\frac{v_x}{v^*} = 8.74\left[\frac{(a - r)v^*}{\nu}\right]^{1/7} \tag{148}$$

Figure 47 shows the velocity profile for fully developed turbulent flow, indicating the regions in which each of the three zones characterise the flow.

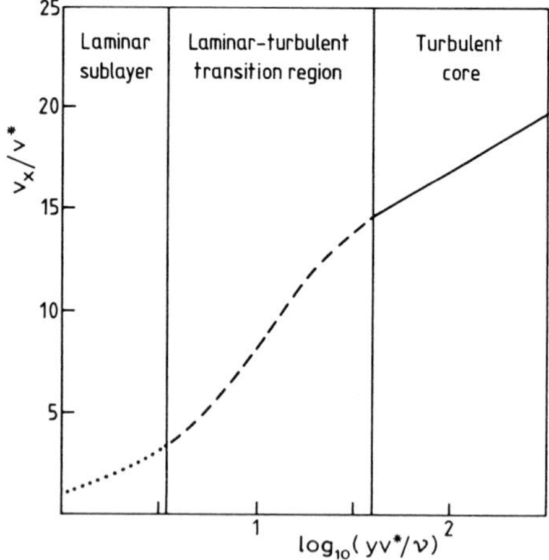

Fig. 47. The three "zones" within the turbulent boundary layer at tubes $y = a - r$.

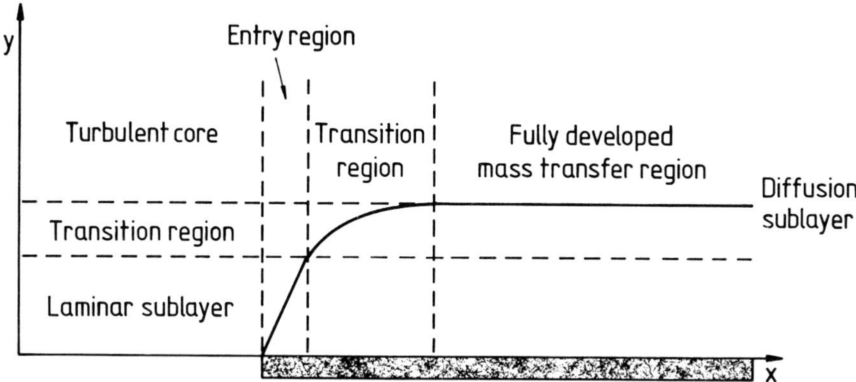

Fig. 48. Schematic diagram delineating the three mass-transport regions for turbulent flow as a function of electrode length.

In solving the problem of mass transfer to tubular electrodes, the contribution of each of the zones within the turbulent boundary layer, as a function of distance down the electrode, has to be assessed. This is illustrated by considering a species which undergoes simple electron transfer at the electrode and comparing the magnitude of the resulting concentration boundary layer relative to the mass transport boundary layer (composed of the three "zones"), as shown schematically in Fig. 48. At short distances from the upstream edge of the electrode, in the so-called "entry region", mass transport to the electrode can be considered in terms of the laminar sub-layer alone. However, as the electrolyte moves downstream, the diffusion sublayer is extended to greater distances from the electrode surface and the contribution of the turbulent eddies to the transport process is seen to be increasingly important. Eventually, at a sufficient distance downstream, mass transfer becomes x-independent and the diffusion sub-layer maintains a constant thickness. This is termed the "fully developed mass transport region".

Dreeson and Vielstich [146] solved the problem of turbulent mass transport to a tubular electrode in both the entry and the fully developed regions. The steady-state mass transport equation to be solved may be written as

$$\bar{v}_x \frac{\partial \bar{c}}{\partial x} = (D + D_t) \frac{\partial^2 \bar{c}}{\partial y^2} \tag{149}$$

where $\bar{v}_x$ and $\bar{c}$ are the time-averaged solution velocity in the x direction and concentration of the species of interest, respectively, and D_t is the coefficient of eddy diffusivity, as distinct from the coefficient of molecular diffusion, D. Experimental data show D_t to be $(a - r)^3$-dependent [115] and, on this basis, Vielstich has suggested that D_t can be represented by [156]

$$D_t = D \left(\frac{y}{\delta_{td}} \right)^3 \tag{150}$$

where δ_{td} is the thickness of the diffusion sub-layer under turbulent conditions, given by

$$\delta_{dt} = \frac{50a}{Re^{7/8}Pr^{1/3}} \tag{151}$$

with the Prandtl number as

$$Pr = \frac{\nu}{D} \tag{152}$$

By substituting eqns. (150) and (151) into eqn. (149), expressing $\bar{v}_x$ in terms of the 1/7-power velocity distribution law, and introducing a dimensionless parameter

$$C = \frac{\bar{c}}{c^{\infty}} \tag{153}$$

Dreeson and Vielstich obtained the normalized convective-diffusion equation

$$25u \frac{\partial C}{\partial s} = \frac{\partial}{\partial u} (1 + u^3) \frac{\partial C}{\partial u} \tag{154}$$

where

$$s = x \frac{Re^{7/8}Pr^{-1/3}}{50a} \tag{155}$$

and

$$u = y \frac{Re^{7/8}Pr^{1/3}}{50a} \tag{156}$$

The boundary conditions appropriate to the solution of eqn. (154) are delineated in Table 13.

TABLE 13

Boundary conditions for the solution of eqn. (154)

Boundary		$C = \frac{\bar{c}}{c^{\infty}}$
s	u	
$0 > s \leqslant \frac{l\,Re^{7/8}Pr^{-1/3}}{50a}$	0	0
0	$\geqslant 0$	1
$0 > s \leqslant \frac{l\,Re^{7/8}Pr^{-1/3}}{50a}$	$> 1\ [y > \delta_{td}]$	1

The flux to an infinitesimal segment of the electrode (i.e. the local flux) under turbulent flow is given by

$$J_L = D \frac{\partial \bar{c}}{\partial y}\bigg|_{y=0} \tag{157}$$

This is analogous to eqn. (19) for the laminar case. Dreeson and Vielstich also calculated the average flux to a point (in the x direction) via

$$J_A = \frac{1}{l}\int_0^l J_L \, dx \tag{158}$$

Both J_L and J_A are conveniently defined in terms of the Nusselt number

$$Nu_L = \frac{2J_L a}{Dc^\infty} \tag{159}$$

and

$$Nu_A = \frac{1}{l}\int_0^l Nu_L \, dx = \frac{2J_A a}{Dc^\infty} \tag{160}$$

Equation (154) was solved by numerical integration to yield [146]

$$\frac{\partial C}{\partial u}\bigg|_{u=0} = \frac{Nu_L \, Re^{-7/8} Pr^{-1/3}}{25} \tag{161}$$

and

$$\frac{1}{s}\int_0^s \frac{\partial C}{\partial u}\bigg|_{u=0} ds = \frac{Nu_A \, Re^{-7/8} Pr^{-1/3}}{25} \tag{162}$$

The rates of mass transfer, in terms of a normalized Nusselt number

$$\mathcal{N}_{L,A} = Nu_{L,A} \, Re^{-7/8} Pr^{-1/3} \tag{163}$$

are shown in Fig. 49 (taken from ref. 146). The vertical lines, assigning the $\mathcal{N}_{L,A}$–s behaviour to be characteristic of either the entry or fully developed regions, are purely arbitrary, but represent zones in which mass transfer can be safely categorised in terms of one of the two types.

Dreeson and Vielstich [146] found that, in both the entry and fully developed regions, mass transfer could be described by

$$Nu \propto Re^a Pr^b \tag{164}$$

In the entry region, $a = \frac{7}{12}$ and $b = \frac{1}{3}$, in agreement with mass transfer experiments [99, 116, 144, 145, 157, 158], which have been shown to follow

$$Nu = 0.276 \, Re^{7/12} Pr^{1/3} \left(\frac{2a}{l}\right)^{1/3} \tag{165}$$

The above equation was given earlier in a slightly different form [eqn. (106)].

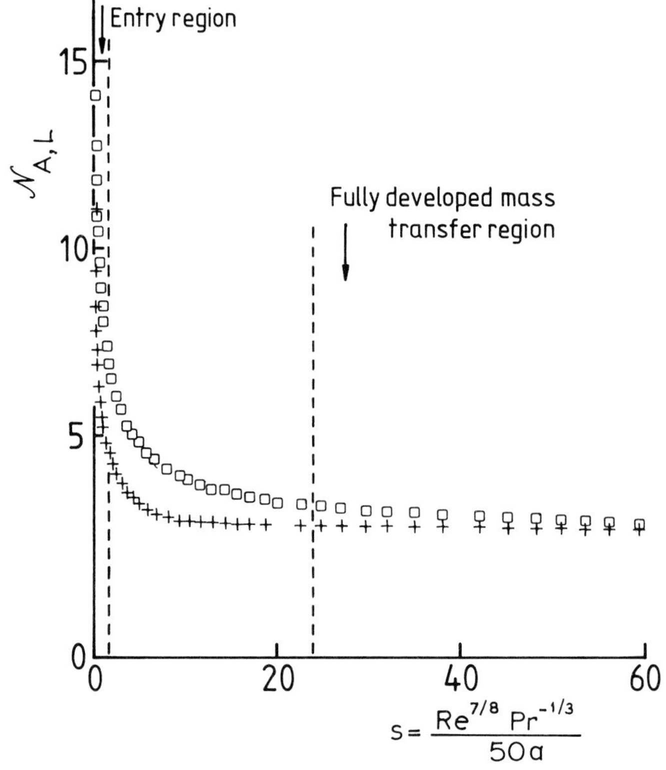

Fig. 49. Rates of mass transfer under a turbulent flow regime [in terms of the normalized Nusselt number $\mathcal{N}_A$ (□) or $\mathcal{N}_L$ (+)] as a function of the dimensionless x direction, s.

In the fully developed region, Dreeson and Vielstich found that mass transfer was accurately represented by $a = \frac{7}{8}$ and $b = \frac{1}{3}$. This compares with a reported range in a of 0.8–0.94 [115, 157, 159, 162] and in b of 0.25–0.35 [156–162].

The above analysis serves as a useful guide to suitable geometrical parameters of the cell and electrodes employed in the study of reaction mechanisms. It is apparent from Fig. 49 that, provided $s \leq 1$, the investigation of electrode reaction mechanisms can be set up so that mass transport to the entire electrode is described in terms of the entry region alone. For a "typical" turbulent tubular electrode cell with $a = 0.11$ cm [120] and an electroactive species of $D = 5 \times 10^{-6}\,\text{cm}^2\,\text{s}^{-1}$ in a solvent with kinematic viscosity of $0.01\,\text{cm}^2\,\text{s}^{-1}$ (e.g. H_2O at 25°C), this dictates that $69.3 \geq x\,Re^{7/8}$. Hence, if solution flow rates characterised by Reynold's numbers up to 10^5 are to be employed, $l \leq 30\,\mu$m in order that the entry region constraint is not violated. Of course, the study of fast electrode kinetics necessitates the employment of short electrodes in any case (vide supra). The above merely serves to illustrate that limiting mass transport to the entry region does not place practical (constructional and operational) constraints on the electrode geometry. In the examples that follow, it is therefore assumed that mass transport is confined to the entry region.

6.2.2 Current–potential relationship for steady-state electron transfer

Vielstich and co-workers [99] modified the analysis of Blaedel and Klatt [66] to derive the steady-state current–potential relationship at the micro-tubular electrode under turbulent flow conditions. Their analysis assumed that the electrode/cell/flow rate parameters were such that mass transport to the electrode could be considered to be controlled by the laminar sub-layer. In terms of the parameter

$$u_t = \frac{0.67 I_{\mathrm{LIM}}^{\mathrm{A}}}{A[\mathrm{Red}]^{\infty}[k_{\mathrm{b}} + (I_{\mathrm{LIM}}^{\mathrm{A}}/I_{\mathrm{LIM}}^{\mathrm{C}})k_{\mathrm{f}}]} \tag{166}$$

where k_f and k_b are the forward and back rate constants for the electron transfer process (defined in terms of the cathodic reaction) given by eqns. (55) and (56), and $I_{\mathrm{LIM}}^{\mathrm{A}}$, $I_{\mathrm{LIM}}^{\mathrm{C}}$ are the anodic and cathodic mass-transport limited currents given by eqn. (106), it was found that the current–potential behaviour relative to a reversible process was again given by eqn. (57), where u_t replaces u, and I_{rev} can be written as

$$I_{\mathrm{rev}} = \frac{I_{\mathrm{LIM}}^{\mathrm{A}}[1 - \exp\{-F(E - E^{\ominus})/RT\}]}{1 + (D_{\mathrm{Red}}/D_{\mathrm{Ox}})^{2/3}\exp\{-F(E - E^{\ominus})/RT\}} \tag{167}$$

Equations (57), (166), and (167) provide a basis for the calculation of the kinetic parameters α and k_s if the steady-state current–potential behaviour characterising the electrode reaction in question is observably different from the behaviour of a reversible process ($I/I_{\mathrm{rev}} < 0.95$) over a reasonable potential range. The evaluation of the ratio I/I_{rev} at several potentials allows the corresponding value of u to be deduced via eqn. (57). Equation (166) then indicates that a plot of $\ln u_t$ vs. $(E - E^{\ominus})$ allows the determination of (a) the transfer coefficient, α, from the limiting slope (where either k_f or k_b is "out-run" by the "reverse" reaction) and (b) the standard rate constant, k, by extrapolating the limiting slope to zero overpotential.

Whilst the kinetic parameters of an electron–transfer reaction can be obtained in an identical fashion under laminar conditions [where u is now given by eqn. (58)] as illustrated by Blaedel [66], it is evident that the dependence of u on the cube root of the solution velocity in the laminar case [eqn. (58)] compared with the $\frac{7}{12}$-dependence under turbulent conditions [eqn. (166)], implies that faster electron–transfer reactions can be investigated via the latter route. This is best illustrated with a practical example. Using flow rates characterised by Reynolds numbers up to 2×10^5 at a tubular electrode 7 μm in length within a tubular cell of radius 5 mm, Vielstich and co-workers [99] were able to measure α and k_s for the ferro–ferricyanide redox couple (at 33.5°C). Their experimental data, in terms of a plot of $\ln u_t$ vs. $(E - E^{\ominus})$, is represented in Fig. 50. The slope of both of the linear

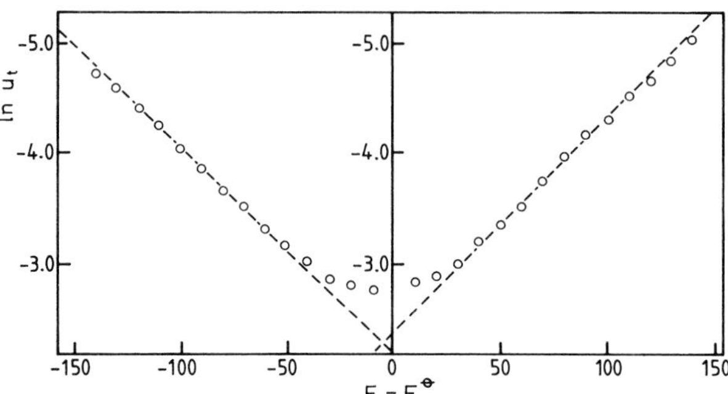

Fig. 50. Plot of $\ln u_t$ vs. $(E - E^{\ominus})$ for the ferro-ferricyanide redox couple as measured at a turbulent tubular electrode 7 μm in length employing a mean solution velocity of 12 m s^{-1}. The data are taken from ref. 99.

portions of the plot yield the expected value for α of 0.50. The standard rate constant, k_s, is evaluated to be 1.3 cm s^{-1} (upon extrapolation of both linear segments to zero overpotential).

The strength of the technique in this application is demonstrated by Fig. 51, which depicts the current–potential behaviour (calculated from the experimental data in Fig. 50) with reference to that of a reversible process. It is clear that the kinetic parameters of the ferrocyanide/ferricyanide couple, a typical reversible electrode reaction [163], can be measured. Vielstich and co-workers have suggested [99] that, with the condition $I/I_{rev} < 0.95$ (over a reasonable potential range), the technique can be used to measure k_s values up to 5 cm s^{-1} with an accuracy of around 10%. This

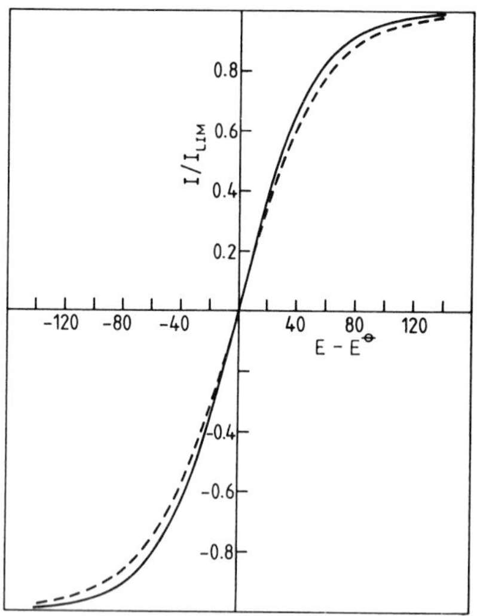

Fig. 51. Comparison of the current–voltage behaviour of ——, a reversible electrode process and – – –, that for the ferro–ferricyanide redox couple as measured at a turbulent tubular electrode, under the conditions outlined in Fig. 50.

TABLE 14

The approximate maximum measurable values of k_s; a comparison of various experimental methods

Method	$k_s/\mathrm{cm\,s^{-1}}$	Ref.
Steady-state RDE voltammetry (laminar hydrodynamics)	0.02	99
Potential-step chronoamperometry	0.1	164
Chronocoulometry	0.1	68
Chronopotentiometry	0.2	165
Cyclic voltammetry	1	166
A.c. impedance	3	68
Micro-tubular electrode (turbulent hydrodynamics)	5	99
Faradaic rectification	10	165
Microelectrode voltammetry	20	98

compares extremely favourably with other experimental techniques, as shown in Table 14.

6.2.3 Electrode kinetics and coupled homogeneous kinetics: the measurement of the rate of fast chemical reactions

Following the discussion in the previous section, it should be evident that the principle allowing the measurement of fast heterogeneous electron transfer reactions also carries over to the investigation of homogeneous chemical reactions coupled to electron transfer steps. Vielstich and co-workers have stated [114, 115] that the technique can be used to measure preceding chemical reactions with first-order rate constants of up to $10^8\,\mathrm{s^{-1}}$; whilst the adoption of the double electrode in turbulent tubular flow allows following chemical reactions with rate constants of up to $4 \times 10^6\,\mathrm{s^{-1}}$ to be characterised [115, 120]. Here, we will briefly review the two approaches.

(a) CE mechanism

This process was discussed both in the Introduction and in Sect. 4.3. A particularly interesting case of this class of mechanism concerns the dissociation of a weak acid, HA, in aqueous solution to yield H^+, which undergoes reduction at the electrode

$$\mathrm{HA} \underset{k_a}{\overset{k_d}{\rightleftharpoons}} \mathrm{H^+} + \mathrm{A^-} \qquad \text{(xxvii)}$$

$$\mathrm{H^+} + \mathrm{e^-} \longrightarrow \tfrac{1}{2}\mathrm{H_2} \qquad \text{(xxviii)}$$

HA and A^- are electro-inactive at the potential of interest.

Whilst it has already been illustrated that hydrodynamic electrodes operated under laminar regimes can be used to investigate the dissociation/association kinetics of acetic acid by measuring the "sub-Levich" behaviour at high rotation speeds (or flow rates) (see Table 1 and Fig. 1), other acids (e.g. monochloroacetic or formic acids) dissociate rapidly on the "laminar flow time scale", so that Levich behaviour is observed at all practical rotation speeds (or solution velocities). By switching to a turbulent regime, however, Vielstich and co-workers have suggested that the resulting rates of mass transfer are high enough to compete with the dissociation kinetics of both formic and monochloroacetic acids, allowing the measurement of the appropriate rates constants.

Under the reaction-layer concept, the relationship between the mass transport-limited flux, influenced by a preceding chemical reaction, J_k, and the transport-limited flux that would result if all HA transported to the electrode reacted, J_d, is [115]

$$\frac{J_k}{J_d} = \frac{1}{1 + J_d(D_{H+}K_{eqm}k_d)^{-1/2}([HA]^{\infty})^{-1}} \tag{168}$$

where

$$K_{eqm} = \frac{K_d}{[A^-]^{\infty}} \tag{169}$$

K_d being the dissociation constant of the weak acid in question. Provided that mass transport is confined to the entry region, the ratio of the two (kinetic, I_k, and mass-transport, I_d) limited currents are given by

$$\frac{I_k}{I_d} = 1 - 2u_{tk} + 2u_{tk}^2 \ln(1 + u_{tk}^{-1}) \tag{170}$$

where

$$u_{tk} = \frac{0.67 I_d}{A[HA]^{\infty}(D_{H+}K_{eqm}k_d)^{1/2}} \tag{171}$$

Hence, provided that the rate of mass transport is such that I_k is observed to be lower than predicted values of I_d, the product $K_{eqm}k_d$ can be evaluated via eqns. (170) and (171). An independent measurement of K_{eqm} then allows k_d (and thus k_a) to be determined. Figure 52 shows the results obtained by Bernstein and Vielstich [114] for the monochloroacetic acid system in terms of a plot of the measured mass transport-limited current against $Re^{7/12}$. The straight line represents the predicted mass-transport limited current behaviour for a kinetically stable species. On the basis of the results in Fig. 52, Bernstein and Vielstich [114] evaluated k_a and k_d for monochloroacetic acid to be $2.5 \times 10^{10}\ M^{-1}\ s^{-1}$ and $4 \times 10^7\ s^{-1}$, respectively, with an error of $\pm 50\%$. Results obtained on formic acid allowed k_a and k_d to be calculated as $2.6 \times 10^{10}\ M^{-1}\ s^{-1}$ and $8.4 \times 10^6\ s^{-1}$, respectively, with an error estimated to be $\pm 15\%$. This degree of accuracy is better than that which can be obtained via RDE steady-state voltammetry [167] which entails errors of up to 50% in this application.

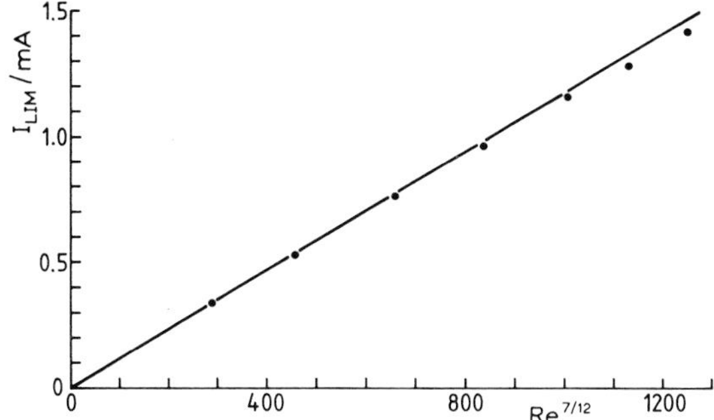

Fig. 52. I_{LIM}–$Re^{7/12}$ characteristics for the reduction of H^+ from a monochloroacetic acid system comprising an aqueous solution of 2×10^{-3} M $ClCH_2COOH$, 2×10^{-2} M $ClCH_2COONa$, and 1 M NaCl. The points represent the behaviour determined experimentally by Bernstein and Vielstich [114], whilst the straight line represents the behaviour predicted for a kinetically stable species.

(b) EC′ mechanism

EC′ reactions (and the rate of the catalytic step) are readily investigated at single channel electrodes under laminar flow conditions by measuring the catalytic current as a function of flow rate, as discussed in Sect. 4.4. Double channel electrodes can also be put to this task by measurement of the collection efficiency (Sect. 6.1.2). In this last case, Vielstich and co-workers [120] have recommended that, when the rate of the homogeneous regenerative catalytic step is very fast, advantages lie in the employment of turbulent flow in order that the ratio I_{det}/I_{gen} is maximised as far as possible. By again setting up the experimental system so that mass transport is confined to the entry region, they modified the parameter Λ_{cat} [derived by Matsuda and given by eqn. (103)] such that

$$\Lambda_{cat}^{turb} = \left(\frac{k}{D^{1/3}}\right)^{1/2} \left[\frac{x_1 \nu^{3/4} (2a)^{1/4}}{0.03325 \bar{U}^{7/4}}\right]^{1/3} \tag{172}$$

$\bar{U}$ can be evaluated in situ via eqn. (106).

Replacing the "laminar parameter" Λ_{cat} with Λ_{cat}^{turb} renders both eqns. (121) and (122) and the working curves in ref. 126 applicable to the determination of k under turbulent flow conditions. Experimental evidence for this has been presented by Vielstich and co-workers [120] from measurements on the catalytic system

upstream electrode: $Fe^{3+} + e^- \longrightarrow Fe^{2+}$ (xxix)

solution: $\frac{1}{2}H_2O_2 + Fe^{2+} + H^+ \xrightarrow{k} H_2O + Fe^{3+}$ (xxx)

downstream electrode: $Fe^{2+} \longrightarrow Fe^{3+} + e^-$ (xxxi)

in the presence of the species $N(C_2H_4OH)_3$, where $[N(C_2H_4OH)_3] \simeq 100[Fe^{3+}]$.

6.3 CHANNEL ELECTRODES IN THE STUDY OF PITTING CORROSION

The resistance of many metals and alloys to corrosion depends critically upon the presence of a thin (10–1000 Å [168]) passive surface film [169]. In "aggressive" environments, this film may become damaged locally via several processes, e.g. surface stress effects (either flow-induced [170, 171] or as a result of anion adsorption [168]), the impingement of small particles on the surface [169], spontaneous depassivation [169]. Retention of the protective film by the metal only results if repassivation of the unprotected area is feasible compared with pit growth.

It is now very well established that the pitting corrosion of a number of metals and alloys is accompanied by the formation of metal salt films [172–181]. In particular, Beck and Alkire [181] predicted that the current densities resulting from pit initiation could be large enough to produce a supersaturated metal salt solution in the vicinity of the metal surface with subsequent, rapid precipitation of a salt film. For some metals, the precursor to the re-establishment of passivity is the precipitation of a metal salt film [172–180]; for others (e.g. certain metals in halide-containing mineral acids [169, 181]), the precipitated metal salt hinders pit repassivation.

It can be readily envisaged that the rate of transfer of fresh solution to a pit will be an important factor in whether the precipitation process occurs and, by extension, in the fate (growth or death) of the pit. It has recently been recognised that channel electrodes are a useful experimental system by which the effect of flow on pitting corrosion can be studied [111–113, 169] for the following reasons.

Firstly, the effect of both laminar and turbulent flow regimes is readily investigated, as illustrated in Sects. 6.1 and 6.2, and secondly, direct observation of the electrode area under study may be achieved by incorporating a window in the channel cell, permitting light microscopy. In this way, electrochemical measurements can be linked directly to observable events on the surface, such as metal salt film precipitation and pit growth and death.

This section provides a brief review of both the available theoretical models and the experiments relating to pitting corrosion at channel electrodes. A thorough coverage of pitting corrosion per se is beyond the scope of this chapter and the reader is referred to ref. 182 for further detail.

6.3.1 Theoretical models

In this sub-section, we present the available theoretical descriptions for the effect of convection on pitting corrosion and metal salt deposition. Three models described in the literature are relevant: these are discussed in the next three sections.

(*a*) *The Alkire and Cangellari model of the effect of flow on salt formation at a channel electrode* [169]

Given that the precipitation of metal salts is an integral part of pitting, the modelling of the effect of flow on this process provides useful information on the pitting process itself. With this aim, Alkire and Cangellari developed a theory for salt formation at a channel electrode undergoing dissolution. The problem was approached with the following assumptions.

(i) The velocity profile under both laminar and turbulent flow is linear close to the electrode.

(ii) The diffusion layer is thin in comparison with the height of the channel.

(iii) Diffusion in the direction of flow is negligible in comparison with convection.

(iv) The height of the channel is such that the working electrode reaction is not affected either by the size of, or reactions at, the counter electrode (located on the opposite wall).

(v) The electrode dissolution reaction is driven by a large over-potential (~ 1 V).

(vi) The dissolved metal ions are not involved in hydrolysis reactions.

(vii) Migration of dissolved metal ions is neglected owing to the presence of sufficient background electrolyte.

Assumption (i) represents the Lévêque approximation (vide supra), whilst (ii) and (iii) embody the assumptions used in the derivation of the Levich equations (Sect. 2.1). Alkire and Cangellari pointed out that, whilst in cases where the metal salt is of high solubility migration due to the metal ions and anions could no longer be neglected, the resulting increase in mass transfer was offset by a decrease of corresponding magnitude in the diffusion coefficients of the diffusing species. With this argument, assumption (vii) was considered to be reasonable.

Under the assumptions outlined above, the model predicted the current, potential, and concentration distributions on the channel electrode surface for both laminar and turbulent flow as a function of a dimensionless average current and an average saturation current (where the saturation current is the current when the concentration of metal ions on the dissolving electrode is both uniform and equal to the saturation concentration), defined by the parameters ζ, N, and f_{sc}. ζ describes the magnitude of the ohmic resistance of the solution to the charge-transfer resistance and is given by

$$\zeta = \frac{nFl\langle i \rangle}{RT\kappa} \tag{173}$$

$\langle i \rangle$ is the current density averaged over the surface of the electrode and κ is the conductivity ($\Omega^{-1}\,\text{cm}^{-1}$) of the solution. N_l and N_t refer to the dimensionless anodic current densities for laminar and turbulent flow, respectively, under the condition that the concentration of the dissolving species on the electrode surface is at the saturation level

$$N_l = 1.5\,\frac{n^2F^2}{RT\kappa}\,\frac{D(c_{\text{sat}} - c^{\infty})}{\Gamma(\frac{4}{3})}\,\frac{6^{1/3}}{9}\left(\frac{l}{2h}\right)^{1/3}\text{Pe}^{1/3} \tag{174}$$

$$N_{\text{t}} = 0.1857 \frac{n^2 F^2}{RT\kappa} \left(\frac{l}{h}\right)^{1/8} \left(\frac{\text{Pe}}{\text{Sc}}\right)^{7/8} D(c_{\text{sat}} - c^{\infty}) \tag{175}$$

with c_{sat} as the saturation concentration with respect to metal salt formation, and Pe representing the Péclet number, given (in this case) by

$$\text{Pe} = \frac{\bar{U}l}{D} = \frac{\text{Re}\,\text{Sc}l}{h} \tag{176}$$

Finally, f_{sc} is defined in terms of the average current density and the average saturation current density by the equation

$$f_{\text{sc}} = \frac{\langle i \rangle}{\langle i_{\text{sat}} \rangle} = \frac{\zeta}{N} \tag{177}$$

The potential distribution, in terms of the applied potential, V_{A}, relative to the ohmic drop, V_0, obtained by Alkire and Cangellari for various values of the parameters ζ and N are shown in Fig. 53. This figure illustrates that, as ζ takes on larger values (corresponding to a large applied current under conditions of low solution conductivity), the potential distribution becomes non-uniform with the upstream and downstream edges of the electrode experiencing the higher potentials.

The current distribution, in terms of $i/\langle i_{\text{sat}} \rangle$, is illustrated by Fig. 54. Large current densities at the upstream and downstream edges of the electrode result because the potential distribution is controlled by the ohmic resistance (large ζ), as illustrated in Fig. 53. As the solution velocity is increased (N_{l} or N_{t} taking on larger values), the current density over the whole of the electrode surface decreases. Eventually, at a sufficiently high rate of mass transfer, the curent density at all points on the surface is lower than that needed for saturation of the adjacent solution.

The theoretical model also allows the surface concentration of the dissolving species to be evaluated. Figure 55 illustrates the concentration

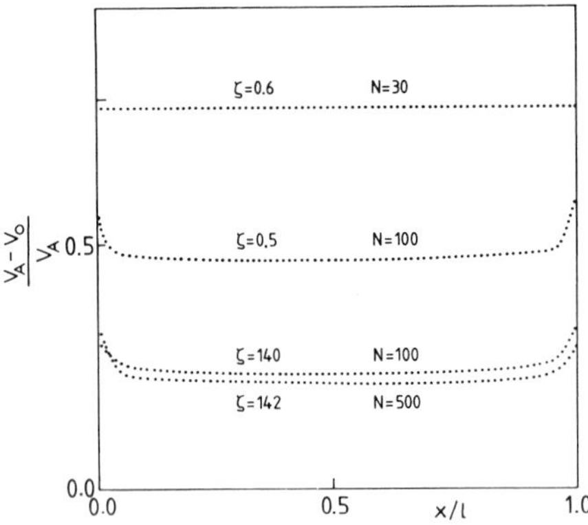

Fig. 53. The potential distribution along a channel electrode surface, as calculated by Alkire and Cangellari [169], for various values of the parameters ζ and N.

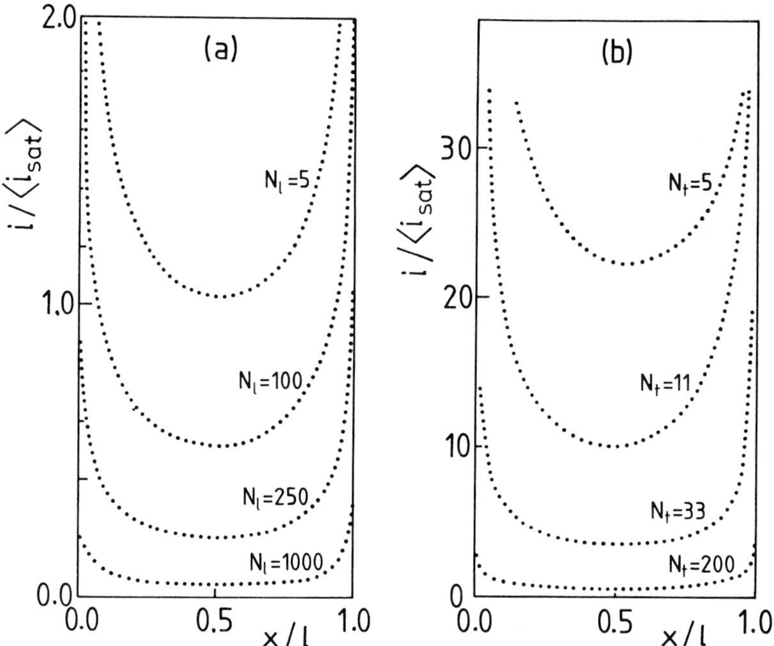

Fig. 54. Current distribution along the channel electrode surface, as calculated by Alkire and Cangellari [169], for various values of N under (a) laminar and (b) turbulent flow regimes. In the former case, the data are for $\zeta = 100$, whilst in the latter, they refer to $\zeta = 300$.

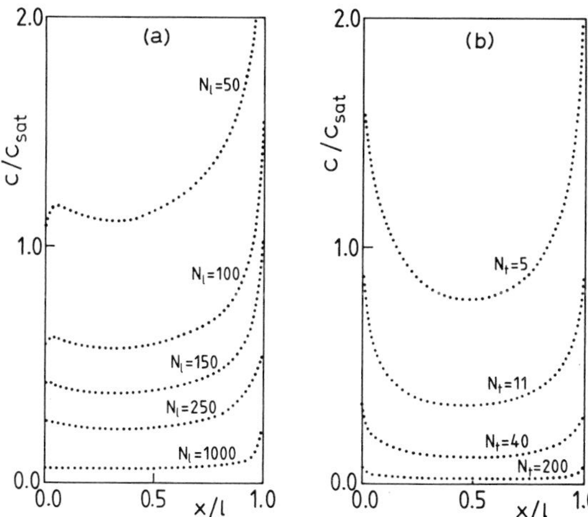

Fig. 55. Concentration distributions along the channel electrode surface, as calculated by Alkire and Cangellari [169], for various values of N under (a) laminar and (b) turbulent flow regimes. The values of ζ employed are as for Fig. 54.

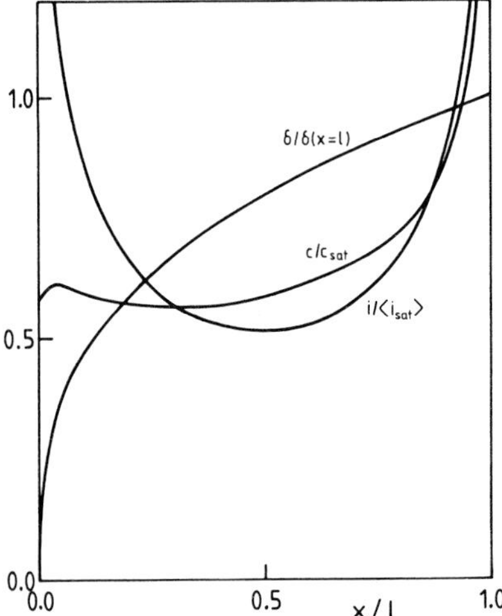

Fig. 56. Concentration, current and diffusion layer variation along the channel electrode under a laminar flow regime for $N_l = 100$ and $\zeta = 100$.

distributions obtained by Alkire and Cangellari at several solution velocities under laminar and turbulent flow. Notice that, whilst switching between laminar and turbulent conditions produced little change in the shape of the current and potential distributions (Figs. 53 and 54), the concentration distributions under the two flow regimes are markedly different. This arises because of the disparity in the shape of the diffusion layers for laminar and turbulent flow. Under laminar conditions, even though the current–x distribution at the upstream and downstream edges of the electrode are very similar, the non-uniform accessibility of the electrode (remember that the diffusion layer thickness varies as $x^{1/3}$) results in species formed at the upstream edge of the electrode escaping into bulk solution more readily than those formed further downstream. Consequently, the surface concentration takes on the shape shown in Fig. 56. Note that this effect is not as pronounced when the flow is turbulent, since the mass transfer boundary layer only varies as $x^{1/3}$ in the entry region, which is confined to the upstream edge of the electrodes employed in this study [169]. A large portion of the electrode then receives mass transport in the (uniformly accessible) fully developed region, as described by Fig. 48 (Sect. 6.2).

Figure 55 demonstrates that the concentration of dissolution products in the solution adjacent to the electrode surface can be in three states: completely saturated, completely unsaturated, or partially saturated (at the downstream and upstream edges), depending upon the solution velocity employed. This led Alkire and Cangellari to derive expressions for the critical (minimum) Reynold's number, Re_{crit}, at which the concentration of dissolving species over the whole electrode surface was below the saturation level. Four cases were considered.

$\zeta < 1$ (*charge transfer control*); *laminar flow*

$$\mathrm{Re}_{\mathrm{crit}} = 0.6328 \frac{h^2 D l}{\nu} \left[\frac{i_0 \exp\{\alpha n F V_A / RT\}}{D(c_{\mathrm{sat}} - c^\infty)} \right]^3 \tag{178}$$

where i_0 is the exchange current density.

$\zeta < 1$; *turbulent flow*

$$\mathrm{Re}_{\mathrm{crit}} = 86.13 \left[\frac{h i_0 \exp\{\alpha n F V_A / RT\}}{nFD(c_{\mathrm{sat}} - c^\infty)\mathrm{Sc}^{1/3}} \right]^{8/7} \tag{179}$$

$\zeta > 1$ (*ohmic control*); *laminar flow*

$$\mathrm{Re}_{\mathrm{crit}} = 10.45 \frac{h^2 D}{l^2 \nu} \left[\frac{\kappa V_A}{f_{\mathrm{sc}} nFD(c_{\mathrm{sat}} - c^\infty)} \right]^3 \tag{180}$$

$\zeta > 1$; *turbulent flow*

$$\mathrm{Re}_{\mathrm{crit}} = 250.6 \left[\frac{h V_A}{lFnf_{\mathrm{sc}} D(c_{\mathrm{sat}} - c^\infty)\mathrm{Sc}^{1/3}} \right]^{8/7} \tag{181}$$

(b) The effect of solution velocity on the growth of a single pit

An explanation of the effect of convective mass transfer on the growth of a single pit was provided by Beck and Chan [112]. They suggested that the current due to the pitting process could be limited by

(i) the conductivity of the solution, in which case the expression for the current density at a concave pit surface (following Vetter and Strehblow [184]) is given by

$$i_\Omega = \frac{\kappa V_0}{b r_p} \tag{182}$$

where b is a geometric constant [112, 184] and r_p is the radius of the pit.

(ii) hemispherical diffusion, so that the limiting current density may be written (approximately) as [112, 184]

$$i_{\mathrm{HD}} = \frac{nFDc_{\mathrm{sat}}}{b r_p t_-} \tag{183}$$

where t_- is the transfer number.

(iii) the rate of convective mass transfer. To obtain an expression for this case, as applied to a small circular pit in channel flow, the equation derived by Levich for the dissolution at an inclusion, in the form of a strip of width d_i at a distance l_i from the upstream edge of a semi-infinite plate [185], was modified

$$i_{\mathrm{CT}} = \frac{0.5nFD^{2/3} c_{\mathrm{sat}} \bar{U}^{1/2}}{\nu^{1/6} d_i^{1/3} l_i b t_-} \tag{184}$$

bt_- may be approximated as unity [112]. For circular pits, d_i was replaced by the diameter of the disc. The term l_i was assumed to be given by the entrance length, l_e, for the establishment of fully developed laminar or turbulent flow [112]. Although Levich derived eqn. (184) for laminar flow only, Beck and Chan applied it in their studies to both laminar and turbulent conditions [112].

Equations (182)–(184) provide a basis for defining, firstly a threshold velocity above which the current density is influenced by convective mass transfer

$$\bar{U}_{thr} = 6.35\, \nu^{1/3} D^{2/3} l_e^{1/3} r_p^{-4/3} \tag{185}$$

and secondly a critical velocity, defining the point at which the current becomes limited by ohmic resistance, rather than the rate of convection

$$\bar{U}_{crit} = \frac{2.52\kappa^2 V_A^2 \nu^{1/3} l_e^{1/3} r_p^{-4/3}}{n^2 F^2 D^{4/3} c_{sat}^2} \tag{186}$$

Beck and Chan pointed out that, for the particular case of stainless steel, $\bar{U}_{crit}$ corresponded to the velocity at which the metal salt film is removed with subsequent repassivation of the metal. Notice that both $\bar{U}_{thr}$ and $\bar{U}_{crit}$ take on lower values as the pit increases in size.

(c) *The effect of convection on the transport of dissolution products from small pits*

Whilst the treatment due to Beck and Chan [112], given in the preceding section, usefully introduced the concept of a critical velocity beyond which salt films are removed, a more sophisticated treatment of the problem has recently been presented by Alkire et al. [183]. Employing the finite element method, these workers investigated the effect of convection on the removal of dissolution products from a number of two-dimensional pit geometries. Two cases were considered: firstly, when the pit surface is coated with a metal salt film, i.e.

$$\left.\frac{c}{c_{sat}}\right|_{\text{pit surface}} = 1 \tag{187}$$

where c is the concentration of dissolution products and secondly that uniform dissolution of the surface occurs, such that the rate is given by

$$\left.\frac{\partial(c/c_{sat})}{\partial n}\right|_{\text{pit surface}} = -\,\text{Sh} \tag{188}$$

where the Sherwood number, Sh, in this application is defined by

$$\text{Sh} = \frac{i r_p}{n F c_{sat} D} \tag{189}$$

In the latter case, Alkire et al. presented relationships between the Péclet number and Sh for various pit geometries as illustrated in Fig. 57. The Pe

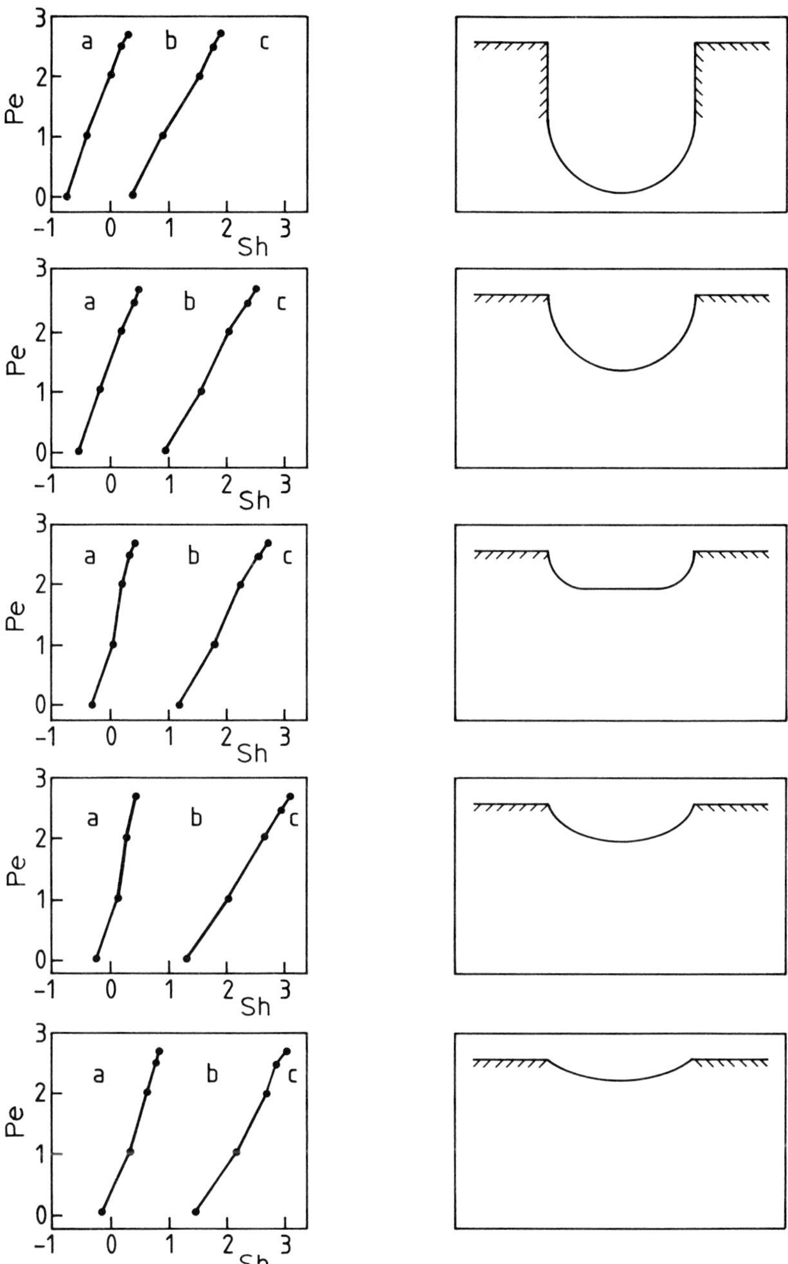

Fig. 57. The theoretical relationships between the Péclet number, Pe, and the Sherwood number, Sh, as calculated by Alkire et al. [183] for various 2-dimensional pit geometries (located adjacent). The labels a, b, and c refer to zones in which the whole surface is below saturation, partially saturated, and completely above saturation with respect to dissolution products.

number is now defined by

$$\mathrm{Pe} = \frac{u_0 r_\mathrm{p}}{D} \tag{190}$$

where u_0 is the velocity at the centre of the pit mouth, which may be calculated numerically [183]. Alternatively, Harb [113] has suggested that, to a good approximation, u_0 can be taken to be 10% of v_x at a position r_p above the wall of the channel, i.e.

$$u_0 = \frac{r_\mathrm{p}}{10} \left.\frac{\partial v_x}{\partial y}\right|_{y=h} \tag{191}$$

Figure 57 shows the general trend that, as the depth of the pit increases relative to the diameter, higher rates of convection are necessary, for a given pit radius, if the concentration of dissolution products in the solution adjacent to the pit surface is to be maintained below the saturation value. Figure 57 thus provides a basis for predicting the critical Péclet number, $\mathrm{Pe}_\mathrm{crit}$, at which the solution in the vicinity of the pit surface is kept below the saturation level with respect to the dissolving species. Notice that, this being the case, analysis in terms of $\mathrm{Pe}_\mathrm{crit}$ using eqn. (191) for u_0 predicts a critical solution velocity dependence on r_p^{-2}, in contrast to the dependence of critical velocity on $r_\mathrm{p}^{-4/3}$ in eqn. (186).

6.3.2 Experimental investigation of pitting corrosion at channel electrodes

(a) Iron

Alkire and Cangellari [169] investigated the pitting corrosion of iron in 0.5 M H_2SO_4 at the channel electrode. Applying the chronoamperometric technique, in which the potential was stepped from a region in which a slight cathodic current was being passed, to anodic potentials of 0.8, 1.0, and 1.2 V vs. NHE (such that pitting corrosion would occur in stationary solution [174]), the resulting chronoamperometric behaviour was measured as a function of flow rate. Typical results, obtained by these workers, are shown in Fig. 58. Evidently, whilst at low Reynolds numbers repassivation of the surface occurs, at higher Reynolds numbers this is no longer feasible. This is attributed to the fact that high solution velocities prohibit the saturation of the solution adjacent to the surface of the iron anode with respect to dissolution products. It is therefore concluded that, in the case of the pitting of iron in a sulphate medium, a salt film is a prerequisite to repassivation of the metal surface. Alkire and Cangellari interpreted their results in terms of their theoretical model, described in Sect. 6.3.1.(a), and found that the data was in excellent agreement with theory if a supersaturated solution, with a concentration 130% of the saturation value, was assumed to be necessary for precipitation of the salt film.

Beck and Chan [112] employed a pulsed-flow technique to study the effect of convection on the pitting corrosion of iron, in which the effect of a flow

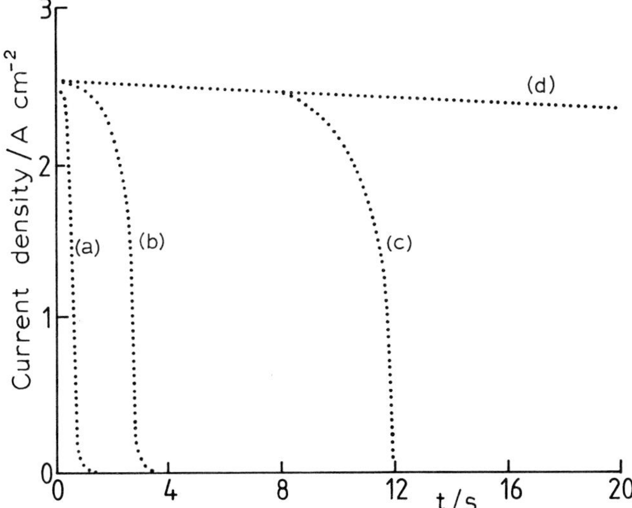

Fig. 58. The effect of Reynolds number on the chronoamperometric behaviour of an iron electrode, stepped to a potential of +1.0 V vs. N.H.E. (a) Re = 1000; (b) Re = 1900; (c) Re = 3400; (d) Re = 4800. The data are taken from ref. 169.

pulse (from stationary conditions to a predetermined velocity and back to stationary conditions) on the pitting current, was measured. The pulse was typically 0.2 s in length (see, for example, Fig. 4 in ref. 112). Figure 59 shows the results for a single artificial pit in 1 M NaCl at pH 1, created in the surface of the iron by masking off all but a 35 mm diam. circular section of the electrode [112]. By observing the surface of the electrode in conjunction with the corresponding electrochemical characteristics of the system, it is possible to show that, during the plateau current–time regions, A, B, and C in Fig. 59, a salt film is present on the surface of the electrode, which is dissipated upon application of the flow pulse whose magnitude is also shown. Beck and Chan suggested that, this being the case, the magnitude of the current at the summit of the peaks preceding the plateaus B and C could be described by eqn. (182), the effect of flow being to increase the curent to the ohmic limit. The plateaus A, B, and C may be described by eqn. (183).

(*b*) *Stainless steel 304*

The effect of convection on the pitting behaviour of stainless steel 304 is in total contrast to that of pure iron, in that removal of the metal salt film leads to the repassivation of the metal surface. Harb [113] has recently investigated the pitting-convection behaviour of single pits in a pH 3.5 solution of 0.1 M Na_2SO_4 and 0.2 M NaCl. Basing his analysis on the concept of a critical Péclet number [113], he showed that high Péclet numbers (Pe > 10) favour removal of the dissolution products from the pit surface and subsequent repassivation, whilst for flow rates and/or pit radii for which

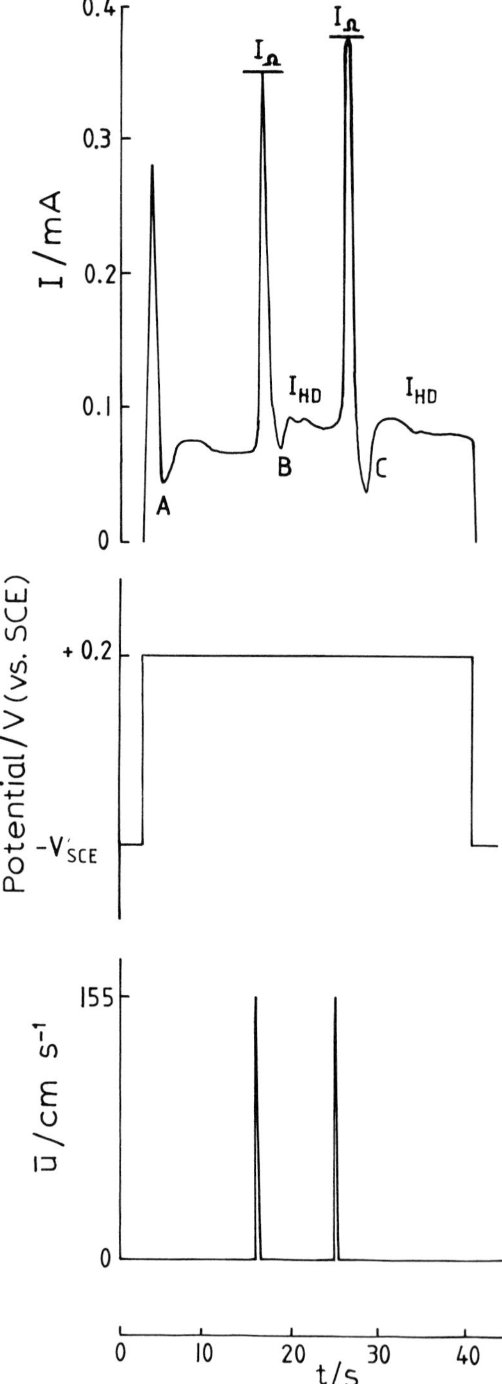

Fig. 59. Chronoamperometric behaviour for a single artificial iron pit subjected to the potential and flow formats shown. The chronoamperometric data are taken from ref. 112.

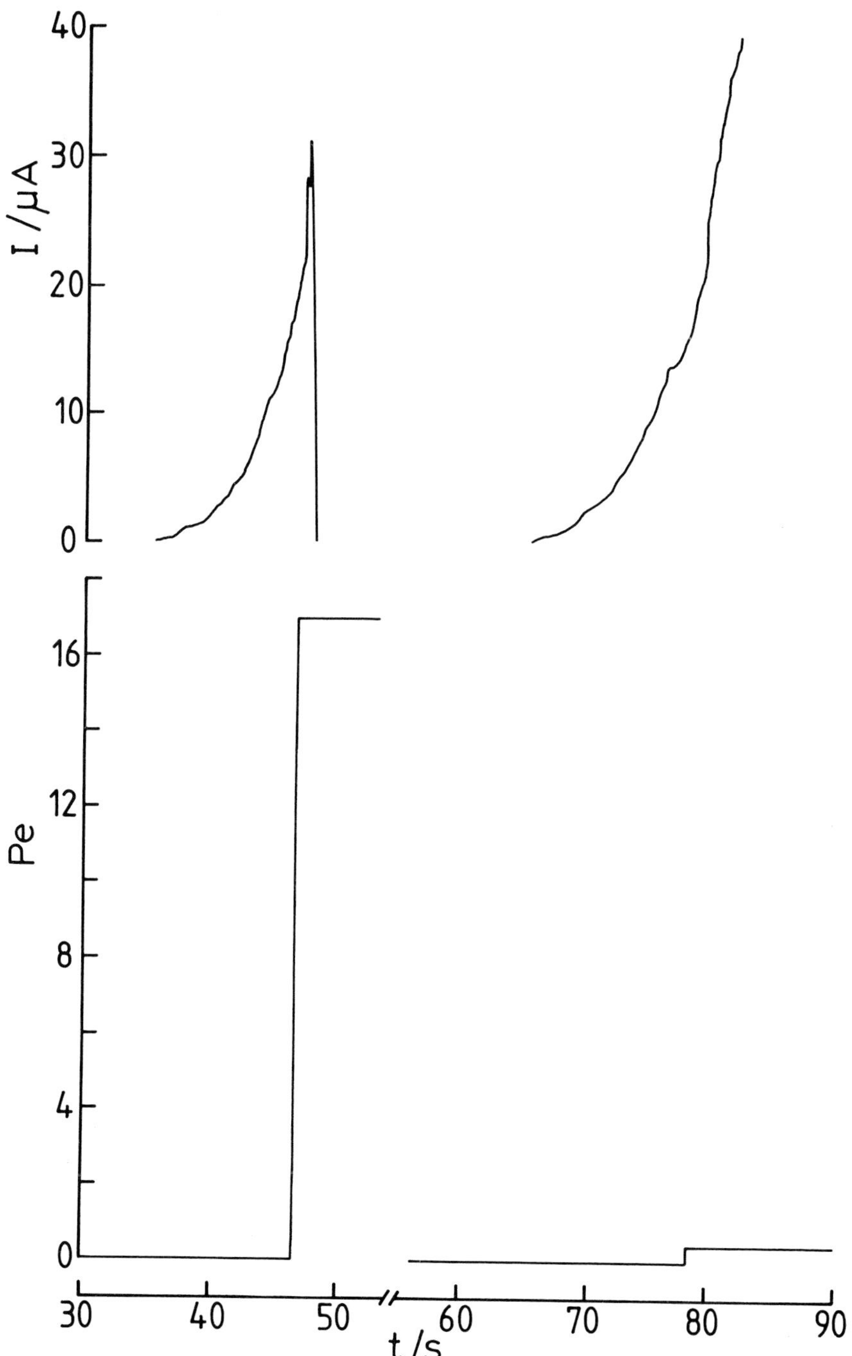

Fig. 60. The effect of convection on the pitting behaviour of stainless steel 304. "Stepping" the flow rate such that Pe = 17 results in pit repassivation, whilst for Pe < 1, continued dissolution is preferred. The data are taken from ref. 113.

Pe < 1, continued dissolution is preferred. Figure 60 serves to illustrate the dramatic effect of flow rate on the pitting behaviour.

The same phenomenon was also observed by Beck and Chan [112] in their investigation of the pitting of stainless steel in a chloride aqueous medium. Analysis in terms of the theory outlined in Sect. 6.3.1.(b) predicted that the critical velocity for repassivation varied as $r_p^{-4/3}$.

6.3.3 Concluding remarks

The above discussion indicates the recent advances that have been made in the understanding of pitting corrosion and, in particular, the effect of convection on this process, through the introduction of the channel electrode technique.

6.4 CHANNEL ELECTRODE METHODOLOGY IN THE STUDY OF THE DISSOLUTION KINETICS OF IONIC SOLIDS

The dissolution kinetics of ionic solids is currently an intense area of research activity [186–212]. This arises from the obvious environmental and technological significance, as well as the intrinsic interest in the underlying fundamental principles. In this section, we establish the merits of employing channel electrode methodology in this general area. The experiments described have general applicability, but will be illustrated by reference to our work on calcite. The extension to other systems is self-evident and involves no new principles.

6.4.1 Calcium carbonate dissolution and channel electrodes

(a) Early work and the need for controlled hydrodynamics

Initial experiments on the dissolution of calcium carbonate were made with suspensions of the powdered material [213–224]. In this way, it was shown that the dissolution rate is increased by the presence of both H^+ and dissolved CO_2. In particular, Plummer et al. [223] were able to write the following equation for the net dissolution rate of calcite, based on their work with crushed Iceland spar crystals

$$\text{Rate} = k_1 a_{H^+} + k_2 a_{H_2O} + k_3 a_{H_2CO_3} - k_4 a_{Ca^{2+}} a_{HCO_3^-} \tag{192}$$

where a_j represents the activity of the species j and k_n are rate constants; k_4 was found to be a function of P_{CO_2}, the partial pressure of CO_2 in the gas phase in equilibrium with a particular concentration of CO_2 in the solution. It was suggested that k_1 is controlled by the rate of transport of H^+ to the crystal surface whilst k_2 was ascribed to the surface reaction [223]

$$CaCO_3 + H_2O \longrightarrow Ca^{2+} + HCO_3^- + OH^- \tag{xxxii}$$

Attempts to separate surface chemical kinetic phenomena from mass transport effects have frequently employed no more experimental sophistication than the variation in the speed of a magnetic stirrer. As the reader will recognise, this approach may be hampered by the ill-defined and indescribable hydrodynamic conditions prevailing in suspensions thus agitated. In order to overcome this problem, Daly and co-workers [209, 210, 227], amongst others [196–199, 228–231], employed the rotating disc technique in which one face of a single crystal forms the disc, thereby ensuring that mass transport to the crystal surface is both well-defined and readily changed by alteration of the disc rotation speed. An added bonus of this approach is that the crystal surface can be pretreated (e.g. polished or etched) so as to provide reproducible surfaces, hence allowing comparisons between different experimental conditions to be made with confidence. Moreover, the influence of different surface treatments on the rate of the surface process is readily facilitated [209, 210].

Tne results of Daly and co-workers allowed the rate equation for the flux of dissolving Ca^{2+}, $j_{Ca^{2+}}$, in the absence of dissolved CO_2 to be given by

$$j_{Ca^{2+}} = k_1[H^+] + k_2 \tag{193}$$

where, in the case of the rotating disc set-up, k_1 is given by

$$k_1 = 1.554 D_{H^+}^{2/3} \nu^{-1/6} W^{1/2} \tag{194}$$

and k_2 is again attributed to the surface process described by eqn. (xxxii).

(b) The channel electrode system

Extension of the hydrodynamic approach in the study of dissolution to encompass channel electrodes represents a significant advance on the rotating-disc technique, both practically and in the comparison of the informational content of corresponding experiments at the two systems (vide infra). Figure 61 shows the simplest experimental channel set-up by which the dissolution kinetics of ionic solids may be studied. It consists of the type of channel electrode flow cell depicted in Fig. 31(a), but with the perspex coverplate (bearing the working electrode) replaced by one of epoxy resin into which a calcite single crystal and either an amperometric or a potentiometric "detector" electrode are cast so as to produce the crystal surface/electrode geometry shown in Fig. 61. The exposed crystal surface can then be pretreated in an analagous manner to that described by Daly and co-workers [209, 210]. Electrolyte flows over the crystal surface and, in the case of amperometric measurement, the mass transport-limited current flowing at the detector electrode due to the reduction of H^+

$$H^+ + e^- \longrightarrow \tfrac{1}{2}H_2$$

is indicative of the rate at which protons are consumed in the dissolution reaction. This is because the flux of protons consumed on the upstream $CaCO_3$ surface reduces the proton flux at the detector electrode which is thus said to be "shielded". Whilst this mode of experiment is, in principle, con-

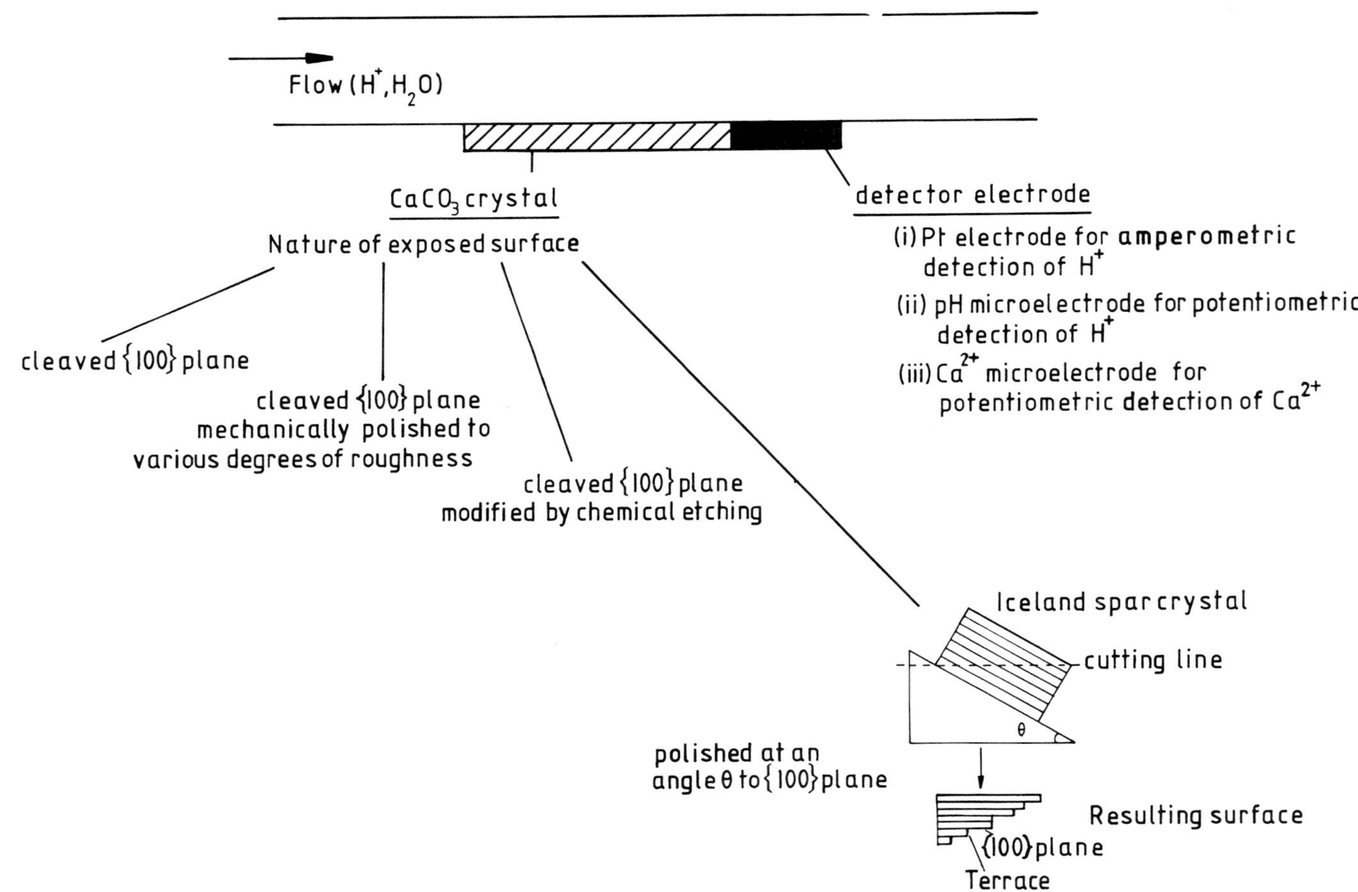

Fig. 61. Channel electrode set-up for dissolution studies.

ceivable in a ring–disc geometry, in which part of a crystal cast in a cylinder of epoxy resin forms the disc which is surrounded by an amperometric electrode, it would be impossibly difficult in practice, particularly since precise cylindrical symmetry with respect to both the ring and the disc needs to be maintained [21].

An additional merit of the channel electrode set-up is that the exposed calcite surface can be viewed in situ, by light microscopy, allowing the crystallographic nature of the dissolution process to be linked directly to kinetic measurements. Furthermore, it should be noted that ex-situ studies on the morphology of crystal surfaces subjected to dissolution in the channel set-up are likely to provide greater information, by nature of the non-uniform accessibility of the crystal surface to protons, than might be anticipated from the same measurements on a crystal surface exposed to a uniform flux of electrolyte (vide infra).

At the present time, there is considerable fundamental and applied interest in explaining the effect of inhibitors on both the dissolution and precipitation mechanisms of calcium carbonate [191, 219, 220, 231–243]. On a simple level, such species are thought to operate via adsorption at kink sites, thereby reducing the rate of dissolution (or growth) [215, 220] which is thought to proceed at these features. The experimental investigation of the kinetics of adsorption, when this is rapid, demands that the inhibitor can be introduced into a flow system, transported to the crystal surface, and monitored downstream of the crystal in a well-defined and reproducible manner. Whilst such methods are readily facilitated in channel electrode systems where inhibitors can either be electrogenerated at an electrode located upstream of the crystal (e.g. Cu^{2+} [213, 244, 245]) or else introduced via "hydrodynamic" flow injection [246–248], both of these principles are ruled out at a rotating disc, where the double electrode and hydrodynamic injection techniques would be difficult to realise practically.

Below, we describe crystal-detector electrode shielding experiments for both the dissolution process alone and in the presence of polymaleic acid. Homo- and co-polymers of maleic acid are well-documented precipitation inhibitors [249–251]. Firstly, however, we consider the theoretical relationship between the shielding current and the heterogeneous rate constant for the reaction of H^+ with the crystal surface.

6.4.2 Dissolution with amperometric detection: the relationship between the shielding of the transport-limited detector electrode current and the rate constant for the heterogeneous process on the crystal surface

Two cases require consideration: mass transport-controlled dissolution and dissolution at a finite rate. The former represents the simpler case in mathematical terms and we shall consider this first.

(a) Mass transport-controlled reaction of H^+ at the crystal surface

Figure 62 shows the crystal–electrode geometry with the appropriate

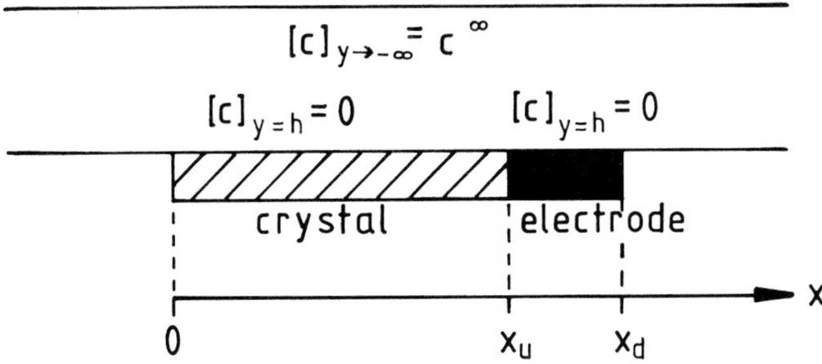

Fig. 62. Crystal–electrode geometry, notation, and boundary conditions for mass transport-controlled reaction of H^+ at the crystal surface.

co-ordinate system. The boundary conditions relevant to this case are delineated in Table 15. Notice that these are analogous to those employed in the solution of the Levich equation (Sect. 2.1). Hence, by extension, the transport-limited shielded current at the detector electrode can be written as

$$I_{\mathrm{LIM}}^{k'\to\infty} = 0.925nFc_{\mathrm{H^+}}^{\infty} D_{\mathrm{H^+}}^{2/3}\left(\frac{V_{\mathrm{f}}}{h^2 d}\right)^{1/3} w[x_{\mathrm{d}}^{2/3} - x_{\mathrm{u}}^{2/3}] \tag{195}$$

where k' denotes the heterogeneous rate constant for the dissolution process.

(b) *Reaction of H^+ on the crystal surface at a finite rate*

The consumption of H^+ on the crystal surface now proceeds at a rate such that

$$D_{\mathrm{H^+}} \left.\frac{\partial[\mathrm{H^+}]}{\partial y}\right|_{y=h} = k'[\mathrm{H^+}]_{y=h} \tag{196}$$

Figure 63 illustrates the surface concentration variation over the crystal and electrode for this case, along with the co-ordinate systems we will adopt to solve the problem. χ_1 represents a dimensionless x direction, normalized such that

$$\chi_1 = \frac{x}{x_{\mathrm{d}}} \tag{197}$$

TABLE 15

Boundary conditions for the mass transport-controlled reaction of H^+ at the crystal surface

Boundary	Condition
$0 < x \leqslant x_{\mathrm{d}};\quad y = h$	$c = 0$
$0 \leqslant x \leqslant x_{\mathrm{d}};\quad y \to -\infty (\simeq -h)$	$c \to c^{\infty}$

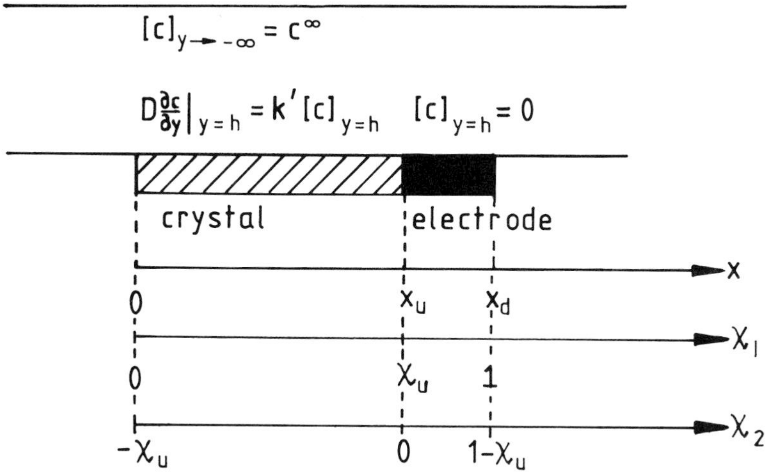

Fig. 63. Crystal–electrode geometry, notation, and boundary conditions for the reaction of H^+ at the crystal surface at a finite rate.

and χ_2 is defined by

$$\chi_2 = \chi_1 - \chi_u \tag{198}$$

where

$$\chi_u = \frac{x_u}{x_d} \tag{199}$$

The dimensionless steady-state convective–diffusion equation requiring solution is

$$0 = \frac{\partial^2 c}{\partial \xi^2} - \xi \frac{\partial c}{\partial \chi_1} \tag{200}$$

where ξ is given by eqn. (40), but l is now replaced by x_d; c denotes $[H^+]$.

In order to solve eqn. (200), c is divided into two parts [252] such that

$$c = c_1 - c_2 \tag{201}$$

and, in the domain $0 \le x \le x_u$

$$c = c_1 \tag{202}$$

The boundary conditions appropriate to the solution of eqn. (200) are delineated in Table 16 where

$$\tilde{k} = k' \left(\frac{h x_d}{3 \bar{U} D^2} \right)^{1/3} \tag{203}$$

We first seek a solution to eqn. (200) in the vicinity of the crystal. Laplace

TABLE 16

Boundary conditions for the solution of eqn. (199)

Boundary	Conditions		
$\xi \to \infty$	$c \to c^{\infty}$;	$c_1 \to c^{\infty}$;	$c_2 \to 0$
$\xi = 0;\quad 0 < \chi_1 \leqslant \chi_u$	$\dfrac{\partial c}{\partial \xi} = \tilde{k}[c]_{\xi=0}$;	$\dfrac{\partial c_1}{\partial \xi} = \tilde{k}[c_1]_{\xi=0}$;	$c_2 = 0$
$\xi = 0;\quad \chi_u < \chi_1 \leqslant 1$	$c = 0$;		$c_2 = c_1$

transformation of eqn. (200) with respect to χ_1 (transform variable p_1) gives

$$\frac{\partial^2 \bar{u}_1}{\partial \xi^2} = p_1 \xi \bar{u}_1 \tag{204}$$

where

$$u_1 = c^{\infty} - c_1 \tag{205}$$

and $\bar{f}$ denotes the Laplace transform of f. The solution to eqn. (204) is

$$\bar{u}_1 = A\mathrm{Ai}(p_1^{1/3}\xi) \tag{206}$$

where A is a constant fixed by the boundary conditions and $\mathrm{Ai}(x)$ is the Airy function. By definition [eqn. (205)], $\bar{u}_1$ is also given by

$$\bar{u}_1 = \frac{c^{\infty}}{p_1} - \bar{c}_1 \tag{207}$$

and hence

$$\frac{c^{\infty}}{p_1} - \bar{c}_1 = A\mathrm{Ai}(p_1^{1/3}\xi) \tag{208}$$

When $\xi \to 0$ (at the electrode surface), eqn. (208) may be written as

$$-\left.\frac{\partial \bar{c}_1}{\partial \xi}\right|_{\xi=0} = Ap_1^{1/3}\mathrm{Ai}'(0) \tag{209}$$

$$= -\tilde{k}\left[\frac{c^{\infty}}{p_1} - \bar{u}_1\right]_{\xi=0} \tag{210}$$

Laplace transformation of eqn. (210) together with eqn. (207) then gives

$$-\left.\frac{\partial \bar{c}_1}{\partial \xi}\right|_{\xi=0} = -\tilde{k}\left[\frac{c^{\infty}}{p_1} - A\mathrm{Ai}(0)\right]_{\xi=0} \tag{211}$$

Equating eqns. (209) and (211) leads to an equation for A

$$A = -\frac{\tilde{k}c^{\infty}}{p_1[p_1^{1/3}\mathrm{Ai}'(0) - \tilde{k}\mathrm{Ai}(0)]} \tag{212}$$

Using this result in eqn. (208) then affords the following result for $\bar{c}_1$

$$\bar{c}_1 = \frac{\tilde{k}\mathrm{Ai}(p_1^{1/3}\xi)c^\infty}{p_1[p_1^{1/3}\mathrm{Ai}'(0) - \tilde{k}\mathrm{Ai}(0)]} + \frac{c^\infty}{p_1} \tag{213}$$

Notice that as $\tilde{k} \to 0$, $c_1 \to c^\infty$, i.e. H^+ is not consumed on the crystal surface, whilst as $\tilde{k} \to \infty$, $[c_1]_{\xi=0} \to 0$, i.e. the surface process becomes mass transport-controlled. In the limit of small $\tilde{k}$, eqn. (213) becomes

$$\bar{c}_1 = \frac{\tilde{k}c^\infty \mathrm{Ai}(p^{1/3}\xi)}{p_1^{4/3}\mathrm{Ai}'(0)} + \frac{c^\infty}{p_1} \tag{214}$$

We next seek a solution to eqn. (200) as applied to the zone of the electrode. The appropriate convective–diffusion equation is

$$0 = \frac{\partial^2 c_2}{\partial \xi^2} - \xi \frac{\partial c_2}{\partial \chi_2} \tag{215}$$

which, upon Laplace transformation with respect to χ_2 (transform variable p_2)

$$\frac{\partial^2 \bar{c}_2}{\partial \xi^2} = p_2 \xi \bar{c}_2 \tag{216}$$

may be solved in Laplace space to give

$$\bar{c}_2 = B\mathrm{Ai}(p_2^{1/3}\xi) \tag{217}$$

where B is a constant. Now, by definition (boundary conditions)

$$[\bar{c}_2]_{\xi=0} = [\bar{c}_1]_{\xi=0} \tag{218}$$

$$= \mathscr{L}_{\chi_2}\mathscr{L}_{\chi_1}^{-1}[\bar{c}_1]_{\xi=0} \tag{219}$$

$$= \mathscr{L}_{\chi_2}\mathscr{L}_{\chi_1}^{-1}\left[\frac{c^\infty}{p_1} - \frac{\tilde{k}\mathrm{Ai}(0)c^\infty}{p_1[\tilde{k}\mathrm{Ai}(0) - p_1^{1/3}\mathrm{Ai}'(0)]}\right] \tag{220}$$

on substituting eqn. (213). Hence

$$\bar{c}_2 = \frac{\mathrm{Ai}(p_2^{1/3}\xi)}{\mathrm{Ai}(0)} \mathscr{L}_{\chi_2}\mathscr{L}_{\chi_1}^{-1}\left[\frac{c^\infty}{p_1} - \frac{\tilde{k}\mathrm{Ai}(0)c^\infty}{p_1[\tilde{k}\mathrm{Ai}(0) - p_1^{1/3}\mathrm{Ai}'(0)]}\right] \tag{221}$$

and upon using the inverse Laplace transform derived in Appendix 2, we find that

$$\bar{c}_2 = \frac{\mathrm{Ai}(p_2^{1/3}\xi)}{\mathrm{Ai}(0)} c^\infty \mathscr{L}_{\chi_2}\mathrm{g}(\chi_1/\beta^3) \tag{222}$$

where

$$\mathrm{g}(\chi_1) = \exp(-\chi_1) + \frac{\chi_1^{2/3}}{\Gamma(\frac{5}{3})} - \frac{\exp(-\chi_1)}{\Gamma(\frac{5}{3})}\int_0^{\chi_1} \lambda^{2/3}\exp(\lambda)\,\mathrm{d}\lambda - \frac{\chi_1^{1/3}}{\Gamma(\frac{4}{3})} + \frac{\exp(-\chi_1)}{\Gamma(\frac{4}{3})}\int_0^{\chi_1} \lambda^{1/3}\exp(\lambda)\,\mathrm{d}\lambda \tag{223}$$

and

$$\beta = -\frac{\text{Ai}'(0)}{\text{Ai}(0)} \times \frac{1}{\tilde{k}} \tag{224}$$

This gives us suitable expressions for both c_1 and c_2 in Laplace space [eqns. (214) and (222), respectively]. Now, the transport-limited current at the detector is given by

$$I_{\text{LIM}} \propto \int_{\chi_u}^{1} \left.\frac{\partial c}{\partial \xi}\right|_{\xi=0} \tag{225}$$

$$\propto \int_{\chi_u}^{1} \left\{ \left.\frac{\partial c_1}{\partial \xi}\right|_{\xi=0} - \left.\frac{\partial c_2}{\partial \xi}\right|_{\xi=0} \right\} d\chi_1 \tag{226}$$

Hence, in order to calculate the current, the appropriate functions to invert from Laplace space are

$$\mathscr{L}_{\chi_2}^{-1} \frac{1}{p_2} \left.\frac{\partial \bar{c}_2}{\partial \xi}\right|_{\xi=0} = \int_{0}^{\chi_2} \left.\frac{\partial c_2}{\partial \xi}\right|_{\xi=0} d\chi_2 \tag{227}$$

and

$$\mathscr{L}_{\chi_1}^{-1} \frac{1}{p_1} \left.\frac{\partial \bar{c}_1}{\partial \xi}\right|_{\xi=0} = \int_{0}^{\chi_1} \left.\frac{\partial c_1}{\partial \xi}\right|_{\xi=0} d\chi_1 \tag{228}$$

If we first consider eqn. (227)

$$\mathscr{L}_{\chi_2}^{-1} \frac{1}{p_2} \left.\frac{\partial \bar{c}_2}{\partial \xi}\right|_{\xi=0} = \frac{\text{Ai}'(0)}{\text{Ai}(0)} c^{\infty} \mathscr{L}_{\chi_2}^{-1} \frac{1}{p_2^{2/3}} \left\{ \mathscr{L}_{\chi_2} \text{g}\left(\frac{\chi_2 + \chi_u}{\beta^3}\right)\right\} \tag{229}$$

Applying the convolution theorem [253] allows us to write

$$\mathscr{L}_{\chi_2}^{-1} \frac{1}{p_2} \left.\frac{\partial \bar{c}_2}{\partial \xi}\right|_{\xi=0} = \frac{c^{\infty} \text{Ai}'(0)}{\Gamma(\frac{2}{3})\text{Ai}(0)} \int_{\chi_1=\chi_u}^{\chi_1=1} (1 - \chi_1)^{-1/3} \text{g}(\chi_1/\beta^3)\, d\chi_1 \tag{230}$$

Integrating by parts gives

$$\mathscr{L}_{\chi_2}^{-1} \frac{1}{p_2} \left.\frac{\partial \bar{c}_2}{\partial \xi}\right|_{\xi=0} = \frac{3c^{\infty} \text{Ai}'(0)}{2\Gamma(\frac{2}{3})\text{Ai}(0)} \left\{ (1 - \chi_u)^{2/3} \text{g}(\chi_u/\beta^3) + \int_{\chi_u}^{1} (1 - \chi_1)^{2/3} \text{h}(\chi_1)\, d\chi_1 \right\} \tag{231}$$

where

$$\text{h}(\chi_1) = \frac{d[\text{g}(\chi_1/\beta^3)]}{d\chi_1} = \frac{1}{\beta^3} g'(\chi_1/\beta^3) \tag{232}$$

and

$$g'(\chi) = \exp(-\chi) + \frac{2\chi^{-1/3}}{3\Gamma(\frac{5}{3})} + \frac{\exp(-\chi)}{\Gamma(\frac{5}{3})}\int_0^{\chi} \lambda^{2/3}\exp(\lambda)\,d\lambda - \frac{\chi^{2/3}}{\Gamma(\frac{5}{3})}$$

$$- \frac{\chi^{-2/3}}{3\Gamma(\frac{4}{3})} - \frac{\exp(-\chi)}{\Gamma(\frac{4}{3})}\int_0^{\chi} \lambda^{1/3}\exp(\lambda)\,d\lambda + \frac{\chi^{1/3}}{\Gamma(\frac{4}{3})} \quad (233)$$

Hence

$$\int_{\chi_u}^{1} \frac{\partial c_2}{\partial \xi}\bigg|_{\xi=0} d\chi_1 = \frac{3c^{\infty}\mathrm{Ai}'(0)}{2\Gamma(\frac{2}{3})\mathrm{Ai}(0)}\left\{(1 - \chi_u)^{2/3}g(\chi_u/\beta^3)\right.$$

$$\left. + \int_{\chi_u}^{1} (1 - \chi_1)^{2/3}\,\frac{g'(\chi_1/\beta^3)}{\beta^3}\,d\chi_1\right\} \quad (234)$$

We next consider an appropriate expression by which the inversion in eqn. (228) may be carried out. We deduce from eqn. (214) that

$$\frac{1}{p_1}\frac{\partial \bar{c}_1}{\partial \xi}\bigg|_{\xi=0} = -\frac{\mathrm{Ai}'(0)c^{\infty}}{\mathrm{Ai}(0)p_1^{5/3}} \times \frac{1}{(1 + \beta p_1^{1/3})} \quad (235)$$

which may be inverted by applying the inverse Laplace transform derived in Appendix 3 to give

$$\int_{\chi_u}^{1} \frac{\partial c_1}{\partial \xi}\bigg|_{\xi=0} d\chi_1 = -\frac{\mathrm{Ai}'(0)c^{\infty}}{\mathrm{Ai}(0)}\left[\frac{\chi_1^{2/3}}{\Gamma(\frac{5}{3})} - \frac{\beta\chi_1^{1/3}}{\Gamma(\frac{4}{3})} + \beta^2 - \beta^2\exp(-\chi_1/\beta^3)\right.$$

$$\left. \times\left\{1 + \frac{1}{\Gamma(\frac{2}{3})}\int_0^{\chi_1/\beta^3} \lambda^{-1/3}\exp(\lambda)\,d\lambda - \frac{1}{\Gamma(\frac{1}{3})}\int_0^{\chi_1/\beta^3} \lambda^{-2/3}\exp(\lambda)\,d\lambda\right\}\right]_{\chi_u}^{1} \quad (236)$$

Hence, from eqns. (226), (234), and (236), we deduce that the transport-limited current at the detector electrode is given by

$$I_{\mathrm{LIM}}^{k'} = -nFw\left(\frac{3\bar{U}D^2x_d^2}{h}\right)^{1/3}\frac{\mathrm{Ai}'(0)}{\mathrm{Ai}(0)}c^{\infty}\left\{\frac{3}{2\Gamma(\frac{2}{3})}\left((1 - \chi_u)^{2/3}\left[\exp(-\chi_u/\beta^3)\right.\right.\right.$$

$$+ \frac{(\chi_u/\beta^3)^{2/3}}{\Gamma(\frac{5}{3})} - \frac{\exp(-\chi_u/\beta^3)}{\Gamma(\frac{5}{3})}\int_0^{\chi_u/\beta^3} \lambda^{2/3}\exp(\lambda)\,d\lambda - \frac{(\chi_u/\beta^3)^{1/3}}{\Gamma(\frac{4}{3})}$$

$$\left. + \frac{\exp(-\chi_u/\beta^3)}{\Gamma(\frac{4}{3})}\int_0^{\chi_u/\beta^3} \lambda^{1/3}\exp(\lambda)\,d\lambda\right] + \int_{\chi_u}^{1} \frac{(1 - \chi)^{2/3}}{\beta^3}\left[\frac{2(\chi/\beta^3)^{-1/3}}{3\Gamma(\frac{5}{3})}\right.$$

$$- \exp(-\chi/\beta^3) + \frac{\exp(-\chi/\beta^3)}{\Gamma(\frac{5}{3})}\int_0^{\chi/\beta^3} \lambda^{2/3}\exp(\lambda)\,d\lambda - \frac{(\chi/\beta^3)^{2/3}}{\Gamma(\frac{5}{3})} - \frac{(\chi/\beta^3)^{-2/3}}{3\Gamma(\frac{4}{3})}$$

$$\left.\left. - \frac{\exp(-\chi_u/\beta^3)}{\Gamma(\frac{4}{3})}\int_0^{\chi_u/\beta^3} \lambda^{1/3}\exp(\lambda)\,d\lambda + \frac{(\chi_u/\beta^3)^{1/3}}{\Gamma(\frac{4}{3})}\right]\right) + \frac{(1 - \chi_u^{2/3})}{\Gamma(\frac{5}{3})}$$

References pp. 290–296

$$- \frac{\beta(1 - \chi_u^{1/3})}{\Gamma(\frac{4}{3})} - \beta^2 \exp(-1/\beta^3) \left(1 - \frac{3}{\Gamma(\frac{1}{3})} \left[\lambda^{1/3} \exp(\lambda) - \int_0^{1/\beta^3} \lambda^{1/3} \exp(\lambda)\, d\lambda \right]_0^{1/\beta^3} + \frac{3}{2\Gamma(\frac{2}{3})} \left[\lambda^{2/3} \exp(\lambda) - \int_0^{1/\beta^3} \lambda^{2/3} \exp(\lambda)\, d\lambda \right] \right)$$

$$+ \beta^2 \exp(-\chi_u/\beta^3) \left(1 - \frac{3}{\Gamma(\frac{1}{3})} \left[\lambda^{1/3} \exp(\lambda) - \int_0^{\chi_u/\beta^3} \lambda^{1/3} \exp(\lambda)\, d\lambda \right]_0^{\chi_u/\beta^3} + \frac{3}{2\Gamma(\frac{2}{3})} \left[\lambda^{2/3} \exp(\lambda) - \int_0^{\chi_u/\beta^3} \lambda^{2/3} \exp(\lambda)\, d\lambda \right] \right) \Bigg\} \qquad (237)$$

Values of $I_{\text{LIM}}^{k'}$ for specified electrode–crystal geometries and $\tilde{k}$ are readily calculated from eqn. (237) via a FORTRAN computer program and the NAG subroutines D01 BAF, D01 BAZ, and D01 GAF (which were used for the evaluation of the integrals). By comparing eqn. (237) with the Levich equation (eqn. (27)), we deduce that the shielding factor may be written as

$$S_f = \frac{I_{\text{LIM}}^{k'}}{I_{\text{LIM}}^{k'=0}} = \frac{0.9025 x_d^{2/3}}{(x_d - x_u)^{2/3}} \, G(\chi_u, \beta) \qquad (238)$$

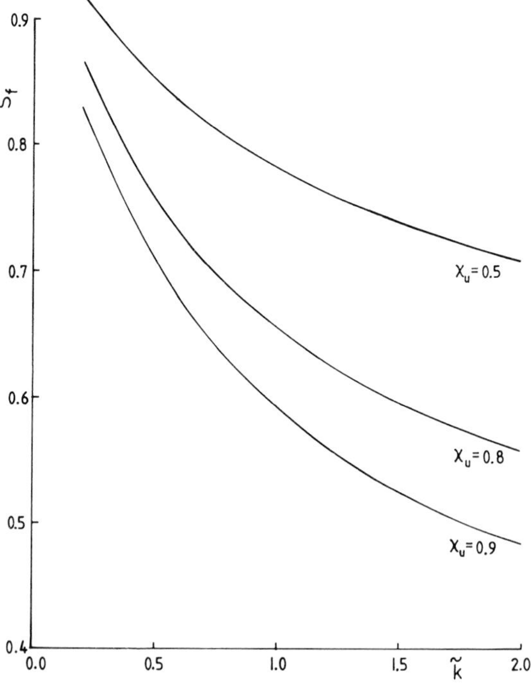

Fig. 64. "Working curves" relating the shielding factor, S_f, to $\tilde{k}$ for crystal–electrode geometries defined by $\chi_u = 0.5$, 0.8, and 0.9.

where $G(\chi_u, \beta)$ denotes the terms within the braces in eqn. (237). Figure 64 shows working curves of S_f vs. $\tilde{k}$ for various electrode–crystal geometries. These curves provide a means of interpreting the transport–limited current at the detector electrode in terms of the rate at which the reaction on the crystal takes place.

6.4.3 The dissolution of calcite at pH 3

Channel electrode methodology has been used to investigate calcite dissolution in aqueous solution at around pH 3 where the process is thought to be limited by the rate of transport of H^+ to the crystal surface [197–199, 209, 220, 223, 225,226, 230, 254–257]. Amperometric detection was used in the form of a platinum electrode held at a potential corresponding to the transport-limited reduction of H^+ [10^{-3} M]. The detector electrode was positioned downstream of a dissolving calcite crystal and, with the notation of the previous section, the cell geometry was defined by the parameters $x_u = 0.469$ cm, $x_d = 0.512$ cm. Figure 65 shows a plot of $(I_{LIM}/D_{H^+}^{2/3}[H^+]^\infty)$ against flow rate. The former quantity is a measure of the average flux of H^+ to the detector electrode. In order to determine an experimental shielding factor (see Sect. 6.4.2), the transport-limited current at the detector electrode for the reduction of ferricyanide [3×10^{-3} M] was measured. The latter does not react at the calcite surface and so

$$S_f = \frac{(I_{LIM}/D_{H^+}^{2/3}[H^+]^\infty V_f^{1/3})_{H^+}}{(I_{LIM}/D_{Fe(CN)_6^{3-}}^{2/3}[Fe(CN)_6^{3-}]^\infty V_f^{1/3})_{Fe(CN)_6^{3-}}} \qquad (239)$$

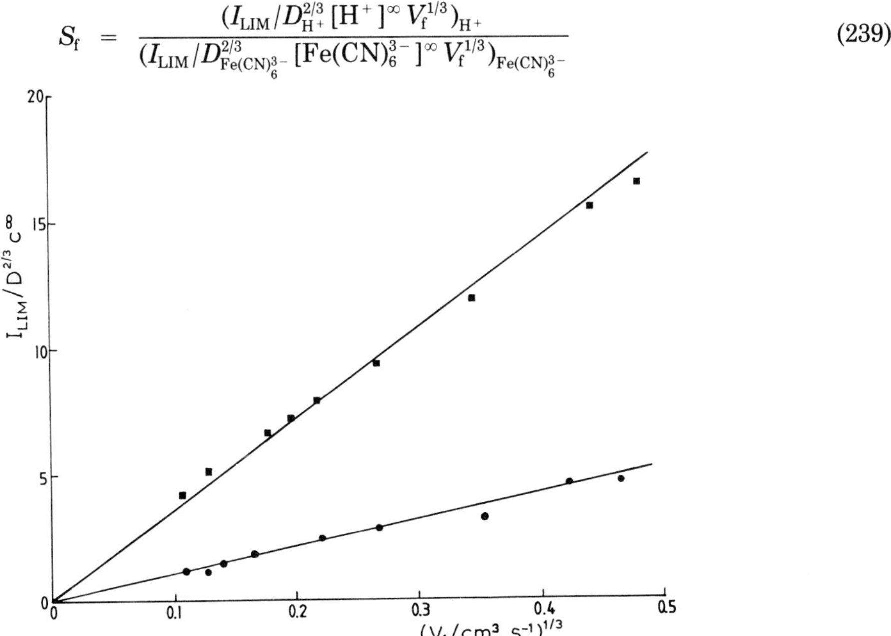

Fig. 65. Transport-limited current–flow rate behaviour for the reduction of H^+ (●) and ferricyanide (■) at a Pt detector electrode for a calcite crystal–electrode geometry defined by $x_u = 0.469$ and $x_d = 0.512$ cm. Data from the authors' laboratory.

where the terms additional to the two experimentally determined transport-limited currents, $(I_{LIM})_{H^+}$ and $(I_{LIM})_{Fe(CN)_6^{3-}}$, are needed to account for the fact that H^+ and $Fe(CN)_6^{3-}$ have significantly different diffusion coefficients ($D_{H^+} = 7.6 \times 10^{-5}\,cm^2\,s^{-1}$; $D_{Fe(CN)_6^{3-}} = 7.6 \times 10^{-6}\,cm^2\,s^{-1}$ at 25°C) and that different concentrations of H^+ and $Fe(CN)_6^{3-}$ were used in the two experiments. Experimentally, S_f is found from the ratio of the gradients of the two lines shown in Fig. 65 and this was measured as 0.298. This compares with a value of 0.296 calculated from eqns. (27) and (195), assuming transport-limited reaction of H^+ on the dissolving calcite surface. The excellent agreement demonstrates the viability of the method and confirms the nature of the kinetics of the dissolution process at this pH.

Additional mechanistic detail was obtained by using light microscopy in conjunction with the flow cell technique. Specifically, the crystal length and channel height were fixed so that, over the length of the crystal surface, there was a transition from a "Levich" regime to "thin-layer cell" behaviour, so that $[H^+]$ became substantially depleted by the downstream edge of the crystal. Plates 1 and 2 show differential interference contrast micrographs of the upstream and downstream sections of the crystal after dissolution for 120 min. Dissolution is clearly proceeding via the formation of pits on the crystal surface. There is, however, a considerable difference in the pit size in the upstream and downstream zones due to the depletion of H^+ over

Plate 1. Differential interference contrast micrograph (DIC) of the upstream zone of a cleaved calcite crystal subjected to dissolution by 10^{-3} M HCl in the channel electrode system. The bar represents 67 μm.

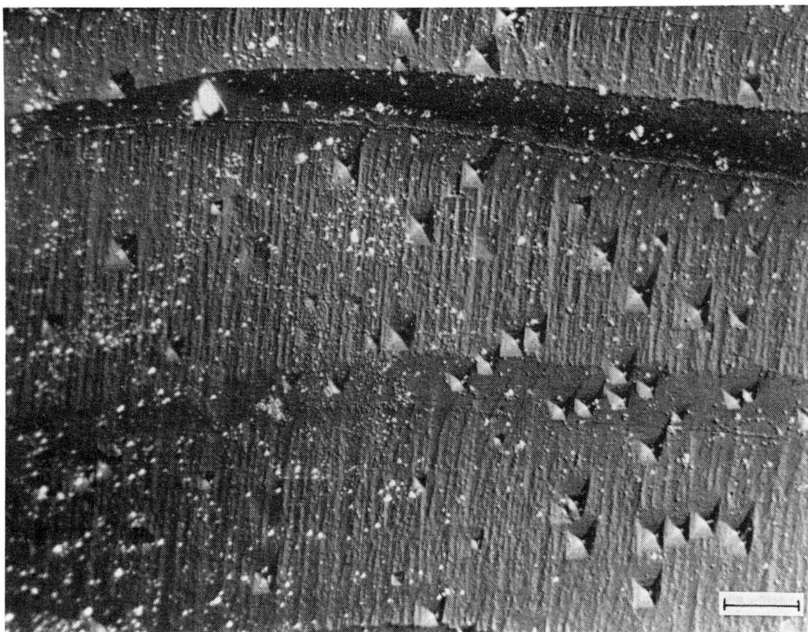

Plate 2. DIC of the downstream zone of the crystal described in Plate 1. The bar represents 67 μm.

the crystal. Importantly, however, the pit density between the two extremes remains unchanged ($1.1–1.2 \times 10^4\,cm^{-2}$). This may be taken as providing unambiguous evidence that instantaneous (rather than progressive) nucleation operates in the dissolution process. In conclusion, the combination of optical surface studies and kinetic measurements via electrochemical methods is a particularly powerful approach to the study of dissolution/precipitation kinetics. Considerably more, and unambiguous, information results than from the application of either type of measurement in isolation.

6.4.4 The dissolution of calcite in aqueous polymaleic acid (PMA) solution at around pH 3

The dramatic inhibitory effect of even trace amounts of PMA on the precipitation of calcium carbonate has long been recognised [249] and it has thus found considerable commercial application as an inhibitor of the formation of scale in hard water systems [251]. However, the mechanism by which this species inhibits the precipitation reaction remains unclear. Furthermore, to our knowledge, there have been no kinetic or morphological studies on the effect of polymaleic acid on the dissolution process.

In this section, we report results on the dissolution of calcite in PMA using the channel electrode technique. Dissolution experiments were carried out with unbuffered polymaleic acid solutions using the channel electrode set-up described in Sect. 6.4.3, a platinum electrode located downstream of the calcite crystal surface again providing a means of measuring the rate of dissolution amperometrically via the reduction of H^+. PMA solutions were made up via hydrolysis of polymaleic anhydride (wt. ave. mol. wt. 1082; no. ave. mol. wt. 891). Typically, these were of around $0.5\,g\,l^{-1}$ in concentration with respect to the polymer, corresponding (approximately) to concentrations of 4.3–8.6 $\times$ 10^{-3} M in the repeating unit

$$\left[\begin{array}{cc} -HC & -CH- \\ | & | \\ CO_2H & CO_2H \end{array}\right]$$

Potassium chloride (0.5 M) served as supporting electrolyte. Specifically, a solution of $0.885\,g\,l^{-1}$ in PMA gave rise to a solution of pH 2.70, suggesting that the average composition of such solutions can be represented in the form

$$R\left(\begin{array}{cccccccc} CH- & CH- & CH- & CH- & CH- & CH- & CH- & CH \\ | & | & | & | & | & | & | & | \\ CO_2H & CO_2H & CO_2H & CO_2^- & CO_2H & CO_2H & CO_2H & CO_2H \end{array}\right)_n R \quad + \; nH^+$$

where $n \approx 2$. Rotating disc studies on the reduction of H^+ in such solutions indicate Levich behaviour at all practical rotation speeds, with the bulk concentration of H^+, thus measured, being equivalent to that determined potentiometrically [258]. This implies that the species PMA^{2-} does not undergo further dissociation on the timescale of the RDE experiment. We may thus reasonably assume that, for the purposes of the interpretation of channel electrode PMA/calcite dissolution data, experimental shielding factors may be calculated in terms of the ratio of the observed transport-limited current for the reduction of H^+ to that predicted on the basis of the bulk concentration of H^+ in the unbuffered solution in the absence of any reaction on the crystal surface. Figure 66 displays the S_f–time behaviour calculated in this way for a crystal–electrode geometry for which transport-controlled reaction of H^+ on the crystal surface is theoretically predicted to give rise to a shielding factor of 0.333. Initially, values of S_f are very much lower than this, indicating that the rate of calcite dissolution is greater than can be explained in terms of the transport-controlled reaction of H^+ with the crystal surface alone. We suggest that this is due to the fact that the species PMA^{2-} also reacts on the crystal surface. The dissolution of $CaCO_3$ in polymaleic acid may therefore be represented by the equations

$$H^+_{(aq)} + CaCO_3 \longrightarrow Ca^{2+}_{(aq)} + HCO^-_{3(aq)} \qquad \text{(xxxiii)}$$

$$PMA^{2-}_{(aq)} + CaCO_3 \longrightarrow |Ca^{2+}PMA^{2-}|_{(aq)} + CO^{2-}_{3(aq)} \qquad \text{(xxxiv)}$$

CO_3^{2-} released according to eqn. (xxxiv) will then undergo rapid reaction with H^+ in solution

$$CO^{2-}_{3(aq)} + H^+_{(aq)} \longrightarrow HCO^-_{3(aq)} \qquad \text{(xxxv)}$$

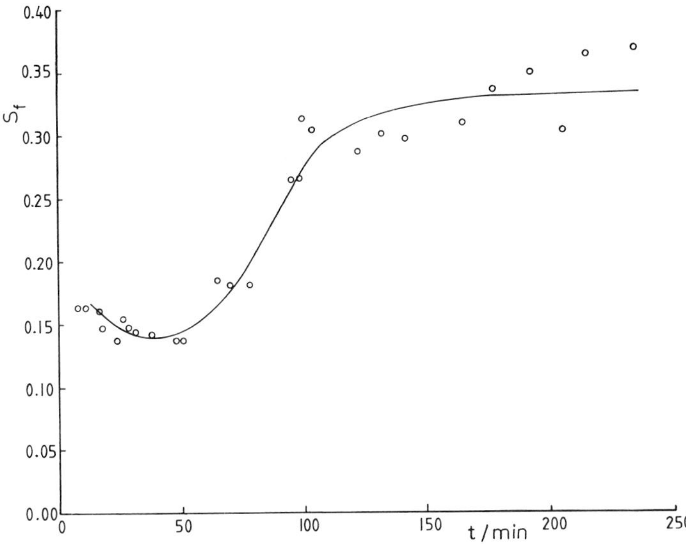

Fig. 66. The variation in shielding factor with time for the reduction of H^+ in an aqueous polymaleic acid system of concentration 7.63 mM in monomer units. The flow rate employed is $1.64 \times 10^{-2}\,cm^3\,s^{-1}$. A theoretical shielding factor of 0.333 is predicted on the basis of the crystal electrode geometry, for mass transport-controlled reaction of H^+ at the crystal surface. Data from the authors' laboratory.

accounting for the observed depletion of H^+ (in the measured shielding factor), beyond that expected for transport-controlled reaction of H^+ with $CaCO_3$ alone. The interpretation of the data in Fig. 66 in this way would suggest that the efficiency of PMA as a scale inhibitor relates, at least in part, to its ability to promote the dissolution of $CaCO_3$. This is perhaps not unexpected in the light of the known efficiency of PMA in complexing Ca^{2+} [250].

At longer times, the increase in S_f, shown in Figs. 66 and 67, may be rationalised if the calcite surface is considered to undergo transport-controlled reaction with H^+ alone, and so PMA^{2-} no longer takes part in the dissolution process. This is consistent with microscopy studies on the morphology of cleaved calcite surfaces etched in PMA [259], which indicate that, at long times, sites on the calcite surface with which PMA^{2-} reacts become exhausted.

6.4.5 *The dissolution of calcite in aqueous maleic acid solution* (3 *mM*): *a morphological study*

In this section, we briefly provide preliminary results on the morphology of etch pits obtained on cleaved calcite crystals by dissolution in unbuffered aqueous solutions of maleic acid (3 mM) under both stationary conditions and through channel electrode experiments. Kinetic and morphological studies on the effect of maleic acid on the dissolution of calcium carbonate

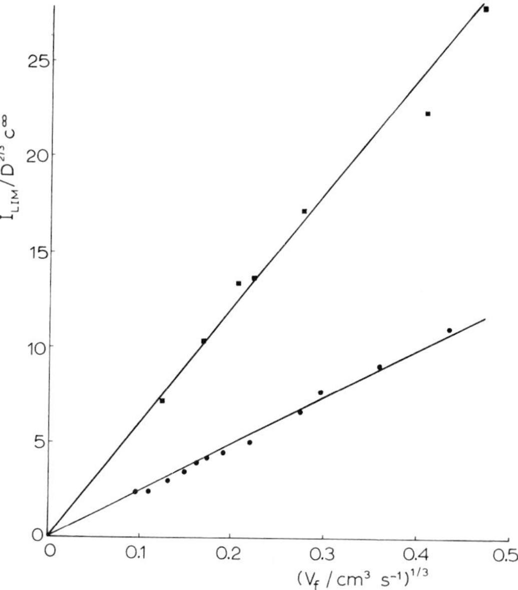

Fig. 67. Transport-limited current–flow rate behaviour for the reduction of H^+ (●) in the system defined in Fig. 66 for $t > 300$ min. Also shown is the measured behaviour for the ferricyanide system (■). Data from the authors' laboratory.

are interesting in that, firstly, they provide a useful system for comparison with the behaviour of PMA and secondly since the maleate dianion, A^{2-}, is a known precipitation inhibitor [260].

In stationary solution, the dissolution of calcite in aqueous maleic acid occurs via the formation of characteristically shaped etch pits. Plates 3 and 4 show etch pits obtained by employing 3 mM aqueous maleic acid which, unbuffered, gives rise to a 1:1 solution in H^+ and the monoanion, HA^- (pK_1 ca. 1.52). Etching times were 15 and 120 min, respectively.

Now, on the time-scale of dissolution in stationary solution, HA^- will undergo virtually complete dissociation

$$HA^- \rightleftharpoons H^+ + A^{2-}$$

where A^{2-} = *cis*-$^-O_2CHC{=}CHCO_2^-$. Hence dissolution may be considered to be in terms of $2\,H^+$ and A^{2-}. Comparison of the morphology of the etch pits in Plates 3 and 4 with those in 1 and 2 highlights the influence of the A^{2-} species on the dissolution process. This effect may be explained in terms of the adsorption of A^{2-} on the calcite surface, resulting in the inhibition of dissolution (in specific crystallographic directions), particularly given that A^{2-} is thought to adsorb readily on $CaCO_3$ [261, 262].

In contrast, channel electrode voltammetric studies at a platinum electrode on unbuffered maleic acid solutions of this concentration indicate that there is little dissociation of HA^- on the timescale of the channel electrode experiment [263]. Thus, the calcite "sees" HA^- and H^+, but little A^{2-} and,

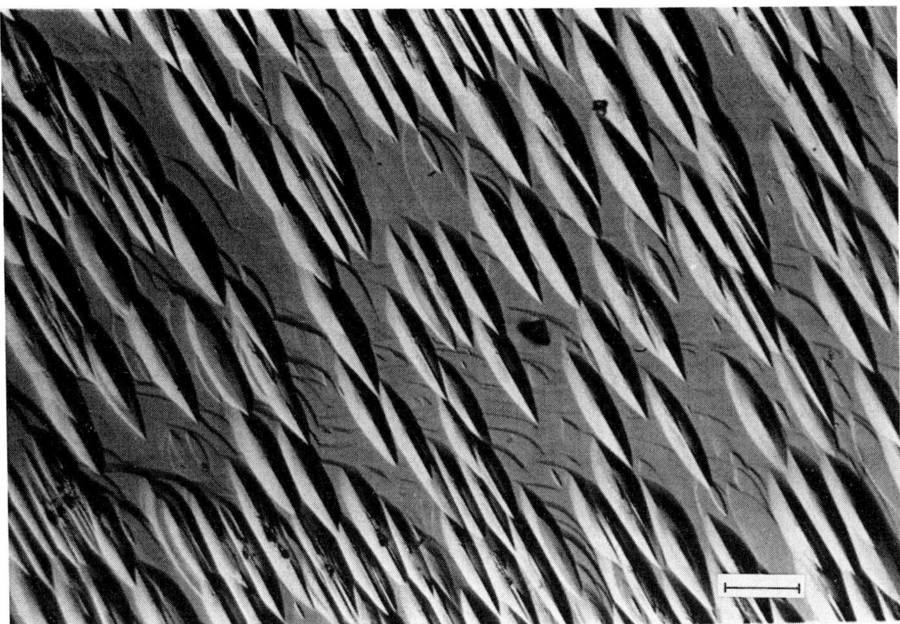

Plate 3. DIC of the etch pits resulting from the dissolution of a calcite cleavage in 3 mM aqueous maleic acid for 15 min under no-flow conditions. The bar represents 56 μm.

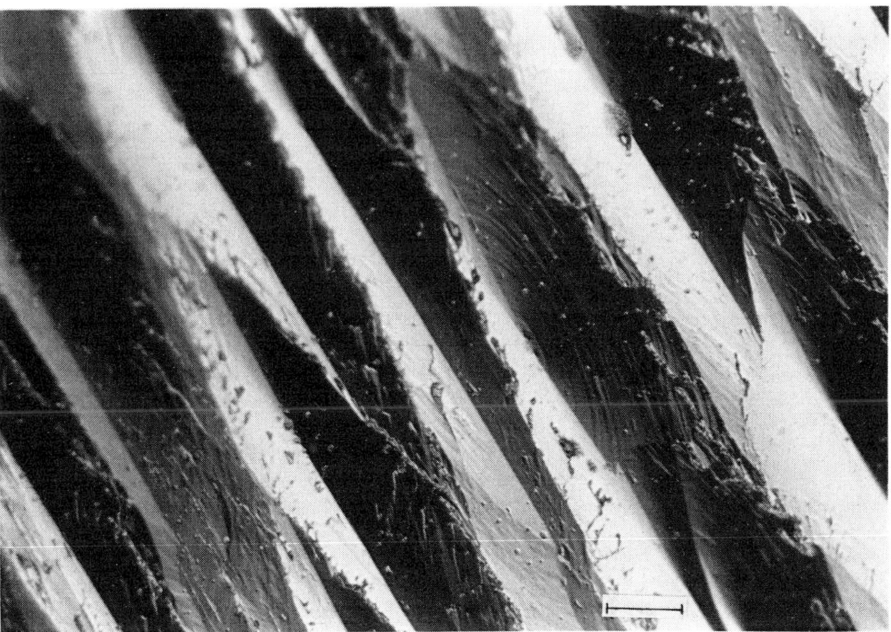

Plate 4. DIC of the etch pits on calcite after dissolution for 120 min in 3 mM aqueous maleic acid under no-flow conditions. The bar represents 56 μm.

References pp. 290–296

Plate 5. DIC of the etch pits at the upstream edge of a cleaved calcite crystal subjected to dissolution in 3 mM maleic acid solution for 45 min in a channel electrode system ($V_f = 10^{-2}$ cm^3 s^{-1}). The bar represents 56 μm.

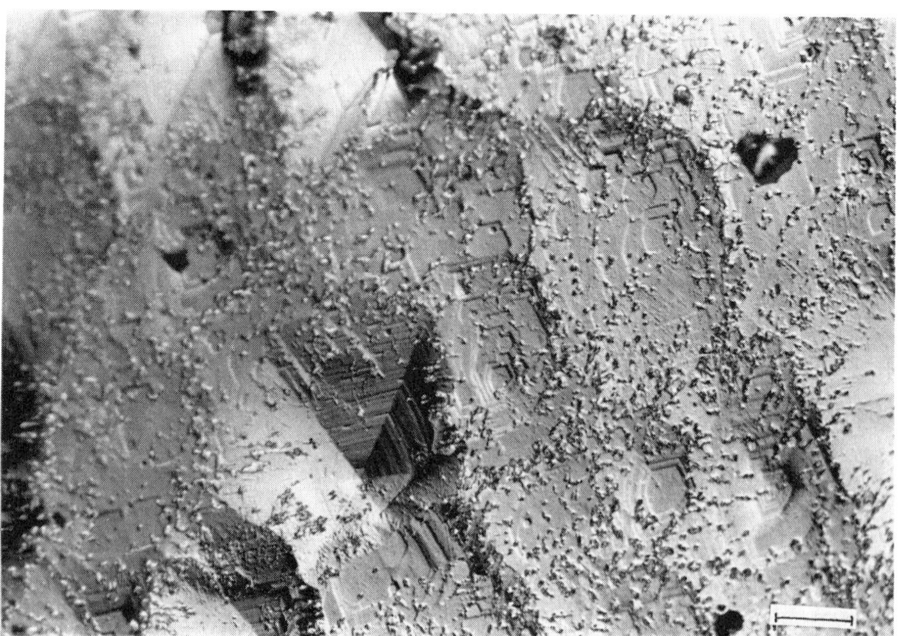

Plate 6. DIC of the downstream zone of the crystal described in Plate 5. The bar represents 56 μm.

consequently, the morphology of etch pits subjected to dissolution by maleic acid solutions in a convective system attain different structures to those resulting from dissolution in stationary solution, as can be seen by reference to Plates 5 and 6. These differential interference contrast micrographs show the etch pit morphology at the upstream and downstream edges of a calcite crystal subjected to dissolution in 3 mM maleic acid solution for 45 min at a flow rate of $10^{-2}\,\mathrm{cm^3\,s^{-1}}$.

Whilst the above results are preliminary, the implications are clear: the study of the effect of organic acid inhibitors on calcite dissolution in a convective system provides greater control than in stationary solution since the degree of acid dissociation, and thus the solution speciation, in the vicinity of the crystal can be controlled by varying the rate of convection. Furthermore, in employing the non-uniformly accessible channel electrode in conjunction with microscopy studies, we anticipate that, as in the case of the study of pure dissolution (Sect. 6.4.3), greater information can be obtained than by using a uniformly accessible system. Finally, tying together morphological studies with the measurement of the dissolution kinetics using the channel electrode system described above constitutes a powerful and comprehensive means of investigating both the kinetics of ionic solid dissolution and the effect of inhibitors on the process. Work in this direction is currently in progress.

Appendix 1

The aim of this appendix is to show how the equation

$$\frac{2v_0\sigma}{h}\frac{\partial c}{\partial x} = D\frac{\partial^2 c}{\partial \sigma^2} \qquad \text{(A1.1)}$$

is reduced to a second-order differential equation by means of the substitution

$$\eta = \left(\frac{v_0}{xh}\right)^{1/3}\sigma \qquad \text{(A1.2)}$$

Now

$$\mathrm{d}\eta = \left(\frac{\partial \eta}{\partial x}\right)_\sigma \mathrm{d}x + \left(\frac{\partial \eta}{\partial \sigma}\right)_x \mathrm{d}\sigma \qquad \text{(A1.3)}$$

and

$$\mathrm{d}c = \left(\frac{\mathrm{d}c}{\mathrm{d}\eta}\right)\mathrm{d}\eta \qquad \text{(A1.4)}$$

From which we may deduce that

$$\left(\frac{\partial c}{\partial \sigma}\right)_x = \left(\frac{\partial \eta}{\partial \sigma}\right)_x \frac{\mathrm{d}c}{\mathrm{d}\eta} = \frac{\eta}{\sigma}\frac{\mathrm{d}c}{\mathrm{d}\eta} \qquad \text{(A1.5)}$$

and

$$\left(\frac{\partial c}{\partial x}\right)_\sigma = \left(\frac{\partial \eta}{\partial x}\right)_\sigma \frac{dc}{d\eta} = -\frac{\eta^4 Dh}{3v_0\sigma^3}\frac{dc}{d\eta} \tag{A1.6}$$

It then follows that

$$\left(\frac{\partial^2 c}{\partial \sigma^2}\right)_x = \frac{\partial}{\partial \sigma}\left(\frac{\eta}{y}\frac{dc}{d\eta}\right) = \frac{\eta^2}{\sigma^2}\frac{d^2 c}{d\eta^2} \tag{A1.7}$$

and substituting eqns. (A1.6) and (A1.7) with eqn. (A1.1) we find that

$$\frac{d^2 c}{d\eta^2} + \frac{2}{3}\eta^2\frac{dc}{d\eta} = 0 \tag{A1.8}$$

which is the equation sought.

Appendix 2

In this appendix, we derive an expression for the inverse Laplace transform

$$\mathscr{L}_\chi^{-1}\frac{1}{p(1+\beta p^{1/3})} = \mathrm{f}(\chi/\beta^n) \tag{A2.1}$$

Consider the inverse Laplace transform

$$\mathscr{L}_\chi^{-1}\frac{1}{p(1+p^{1/3})} = \mathrm{f}(\chi) \tag{A2.2}$$

from which, by definition

$$\frac{1}{p(1+p^{1/3})} = \int_0^\infty \exp(-p\chi) \times \mathrm{f}(\chi)\,d\chi \tag{A2.3}$$

Now, let $y = (\chi/\beta^n)$, then

$$\int_0^\infty \exp(-p\chi) \times \mathrm{f}(\chi/\beta^n)\,d\chi = \beta^n \int_0^\infty \exp(-p\beta^n y) \times \mathrm{f}(y)\,dy \tag{A2.4}$$

$$= \frac{\beta^n}{p\beta^n(1+p^{1/3}\beta^{n/3})} \tag{A2.5}$$

$$= \frac{1}{p(1+\beta p^{1/3})} \tag{A2.6}$$

for $n = 3$. We may thus conclude that

$$\mathscr{L}_\chi^{-1}\frac{1}{p(1+\beta p^{1/3})} = \mathrm{f}(\chi/\beta^3) \tag{A2.7}$$

Now, considering eqn. (A2.2)

$$p^{-1}(1 + p^{1/3})^{-1} = p^{-1}(1 + p)^{-1} - p^{-5/3} + p^{-5/3}(1 + p)^{-1} + p^{-4/3} - p^{-4/3}(1 + p)^{-1} \quad \text{(A2.8)}$$

All the terms on the right-hand side of eqn. (A2.8) may now be inverted [264]

$$\mathscr{L}^{-1}p^{-(v+1)} = \frac{\chi^{v}}{\Gamma(v + 1)} \quad \text{(A2.9)}$$

$$\mathscr{L}^{-1}p^{-1}(1 + p)^{-1} = 1 - \exp(-\chi) \quad \text{(A2.10)}$$

$$\mathscr{L}^{-1}p^{-5/3}(1 + p)^{-1} = \frac{\exp(-\chi)}{\Gamma(\frac{5}{3})} \int_0^{\chi} \lambda^{2/3} \exp(\lambda)\, d\lambda \quad \text{(A2.11)}$$

$$\mathscr{L}^{-1}p^{-4/3}(1 + p)^{-1} = \frac{\exp(-\chi)}{\Gamma(\frac{4}{3})} \int_0^{\chi} \lambda^{1/3} \exp(\lambda)\, d\lambda \quad \text{(A2.12)}$$

Hence the required expression for $f(\chi)$ is

$$f(\chi) = 1 - \exp(-\chi) - \frac{\chi^{2/3}}{\Gamma(\frac{5}{3})} + \frac{\exp(-\chi)}{\Gamma(\frac{5}{3})} \int_0^{\chi} \lambda^{2/3} \exp(\lambda)\, d\lambda + \frac{\chi^{1/3}}{\Gamma(\frac{4}{3})} - \frac{\exp(-\chi)}{\Gamma(\frac{4}{3})} \int_0^{\chi} \lambda^{1/3} \exp(\lambda)\, d\lambda \quad \text{(A2.13)}$$

Appendix 3

In this appendix, we derive an expression for the inverse Laplace transform

$$\mathscr{L}_{\chi}^{-1} \frac{1}{p^{5/3}(1 + \beta p^{1/3})} = f(\chi) \quad \text{(A3.1)}$$

This is more conveniently written as

$$\mathscr{L}_{\chi}^{-1} \frac{1}{p^{5/3}(1 + \beta p^{1/3})} = \beta^{2} \mathscr{L}_{\chi/\beta^{3}}^{-1} \frac{1}{p^{5/3}(1 + p^{1/3})} \quad \text{(A3.2)}$$

Now

$$p^{-5/3}(1 + p^{1/3})^{-1} = p^{-5/3} - p^{-4/3} - p^{-1} + p^{-1/3}(1 + p)^{-1} - (1 + p)^{-1} - p^{-2/3}(1 + p)^{-1} \quad \text{(A3.3)}$$

All of the terms on the right-hand side of eqn. (A3.3) may be inverted via eqn. (A2.9) and

$$\mathscr{L}^{-1}(1 + p)^{-1} = \exp(\chi) \quad \text{(A3.4)}$$

$$\mathscr{L}^{-1}p^{-1} = 1 \tag{A3.5}$$

$$\mathscr{L}^{-1}p^{-1/3}(1+p)^{-1} = \frac{\exp(-\chi)}{\Gamma(\frac{1}{3})}\int_0^{\chi}\lambda^{-2/3}\exp(\lambda)\,\mathrm{d}\lambda \tag{A3.6}$$

$$\mathscr{L}^{-1}p^{-2/3}(1+p)^{-1} = \frac{\exp(-\chi)}{\Gamma(\frac{2}{3})}\int_0^{\chi}\lambda^{-1/3}\exp(\lambda)\,\mathrm{d}\lambda \tag{A3.7}$$

Hence

$$\mathrm{f}(\chi) = \frac{\chi^{2/3}}{\Gamma(\frac{5}{3})} - \frac{\beta^{1/3}\chi^{1/3}}{\Gamma(\frac{4}{3})} + \beta^{2/3} - \beta^{2/3}\exp(-\chi/\beta)$$

$$\times\left\{1 - \frac{1}{\Gamma(\frac{1}{3})}\int_0^{\chi/\beta}\lambda^{-2/3}\exp(\lambda)\,\mathrm{d}\lambda + \frac{1}{\Gamma(\frac{2}{3})}\int_0^{\chi/\beta}\lambda^{-1/3}\exp(\lambda)\,\mathrm{d}\lambda\right\} \tag{A3.8}$$

Acknowledgements

We thank our collaborators both at Oxford (Tony Barwise, Roger Bowler, Matt Pilkington, Keith Pritchard, Geoff Stearn and Andy Waller) and at Ciba–Geigy, Manchester (Alan Gerrard, Gary Grigg, Michael Lees, and Philip Sylvester) for interesting and stimulating discussions on various aspects of this work. We are also grateful to Prof. J.L. Anderson for providing us with preprints of his work and a copy of his BI program. We thank Pat Dunn and Marjorie Wallace for typing the manuscript and David Kozlow for preparing the figures.

References

1 P. Bindra, A.P. Brown, M. Fleischmann and D. Pletcher, J. Electroanal. Chem., 58 (1975) 31.
2 V.G. Levich, Physicochemical Hydrodynamics, Prentice–Hall, Englewood Cliffs, 1962, pp. 60–70.
3 C.M.A. Brett and A.M.C.F. Oliveira Brett, in C.H. Bamford and R.G. Compton, (Eds.), Comprehensive Chemical Kinetics, Vol. 26, Elsevier, Amsterdam, 1986.
4 J. Koutecky and V.G. Levich, Dokl. Akad. Nauk SSSR, 117 (1957) 441.
5 J. Koutecky and V.G. Levich, Zh. Fiz. Khim., 32 (1958) 1565.
6 W. Vielstich and D. Jahn, Z. Elektrochem., 64 (1960) 43.
7 Z. Galus and R.N. Adams, J. Electroanal. Chem., 4 (1962) 248.
8 L.K.J. Tong, K. Liang and W.R. Ruby, J. Electroanal. Chem., 13 (1967) 245.
9 R.G. Compton, D. Mason and P.R. Unwin, J. Chem. Soc. Faraday Trans. 1, 84 (1988) 473.
10 R.G. Compton, R.G. Harland, P.R. Unwin and A.M. Waller, J. Chem. Soc. Faraday Trans. 1, 83 (1987) 1261.
11 W.J. Blaedel, C.L. Olson and L.R. Sharma, Anal. Chem., 35 (1963) 2100.
12 J. Ruzicka and E.H. Hansen, Anal. Chim. Acta, 179 (1986) 1.
13 J. Ruzicka and E.H. Hansen, Anal. Chim. Acta, 114 (1980) 14.
14 J. Ruzicka and E.H. Hansen, Flow Injection Analysis, Wiley, New York, 1981.
15 J. Ruzicka, Anal. Chem., 55 (1983) 1040A.
16 D.A. Roston, R.E. Shoup and P.T. Kissinger, Anal. Chem., 54 (1982) 1417A.

17 R. Appelqvist, G.R. Beecher, H. Bergamin, F. G. Denboef, J. Emnéus, Z. Fang, L. Gorton, E.H. Hansen, P.E. Hare, J.M. Harris, J.J. Harrow, N. Ishibash, J. Janata, G. Johansson, B. Karlberg, F.J. Krug, W.E. Van der Linden, M.D. Luque de Castro, G. Markovarga, J.N. Miller, H.A. Mottola, H. Muller, G.E. Pacey, C. Riley, J. Ruzicka, R.C. Schothorst, K.K. Stewart, A. Townsend, J.F. Tyson, K. Ueno, M. Valcárcel, J. Vanderslice, P.J. Worsfold, N. Yoza and E.A.G. Zagatto, Anal. Chim. Acta, 180 (1986) 1.
18 D.C. Johnson, S.G. Weber, A.M. Bond, R.M. Wightman, R.E. Shoup and I.S. Krull, Anal. Chim. Acta, 180 (1986) 187.
19 M. Fleischmann, F. Lasserre and J. Robinson, J. Electroanal. Chem., 177 (1984) 115.
20 W.J. Albery and S. Bruckenstein, J. Electroanal. Chem., 144 (1983) 105.
21 W.J. Albery and M.L. Hitchman, Ring–Disc Electrodes, Clarendon Press, Oxford, 1971.
22 T. Singh and J. Dutt, J. Electroanal. Chem., 182 (1985) 259.
23 T. Singh and J. Dutt, J. Electroanal Chem., 190 (1985) 65.
24 T. Singh and J. Dutt, J. Electroanal Chem., 196 (1985) 35.
25 J. Dutt and T. Singh, J. Electroanal. Chem., 207 (1986) 41.
26 V. Yu. Filinovskii and Yu. V. Pleskov, in E. Yeager, J. O'M. Bockris, B.E. Conway and S. Sarangapani (Eds.), Comprehensive Treatise of Electrochemistry, Vol. 9, Plenum Press, New York, 1984, p. 298.
27 F. Opekar and P. Beran, J. Electroanal. Chem., 69 (1975) 1.
28 R.B. Bird, W.E. Stewart and E.N. Lightfoot, Transport Phenomena, Wiley, New York, 1960, p. 47.
29 V.G. Levich, Physicochemical Hydrodynamics, Prentice–Hall, Englewood Cliffs, 1962, p. 32.
30 J.S. Newman, Electrochemical Systems, Prentice–Hall, Englewood Cliffs, 1963.
31 R.G. Compton and B.A. Coles, J. Electroanal. Chem., 144 (1983) 87.
32 V.G. Levich, Physicochemical Hydrodynamics, Prentice–Hall, Englewood Cliffs, 1962, pp. 112–116.
33 M.A. Lévêque, Ann. Mines Mem. Ser. 12, 13 (1928) 201.
34 M. Abramowitz and I.A. Stegun, Handbook of Mathematical Functions, Dover, New York, 1970, p. 255.
35 J.B. Flanagan and L. Marcoux, J. Phys. Chem., 78 (1974) 718.
36 S. Moldoveanu and J.L. Anderson, J. Electroanal. Chem., 175 (1984) 67.
37 J.L. Anderson and S. Moldoveanu, J. Electroanal. Chem., 179 (1984) 107.
38 J.L. Anderson and S. Moldoveanu, J. Electroanal. Chem., 179 (1984) 119.
39 S.W. Feldberg, in A.J. Bard (Ed.), Electroanalytical Chemistry, Vol. 3, Marcel Dekker, New York, 1969, p. 199.
40 J. Heinze, J. Electroanal. Chem., 124 (1981) 73.
41 R.G. Compton and P.R. Unwin, J. Electroanal. Chem., 205 (1986) 1 and refs. cited therein.
42 J. Lankelma and H. Poppe, J. Chromatogr., 125 (1976) 375.
43 M.R. Goldman and L.R. Barnett, Trans. Inst. Chem. Eng., 47 (1969) T29.
44 A.A. Wragg and T.K. Ross, Electrochim. Acta, 12 (1967) 1421.
45 A.A. Wragg, Electrochim. Acta, 16 (1971) 373.
46 D.J. Robinson, Electrochim. Acta, 17 (1972) 791.
47 R. Alkire and A.A. Mirarefi, J. Electrochem. Soc., 124 (1977) 1043.
48 K.B. Oldham, J. Electroanal. Chem., 122 (1981) 1.
49 K. Aoki, K. Tokuda and H. Matsuda, J. Electroanal. Chem., 217 (1987) 33.
50 M. Abramowitz and I.A. Stegun, Handbook of Mathematical Functions, Dover, New York, 1970, p. 450.
51 D.K. Cope and D.E. Tallman, J. Electroanal. Chem., 188 (1985) 21.
52 V. Yu. Filinovskii, Electrochim. Acta, 25 (1980) 309.
53 D.E. Weisshaar, D.E. Tallman and J.L. Anderson, Anal. Chem., 53 (1981) 1809.
54 D.E. Weisshaar and D.E. Tallman, Anal. Chem., 55 (1983) 1146.
55 D.E. Tallman and D.E. Weisshaar, J. Liq. Chromatogr., 6 (1983) 2157.
56 J.L. Anderson and S. Moldoveanu, J. Electroanal Chem., 185 (1985) 239.

57 J.L. Anderson, T.-Y. Ou and S. Moldoveanu, J. Electroanal. Chem., 196 (1985) 213.
58 J.L. Anderson, K.K. Whiten, J.D. Brewster, T.-Y. Ou and W.K. Nonidez, Anal. Chem., 57 (1985) 1366.
59 L.E. Fosdick and J.L. Anderson, Anal. Chem., 58 (1986) 2481.
60 L.E. Fosdick, J.L. Anderson, T.A. Baginski and R.C. Jaeger, Anal. Chem., 58 (1986) 2750.
61 R.G. Compton, M.B.G. Pilkington, G.M. Stearn and P.R. Unwin, J. Electroanal. Chem., 238 (1987) 43.
62 K. Aoki, K. Tokuda and H. Matsuda, J. Electroanal. Chem., 209 (1986) 247.
63 R.G. Compton, D.J. Page and G.R. Sealy, J. Electroanal. Chem., 161 (1984) 129.
64 R.G. Compton, P.J. Daly, P.R. Unwin and A.M. Waller, J. Electroanal. Chem., 191 (1985) 15.
65 W.J. Blaedel and L.N. Klatt, Anal. Chem., 38 (1966) 879.
66 W.J. Blaedel and L.N. Klatt, Anal. Chem., 39 (1967) 1065.
67 R.G. Compton and P.R. Unwin, J. Electroanal. Chem., 206 (1986) 57.
68 M. Sluyters-Rehbach and J.H. Sluyters, in C.H. Bamford and R.G. Compton (Eds.), Comprehensive Chemical Kinetics, Elsevier, Amsterdam, 1986, pp. 203–354.
69 R.G. Compton and G.R. Sealy, J. Electroanal. Chem., 145 (1983) 35.
70 R.G. Compton, M.E. Laing and P.R. Unwin, J. Electroanal. Chem., 207 (1986) 309.
71 See, for example, Southampton Electrochemistry Group, Instrumental Methods in Electrochemistry, Ellis Horwood, Chichester, 1985, p. 265.
72 D.D. MacDonald, Transient Techniques in Electrochemistry, Plenum Press, New York, 1977, pp. 119–184.
73 K. Aoki and H. Matsuda, J. Electroanal. Chem., 90 (1978) 333.
74 F.G. Cottrell, Z. Phys. Chem., 42 (1902) 385.
75 R.G. Compton and P.J. Daly, J. Electroanal. Chem., 178 (1984) 45.
76 S. Bruckenstein and S. Prager, Anal. Chem., 39 (1967) 1161.
77 T.M. Florence, J. Electroanal. Chem., 168 (1984) 207.
78 W.R. Seitz, R. Jones, L.N. Klatt and W.D. Mason, Anal. Chem., 45 (1973) 840.
79 S.H. Lieberman and A. Zirino, Anal. Chem., 46 (1974) 20.
80 R.W. Andrews and D.C. Johnson, Anal. Chem., 48 (1976) 1057.
81 D.C. Johnson, J.A. Polta, T.Z. Polta, G.G. Neuburger, J. Johnson, A.P.-C. Tang, I.-H. Yeo and J. Baur, J. Chem. Soc. Faraday Trans. 1, 82 (1986) 1081 and refs. cited therein.
82 D.S. Austin, J.A. Polta, T.Z. Polta, A.P.-C. Tang, T.D. Cabelka and D.C. Johnson, J. Electroanal. Chem., 168 (1984) 227.
83 J.A. Polta and D.C. Johnson, Anal. Chem., 57 (1985) 1373.
84 B. Miller and S. Bruckenstein, Anal. Chem., 46 (1974) 2026.
85 T. Mizushina, T. Maruyama, S. Ide and Y. Mizukami, J. Chem. Eng. Jpn., 6 (1973) 152.
86 R.D. Patel, J.J. McFeeley and K.R. Jolls, AIChE J., 21 (1975) 259.
87 T. Maruyama and T. Mizushina, AIChE J., 21 (1975) 1035.
88 W.J. Blaedel and S.C. Boyer, Anal. Chem., 43 (1971) 1583.
89 W.J. Blaedel and D.G. Iveson, Anal. Chem., 49 (1977) 1563.
90 W.J. Blaedel and Z. Yim, Anal. Chem., 52 (1980) 565.
91 C. Amatore and J.M. Savéant, J. Electroanal. Chem., 85 (1977) 27.
92 C. Amatore, M. Gareil and J.M. Savéant, J. Electroanal. Chem., 147 (1983) 1.
93 C.P. Andrieux and J.M. Savéant, in C.F. Bernasconi (Ed.), Investigation of Rates and Mechanisms of Reactions Part II, Vol. 6, Wiley, New York, 1986, pp. 305–390.
94 L.S. Marcoux, R.N. Adams and S.W. Feldberg, J. Phys. Chem., 73 (1969) 2611.
95 Yu. V. Pleskov and V. Yu Filinovskii, The Rotating Disc Electrode, Consultants Bureau, 1976, p. 374.
96 A.J. Bard and L.R. Faulkner, Electrochemical Methods, Wiley, New York, 1980, p. 298.
97 M. Fleischmann, F. Lasserre, J. Robinson and D. Swan, J. Electroanal. Chem., 177 (1984) 97.
98 J. Robinson, This volume, Chap. 5.
99 Ch. Bernstein, A. Heindrichs and W. Vielstich, J. Electroanal. Chem., 87 (1978) 81.

100 R.G. Compton and B.A. Coles, J. Electroanal. Chem., 127 (1981) 37.
101 H. Matsuda, J. Electroanal. Chem., 15 (1967) 325.
102 K. Tokuda, K. Aoki and H. Matsuda, J. Electroanal. Chem., 80 (1977) 211.
103 L.N. Klatt and W.J. Blaedel, Anal. Chem., 40 (1968) 512.
104 K. Aoki, K. Tokuda and H. Matsuda, J. Electroanal. Chem., 76 (1977) 217.
105 R.E. Meyer, M.C. Banta, P.M. Lantz and F.A. Posey, J. Electroanal. Chem., 30 (1971) 345.
106 A.M. Waller and R.G. Compton, This volume, Chap. 7.
107 W.J. Albery, B.A. Coles and A.M. Couper, J. Electroanal. Chem., 65 (1975) 901.
108 P. Wood, D. Phil. Thesis, University of Oxford, 1979.
109 K. Stulik and V. Hora, J. Electroanal. Chem., 70 (1976) 253.
110 P.L. Meschi and D.C. Johnson, Anal. Chem., 52 (1980) 1304.
111 T.R. Beck, Corrosion, 33 (1977) 9.
112 T.R. Beck and S.G. Chan, Corrosion, 37 (1981) 665.
113 J.N. Harb, J. Electrochem. Soc., 133 (1986) 439C.
114 Ch. Bernstein and W. Vielstich, Proc. Electrochem. Soc., 9 (1980) 350.
115 F. Barz, Ch. Bernstein and W. Vielstich, in H. Gerischer (Ed.), Advances in Electrochemistry and Electrochemical Engineering, Vol. 13, Interscience, New York, 1984, pp. 261–353.
116 P. Van Shaw and T.J. Hanratty, AIChE J., 10 (1964) 475.
117 A.N. Frumkin and L.I. Nekrasov, Dokl. Akad. Nauk SSSR, 126 (1959) 115.
118 H. Gerischer, I. Mattes and R. Braun, J. Electroanal. Chem., 10 (1965) 553.
119 G.W. Schieffer and W.J. Blaedel, Anal. Chem., 49 (1977) 49.
120 J. Herrmann, H. Schmidt and W. Vielstich, Z. Phys. Chem., 139 (1984) 83.
121 H. Matsuda, J. Electroanal. Chem., 16 (1968) 153.
122 R. Braun, J. Electroanal. Chem., 19 (1968) 23.
123 R.N. Cook, D. Phil. Thesis, University of Oxford, 1973, p. 313.
124 W.J. Albery and S. Bruckenstein, Trans. Faraday Soc., 62 (1966) 1920.
125 K. Aoki and H. Matsuda, J. Electroanal. Chem., 94 (1978) 157.
126 K. Aoki, K. Tokuda and H. Matsuda, J. Electroanal. Chem., 79 (1977) 49.
127 W.H. Smyrl, in J.O'M. Bockris, B.E. Conway, E. Yeager and R.E. White (Eds.), Comprehensive Treatise of Electrochemistry, Vol. 4, Plenum Press, New York, 1981, pp. 97–149, and refs. cited therein.
128 S. Bruckenstein and D.T. Napp, J. Am. Chem. Soc., 90 (1968) 6303.
129 T. Tsuru, T. Nishimura and S. Haruyama, in M. Duprat (Ed.), Electrochemical Methods in Corrosion Research, Materials Science Forum, Vol. 8, Trans. Tech., Switzerland, 1986, pp. 429–438.
130 T. Tsuru, T. Nishimura and S. Haruyama, Denki Kagaku, 52 (1984) 532.
131 T. Tsuru, T. Nishimura, K. Aoki and S. Haruyama, Denki Kagaku, 50 (1982) 712.
132 K. Aoki, K. Tokuda and H. Matsuda, J. Electroanal. Chem., 195 (1986) 229.
133 M. Abramowitz and I.A. Stegun, Handbook of Mathematical Functions, Dover, New York, 1970, p. 227.
134 J. O'M. Bockris, D. Drazic and A. Despic, Electrochim. Acta, 4 (1961) 325.
135 K.E. Heusler, Z. Elektrochem., 62 (1958) 582.
136 F. Hilbert, Y. Miyoshi, G. Eichkorn and W.J. Lorenz, J. Electrochem. Soc., 118 (1971) 1919.
137 F. Hilbert, Y. Miyoshi, G. Eichkorn and W.J. Lorenz, J. Electrochem. Soc., 118 (1971) 1927.
138 D. Geana, A.A. El-Miligy and W.J. Lorenz, Corros. Sci., 14 (1974) 657.
139 D. Geana, A.A. El-Miligy and W.J. Lorenz, Electrochim. Acta, 20 (1975) 273.
140 H. Schweickert, W.J. Lorenz and H. Friedburg, J. Electrochem. Soc., 127 (1980) 1693.
141 G.J. Bignold and M. Fleischmann, Electrochim. Acta, 19 (1974) 363.
142 J.A. Harrison and W.J. Lorenz, Electrochim. Acta, 22 (1977) 205.
143 W.J. Lorenz and J.R. Vilche, Corros. Sci., 12 (1972) 785.
144 P. Van Shaw, L.R. Kerss and T.J. Hanratty, AIChE J., 9 (1963) 362.

145 E.W. Dreeson and W. Vielstich, Ber. Bunsenges Phys. Chem., 79 (1975) 6.
146 E.W. Dreeson and W. Vielstich, Ber. Bunsenges Phys. Chem., 79 (1975) 12.
147 W.J. Blaedel and G.W. Schieffer, Anal. Chem., 46 (1974) 1564.
148 W.J. Blaedel and G.W. Schieffer, Anal. Chem., 49 (1977) 259.
149 H. Schlichting, Boundary-Layer Theory, McGraw-Hill, New York, 1979, p. 555.
150 H. Schlichting, Boundary-Layer Theory, McGraw-Hill, New York, 1979, p. 599.
151 J. Nikuradse, Forsch. Geb. Ingenieurwes., 356 (1932).
152 W. Szablewski, Ing. Arch., 21 (1953) 323.
153 T. von Kárman, Z. Angew. Math. Mech., 1 (1921) 244.
154 H. Schlichting, Boundary-Layer Theory, McGraw-Hill, New York, 1979, p. 537.
155 H. Reichardt, Z. Angew. Math. Mech., 31 (1951) 208.
156 W. Vielstich, Ber. Bunsenges. Phys. Chem., 57 (1953) 646.
157 W.L. Friend and A.B. Metzner, AIChE J., 4 (1958) 393.
158 G. Schütz, Int. J. Heat Mass Transfer, 7 (1964) 1077.
159 J.S. Son and T.J. Hanratty, AIChE J., 13 (1967) 689.
160 D.W. Hubbard and E.N. Lightfoot, Ind. Eng. Chem. Fundam., 5 (1966) 370.
161 E.S.C. Meyerink and S.K. Friedlander, Chem. Eng. Sci., 17 (1962) 121.
162 D.A. Shaw and T.J. Hanratty, AIChE J., 23 (1977) 28.
163 W.J. Albery, Electrode Kinetics, Oxford University Press, Oxford, 1975, p. 70.
164 Southampton Electrochemistry Group, Instrumental Methods in Electrochemistry, Ellis Horwood, Chichester, 1985, p. 53.
165 P. Delahay, in P. Delahay and C.W. Tobias (Eds.), Advances in Electrochemistry and Electrochemical Engineering, Vol. 1, Interscience, New York, 1961, p. 233.
166 Southampton Electrochemistry Group, Instrumental Methods in Electrochemistry, Ellis Horwood, Chichester, 1985, p. 189.
167 B. Gostisa-Michelcic and W. Vielstich, Ber. Bunsenges. Phys. Chem., 77 (1973) 476.
168 N. Sato, Electrochim. Acta, 16 (1971) 1683.
169 R. Alkire and A. Cangellari, J. Electrochem. Soc., 130 (1983) 1252.
170 B.T. Ellison and C.J. Wen, AIChE J. Symp. Ser., 77 (1981) 161.
171 B.C. Syrett, Corrosion, 32 (1976) 242.
172 I.L. Rosenfeld, I.S. Danilov and R.N. Oranskaya, J. Electrochem. Soc., 125 (1978) 1729.
173 H.J. Engell, Electrochim. Acta, 22 (1977) 987.
174 T.R. Beck, J. Electrochem. Soc., 120 (1973) 1317.
175 H.S. Isaacs, J. Electrochem. Soc., 120 (1973) 1454.
176 H.H. Strehblow and J. Wenners, Z. Phys. Chem., 98 (1975) 199.
177 Ya. M. Kolotyrkin and Yu. A. Popov, Elektrokhimiya, 12 (1976) 406.
178 Ya. M. Kolotyrkin and Yu. A. Popov, Elektrokhimiya, 12 (1976) 527.
179 R. Alkire, D. Ernsberger and T.R. Beck, J. Electrochem. Soc., 125 (1978) 458.
180 T.R. Beck, J. Electrochem. Soc., 129 (1982) 2412.
181 T.R. Beck and R.C. Alkire, J. Electrochem. Soc., 126 (1979) 1662.
182 M. Janik-Czachor, J. Electrochem. Soc., 128 (1981) 513C.
183 R.C. Alkire, D.B. Reiser and R.L. Sani, J. Electrochem. Soc., 131 (1984) 2795.
184 K.J. Vetter and H.H. Strehblow, Ber. Bunsenges. Phys. Chem., 74 (1970) 1024.
185 V.G. Levich, Physicochemical Hydrodynamics, Prentice-Hall, Englewood Cliffs, 1962, p. 344.
186 J.W. Morse, Rev. Mineral., 11 (1983) 227.
187 E.L. Sjöberg, Sci. Geol. Mem., 71 (1983) 119.
188 R.A. Berner, Crist. Deform., Dissolution Carbonates, Reun., (1980) 33.
189 N. Litsakes and P. Ney, Fortschr. Mineral., 63 (1985) 135.
190 N. Litsakes and P. Ney, Fortschr. Mineral., 63 (1985) 155.
191 N. Litsakes, Fortschr. Mineral., 63 (1985) 165.
192 A. Ludwig, S.C. Dave, W.I. Higuchi and J.L. Fox, Expo.-Congr. Int. Technol. Pharm. 3rd., 1 (1983) 97.
193 R. Yushan, T. Qiu and Z. Yuanmou, Process Technol. Proc., 2 (1984) 289.

194 J. Christoffersen and M.R. Christoffersen, Faraday Discuss. Chem. Soc., 77 (1984) 235.
195 V.K. Cheng, B.A.W. Coller and J.L. Powell, Faraday Discuss. Chem. Soc., 77 (1984) 243.
196 E.L. Sjöberg and D. Rickard, Geochim. Cosmochim. Acta, 48 (1984) 485.
197 E.L. Sjöberg and D. Rickard, Chem. Geol., 49 (1985) 405.
198 D. Rickard and E.L. Sjöberg, Am. J. Sci., 283 (1983) 815.
199 E.L. Sjöberg and D. Rickard, Chem. Geol., 42 (1984) 119.
200 N.N. Kruglitskii, G.A. Kataev, B.P. Zhantalai, K.A. Rubezhanskii and A.A. Kolomiets, Kolloidn. Zh., 47 (1985) 493.
201 A.E. Smirnov, A.A. Urusovskaya and V.R. Regel, Dokl. Akad. Nauk SSSR, 280 (1985) 1122.
202 G.R. Gobran and S. Miyamoto, Soil Sci., 140 (1985) 89.
203 J. Ferret and R. Gout, Mater. Sci. Monogr., 28A (1985) 523.
204 O. Fruhwirth, G.W. Herzog and J. Poulios, Surf. Technol., 24 (1985) 293.
205 A.J. Shah and J.R. Pandya, Cryst. Res. Technol., 20 (1985) 813.
206 L.R. Bhagia and J.R. Pandya, Indian J. Pure Appl. Phys., 23 (1985) 27.
207 C. Amrhein, J.J. Jurinak and W.M. Moore, Soil Sci. Soc. Am. J., 49 (1985) 1393.
208 I. Nicoara, O.F.G. Aczel, D. Nicoara and Z. Schlett, Cryst. Res. Technol., 21 (1986) 647.
209 R.G. Compton and P.J. Daly, J. Colloid Interface Sci., 101 (1984) 159.
210 R.G. Compton, P.J. Daly and W.A. House, J. Colloid Interface Sci., 113 (1986) 12.
211 B. Capron, A. Girou and L. Humbert, Bull. Soc. Chim. Fr., 2 (1987) 280.
212 B. Capron, A. Girou, L. Humbert and E. Puech-Costes, Bull. Soc. Chim. Fr., 2 (1987) 288.
213 O. Erga and S.G. Terjesen, Acta Chem. Scand., 10 (1956) 872.
214 S.G. Terjesen, O. Erga, G. Thorsen and A. Ve, Chem. Eng. Sci., 74 (1961) 277.
215 I. Nestaas and S.G. Terjesen, Acta Chem. Scand., 23 (1969) 2519.
216 M.N.A. Peterson, Science, 154 (1966) 1542.
217 R.A. Berner and P. Wilde, Am. J. Sci., 272 (1972) 826.
218 J.W. Morse and R.A. Berner, Am. J. Sci., 272 (1972) 840.
219 J.W. Morse, Am. J. Sci., 274 (1974) 97.
220 R.A. Berner and J.W. Morse, Am. J. Sci., 274 (1974) 108.
221 J.W. Morse, Am. J. Sci., 274 (1974) 638.
222 L.N. Plummer and T.M.L. Wigley, Geochim. Cosmochim. Acta, 40 (1976) 191.
223 L.N. Plummer, T.M.L. Wigley and D.L. Parkhurst, Am. J. Sci., 278 (1978) 179.
224 L.N. Plummer, T.M.L. Wigley and D.L. Parkhurst, in E.A. Jenne (Ed.), Chemical Modelling of Aqueous Systems, American Chemical Society Symposium Series, Vol. 93, 1979, pp. 538–573.
225 E.L. Sjöberg, Stockholm Contrib. Geol., 32 (1978) 1.
226 E.L. Sjöberg, Geochim. Cosmochim. Acta, 40 (1976) 441.
227 R.G. Compton and P.J. Daly, J. Colloid Interface Sci., 115 (1987) 493.
228 D.R. Boomer, C.C. McCune and H.S. Fogler, Rev. Sci. Instrum., 43 (1972) 225.
229 K. Lund, H.S. Fogler and C.C. McCune, Chem. Eng. Sci., 28 (1973) 691.
230 K. Lund, H.S. Fogler, C.C. McCune and J.W. Ault, Chem. Eng. Sci., 30 (1975) 825.
231 G.H. Nancollas, T.F. Kazmierizak and E. Schuttringer, Corrosion, 37 (1981) 76.
232 Z. Amjad, Langmuir, 3 (1987) 224.
233 Z. Amjad and J.J. Hooley, J. Colloid Interface Sci., 111 (1986) 496.
234 Z. Amjad, Desalination, 54 (1985) 263.
235 A.G. Xyla and P.G. Koutsoukos, J. Chem. Soc. Faraday Trans. 1, 83 (1987) 1477.
236 W.A. House and L. Donaldson, J. Colloid Interface Sci., 112 (1986) 309.
237 W.A. House, H. Casey, L. Donaldson and S. Smith, Water Res., 20 (1986) 917.
238 W.A. House, H. Casey and S. Smith, Water Res., 20 (1986) 923.
239 L.M. Walter and E.A. Burton, Chem. Geol., 56 (1986) 313 and refs. cited therein.
240 A. Girou, Rev. Fr. Sci. Eau, 4 (1985) 129.
241 M.M. Reddy, J. Cryst. Growth, 41 (1977) 287.
242 M.M. Reddy and K.K. Wang, J. Cryst. Growth, 50 (1980) 470.
243 A. Glaser and D. Weiss, J. Inorg. Nucl. Chem., 42 (1980) 655.

244 P.J. Daly, Ph.D. Thesis, University of Liverpool, 1985.
245 E.A. Pashkova, Probl. Osad. Geol. Dokembr., (1985) 20.
246 J. Ruzicka and E.H. Hansen, Anal. Chim. Acta, 99 (1978) 37.
247 J. Ruzicka and E.H. Hansen, Anal. Chim. Acta, 145 (1983) 1.
248 J.T. Vanderslice, A.G. Rosenfeld and G.R. Beecher, Anal. Chim. Acta, 179 (1986) 119.
249 F.V. Williams and R.A. Ruehrwein, J. Am. Chem. Soc., 79 (1957) 4898.
250 M. Okamoto, Nippon Kagaku Kaishi, (1986) 1153.
251 Ciba-Geigy A.G., Br. Pat. 1,369,429 (1972). U.S. Pats. 3,810,834 (1974); 3,897,209 (1975); 3,919,258 (1975); 3,963,636 (1976); 4,089,796 (1978); 4,212,788 (1980). F.R.G. Pats. 2,159,172 (1972); 2,412,926 (1973); 2,518,880 (1975); 2,749,719 (1978). Fr. Pat. 2,167,594 (1973). Belg. Pat. 856,978 (1978). Jpn. Pat. 39,485 (1979).
252 R.G. Compton, D. Phil. Thesis, University of Oxford, 1980.
253 J. Miles, Integral Transforms in Applied Mathematics, Cambridge University Press, London, 1971, p. 17.
254 C.V. King and C.L. Liu, J. Am. Chem. Soc., 55 (1933) 928.
255 H. Tominaga, H. Adzum and T. Isobe, Bull. Chem. Soc. Jpn., 14 (1939) 348.
256 D.E. Nierode and B.B. Williams, Soc. Pet. Eng. J., 11 (1971) 406.
257 P. Barton and T. Vatanatham, Environ. Sci. Technol., 10 (1976) 262.
258 R.G. Compton and P.R. Unwin, unpublished data.
259 R.G. Compton, A.F. Gerrard, G. Grigg, M. Lees, P. Sylvester and P.R. Unwin, unpublished data.
260 C.M. Hwa (W.R. Grace and Co.) Ger. Offen. 1,813,717 (1967).
261 J. Farukawa, S. Yamashika and H. Niwa, Nippon Gomu Kyokaishi, 36 (1963) 295.
262 K.L. Pritchard and W.A. House, private communication, 1987.
263 R.G. Compton and P.R. Unwin, unpublished results.
264 J. Miles, Integral Transforms in Applied Mathematics, Cambridge University Press, London, 1971, p. 84.

Chapter 7

In-Situ Electrochemical ESR

ANDREW M. WALLER and RICHARD G. COMPTON

1. Introduction

The existence of radicals as intermediates in a great variety of electrode reactions is well established. Such paramagnetic species may be detected and identified using electron spin resonance spectroscopy (ESR) and so for nearly thirty years electrochemists have used ESR as a method for the investigation of those complex electrode reactions that proceed via radical intermediates. ESR affords several advantages in comparison with other spectroscopic techniques; the spectra have a high information content, the technique is very sensitive (standard equipment routinely allows concentrations of the order of 10^{-8} M to be observed [1]), and because of the low energies employed, ESR is non-perturbing.

The first application of ESR in electrochemistry is attributed to the work of Ingram and his co-workers in 1958 [2]. Ingram demonstrated the formation of radicals in the controlled potential electrolysis of the aromatic compounds anthracene, benzophenone, and anthraquinone in dimethylformamide solvent with 0.1 M tetraethylammonium iodide as supporting electrolyte. Spectra were obtained from samples which had been withdrawn during the electrolysis, sealed in tubes, and frozen in liquid nitrogen. These tubes were transferred to the ESR spectrometer and the spectra recorded at low temperature. The spectra observed exhibited broad single lines, characteristic of solid-state samples of organic radicals, but the lack of the hyperfine splitting characteristic of solution-phase radicals precluded any positive identification of the radicals involved. The real possibilities of the coupling of ESR with electrochemistry was first shown by Maki and Geske [3, 4] who obtained the first solution-phase spectra of an electrogenerated species, their electrolysis being carried out within the cavity of the ESR spectrometer. A more detailed description of the work of Maki and Geske will be given in Sect. 3.1, but of note already is the fact that there are two distinct methodologies to the electrochemical ESR experiment.

(1) "In-situ" experiments where the generation of radicals is carried out *within* the cavity of the ESR spectrometer.

(2) "Ex-situ" experiments where the generation of radicals occurs in an electrochemical cell outside the ESR cavity and radicals produced are transferred to the cavity by flow, by pumping, or by physical transfer.

A critical comparison of the merits of these two methodologies will occur

References pp. 349–352

in later sections, although the main concern of this chapter will be to give a detailed description of contemporary in-situ techniques.

2. Electron spin resonance: an introductory survey

Before commencing a detailed discussion of the electrochemical ESR experiment, it will now be helpful to give an outline description of the theory and practice of ESR so that, in later sections, spectra may be presented and their interpretation understood.

2.1 ESR THEORY

The origin of ESR spectroscopy lies in the spin angular momentum possessed by the unpaired electron. Associated with this spin is a magnetic moment, $\boldsymbol{\mu}$, which is proportional in size to the spin vector (**S**).

$$\begin{aligned} \boldsymbol{\mu} &= -g_e \beta \mathbf{S} \\ |\mathbf{S}| &= \sqrt{\tfrac{1}{2}(\tfrac{1}{2}+1)}\hbar \end{aligned} \tag{1}$$

where g_e is the dimensionless constant known as the electronic g factor and β is the electronic Bohr magneton equal to $(e\hbar/2mc)$, e and m being the charge and mass of the electron. In a steady magnetic field **B** applied in what is conventionally labelled the z direction it is found that the vector **S** precesses around the z axis only in certain allowed orientations so that the component of the angular momentum in this direction is given by $\pm \hbar/2$. Alternatively, we say that the quantum number M_s representing the allowed eigenvalues of the projection on to the z axis of the spin angular momentum (in units of $\hbar$) has the two possible values $+\frac{1}{2}$ or $-\frac{1}{2}$. This space quantisation of **S** and hence $\boldsymbol{\mu}$ means that the energy of the interaction between the magnetic moment and **B** is limited to certain values.

$$E = -\boldsymbol{\mu}\cdot\mathbf{B} \tag{2}$$

$$= -g_e \beta M_s B_z \qquad M_s = \pm\tfrac{1}{2} \tag{3}$$

It is clear from eqns. (2) and (3) that, the absence of the applied magnetic field, the two spin states $M_s = \pm\frac{1}{2}$ are degenerate but that, in the presence of the applied field, this degeneracy is removed. The energy level separation between the two spin states has the value $g\beta B$ and increases linearly with the magnetic field as shown in Fig. 1. Transitions between the two spin states are induced when electromagnetic radiation of the appropriate frequency ν is applied and where $h\nu$ matches the energy level separation between the two spin states

$$\Delta E = g_e \beta B = h\nu \tag{4}$$

The frequency necessary to induce the transition from one spin state to the

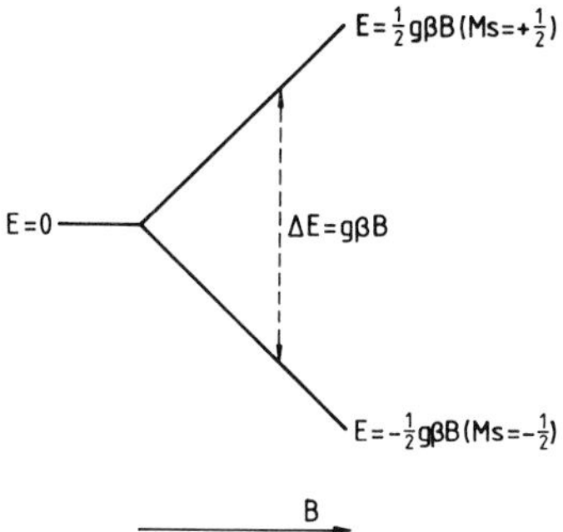

Fig. 1. Energy levels for the two spin states of a free electron in an applied magnetic field. Transitions may be induced between the levels by electromagnetic radiation of the correct frequency.

other is then given by

$$\nu = \frac{g_e \beta B}{h} \tag{5}$$

It is apparent that two distinct experimental possibilities exist for observation of the ESR spectrum; the frequency may be altered at a fixed magnetic field or, alternatively, the magnetic field may be varied at constant frequency. In practice, the ESR spectrum is invariably obtained using the second of these two methods. For an applied field of 3400 Gauss (0.34 Tesla), one finds, by substituting the appropriate physical constants into eqn. (5), that the resonance frequency for a free electron approximates to 9.4 GHz, which is in the microwave region of the spectrum and relates to a wavelength of about 3 cm. In contrast to NMR spectrometers, ESR spectrometers record the spectrum as the first derivative of the absorption spectrum, dS/dB where S is the signal and B the applied field, as shown in Fig. 2. The reasons for this lie in the phase-sensitive methods of detection employed in ESR in order to achieve greater sensitivity and better resolution.

ESR spectra are characterised by three parameters, the g factor, the hyperfine coupling, and the line width. It is these parameters that allow identification of the nature and environment of radicals.

2.1.1 The g factor

If eqn. (5) represented the resonance condition for all electrons, one would predict that they would all resonate at the same applied field for a given applied frequency. Consequently, the g factor would always be equal to that of the free electron. However, as well as possessing spin angular momentum, the unpaired electron also possesses electronic orbital angular momentum. The interaction between the two (via spin–orbit coupling) ensures that the electron has an effective magnetic moment different from that of the free electron. Hence, the resonance condition for an unpaired electron in a

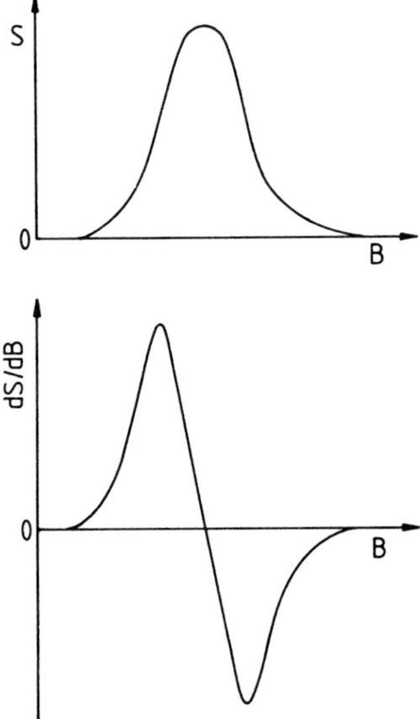

Fig. 2. (a) Absorption spectrum, S, and (b) first-derivative spectrum, $\mathrm{d}S/\mathrm{d}B$, observed in ESR.

molecule is written as

$$h\nu \quad = \quad g\beta B \tag{6}$$

where g is the g factor for the species and is a unique property of the molecule. In the case of organic radicals, g is usually very close to g_e. As g depends upon the electronic structure of the molecules, deductions may be made of the nature of the molecule. The spin–orbit coupling is dependent upon the orientation of the radical within the applied field, making g anisotropic, and thus the position of the observed ESR spectrum is dependent upon the orientation of the radical. Fortunately, for radicals in solution, such as those observed commonly in electrochemical ESR, there is rapid tumbling of the radicals and, effectively, an average g factor is observed.

2.1.2 Hyperfine coupling

We have seen that one result of putting an unpaired electron into a molecule is to alter the value of the g factor away from that of the free electron. However, the most useful characteristic of ESR spectra in the identification of the radical being studied is the splitting due to the hyperfine coupling. This arises because of the electron spin interacting with any magnetic nuclei in the molecule and this interaction will depend on the orientation of the magnetic nuclei whose spin angular moment, **I**, will also be space quantized in the magnetic field; $M_I = I, (I - 1) \ldots 0 \ldots - (I - 1), - I$. [Just as the electron spin angular momentum can be viewed as precessing in the appied field, so can the nuclear spin vector, $|\mathbf{I}| = \sqrt{I(I + 1)}\hbar$.] The result is that the applied field necessary for the resonance condition is

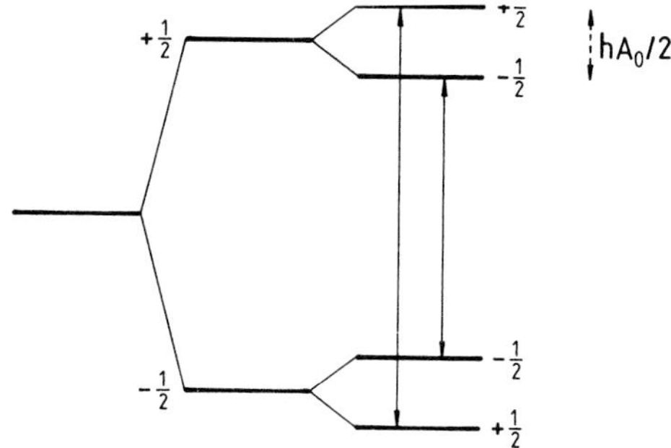

Fig. 3. Energy levels for the hydrogen atom at constant magnetic field. The allowed transitions having energy $= g\beta B \pm (ha)/2$, where a is the hydrogen hyperfine constant.

dependent upon the orientation of the nuclei in the z direction. Therefore, for each spin state of the nucleus, a separate transition is observed in the ESR spectrum. The simplest example is the hydrogen atom where the electron interacts with the proton, which has a nuclear spin equal to $\frac{1}{2}$. The possible energy levels resulting from this interaction can be seen in Fig. 3. During the electronic transition, the nuclear spin quantum number, M_I, does not change, hence two transitions are allowed and the ESR spectrum for the hydrogen atom consists of two lines of equal intensity. The hyperfine coupling constant, a, is the separation between the two lines.

The electronic energy levels for the one electron one nucleus system are given quantitatively by the equation

$$E = M_s[g\beta B + aM_I] \tag{7}$$

where a is the hyperfine coupling constant. The energy levels for a system with $s = \frac{1}{2}$ and $I = 1$ are given in Fig. 4(a), with the allowed transitions which result from the "first-order" selection rules $\Delta M_s = 1$; $\Delta M_I = 0$. Figure 4(b) shows the ESR spectrum for such a system, the nitroxide spin probe

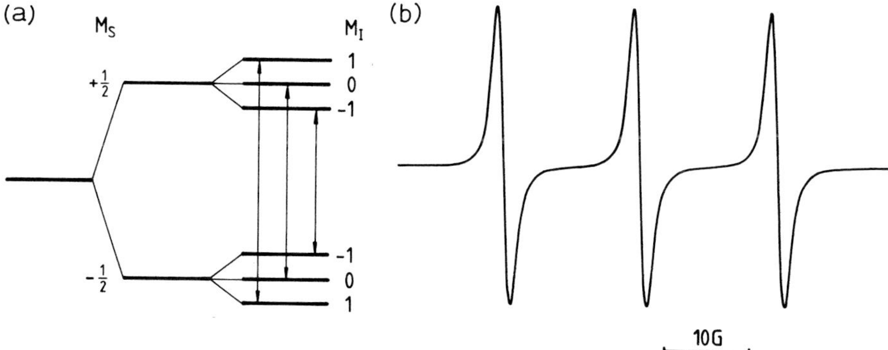

Fig. 4. (a) Energy levels for the system $S = \frac{1}{2}$, $I = 1$ and allowed transitions. (b) ESR spectrum observed from the nitroxide spin probe TEMPO.

TEMPO

Coupling of the spin, which is largely localised on the oxygen nucleus, with the nitrogen nucleus ($I = 1$) results in the characteristic 1:1:1 triplet. Splitting from the protons in the molecule is not resolvable.

The extension of hyperfine coupling to involve more than one magnetic nucleus does not introduce any complications. The ESR spectra of nitroxide radicals produced in spin-trapping experiments with α-phenyl *t*.-butylnitrone (see Sect. 5.7) generally consist of a triplet of doublets (see Fig. 5). The triplet arises from hyperfine coupling with the nitrogen nucleus, $I = 1$, and the doublet from interaction with the hydrogen on the α carbon atom. Such nitroxide spin adducts have two hyperfine coupling constants of different magnitudes and the resultant spectrum is simple; but it can happen that two or more nuclei in a radical may have identical coupling constants. For such species, one will see a reduction in the number of transitions, as some will occur at the same applied field. In general, the energy levels for one electron interacting with two magnetic nuclei are given by

$$E = M_s(g\beta B + a_A M_A + a_B M_B) \tag{9}$$

If the nuclei were equivalent protons with $a_A = a_B$, then the resultant energy levels would be as shown in Fig. 6. The observed spectrum reveals a triplet of intensities 1:2:1. The extra intensity of the middle line, which arises from the transition ($M_s = -\frac{1}{2}$, $M_I = 0$) to ($M_s = \frac{1}{2}$, $M_I = 0$), is because the transition is between doubly degenerate energy levels ($M_A = +\frac{1}{2}$, $M_B = -\frac{1}{2}$, $M_A = -\frac{1}{2}$, $M_B = +\frac{1}{2}$) whereas the outer lines represent transitions between non-degenerate nuclear energy levels. Generally, for an electron interacting with n nuclei of spin $\frac{1}{2}$, the number of ESR transitions observed is equal to

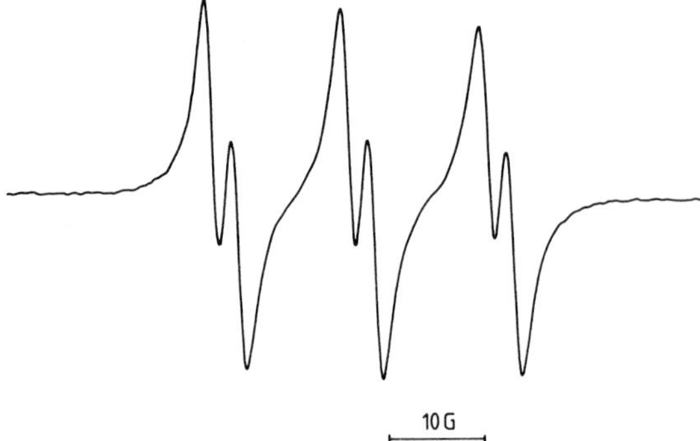

Fig. 5. ESR spectrum of the spin trap adduct H–C(Ph)(R)–N(O$^\cdot$)–Bu^t.

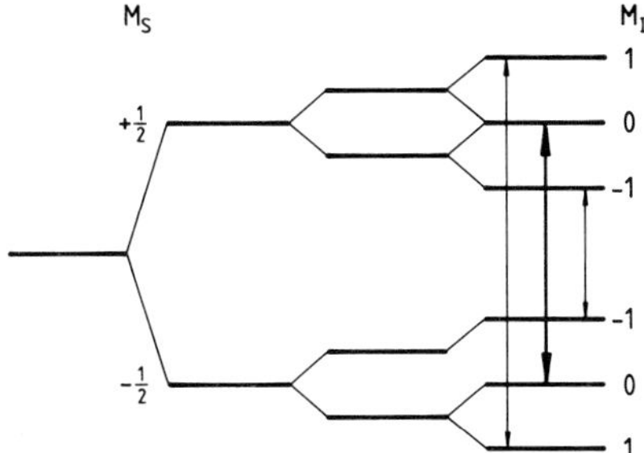

Fig. 6. Energy levels for a system containing a single set of equivalent protons.

$(1 + n)$ and the relative intensity of such transitions are found from the coefficients of the binomial expression $(1 + x)^n$. The successive sets of coefficients are readily found from Pascal's triangle (see Fig. 7).

The examples considered above are relatively simple, but most free radicals contain several magnetic nuclei. The hyperfine coupling from radicals with many magnetic nuclei will produce ESR spectra which are rich in hyperfine lines. The analysis of such spectra can be relatively straightforward and gives useful information regarding the identity of the radical or radicals involved in an electrode reaction.

2.1.3 The ESR linewidth

The linewidth, or lineshape, is the third aspect of the ESR spectrum important in any study of radicals. The two most frequently encountered lineshapes observed are the Lorentzian and Gaussian. The Gaussian lineshape is observed when the line is the superposition of a large number of unresolved individual components, such a line being referred to as inhomogeneously broadened. More important is the Lorentzian lineshape which is often observed for radicals in solution, when complications due to unresolved hyperfine splittings do not occur. The Lorentzian lineshape may be defined by

$$Y = \frac{\Gamma}{\pi}[\Gamma^2 + (B - B_r)^2]^{-1} \tag{10}$$

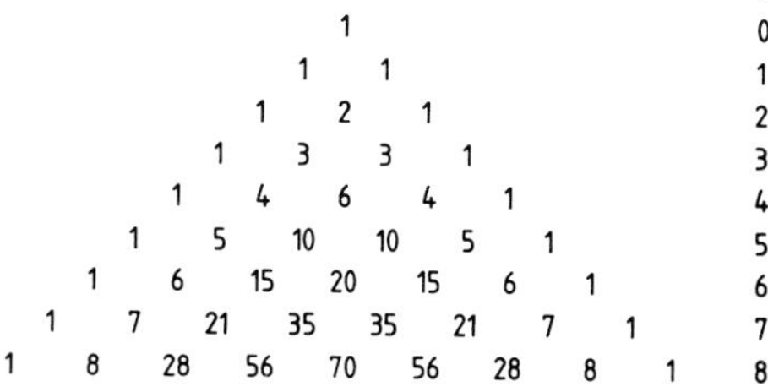

Fig. 7. Pascal's triangle, representing coefficients in the binomial expansion of $(1 + x)^n$.

where Y is the amplitude of the line, B_r the magnetic field at resonance, and Γ is half the linewidth at half-height. The parameter Γ an be used in the definition of a relaxation time τ' where

$$\frac{1}{\tau'} = y_e \Gamma \tag{11}$$

where y_e is the magnetogyric ratio for the electron. The value of the relaxation time τ' is given by

$$\frac{1}{\tau'} = \frac{1}{2\tau_1} + \frac{1}{\tau_2} \tag{12}$$

where τ_1 is the spin–lattice relaxation time and τ_2 the spin–spin relaxation time. The spin–spin relaxation time, τ_2, is approximately equal to τ' since, for most stable free radicals, $\tau_1 \gg \tau_2$. Hence, the linewidth of the Lorentzian line is a measure of $1/\tau_2$. The effect of the relaxation processes is not significant for organic radicals and linewidths observed are narrow, of the order 50 mG, giving well-resolved spectra. However, exchange interactions may occur at high radical or high parent concentration, leading to the linewidth being broadened. At high radical concentrations, exchange of spin can occur between colliding radicals, thus the spin sees an average environment as it moves from radical to radical. At high parent concentration, exchange may occur between radical and parent compound; the extent of broadening may be used to calculate a rate constant for the chemical exchange. Such a process has been carried out for the exchange between benzonitrile and benzonitrile radical anion electrochemically generated in situ [5] giving a rate constant of $2 \times 10^8\,\mathrm{dm^3\,mol^{-1}\,s^{-1}}$.

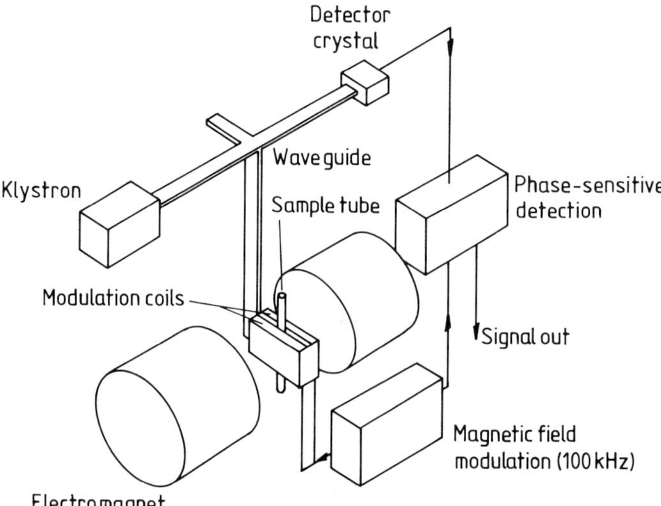

Fig. 8. Basic features of the ESR spectrometer.

2.2 ESR INSTRUMENTATION

The basic features of the ESR spectrometer are shown in Fig. 8. The applied field comes from an electromagnet and should be stable and uniform over the sample. This stability is achieved using a regulated power supply, any variation in the field being detected by a Hall effect device. The magnetic field is modulated at a frequency of 100 kHz in order to facilitate detection by a phase-sensitive method, which limits noise contributions. Microwaves are generated in a klystron, the frequency of which is determined by the voltage applied to the klystron and the stability of the microwave frequency achieved with an automatic frequency control system. The microwaves travel through a waveguide to the ESR cavity, which holds the sample. Within the ESR cavity, electromagnetic waves are set up as shown in Fig. 9 for the ESR cavity commonly used in electrochemical ESR experiments, namely the rectangular TE_{102} cavity. The standing wave results in a spatial separation of the electric and magnetic parts of the radiation. The result is that the magnetic part is at a maximum intensity at the cavity centre and hence concentrated on the sample, whereas the electrical component is at a minimum at the cavity centre. An advantage of this is that aqueous samples, which show considerable dielectric loss at the microwave frequencies, may be still investigated providing the sample is contained within a thin flat cell at the nodal plane where there is no interaction with the electrical component of the microwave field. The cells used are fabricated from silica in preference to glass, which contains paramagnetic impurities observable in the ESR spectrum.

More detailed descriptions of the instrumentation and theory of ESR may be found in several books [1, 6–11]. As an introduction, the book by Symons [8] is particularly useful and for a detailed discussion of instrumentation the work by Poole [11] can be recommended.

3. Electrochemical ESR

3.1 HISTORICAL DEVELOPMENT

As has already been described in the Introduction, the first electrochemical ESR experiment was performed by Ingram and co-workers [2] who demonstrated the formation of aromatic radical ions in electroreductions. However, the real potential of the joint electrochemical ESR technique was demonstrated first by the work of Maki et al. [3, 4, 12–14]. They obtained the first solution spectra of electrochemically generated radicals and by doing so performed the first in-situ electrochemical ESR experiment. This was performed in a two-electrode electrolysis cell, the anode being a platinum wire within a 3 mm o.d. capillary tube, the lip of which was positioned centrally, along the axis of a cylindrical cavity. The cathode was separated from the anode by sinters and was outside the ESR cavity. The system

References pp. 349–352

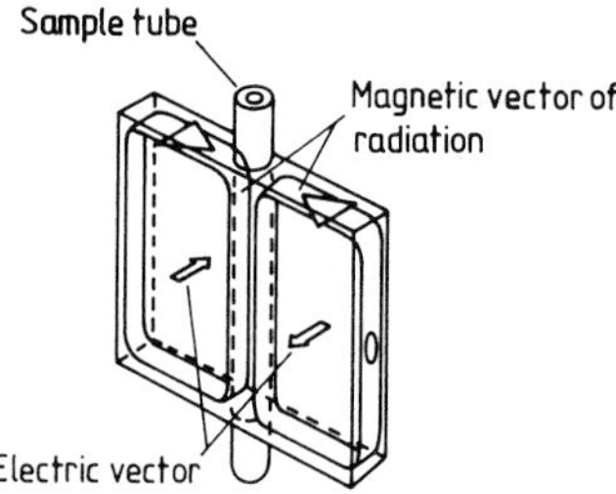

Fig. 9. The TE_{102} ESR cavity showing the electric and magnetic vectors of the electromagnetic radiation.

studied in this first in-situ experiment was the electro-oxidation of $LiClO_4$ in acetonitrile, the spectrum obtained exhibited a hyperfine structure of four equally spaced lines of approximately equal intensity. This was interpreted as being a radical species containing a single chlorine nucleus (both ^{35}Cl and ^{37}Cl have a nuclear spin of $\frac{3}{2}$) and was attributed to the perchlorate radical, $ClO_4\cdot$ [3] although later work showed the spectrum observed to be, in fact, due to the radical $ClO_2\cdot$ [15] formed via the mechanism

$$ClO_4^- \rightarrow ClO_4\cdot + e$$

$$ClO_4\cdot \rightarrow ClO_2\cdot + O_2$$

The in-situ cell was then used by its inventors to investigate the reduction of nitrobenzene [4], *o*-, *m*-, and *p*-dinitrobenzenes [12] and substituted nitrobenzenes [13] in non-aqueous media.

Optimized versions of the Maki and Geske cell were developed by Adams and co-workers [16, 17] who utilised what were essentially (suitably modified) conventional ESR flat cells. These employed a mercury cathode for reductions and a platinum gauze anode for oxidations. Figure 10 shows the platinum gauze cell used for oxidations, the gauze providing a high surface area for the production of large concentrations of radicals within the cell. Adams' cells were used in conjunction with a three-electrode system, a counter electrode being located below the ESR cavity and a reference electrode above it. Initial experiments performed in these cells included a study of the oxidation of aromatic anions and the reduction of both aliphatic and aromatic amines [16]. There have been several variations on the Adams cell used to generate radical cations and anions and modified cells have been used for low-temperature [18] and high-temperature work [19, 20], but the measure of how successful this cell's design is for generating large concentrations of radicals is that cells marketed even today by commercial ESR manufacturers are basically of the same design as that of Adams' cell.

The success of Adams' cell in generating large quantities of relatively long-lived radicals has proved of value to the spectroscopist, but the electrochemist finds that the cell is unsuited for performing experiments where well-defined voltammetric measurements must be made and that the study of

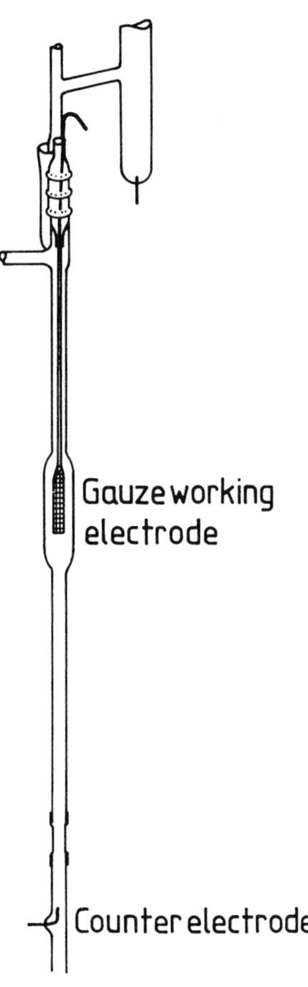

Fig. 10. The Adams' in-situ cell with platinum gauze working electrode.

electrode processes with short-lived paramagnetic intermediates (< seconds) is not possible. The lack of well-defined voltammetric behaviour can be attributed to natural convection within the cell and also to the fact that the thinness of the cell produces a large change in electrical resistance over the surface of the electrode, which results in distortion of the current–voltage curves due to "ohmic drop". The finite radical lifetime accessible in the Adams' cell is limited by the depletion of electroactive species at the electrode as the electrolysis proceeds and this results in a drop in the flux of radicals generated. This depletion is due to there being no flow to bring fresh electroactive species to the electrode. As a consequence, only when the radical produced is sufficiently long-lived for the time necessary to perform the ESR experiment will the cell show sufficient sensitivity.

Finally, it must be pointed out that the Adams cell precluded the measurements of kinetic data. It would be possible, in principle, to open-circuit the working electrode and observe the decay of the radical with time, but the

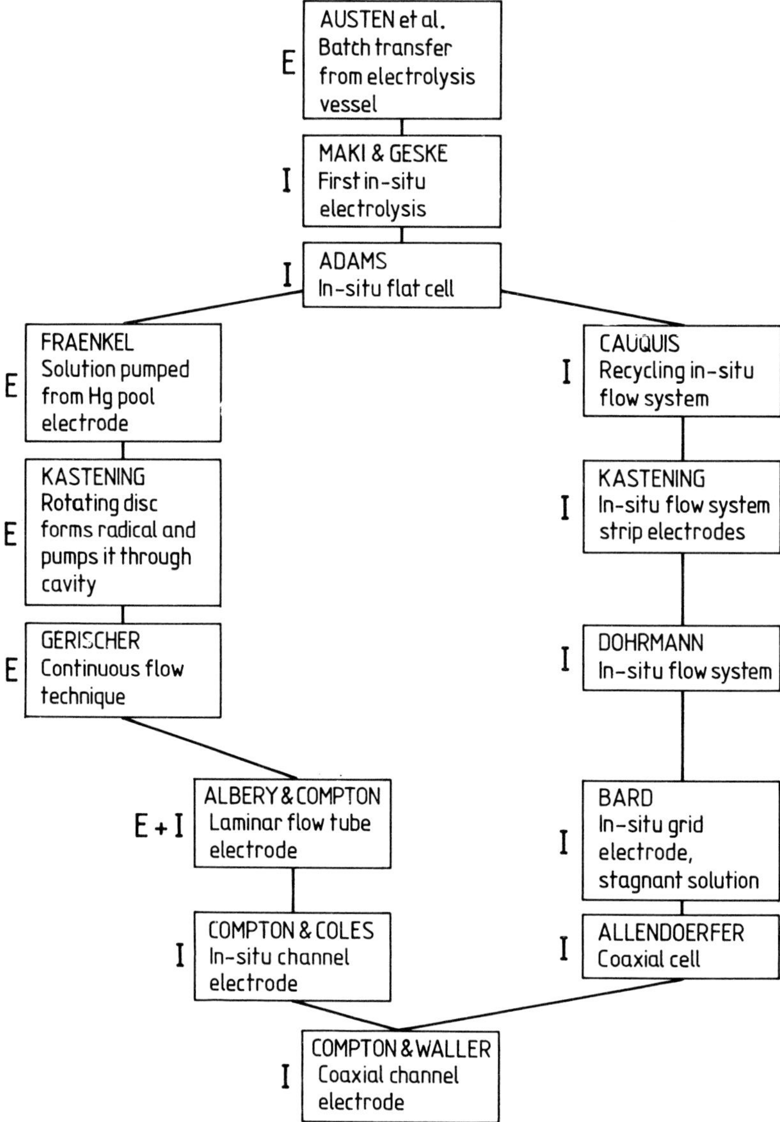

Fig. 11. Flow diagram of the development of electrochemical ESR techniques.

ESR cavity has a non-uniform sensitivity along its length and hence is spatially discriminating. The consequence is that the concentration profile of the electrogenerated radical needs to be calculable in space and time. This is not feasible for such a cell as that developed by Adams due to the presence of natural convection. The latter can be demonstrated by generating a highly coloured radical ion in the cell [21] and observing the irregular motions within the solution.

Further developments in electrochemical ESR were carried out by electrochemists, who attempted to improve the electrochemical behaviour of the cell whilst maintaining the ability to generate large numbers of radicals. Also, attempts were made to determine the lifetimes of radical intermediates and their kinetic modes of decomposition. The subsequent historical development of electrochemical ESR is shown schematically in Fig. 11. As one can see, two alternative approaches were utilised dependent on whether the working electrode was situated inside or outside the cavity.

3.1.1 Ex-situ methods

It is not the purpose of this chapter to give a detailed account of ex-situ methods, but mention must be made of several methods utilized because such cells continue to find applications to date. The ex-situ method is based upon the generation of radicals under conventional electrochemical conditions followed by transfer to the ESR cavity for detection. As has been mentioned, Ingram utilized the ex-situ approach in performing the first electrochemical ESR experiments. Another early, crude approach was that of Fraenkel et al. [22–24], who generated radicals at a mercury pool cathode (of large area) under controlled electrochemical conditions after electrolysis the solution was tipped into a side arm which was used for the ESR measurements. Fraenkel's cell was used to investigate radicals whose lifetime ranged from minutes to several hours, but the value of the electrochemical ESR technique can be judged from his mechanistic work on the electrochemical reduction of various aromatic nitriles, e.g. the reduction of 4-aminobenzonitrile [23] which exhibited an ESR spectrum consistent with the radical anion of 4,4′-dicyanobiphenyl, from which was concluded the mechanism

$$4\text{-}NH_2C_6H_4CN \xrightarrow{+e} [4\text{-}NH_2C_6H_4CN]^{-\bullet}$$

$$[4\text{-}NH_2C_6H_4CN]^{-\bullet} \longrightarrow \dot{C}_6H_4CN + NH_2^{-}$$

$$2\,\dot{C}_6H_4CN \longrightarrow NC\text{-}C_6H_4\text{-}C_6H_4\text{-}CN$$

$$NC\text{-}C_6H_4\text{-}C_6H_4\text{-}CN \xrightarrow{+e} [NC\text{-}C_6H_4\text{-}C_6H_4\text{-}CN]^{-\bullet}$$

The lifetime of radicals can be extended to more unstable radicals investigated if the cell is coupled to a flow system. It was the work of Kastening which first demonstrated this. Kastening [25] generated radicals at a large rotating working electrode, the rapid rotation speed producing hydrostatic pressure which circulated solution in a flow system to the ESR cavity. The flow of solution enabled radicals to be generated throughout the course of the ESR experiment. In later work, the flow of radicals from the working electrode through a capillary tube into the cavity was effected by gravity [26, 27], varying the flow of solution or the point of the capillary at which the ESR signal measured allowed the determination of kinetic data. In particular, the lifetime of the nitrobenzene radical anion at various values of pH was investigated. Similar to the gravity fed flow system used by Kastening was the continuous flow system of Gerischer et al. [28].

The problem associated with the ex-situ technique is the fact that a finite time exists between radical generation at the electrode and its detection within the ESR cavity. The time for transfer of solution restricts the technique to long-lived radical species. This problem was alleviated, to some extent, without resorting to in-situ techniques by Albery et al. [29–33] who made use of the tubular electrode shown in Fig. 12. The electrode consisted of an annular ring which formed part of the wall of a tube through which solution flowed. The tube was situated centrally within the ESR cavity with the electrode immediately upstream outside the cavity.

The flow of solution over the electrode was constrained to be laminar, within the range of flow rates used, enabling the distribution of radicals as a function of flow rate, electrode current, and cell geometry to be calculated. Theory was produced for stable radicals [30] and for radicals decaying by both first-order [31] and second-order kinetics [32], from which information of radical decay mechanisms was obtained by studying the ESR signal–current behaviour at constant flow rate.

The laminar flow tube of Albery perhaps represents the best possible ex-situ generation technique, but nevertheless a finite transit time still exists between the electrode and the sensitive part of the ESR cavity. This limits the lifetimes of radicals that may be studied; the upper limits for second- and first-order reactions have been estimated as $10^4\ dm^3\ mol^{-1}\ s^{-1}$ and $10^2\ s^{-1}$, respectively, [34] but these are certainly very optimistic estimates. The range of lifetimes may be extended if the electrode is placed within the ESR cavity, that is by adopting in-situ techniques, as will be described next.

3.1.2 In-situ methods

Several methods have been employed towards improving the design of an in-situ cell. Cauquis et al. modified an earlier ex-situ design so that in-situ experiments could be performed at a grid electrode located in a flat cell [35–39]. Circulation of solution in a closed loop was effected by bubbling nitrogen through the sidearm, the closed loop making this cell well suited to exhaustive electrolyses. The unfavourable geometry of this and earlier in-

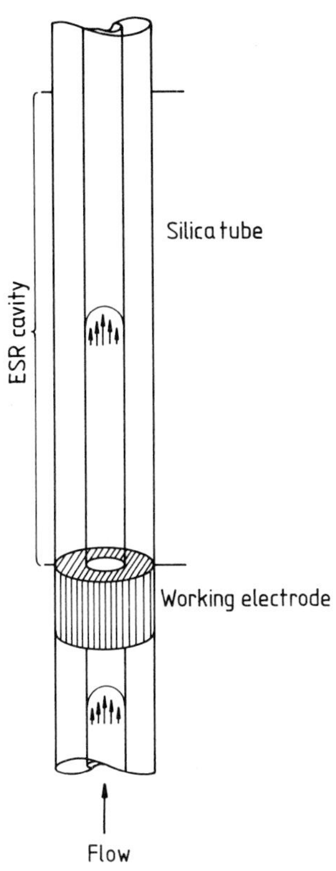

Fig. 12. The ex-situ flow cell of Albery et al., based upon the tube electrode, and located upstream of the ESR cavity.

situ cells allowed for only limited currents to be passed at the working electrode due to the considerable ohmic drop developing across the electrode. To surmount this problem, Kastening et al. [40, 41] developed a cell where the working electrode consisted of a set of strips within a flat cell, each strip having its own current supply. This enabled large currents to be used and by adjusting the potential of each strip, the ESR signal can be maximised. Using such a cell, the reaction of electrogenerated $SO_2^{-\cdot}$ with benzyl bromide [40] was investigated

$$\mathrm{BzBr} + \mathrm{SO_2^{-\cdot}} \xrightarrow{k} \mathrm{Bz{-}SO_2\cdot} + \mathrm{Br^-}$$

which is followed by a fast homogeneous reaction

$$\mathrm{BzSO_2\cdot} + \mathrm{SO_2^{-\cdot}} \longrightarrow \mathrm{Bz{-}SO_2^-} + \mathrm{SO_2}$$

Monitoring the decay of the ESR signal due to $SO_2^{-\cdot}$ enabled a rate constant of 180 $dm^3\,mol^{-1}\,s^{-1}$ to be determined. Unfortunately, doubts occur about the hydrodynamics of flow over the grid electrode and the imperfect agreement between theory and experimental results can be attributed to this.

Dohrmann et al. [42–45] developed a cell which yielded kinetic informa-

References pp. 349–352

tion by interrupting the current at the electrode and observing the ESR signal transient. This cell consisted of a foil electrode placed centrally in an ESR flat cell. Solution was then flowed over the electrode. Dohrmann's cell, and also Kastening's, should be amenable to a steady-state kinetic analysis, as in Albery's ex-situ tube electrode, but Dohrmann preferred to obtain kinetic data from transients, which are inherently less sensitive, unless signal averaging is performed.

The in-situ techniques of Cauquis, Kastening, and Dohrmann described above, although showing advantages over ex-situ methods, did not contribute significant advances towards a kinetic analysis of electrode mechanisms. Improvements upon this are shown by cells developed by several groups of workers for in-situ electrochemical ESR and are described in the following section.

4. Contemporary in-situ electrochemical ESR

This section deals with contemporary in-situ electrochemical ESR methodology and concentrates on the use of such techniques for the investigation of kinetics and their decay mechanism.

4.1 THE BARD AND GOLDBERG CELL

Goldberg et al. [46–48] developed an in-situ stationary solution cell similar to that developed by Adams. A platinum mesh working electrode of grid 1 mm was placed in a flat cell. Along the edges of the cell, a U-shaped tungsten rod was located which served as a counter electrode. A silver wire reference electrode was placed in the flat cell above the working electrode. Preliminary experiments were on substances which produce stable radicals in DMF, e.g. 9,10-anthraquinone, nitrobenzene, and azobenzene [46] were reduced to examine the electrochemical and ESR merits of the cell design. It was observed that the ohmic drop problems associated with Adams' cell were removed by having the counter and reference electrodes within the flat cell and improved electrochemical behaviour was observed. Locating the counter electrode within the ESR cavity presented problems when perchlorate background electrolyte solutions were employed, as $ClO_2\cdot$ radicals may be produced at the counter electrode if reductions occur at the working electrode. No radicals were observed, however, if tetrabutylammonium iodide (TBAI) was used as supporting electrolyte since, during reductions, iodine is formed at the counter electrode. Bard and Goldberg concluded that, where radical formation is possible, it may be necessary to add a non-radical-producing electroactive substance so that the production of secondary radicals is prevented. Later work [48] confirmed that problems arose from radicals produced at the counter electrode. The experiments with the stable radicals also indicated that a stable signal was observed for 20–30 s after the radical generation by a current step, after which natural convection set in.

The ability to maintain a constant steady-state ESR signal suggested that kinetic measurements could be made with the cell within this 20 s time limit.

Goldberg et al. [48] showed that homogeneous kinetics of radical ions generated electrochemically in their cell could be calculated using a current pulse method. The ESR signal was measured during and following the current pulse, the signal transients obtained analysed in terms of working curves for first-order radical decomposition, second-order radical ion dimerisation, and second-order radical ion-parent dimerisation. This technique was used in mechanistic studies of the reduction of olefins and can be illustrated best by the dimerisation of dimethyl fumarate radical anions. Analysis of ESR signal transients generated by a current pulse yielded rate constants of $160 \pm 26\,dm^3\,mol^{-1}\,s^{-1}$ for radical anion dimerisation and $50 \pm 30\,dm^3\,mol^{-1}\,s^{-1}$ for radical anion–parent dimerisation. Comparison of this data with conventional electrochemical measurements [49] in the form of cyclic voltammetry gives a rate constant of $160 \pm 40\,dm^3\,mol^{-1}\,s^{-1}$, which suggests that the mechanism is, in fact, a radical anion dimerisation

$$R + e \longrightarrow R^{-\cdot}$$

$$R^{-\cdot} + R^{-\cdot} \xrightarrow{k} R_2^{2-}$$

The work of Bard and Goldberg clearly showed how an in-situ electrochemical ESR cell with stagnant electrolyte can be used in the determination of reaction mechanisms and rate constants.

4.2 THE BOND CELL FOR IN-SITU ESR AT LOW TEMPERATURES

Very recently Bond and co-workers have described an in-situ cell for use at low temperatures [50]. The cell has a small working electrode and requires sample volumes of less than $0.2\,cm^3$, making it a microcell and benefitting from the advantages of this feature [51, 52]. The cell, a 4 mm o.d. quartz sample tube, contains a working electrode in the form of a $60\,\mu m$ diameter platinum wire sealed into a quartz capillary tube, the end of which is attached a small platinum sphere of diameter 0.5 mm placed centrally within the ESR cavity. An Ag/AgCl reference electrode is positioned 0.5 mm away from the tip of the working electrode and the counter electrode, a platinum wire of 0.5 mm diameter, is located 2 cm away from the centre of the cavity, where the sensitivity of ESR detection approximates to zero. This arrangement was shown to be adequate for the prevention of interference from counter electrode products and also in the fact that the sensitivity of the ESR was only slightly perturbed.

Variable temperature operation was achieved by placing the quartz sample tube within a standard quartz dewar insert through which nitrogen flowed at a controlled temperature in the range -130 to 30°C. Compared with the majority of previous cell designs, the Bond cell offers several advantages in that variable temperature experiments are possible, only small volumes of solution are used (helpful where materials are precious), and, finally, the cell does not suffer from severe ohmic drop problems leading to the recording of undistorted cyclic voltammograms. The cell has been initially used in studies of the oxidation of *mer* and *fac* $Cr(CO)_3P_3$ complexes, where P is a phosphine ligand. ESR spectra of the seventeen-electron *mer*-$[Cr(CO)_3P_3]^+$ isomer were produced in both cases. The rate of isomerisation of *fac*-$[Cr(CO)_3P_3]^+$ and *mer*-$[Cr(CO)_3P_3]^+$ was calculated from cyclic voltammetry and the time dependence of the ESR spectra produced was shown to be consistent with this data, showing that such an in-situ cell may be used for mechanistic studies of electrode reactions.

4.3 THE ALLENDOERFER CELL

Probably the most sensitive in-situ cell for the detection of short-lived species in ESR is the cell due to Allendoerfer and co-workers [53–55]. This cell is based upon the TE_{011} cylindrical ESR cavity shown schematically in Fig. 13. A commercially available cylindrical cavity is modified by placing a cylindrical metallic conductor centrally along the axis of the cavity. A quartz cell contains the metallic conductor and the liquid sample is contained within the unfilled volume of the cell. For conventional ESR, the sensitivity of such a coaxial cavity was shown to be as least as good as that of a rectangular TE_{102} cavity for solvents of high microwave loss; but as an in-situ cell for electrochemistry, a cell based upon the coaxial design would exhibit greater sensitivity compared with in-situ cells based upon the conventional ESR flat cell due to the significantly larger electrode area which can be accommodated when the central conductor of the coaxial cavity acts as the electrode.

Initial experiments by Allendoerfer et al. [53] involved a central conduc-

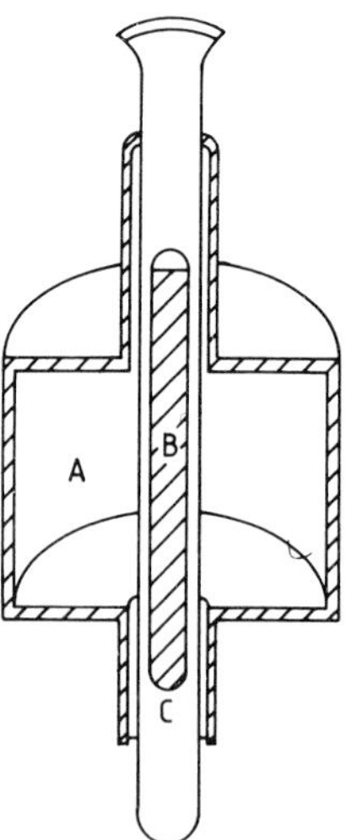

Fig. 13. Schematic diagram of the Allendoerfer cell, based upon the TE_{011} cylindrical cavity, A, with a metal cylinder, B, in a silica sample tube located along the axis of the cavity.

tor of length 50 mm and diameter 6 mm placed within a commercial cylindrical cavity. One problem that arose was that the resonant frequency of the coaxial cavity was greater than the 9.0–9.8 GHz range of the klystron employed. To overcome this, the resonant frequency was lowered by placing cylindrical inserts of a suitable dielectric, e.g. Teflon, around the central conductor. It was shown that, when the central conductor was solid and constructed of good conducting material, the cavity Q (a direct measure of the ESR sensitivity) was not significantly different from that of the cylindrical cavity.

To overcome some problems associated with using a solid central conductor, the inner part of the coaxial cavity was constructed from a finely wound shallow pitched helix, which stands freely and snugly against the inner wall of a quartz test tube of i.d. 6 mm. In this form, the microwaves only "observe" the portion of solution between the helix and the internal wall of the quartz tube and do not penetrate the inside of the helix, as shown in Fig. 14. A result of there being no penetration of microwaves within the helix meant that any material located here would not perturb the ESR experiment and so, in the Allendoerfer in-situ cell, the counter and reference electrodes are located here. The counter electrode is a platinum cylinder placed centrally along the axis of the cavity, such that the current flow is radial, and this geometry ensures a uniform current density over the whole surface of the cylindrical working electrode. The uniform current density meant that potential control of the whole electrode could be achieved by monitoring a single point with a Luggin capillary.

Allendoerfer et al. estimated the surface area of the working electrode to be close to 22 cm^2, significantly larger than previous designs, and found that the cell resistance was 13 Ω when filled with 0.1 M tetrabutylammonium perchlorate (TBAP) in dimethylformamide, which compares well with the

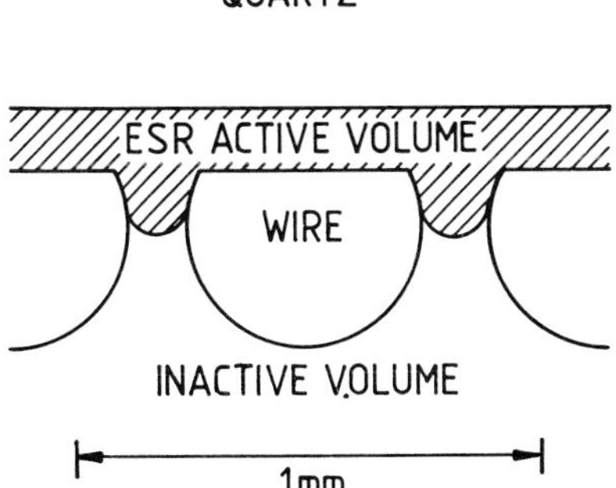

Fig. 14. The ESR active and inactive parts of the Allendoerfer cell where the central conductor is the finely wound helix. (Reproduced from ref. 55 with permission.)

References pp. 349–352

5–10 kΩ estimated by Goldberg and Bard [46] for an in-situ cell based upon a conventional flat cell. Electrochemical measurements made with this cell design were shown to be free of any distortions due to ohmic drop, even when used with poorly conducting electrolyte solutions, and that cyclic voltammograms observed from the in-situ cell were indistinguishable from those obtained from normal polarographic cells, except for the large increase in current due to the large surface area. Allendoerfer et al. estimated that the uncompensated solution resistance was less than 1 Ω, even when highly resistive non-aqueous solvents were used. It was estimated that, using such an in-situ cell, radicals as short-lived as 10^{-5} s could be studied provided a high concentration (approx. 10^{-1} M) of electroactive material was used. This agrees with previous estimates [43] of the shortest-lived radical it would be possible to study using in-situ techniques.

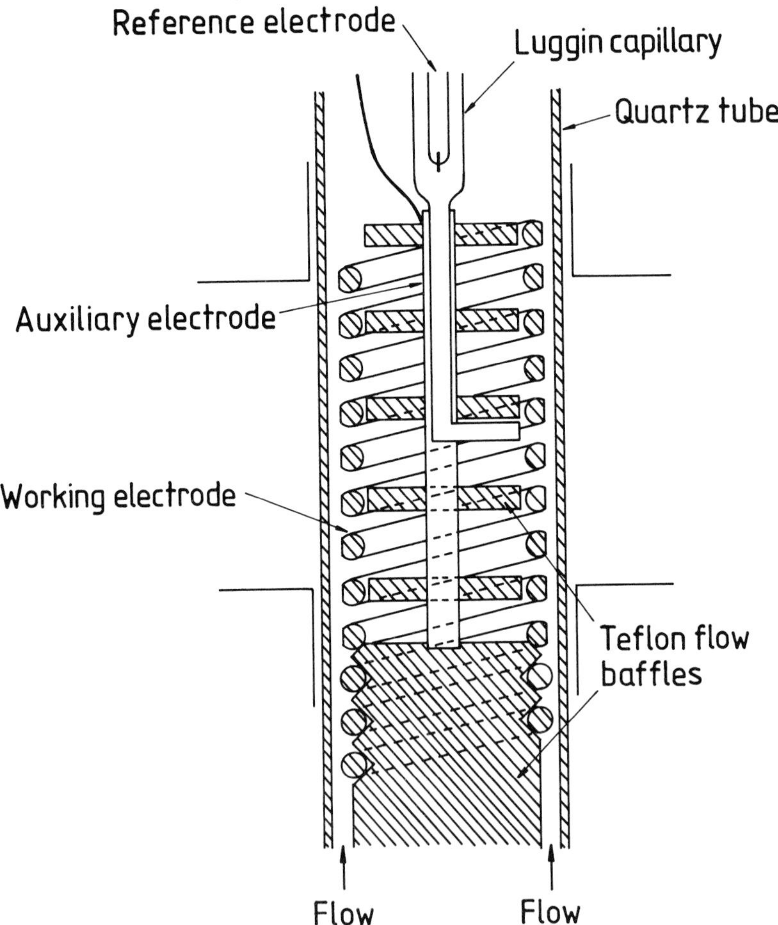

Fig. 15. The Allendoerfer cell, modified by Caroll, for the capability of solution flow over the electrode. (Reproduced from ref. 55 with permission.)

A modification of the Allendoerfer cell has been described by Ohya-Nishiguchi et al. [56–58], whereby low-temperature studies were undertaken on the electrochemistry of 20 aromatic compounds. The low temperature was achieved by placing the electrolysis cell within a temperature control dewar. It should be noted, however, that, in order to simplify the apparatus, the low-temperature in-situ cell did not employ a reference electrode.

In the study of short-lived radicals, the presence of efficient hydrodynamic flow is essential to sustain a constant supply of electroactive material to the electrode surface and hence ensure a steady flux of radicals. For this reason, Carroll adapted the in-situ cell of Allendoerfer to provide the capability of flow [54, 55]. The changes to the helical arrangement can be seen in Fig. 15. Solution is prevented from flowing in the central ESR inactive part of the cell by a complex series of baffles. The value of flow coupled to the Allendoerfer cell was shown by the electroreduction of nitromethane in aqueous conditions; the radical anion of nitromethane was observed and this was shown to have a lifetime in the order of 10 ms. To data, this is the shortest radical lifetime observed. Calculations [56] suggested that such a cell should be capable of observing radicals with lifetimes of 10^{-5} s.

4.4 THE COMPTON–COLES CELL

The high sensitivity of the Allendoerfer cell makes it of great value in the detection of unstable radicals but, for the study of the kinetics and mechanism of radical decay, the use of a hydrodynamic flow is required. The use of a controlled, defined, and laminar flow of solution past the electrode allows the criteria of mechanism to be established from the solution of the appropriate convective diffusion equation. The uncertain hydrodynamics of earlier in-situ cells employing flow, e.g. Dohrmann [42–45] and Kastening [40, 41], makes such a computational process uncertain and difficult. Similarly, the complex flow between helical electrode surface and internal wall of the quartz cell in the Allendoerfer cell [54, 55] means that the nature of the flow cannot be predicted and so the convective diffusion equation cannot be readily written down, let alone solved! Such problems are not experienced by the channel electrode [59], which has well-defined hydrodynamic properties. Compton and Coles [60] adopted the channel electrode as an in-situ ESR cell.

The design of the Compton–Coles in-situ cell, a demountable channel electrode constructed from synthetic silica, is shown in Fig. 16. Metal-foil electrodes are cemented to the silica cover plate and polished flat. Electrical connection is achieved via the hole in the cover plate, either by 0.12 mm copper wire and conductive silver paint or by direct connection to the electrode. The channel unit and cover plate are mounted together using a low-melting wax. The cell is connected to Teflon tubing for connection into a flow system. The assembled cell with relevant connections is held within a TE_{102} rectangular ESR cavity, this being achieved by placing the cell within a silica tube and fixing its position using Teflon spacers. The position of this tube is fixed by nylon collars. Movement of the silica tube then

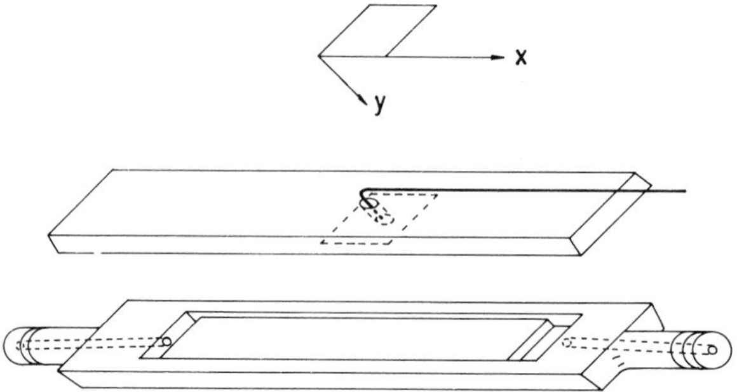

Fig. 16. The Compton–Coles in-situ flow cell.

adjusts the position of the cell within the ESR cavity, Optimally, the cell is positioned at the point of maximum ESR sensitivity and hence centrally within the ESR cavity.

The flow system used in conjunction with the channel cell is capable of a variable flow rate in the range 10^{-1} to 10^{-4} $cm^3\,s^{-1}$. This is achieved either via a gravity feed system, utilising capillaries of different diameter and changes in reservoir height, or by a mechanically driven pumping system. A counter electrode platinum gauze is placed downstream of the working electrode *outside* the ESR cavity and a reference electrode, normally calomel or Ag/AgCl, located upstream. When correctly positioned within the ESR cavity, the in-situ cell was found to be compatible with the cavity's microwave field if the electrode was in a plane parallel to the electric field direction. Adjusting the vertical and horizontal position of the electrode allowed optimisation of ESR sensitivity. The latter was shown to be comparable with that for a conventional ESR flat cell used for homogeneous chemistry.

For the channel electrode, Levich [61] calculated the diffusion-limited current–flow rate behaviour to be

$$I_{\mathrm{lim}} = 1.65\,nFCD^{2/3}\left[\frac{\bar{u}}{b^2 D}\right]^{1/3} w x_{\mathrm{E}}^{2/3} \tag{13}$$

where $\bar{u}$ is the mean solution velocity, $2b$ the depth of the channel, w the electrode width, and x_{E} the electrode length. The theoretically predicted limiting current dependence upon the cube root of the flow rate has been observed from current–voltage curves when the Compton–Coles cell has been used to investigate a range of electroactive species [60, 62–64]. Thus, the Compton–Coles cell can be seen to act as a satisfactory hydrodynamic electrode.

The dependence of the steady-state ESR signal upon diffusion-limiting current and flow rate was predicted to obey the equation

$$S \propto \frac{I}{\bar{u}^{2/3}} \tag{14}$$

for a stable radical. This equation was shown to hold [60] for radicals produced from electrochemically reversible and irreversible systems; typical results are shown in Fig. 17, this data being the one-electron reduction of fluorescein to semi-fluorescein radical anion [60].

The successful verification of the predicted behaviour for the channel electrode's use as an in-situ cell with stable species subsequently led to a study of unstable radical species. For the determination of the kinetics and mechanism of radical decay, two possible strategies exist, viz.

(1) measurement of the steady-state ESR signal as a function of the electrode current and flow rate and using variable flow rate to probe radical lifetimes and

(2) recording the transient ESR signal, when the working electrode is open circuited, after a steady state has been achieved.

The use of both methods has been shown to be successful with the Compton–Coles cell [65, 66].

The steady-state method can be illustrated by the case of a radical reacting via first-order kinetics [65]. Solution of the appropriate convective diffusion equation allowed Compton et al. to find how the ESR signal, S, varied as a function of the rate constant, K, and the geometry of the electrode. This was expressed in terms of the "ESR detection efficiency", M_K, given by

$$M_K = \frac{S\bar{u}^{2/3}}{I} \tag{15}$$

where K is the normalised rate constant for the radical decay

$$K = k\left(\frac{b^2 l^2}{9\bar{u}^2 D}\right)^{1/3} \tag{16}$$

where l is the length of the ESR cavity. For a stable radical, M_0 is constant. Otherwise, for an electrode positioned in the centre of the ESR cavity, M_K

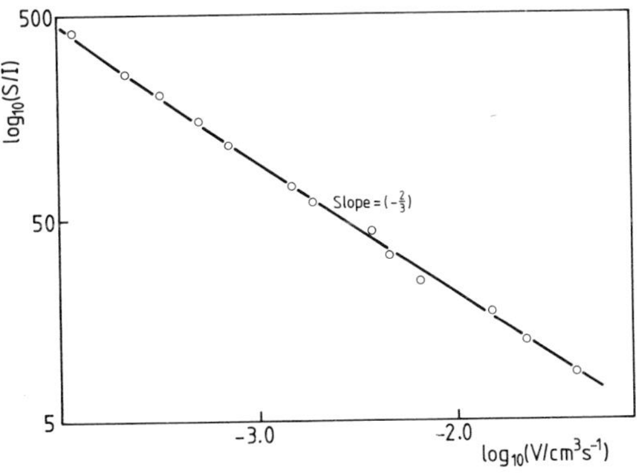

Fig. 17. ESR signal/current/volume flow rate behaviour for the reduction of fluorescein at pH 13 in the Compton–Coles cell.

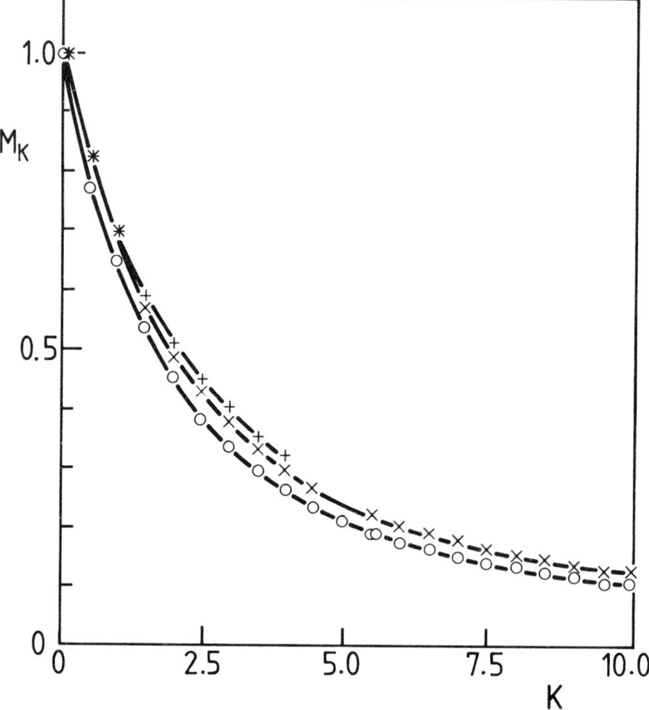

Fig. 18. Variation of ESR detection efficiency, M_K, with the normalised first-order rate constant, K, for different electrode lengths. +, 2 mm; ×, 3 mm; ○, 4 mm.

varies with K, as shown in Fig. 18, for electrodes of different lengths. Experimentally, the ESR signal, S, and the corresponding current, I, are measured for a wide range of flow rates allowing the calculation of the parameter M_K from eqn. (15). Figure 18 allows the deduction of the values of K corresponding to the measured M_K and eqn. (16) shows that the rate constant, k, may be deduced from a plot of K against $\bar{u}^{-2/3}$.

This method was used in a study of the reduction of 2-nitropropane in buffered aqueous solutions (pH 10.2) [65]. It was shown that the parent compound undergoes reduction to the nitropropane radical anion, which undergoes a first-order decay with rate constant $0.36\,s^{-1}$. Combining the in-situ experiments with voltammetric experiments led to the conclusion that the reduction of 2-nitropropane occured via an ECE mechanism, where the rate-determining chemical step was proton uptake by the radical anion.

As described, the alternative method for the deduction of radical kinetics is the use of ESR transient signals [65, 66]. Transient signals are obtained by open-circuiting an electrode which was previously at a potential corresponding to the diffusion-limited current. The electrode is open circuited by opening a switch, rather than stepping the potential between two defined values since the latter method may produce contributions to the radical decay from, e.g. re-oxidation of a radical anion, if the latter is generated by an electrochemically reversible reduction. For a radical decaying by first-order kinetics, the transient is described by

$$S(t) = S(t = 0) \exp(-kt) \quad (17)$$

provided the radical decay is sufficiently fast.

A decay transient for the 2-nitropropane radical anion, produced in the system described above, is shown in Fig. 19 together with the calculated transient. It can be seen that, for the low flow rate used, the transient is exponential, as predicted for the first-order decay, but at higher flow rates (or slower radical decay), one finds the transients are not perfectly exponential. The cause of this is convection of radical from the ESR cavity combining with the decay signal. This problem was shown to be easily overcome by a simple method of analysis [65] whereby the measured transients are treated as exponentials, as in eqn. (17), and an "apparent" rate constant is deduced for the flow rates used. A plot of "apparent" rate constant against $\bar{u}^{2/3}$ allowed extrapolation to zero flow rate and hence the deduction of the true value for the rate constant. Results obtained in this manner for the 2-nitropropane system were shown to be in agreement with the steady-state measurements.

Use of the Compton–Coles cell in quantitative studies has been presented for EC [65], ECE [65, 67], and DISP1 [67] processes. The use of this in-situ cell in the study of radical kinetics and mechanism is well illustrated by the

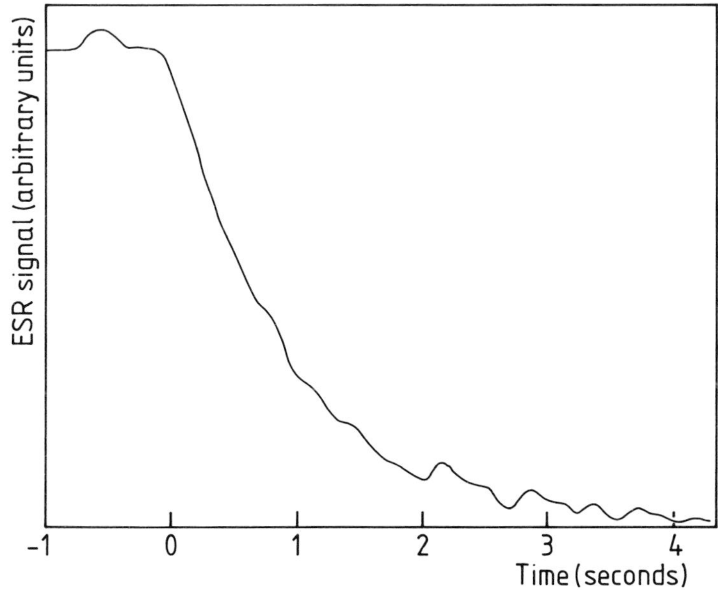

Fig. 19. ESR decay transient for the 2-nitropropane radical anion produced by open-circuiting the working electrode.

discrimination between ECE and DISP1 mechanisms, which is discussed in Sect. 5.4.

4.5 THE ALBERY IN-SITU CELL

The use of a tubular electrode by Albery for ex-situ experiments [29–33] has been described earlier. The tubular electrode is equivalent to the channel electrode in all respects except that the cross-section is circular rather than rectangular. In particular, the convective diffusion equations are equivalent for both electrode designs. The use of the tubular electrode for in-situ ESR has been investigated [34]. Moving the tubular electrode within the ESR cavity did not produce satisfactory results, the $\sin^2$ sensitivity of the cavity was grossly distorted and there was negligible sensitivity at the electrode surface, wherever it was placed within the cavity. Albery and co-workers found that the sensitivity of the ESR cavity was not distorted if the tubular electrode was replaced by a semi-annular tube electrode, as shown in Fig. 20. With the semi-annular tube electrode positioned centrally within the ESR cavity, the expected $\sin^2$ sensitivity profile was observed.

For a semi-annular tube electrode, the limiting current is given by the appropriate form of the Levich equation

$$I_{\text{lim}} = 2.75 nFD^{2/3} x_E^{2/3} V^{1/3} C \tag{18}$$

where x_E is the electrode length and V the volume flow rate. Equation (18) was verified by experimental observations on the oxidation of N,N,N',N'-tetramethyl-p-phenylenediamine in aqueous solution. The diffusion coefficient calculated from this data was in agreement with literature values. This showed the semi-annular tube electrode to be a satisfactory hydrodynamic electrode.

As with the Compton–Coles channel electrode, Albery's cell should obey

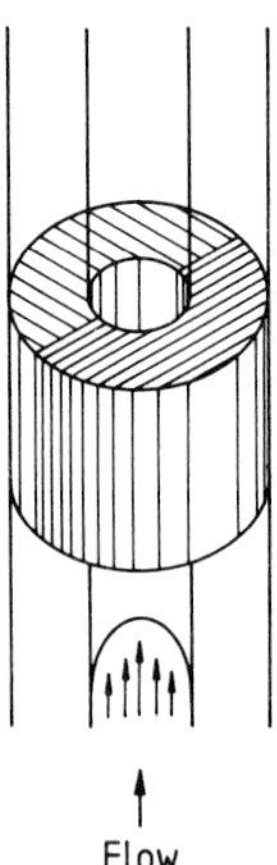

Fig. 20. The in-situ semi-annular tube electrode of Albery and co-workers.

eqn. (14) for the variation of ESR signal, S, with diffusion-limited current and flow rate. Experimental results verified this. Hence, the semi-annular tube electrode of Albery can be regarded as a satisfactory hydrodynamic cell for in-situ ESR experiments, comparable with the Compton–Coles cell. Practically, however, it suffers from several disadvantages compared with the Compton–Coles cell; firstly, the Albery cell is not readily demounted, making examination and polishing of the electrode difficult; secondly, it is uncertain whether eddy currents are induced in the electrode by the 100 kHz modulation rendering areas of the electrode ESR insensitive; thirdly, the channel electrode does not suffer as large an ohmic drop; and finally, irradiation of the electrode is not possible with Albery's design, making this in-situ cell useless for the study of photoelectrochemical phenomena.

4.6 THE COAXIAL IN-SITU CELL

Of the in-situ cells described previously in this section, two designs represent the best methodology for simultaneous electrochemical ESR experiments. The cell developed by Allendoerfer shows the highest sensitivity towards unstable radicals. However, the channel electrode developed by Compton and Coles [60] is more suited to the rigorous study of the kinetics and mechanism of radical decay, although the lifetimes of radicals amenable to study using the channel electrode (greater than 10–100 ms) are rather larger than those accessible with the Allendoerfer cell. To improve upon this, Compton and Waller have described an in-situ cell [68] which adopts the better features of the Allendoerfer cell and the channel electrode. This in-situ cell was shown to utilise a coaxial TE_{011} cavity, allowing a large area of electrode, but retained the well-characterised hydrodynamic flow essential for the rigorous study of electrode reaction mechanisms.

The design of this improved "coaxial cell" is shown in Fig. 21. The basic design encompasses a TE_{011} cylindrical cavity which has been converted into a coaxial cavity by the insertion of a copper rod. The copper rod is smooth and polished and positioned centrally between the stacks of the cylindrical cavity. The diameter of the rod is such that, when placed within the 9 mm precision bore silica tube surrounding it, there is a gap of size $2b$ available through which solution flows. Typically, the annular gap has $2b \approx 100\,\mu$m, but this may be adjusted by polishing the electrode. The annulus through which solution flows is constrained to uniform thickness by nylon thread running throughout the length of the cavity. The central 4 mm portion of the copper rod was insulated from the rest. It is this portion that acts as the working electrode of the cell, at the same time a forming the internal wall of the coaxial cavity. The electrode is plated with mercury so as to achieve the maximum cathodic potential range. The coaxial cell is connected into a flow system with a reference electrode positioned upstream and a counter electrode downstream and a variable flow rate provided by a gravity fed flow system.

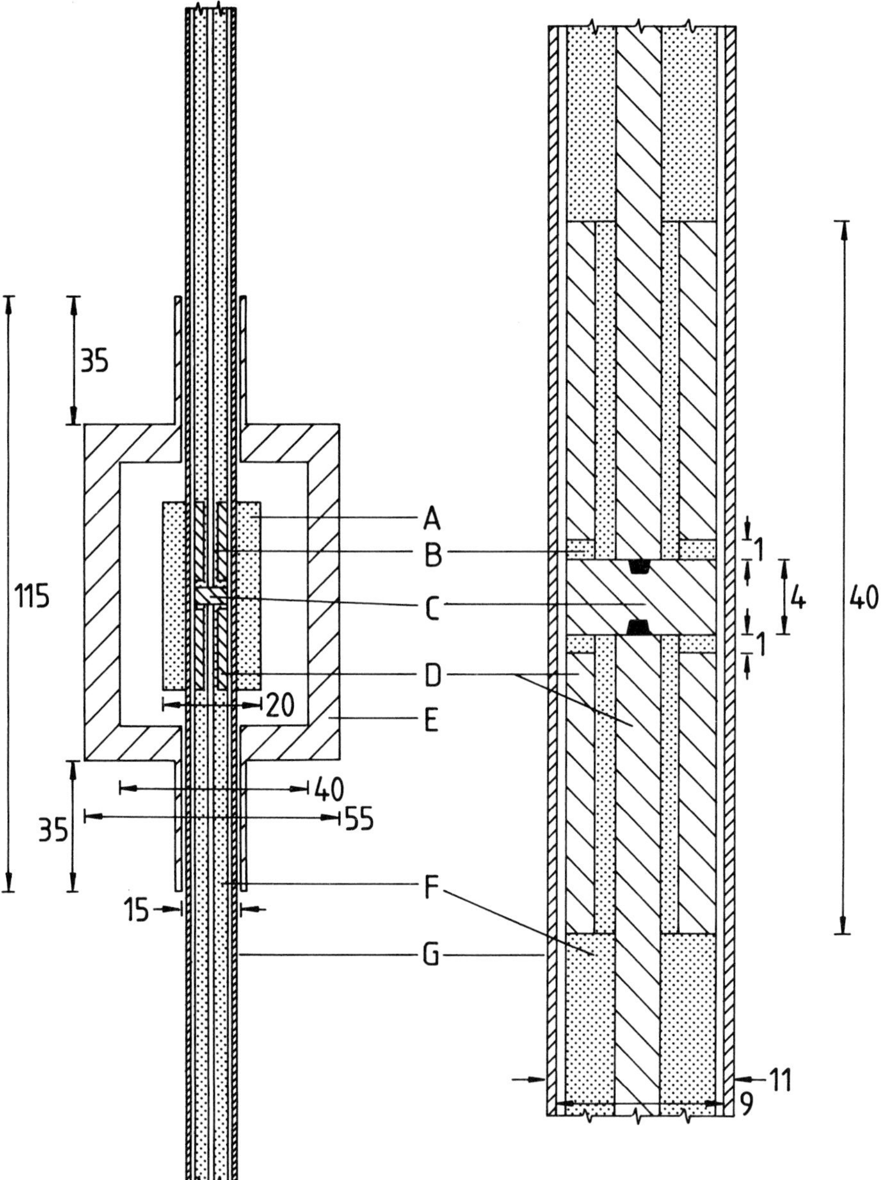

Fig. 21. The Compton–Waller cell. A, PTFE annulus; B, PTE insulation; C, mercury-plated copper electrode; D, copper; E, TE_{011} cylindrical cavity; F, PTFE sheath; G, precision bore silica tubing. Numbers refer to dimensions in mm.

The ESR performance of the coaxial cell was similar to that observed with the Allendoerfer cell. The insertion of the copper rod caused a reduction in the effective size of the cavity and hence the resonant frequency was observed to shift above that of the empty cylindrical cavity. With copper rods of diameter ~ 9 mm, this was found to move the resonant frequency outside the tuning range of the klystron (9–10 GHz). As described in Sect. 4.3, the frequency can be returned to this range by partially filling the cavity with suitable dielectric; for example, an annulus of Teflon can be placed around the silica tubing. The sensitivity of the coaxial cavity was investigated both with distance along its length and also around the cavity at a fixed distance into the cavity; this was performed by moving a minute crystal of the stable radical α,α′-diphenyl-β-picrylhydrazyl down the length of the cavity and around it. Theory [11] would predict a $\sin^2$ sensitivity profile in the axial direction and the experimental results were shown to verify this, as shown in Fig. 22. The experimental results for the sensitivity of the coaxial cavity around the copper rod at a fixed distance are shown in Fig. 23. The observed $\cos^2$ behaviour was not caused by the standing microwave field pattern within the cavity, which would be expected to be symmetrical about the axis of the cylindrical cavity, but was attributed to a perturbation of the component of the magnetic field modulated at 100 kHz. This is caused by eddy currents induced in the copper rod by the 100 kHz component, which induced a magnetic field which opposes the applied field. When the applied field is perpendicular to the copper surface, the effect is maximal but zero when it is parallel to the surface. This phenomenon produced the observed $\cos^2$ variation of sensitivity, which has also been reported by Carroll [55]. Clearly, with the coaxial cell, radicals in solution will experience differing amplitudes of the modulated field depending on their location, but it was shown that this effect will only reduce the ESR signal by a factor of one half from that observed in the absence of the effect. Compton and Waller concluded

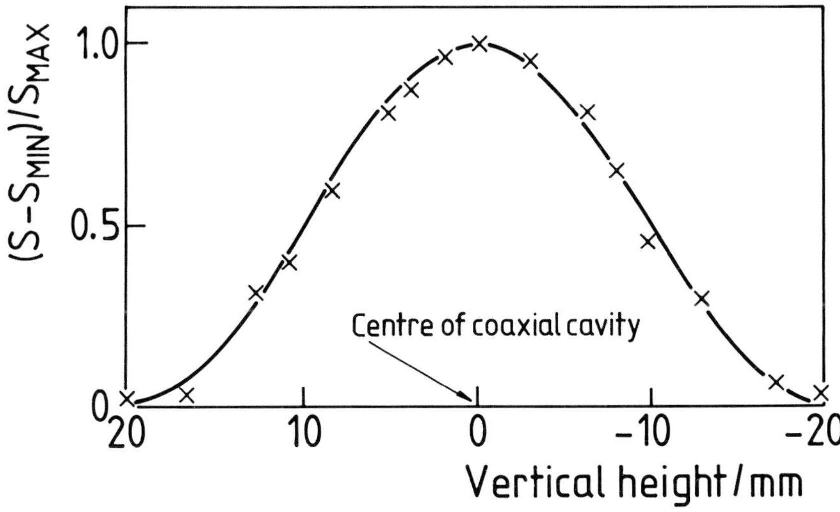

Fig. 22. The sensitivity of the Compton–Waller cell as a function of distance along its axis. The line drawn represents the theoretical $\sin^2$ behaviour.

References pp. 349–352

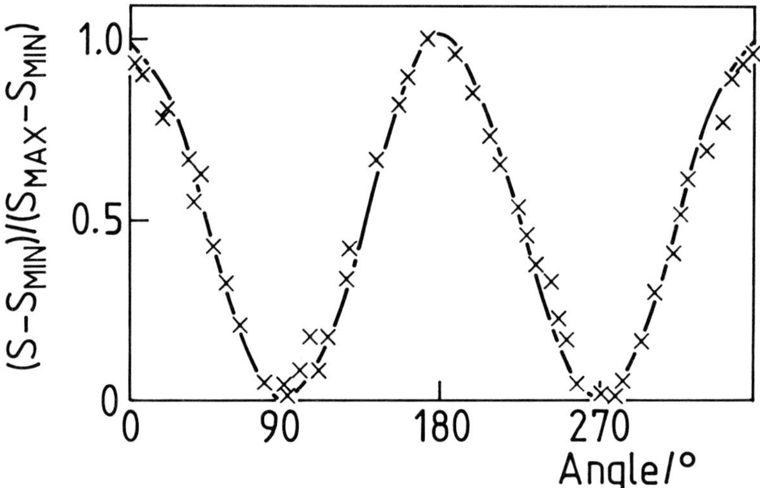

Fig. 23. The sensitivity around the central copper rod at a fixed distance into the cavity.

that the coaxial cell behaved in a satisfactory and understandable manner as regards ESR [68].

The hydrodynamics of flow of solution past the electrode, which is essential to the cell design, was rigorously investigated. In the range of flow rates used (10^{-4} to $10^{-1}\,cm^3\,s^{-1}$), the flow was laminar (Reynolds Number, Re < 10) and hence, beyond a lead-in section of length $0.1\,\mathrm{Re}\cdot b$, a parabolic velocity profile developed across the narrow channel. Thus, the hydrodynamics of the coaxial cell were equivalent to those of the conventional channel electrode [59]. It was predicted that the diffusion-limited current would obey the Levich equation

$$I_{\mathrm{lim}} \quad = \quad 1.165\,nFCD^{2/3}\bar{u}^{1/3}b^{-1}\sigma x_{\mathrm{E}}^{2/3} \tag{19}$$

where σ is the circumference of the electrode. The applicability of eqn. (19) to the coaxial cell was tested using, as a test system, the one-electron reversible reduction of fluorescein to the radical anion semi-fluorescein in 0.1 M NaOH at the mercury surface. Current–voltage curves obtained at the coaxial cell were shown to have a half-wave potential of -1.03 V (vs. saturated calomel), which is in good agreement with literature values for the fluorescein reduction deduced from conventional electrochemical measurements. The limiting current–flow rate behaviour obtained is shown in Fig. 24. The gradient of $\frac{1}{3}$ expected from eqn. (19) was observed. Calculation of a diffusion coefficient, D, from the Levich plot gave a value of $3.0 \times 10^{-6}\,cm^2\,s^{-1}$, again in close agreement with the value of $3.2 \times 10^{-6}\,cm^3\,s^{-1}$ obtained at a mercury-plated rotating-disc electrode [69]. Tafel analysis of the current–voltage curves obtained at the coaxial cell gave satisfactory results. Such a Tafel plot is shown in Fig. 25 where the line

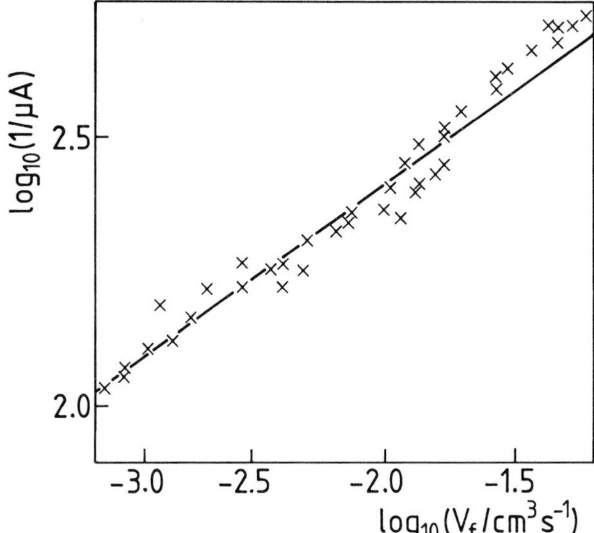

Fig. 24. Limiting current–flow rate behaviour for the reduction of fluorescein (pH 13) at the Compton–Waller cell.

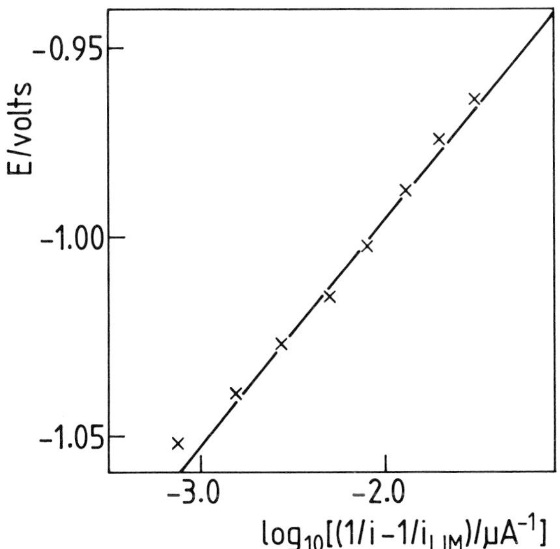

Fig. 25. Tafel analysis of current–voltage curve obtained for the reduction of fluorescein at the Compton–Waller cell.

drawn has the theoretical value of 59 mV $decade^{-1}$ for a reversible one-electron reduction. Thus, the electrochemical behaviour of the coaxial cell was shown to be entirely satisfactory.

Since the coaxial cell has hydrodynamics equivalent to those of the

References pp. 349–352

channel electrode, it was deduced that the behaviour of the steady-state ESR signal with limiting current and flow rate should be identical to that of the channel electrode given by eqn. (14) for a stable radical. This was investigated with the fluorescein system and the data obtained is shown in Fig. 26. For a wide range of flow rates, the data were shown to obey eqn. (14) with the line drawn having the theoretical slope of $-\frac{2}{3}$. But at slow rates ($< 4 \times 10^{-3}\,cm^3\,s^{-1}$), the data were shown to deviate slightly from that theoretically expected. This was explained by the diffusion layer becoming comparable in size to b, hence less radical is produced than expected on the basis of the Levich equation, i.e. by assuming a wide channel [70–72]. At the fast flow rates, the electrode was shown to obey the Levich equation, eqn. (19), whereas at the slow flow rates, there is approaching exhaustive electrolysis of solution as it passes over the electrode, hence the coaxial cell is behaving as a "thin layer cell" [73]. At the slow flow rates, the current is given by

$$I \;=\; 2nF\bar{u}b\sigma C \tag{20}$$

and is less than that predicted by the Levich equation. It was shown that the transition between the two limits in the current corresponds to a flow of about $10^{-3}\,cm^3\,s^{-1}$ for electrodes of the dimensions used in the coaxial cell. The effect was noticeable at slightly higher flow rate in the ESR signal–flow rate–current data (see Fig. 26) because the measured quantity was the concentration of radicals throughout the channel, not just the concentra-

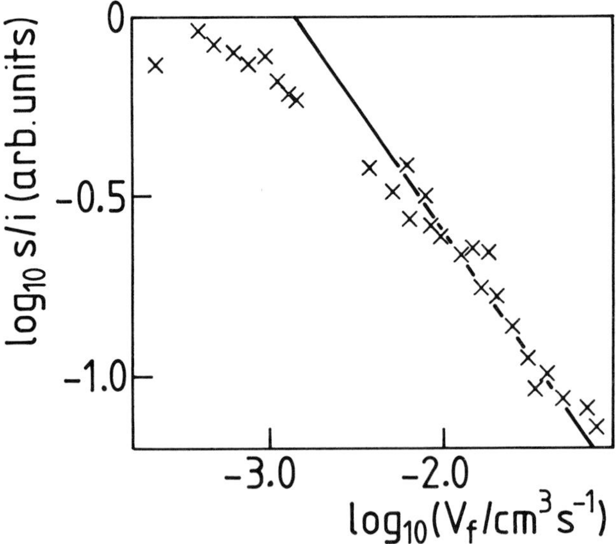

Fig. 26. Steady-state ESR signal/current/flow rate behaviour at the Compton–Waller cell.

tion gradient at the electrode surface as for the current–flow rate data. As a result, the ESR signal–flow rate–current behaviour is more sensitive to depletion effects. Despite this, it was concluded that the coaxial cell behaved in a well-defined calculable manner, apart from at the lowest flow rates.

The use of the coaxial cell developed by Compton and Waller as an in-situ ESR cell gives improvement over previous designs in that it retains the satisfactory hydrodynamic electrode necessary for the study of kinetics and mechanism, alongside the high sensitivity derived from the high area of the working electrode. The coaxial cell thus represents the current state-of-the-art methodology as regards in-situ cells, with changes to the channel thickness $2b$ and flow rate making it amenable to the study of either stable or unstable radicals.

5. Applications

This section describes typical applications of the in-situ electrochemical ESR methodology. Specific examples of radical identification, determination of radical decay mechanisms, polymer-coated electrodes, and spin trapping will be included.

5.1 THE MECHANISM OF THE REDUCTION OF DICYANOBENZENE

The use of electrochemical ESR for radical identification is well illustrated by the reduction of 1,2-dicyanobenzene at a mercury electrode. The reduction has been studied by Gennaro et al. [74], who observed two waves during the first scan of their cyclic voltammetric experiments. The first wave, at -1.32 V (vs. SCE), was attributed to the formation of the radical anion of dicyanobenzene. The second wave, at -2.35 V, was assigned to an ECE process in which the anion formed during the first wave was further reduced to the dicyanobenzene dianion, which was suggested to undergo reaction with the solvent, producing benzonitrile. As benzonitrile is known to be reduced at potentials corresponding to the second wave, a further electron transfer takes place forming the benzonitrile radical anion. Compton and Waller [75], using the channel electrode cell described in Sect. 4.4, reduced 1,2-dicyanobenzene in acetonitrile solvent. Potentiostatting the electrode at potentials corresponding to the first reduction wave, they observed the ESR spectrum shown in Fig. 27(a), which was assigned to the dicyanobenzene radical anion. When the electrode was potentiostatted at the second wave, a different ESR spectrum, shown in Fig. 27(b) was observed. This was found to be indistinguishable (in terms of hyperfine coupling constant and g value) from the spectrum obtained by direct reduction of benzonitrile itself at these potentials. Hence, the mechanism of the reduction inferred indirectly from the cyclic voltammetric experiments of Gennaro et al. [74] was shown to be vindicated by in-situ electrochemical ESR.

Similar observations have been reported by Fraenkel et al. [23] using dimethylformamide as solvent.

5.2 TRIPHENYLACETIC ACID

Considerable effort has been expended by electrochemists in the quest for radical intermediates in the Kolbé reaction, i.e. the anodic oxidation of carboxylate anions in solution. This reaction produces a variety of products dependent upon several factors such as solvent, pH, electrode material, etc.

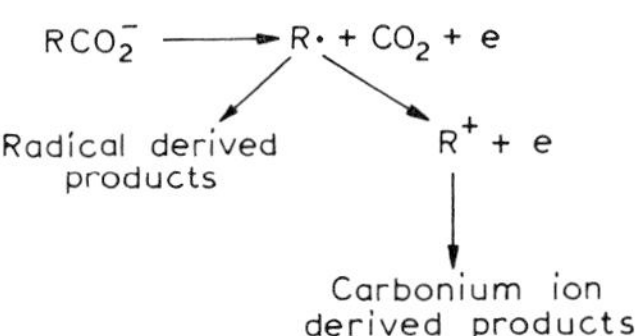

The possible involvement of $R^\cdot$ in the Kolbé reaction has been regularly investigated using ESR but, to date, positive evidence for radical intermediates has been lacking apart from two cases, firstly the photo-Kolbé reaction at irradiated semiconductor electrodes [76] and, secondly, the anodic oxidation of triphenylacetic acid at platinum electrodes in acetonitrile. The latter will be discussed in this section.

The first investigations of the oxidation of triphenylacetic acid were performed by Kondrikov et al. [77, 78] using an in-situ cell adapted from the design of Adams. In this design, the counter and working electrodes were in close proximity within the sensitive part of the ESR cavity. Upon oxidation of triphenylacetic acid, a spectrum was obtained which was attributed to the triphenylacetoxyl radical $Ph_3CCO_2^\cdot$. The spectrum was interpreted in terms of hyperfine interaction with four ortho-protons (a_o = 2.2 G) and two para-protons (a_p = 3.5 G). This suggestion required that one of the phenyl groups does not contribute to the hyperfine splitting and this was explained by the authors as being due to restricted rotation of the CO_2 group around its carbon–carbon bond. The explanation of the observed spectrum is open to question since, firstly, the measured coupling constants are large for such long range interactions, secondly, evidence [79] suggests that the triphenylacetoxyl radical undergoes rapid decarboxylation, and finally, the published coupling constants are not consistent with the published spectrum [78]. The latter appears to be a 1:4:6:4:1 quintet of $a_M \approx$ 2 G, with further small unresolvable splittings.

Bard and co-workers, using the in-situ cell described in Sect. 4.1, obtained rather different results [80] from Kondrikov. Direct oxidation of triphenylacetic acid was found not to result in an ESR spectrum, but spectra attributed to triphenylmethyl radicals $\phi_3C^\cdot$ were observed if the electrode was held at a potential corresponding to TPA oxidation (2.0 V vs. Ag pseudoreference electrode) and then stepped to a value negative of about 0.35 V.

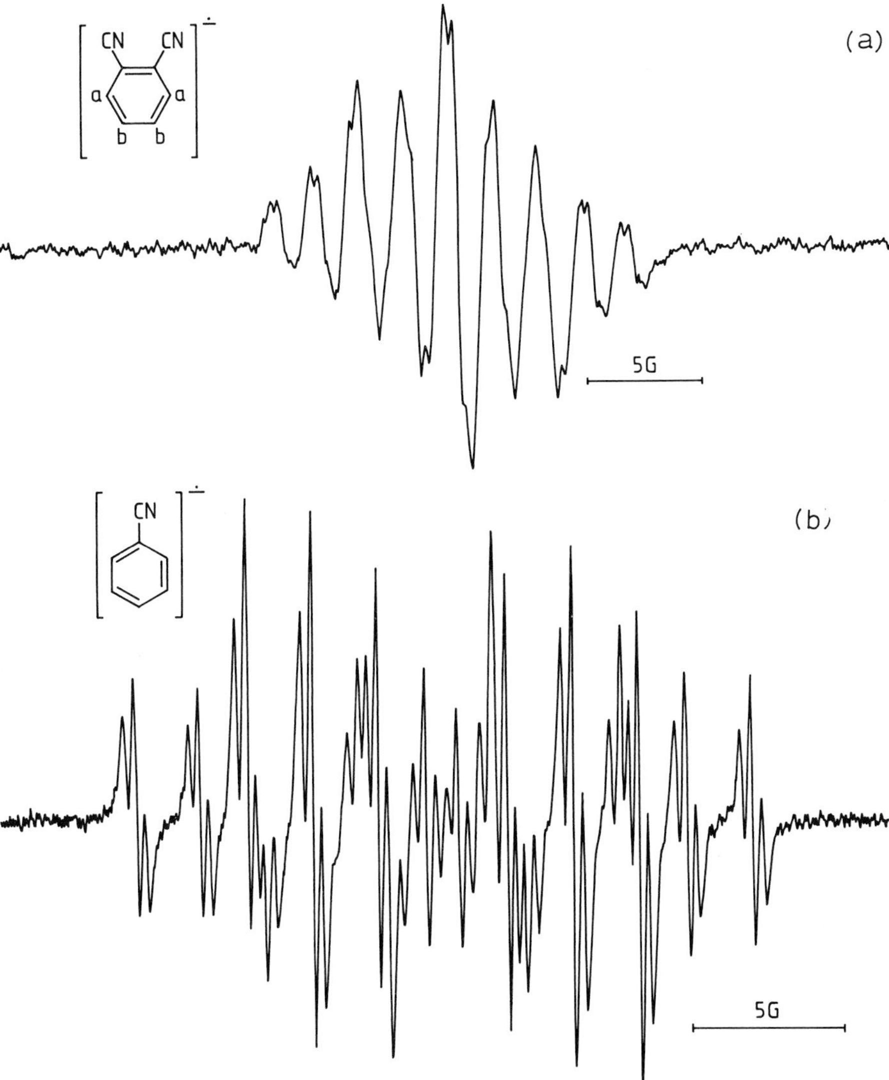

Fig. 27. ESR spectrum obtained by in-situ electrolysis of 1,2-dicyanobenzene at (a) −1.32 V, where the spectrum was shown to be that of 1,2-dicyanobenzene radical anion and (b) −2.35 V where the spectrum was shown to be that of benzonitrile radical anion.

The observation of triphenylmethyl radicals was interpreted by the scheme

$$Ph_3CCO_2H \xrightarrow{2.0\,V} Ph_3C^+ + CO_2 + H^+ + 2e$$

$$Ph_3C^+ + e \xrightarrow{0.35\,V} Ph_3C^{\cdot}$$

The clear discrepancies in the two studies of TPA oxidation led to a further study of this reaction by Day and co-workers [81] using the in-situ

References pp. 349–352

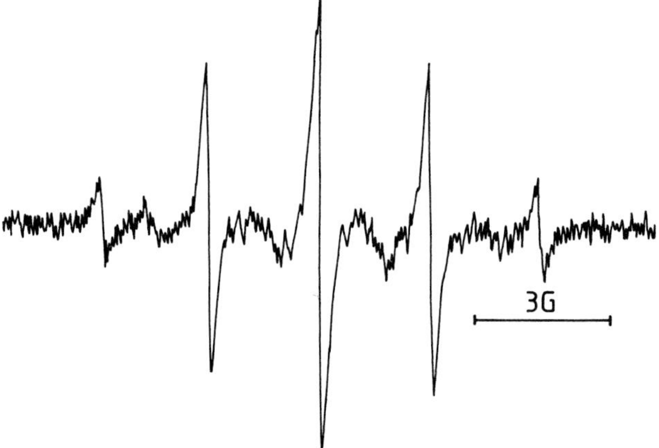

Fig. 28. Spectrum obtained by oxidation of triphenylacetic acid at + 2.0 V before stepping to − 1.9 V in wet acetonitrile.

channel electrode described in Sect. 4.4. They showed that ESR spectra resembling those obtained by Kondrikov could be obtained by an oxidising–reducing potential sequence similar to that used by Bard in wet acetonitrile. TPA was exhaustively oxidised at + 2.025 V (vs. SCE) and then stepped to a negative potential of − 1.80 V (vs. SCE). The spectrum observed from such a potential sequence, shown in Fig. 28(a), was a 1:4:6:4:1 quintet with a coupling constant of 2.41 G which closely resembled the spectrum of the benzoquinone radical anion produced by direct reduction of benzoquinone in acetonitrile. The coupling constants of spectra from benzoquinone reduction and that shown in Fig. 28, and the G value of the two spectra were shown to

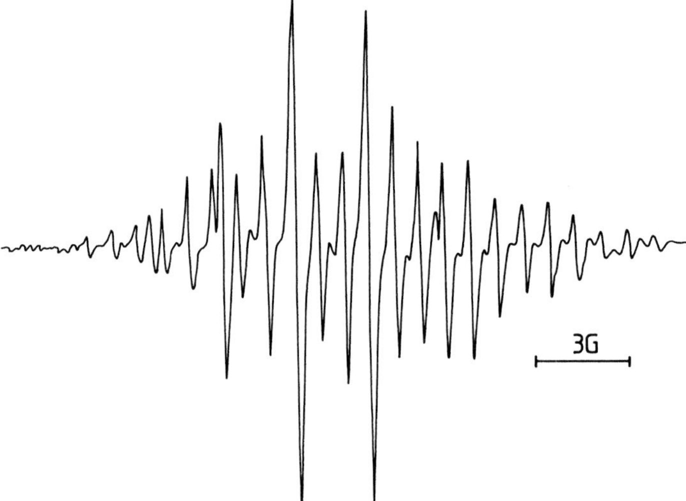

Fig. 29. Spectrum obtained by the oxidation of triphenylacetic acid at + 2.0 V before stepping to − 1.9 V in slightly moist acetonitrile.

be identical. This suggested that the ESR signals observed by Kondrikov et al. [77, 78] were due to the formation of benzoquinone radicals through reduction of working electrode products, this being promoted by the close proximity of the two electrodes within the cell used.

The oxidising–reducing potential sequence was repeated on TPA in dry acetonitrile (~ mM water present); the spectrum observed was different from that obtained in wet solvent and was composed of the superposition of two radicals (see Fig. 29). The relative amounts of the two radicals observed was strongly dependent on the amount of water present. In very wet solvent, only benzoquinone radical anion was observed as described above. Day showed that the spectra was due to both the benzoquinone radical anion and the benzophenone radical anion. It was possible to simulate accurately the measured spectra from the known hyperfine coupling constants and *g* values of the two anion radicals, assuming varying relative amounts of the two species. Day concluded that the spectrum observed by the Russian workers was, in fact, due to the mixture of radicals produced, in their case, by mixing and reaction of counter and working electrode products. The mechanism by which benzoquinone and benzophenone radical anions were formed was rationalized by the scheme

Oxidation

$$Ph_3C\text{–}CO_2H \longrightarrow Ph_3C^+ + CO_2 + H^+ + 2e$$

$$H_2O \longrightarrow 2H^+ + \tfrac{1}{2}O_2 + 2e$$

Reduction

$$Ph_3C^+ + e \longrightarrow Ph_3C^{\cdot}$$

$$Ph_3C^{\cdot} + O_2 \longrightarrow Ph_3C\text{–}O\text{–}O^{\cdot}$$

$$Ph_3C\text{–}O\text{–}O^{\cdot} + 2H^+ + e \longrightarrow Ph_3C\text{–}O\text{–}\overset{+}{O}\text{–}H_2$$

$$Ph_3C\text{–}O\text{–}\overset{+}{O}H_2 \longrightarrow Ph_3CO^+ + H_2O$$

$$Ph_3CO^+ \longrightarrow Ph_2\overset{+}{C}OPh$$

$$Ph_2\overset{+}{C}OPh + H_2O \longrightarrow Ph_2C(OH)\text{–}O\text{–}Ph + H^+$$

$$Ph_2C(OH)\text{–}O\text{–}Ph \longrightarrow Ph_2C{=}O + Ph\text{–}OH$$

$$Ph\text{–}OH \xrightarrow[+H_2,\ -H^+]{\text{"ox"}} O{=}Ph{=}O$$

$$O{=}Ph{=}O \xrightarrow[+e]{E_{1/2} = -0.55\ V} \text{Benzoquinone radical anion}$$

$$Ph_2C{=}O \xrightarrow[+e]{E_{1/2} = -1.75\ V} \text{Benzophenone radical anion}$$

References pp. 349–352

The scheme described involves the well-established rearrangement of Ph_3CO^+ cations to form $Ph_2\overset{+}{C}OPh$, which is known to react with water to form phenol and benzophenone. The oxidation of phenol to form benzoquinone by a residual product of the oxidation step ("ox") is not unreasonable since Day observed that phenol in moist acetonitrile subjected to a similar oxidising–reducing sequence reveals signals attributable to benzoquinone radical anion. The mechanism as suggested gained additional evidence for its support from further work by Day and co-workers on the anodic oxidation of triphenylmethanol in acetonitrile [82], which will be discussed in the following section.

The oxidation of TPA is a good example of the use of in-situ electrochemical ESR to provide details of electrode mechanisms but, at the same time, the above reveals how the results obtained and their subsequent interpretation are easily affected by poor cell design, as is observed from the work of Kondrikov.

5.3 TRIPHENYLMETHANOL

Day and co-workers, in an extension to his work on TPA, investigated the anodic oxidation of triphenylmethanol (TPM) in acetonitrile [82]. Initial cyclic voltammetric experiments revealed that TPM underwent oxidation at a potential of +2.18 V (vs. SCE). If, during the cyclic voltammogram, the potential was held at +2.20 V for several minutes before reversing the scan, a voltammogram such as that shown in Fig. 30 was observed with peaks at +0.25 and +0.05 V. The peak at +0.05 V was attributed to the reduction of H^+ ions formed during the oxidative sweep whilst the peak at +0.25 V was tentatively assigned to the reduction of Ph_3C^+. To confirm this assignation, Day and co-workers performed in-situ electrochemical ESR experiments using the same oxidising–reducing sequence at a channel electrode as in the previous section and this produced the ESR spectrum in Fig. 31, which is identical to the previously reported spectra of the $Ph_3C^{\cdot}$ radical [80]. The production of both H^+ and Ph_3C^+ was rationalised by the scheme

$$Ph_3COH \xrightarrow{2.20\ V} [Ph_3COH]^{+} + e$$

$$[Ph_3COH]^{+\cdot} \begin{cases} \rightarrow Ph_3C^+ + OH \\ \rightarrow Ph_3CO^{\cdot} + H \end{cases}$$

two routes being suggested for the decomposition of $[Ph_3COH]^+$. Day then proceeded to investigate the fate of the triphenylmethoxyl radical $Ph_3CO^{\cdot}$ and suggested two reasonable possibilities

$$\text{A: } Ph_3CO^{\cdot} \rightarrow Ph_3CO{-}COPh_3 \xrightarrow{-2e} Ph_3C^+ + C_2 + CPh_3$$

$$\text{B: } Ph_3CO^{\cdot} \xrightarrow{-e} Ph_3CO^+ \xrightarrow{H^+} Ph_2CO + PhOH$$

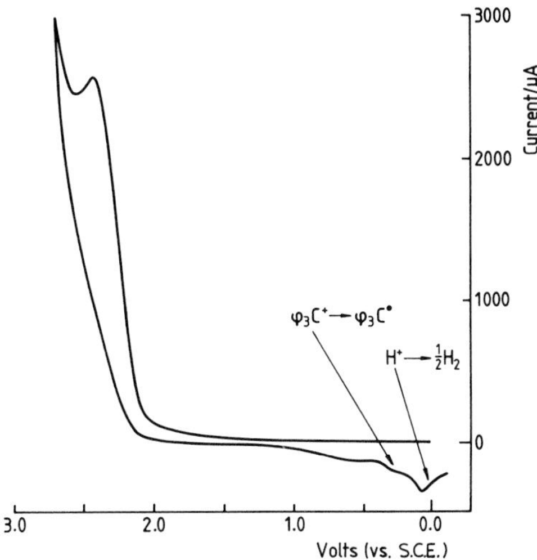

Fig. 30. Cyclic voltammogram showing the oxidation of triphenylmethanol (from ref. 82). The potential was held at +2.20 V for several minutes before reverse scan.

That the fate of the $Ph_3CO^{\cdot}$ radical was further oxidation and not dimerization was confirmed by in-situ electrochemical ESR experiments. Day performed oxidising–reducing potential sequences as carried out with TPA. When the potential was held at +2.20 V before stepping to −1.80 V, the resultant ESR spectrum was identical to that shown in Fig. 29, i.e. a mixture of benzophenone and benzoquinone radical anions produced by the mechanism as detailed earlier for the fate of Ph_3CO^+ species from TPA oxidation.

The oxidation of TPM shows how mechanistic detail can be confirmed by coupling electrochemistry with ESR, when looking at reactions where electrochemistry alone can only tentatively suggest details for the electrode mechanism.

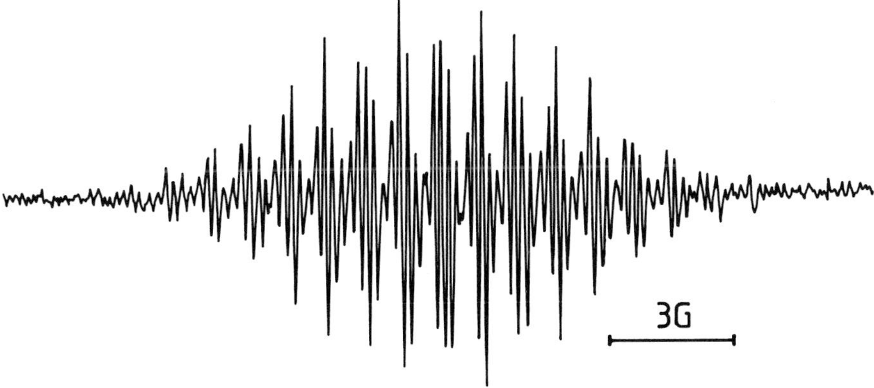

Fig. 31. ESR spectrum by the oxidation of triphenylmethanol at +2.20 V, before stepping to +0.20 V, attributed to $Ph_3C^{\cdot}$.

5.4 ECE/DISP1: THE REDUCTION OF FLUORESCEIN

The three examples described above show the use of in-situ electrochemical ESR in the investigation of the mechanism of electrode reactions. The work of Compton et al. [67] upon the reduction of fluorescein in aqueous media illustrates how in-situ techniques can provide information of both electrode mechanism and kinetics. In buffered solutions in the pH range 9–10 Compton et al. observed an apparent two-electron reduction of fluorescein (F) to leuco-fluorescein (L). In-situ electrochemical ESR experiments using the channel electrode revealed an ESR spectra as shown in Fig. 32 which was attributed to semi-fluorescein (S) where

F =

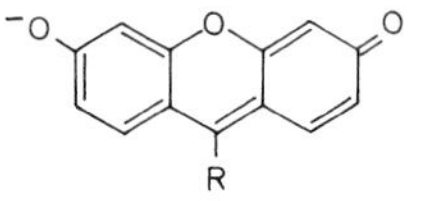

S =

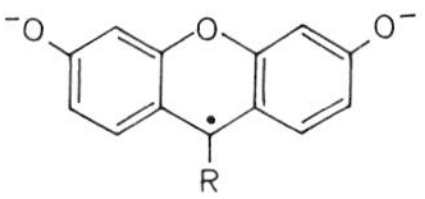

L =

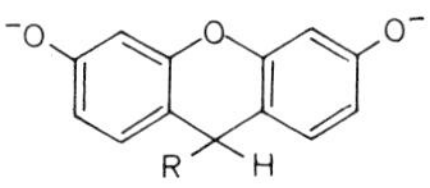

R =

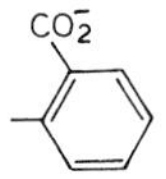

The diffusion-limited current–solution flow rate behaviour for the reduction of fluorescein at about pH 10 was studied. The observed behaviour, shown in Fig. 33, revealed a transition from two-electron transfer at low flow rates to one-electron transfer at fast flow rates. This behaviour is characteristic of an ECE or DISP process (vide infra), where, at fast flow rates, the product of the first electron transfer is swept away from the electrode before the chemical step takes place and one-electron behaviour is observed.

The difference between ECE and DISP mechanisms is that, for ECE, there are two heterogeneous electron transfers, but a DISP process involves a homogeneous second-electron transfer, via disproportionation. Compton described the kinetic scheme

(i) $F + e \rightleftharpoons S^{\cdot}$

(ii) $S^{\cdot} + H^{+} \xrightarrow{k} SH^{+}$

(iii) $SH^{+} + e \rightleftharpoons L$

(iv) $SH^{+} + S^{\cdot} \rightleftharpoons F + L$

for the reduction of fluorescein. There are two possibilities. Firstly, the

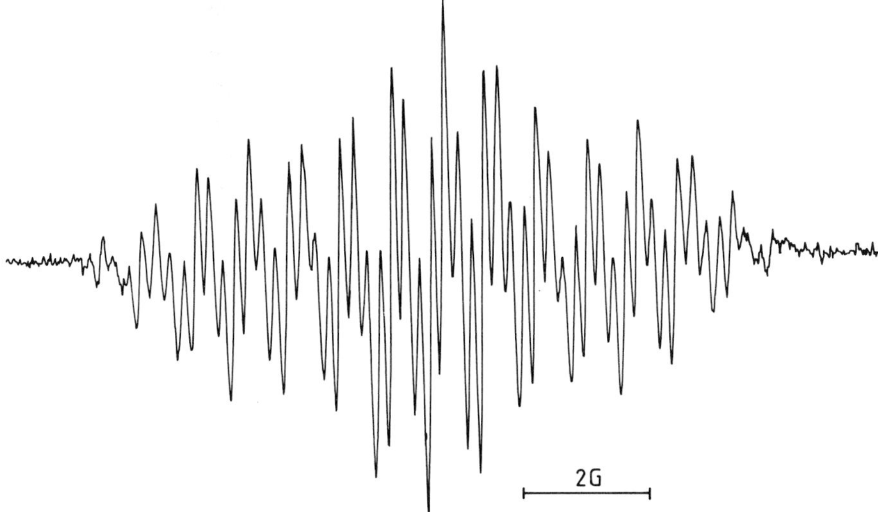

Fig. 32. ESR spectrum of fluorescein radical anion produced by the reduction of fluorescein at pH 9–10.

reaction could proceed via reactions (i), (ii), and (iii) and hence could be termed ECE. Secondly, it could proceed via reactions (i), (ii), and (iv) and be termed DISP. Within the DISP mechanism, two distinct cases occur upon whether reaction (ii) or (iv) is rate-determining. Where reaction (ii) is rate-determining, the mechanism is termed DISP1, as it is a (pseudo-) first-order reaction, whereas when reaction (iv) is rate-determining, the process is

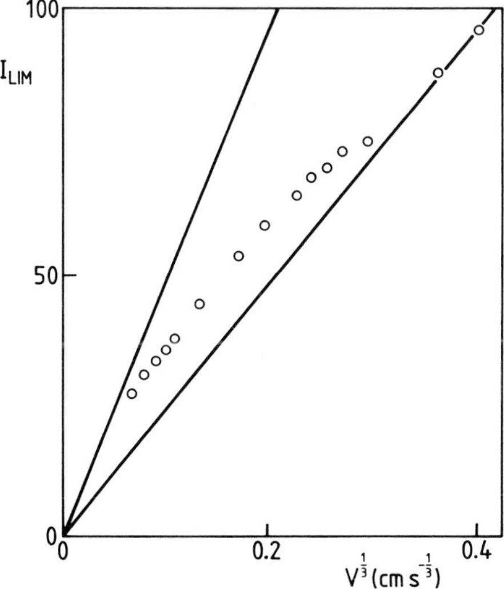

Fig. 33. Diffusion-limited current–solution flow rate behaviour for the reduction of fluorescein at pH 10.05 in the Compton–Coles in-situ cell.

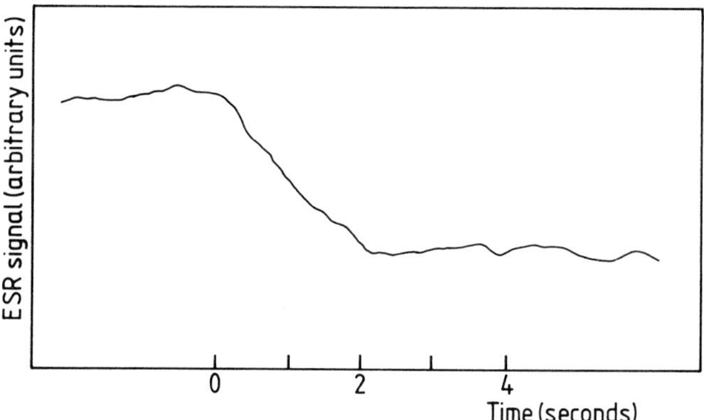

Fig. 34. ESR transient for the decay of semi-fluorescein radical anion.

termed DISP2. Discrimination of DISP2 mechanisms from ECE and DISP1 is readily achieved using conventional electrochemical experiments, but discrimination between ECE and DISP1 mechanisms is not generally possible except with a few methods, e.g. double potential step chronoamperometry [83] and microelectrodes [84]. Compton et al. were able to show that a combination of in-situ electrochemical ESR transient experiments and electrochemical data could be used to distinguish ECE from DISP1.

As discussed earlier, in-situ electrochemical ESR revealed signals that were attributed to semi-fluorescein radicals. By fixing the magnetic field at a peak in the spectrum and then observing the ESR signal decay when the working electrode was open circuited gave ESR transients. An example is shown in Fig. 34. Transients were obtained for various flow rates and values of k_{app} obtained from analysis of the transient in terms of first-order kinetics. To remove any contribution to the transient from radical convection from the cavity, an extrapolation to zero flow rate by plotting k_{app} vs. (flow rate)$^{2/3}$ (vide supra) gave a value for the true rate constant for fluorescein reduction as $1.05\,s^{-1}$ at pH 10.05. The fact that the transients analysed in terms of first-order kinetics ruled out DISP2 as a possible electrode mechanism, but the ESR data alone did not distinguish ECE from DISP1.

Consequently, the electrochemical behaviour, as in Fig. 33, was analysed to give values of k according to either of the two mechanisms chosen. This was possible because of the well-defined and known hydrodynamics of the channel cell. Compton showed that the dependence of the transport-limited current on the flow rate can be calculated for both ECE and DISP1 mechanisms, the predicted behaviour governed by a normalised rate constant, K', where

$$K' = k'\left(\frac{b^2X_E^2}{9u^2D}\right)^{1/3} \tag{21}$$

where k' is the first-order rate constant and the other parameters are as described previously in Sect. 4.4. How the effective number of electrons

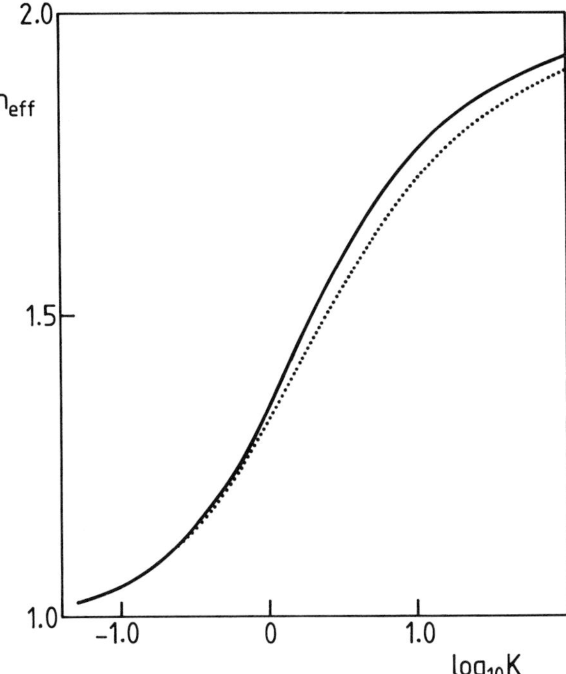

Fig. 35. Dependence of the effective number of electrons transferred upon the normalised rate constant for ECE and DISP1 mechanisms at the channel electrode.

transferred for ECE and DISP1 mechanisms depends on K is shown in Fig. 35. Analysis of the electrochemical data using the working curves in Fig. 35 gave different values of k depending on whether ECE or DISP1 mechanism was chosen. It was found that only values calculated assuming the DISP1 mechanism agreed with the rate constant obtained from the ESR transient data.

Compton and co-workers obtained values of the rate constant for the two-electron reduction of fluroescein at various pH values by ESR transients and electrochemical data. The dependence of the rate constant on pH is shown in Fig. 36 with data from collection efficiency experiments at a rotating ring–disc electrode. The application of electrochemical ESR clearly demonstrates the reduction mechanism to be DISP1. Electrochemical or ESR experiments alone could not make the ECE/DISP1 discrimination independently.

5.5 POLYMER-COATED ELECTRODES

Recently, in-situ electrochemical ESR has been applied to the field of polymer-coated electrodes and conducting polymers. The sensitivity of such cells is sufficient to detect radicals in thin polymer films on electrode surfaces. Various examples are given below.

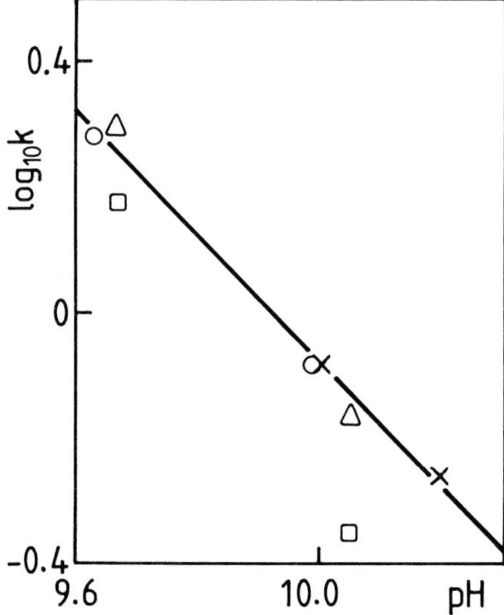

Fig. 36. Dependence of the rate constant on pH for the reduction of fluorescein. Data shown from □, limiting current–flow rate behaviour for the ECE mechanism; △, limiting current–flow rate behaviour for the DISP1 mechanism; ○, ESR decay transients; and ×, rotating ring–disc collection efficiency experiments.

5.5.1 Poly(N-vinylcarbazole)

The electrochemistry of poly(*N*-vinylcarbazole)-modified platinum electrodes has been investigated by Davis and co-workers [85] using the in-situ cell of Compton and Coles [60]. On oxidation of the modified electrode, the ESR spectrum as shown in Fig. 37 was observed. The broad symmetrical single line produced with peak-to-peak linewidth of 3.8 G is indicative of an organic radial "powder" spectrum; strong exchange interactions between

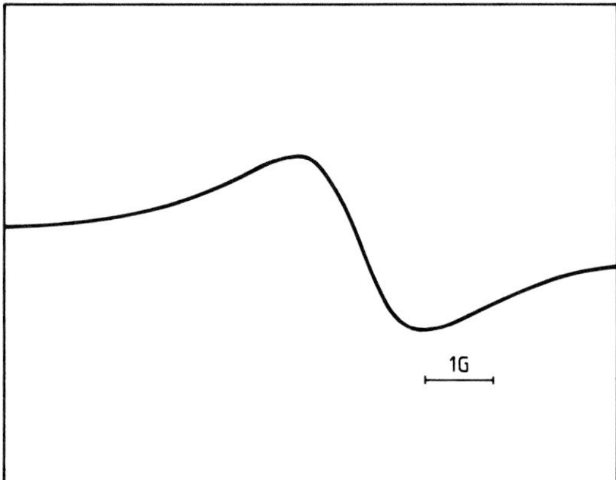

Fig. 37. ESR spectrum obtained from the oxidation of an electrode coated with poly(*N*-vinylcarbazole).

radicals in the polymer coat have removed all hyperfine structure and the effects of dipolar interactions. Quantitative ESR measurements performed on the oxidised polymer showed that the radical species was only a minor component (representing between 3 and 10%) of the pendant groups. The observation that the major component of the film was diamagnetic, together with potential step chromoamperometric data, revealed that the oxidation process involved dimerisation of initially formed cation radicals and further oxidation of the bicarbazolyl group as shown in the scheme

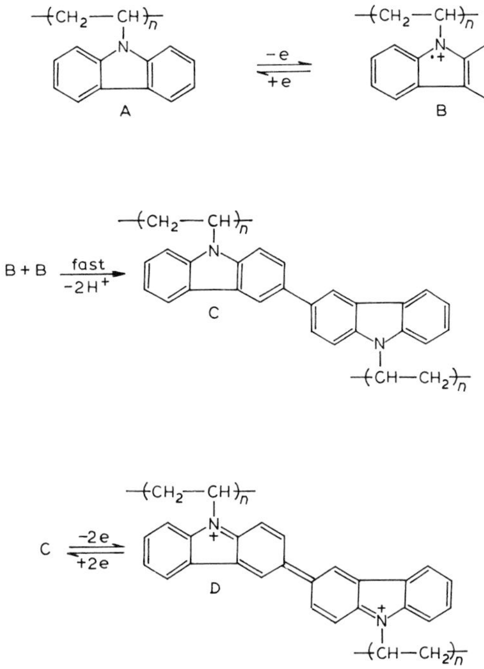

5.5.2 Polypyrrole

An interesting class of polymer-coated electrodes is that in which the polymer is an electronically conducting material. One of the best known examples is polypyrrole, which can be electrochemically grown by oxidation of the monomer on metal electrodes [86–88]. Polypyrrole may be electrochemically switched between its oxidised conducting form (potentials $>$ 0.25 V vs. saturated calomel) and neutral insulating (potentials < -0.75 V vs. saturated calomel). Both in the oxidised conducting and neutral insulating forms polypyrrole exhibits a strong ESR signal. Several workers have used ESR techniques in order to obtain information regarding the charge-carrying species and changes produced by electrochemical cycling [89–92].

The lineshape of the spectrum obtained from oxidised and neutral forms of polypyrrole has aroused much interest. Albery and Jones [89], using the in-situ semi-annular tube, have presented a spectrum which they asserted to

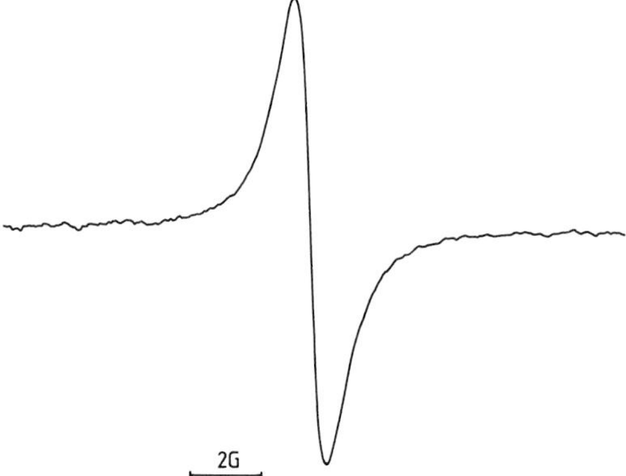

Fig. 38. ESR spectrum from the oxidation of a poly(pyrrole)-coated electrode.

be Dysonian in shape for the oxidised conducting form. This form of ESR lineshape has previously been found in ESR spectra of metals, of polyacetylene, and of radicals adsorbed on surfaces. The Dysonian lineshape is, in general, indicative of metal-like conductivity. Albery and Jones suggested that lineshape analysis could be of use in the determination of the conduction, i.e. as to whether the polymer is a true organic conductor with a developed bond structure or an organic insulator where charge propagation is limited by inefficient electron trapping. Their observations of a Dysonian lineshape [89] for oxidised polypyrrole contrast with the independent results of Pfluger and co-workers [90] and Compton and Waller [92] (the latter using the in-situ channel electrode). The spectrum observed by Compton and Waller is shown in Fig. 38. Pfluger and co-workers also observed a Lorentzian lineshape (with some asymmetry) but noted that the sample thickness and conductivity were too small to give a Dysonian lineshape. The asymmetry of the Lorentzian line, which persisted at low temperatures where the conductivity would be decreased, was attributed to an anisotropy of the g tensor.

Pfluger and co-workers [90] with ex-situ experiments, observed that highly conducting electrochemically cycled films showed little or no ESR signal. They concluded that the paramagnetic species observed in uncycled films were not involved with the carriage of charge within the conducting polymer film and suggested that the charge carrier in oxidised polypyrrole is the bipolaron (see Fig. 39). The non-magnetic bipolaron arises from the combination of two positive paramagnetic species (polarons). But the idea that the basic species responsible for conduction within polypyrrole is the bipolaron was disputed by Nechtstein and co-workers [91]. They performed both steady-state and transient experiments using an in-situ electrochemical ESR cell. Using electropolymerised films of 2000 Å thickness, Nechtstein and co-workers performed experiments where the ESR signal was monitored as a function of the potential applied to the polymer-coated electrode. It was

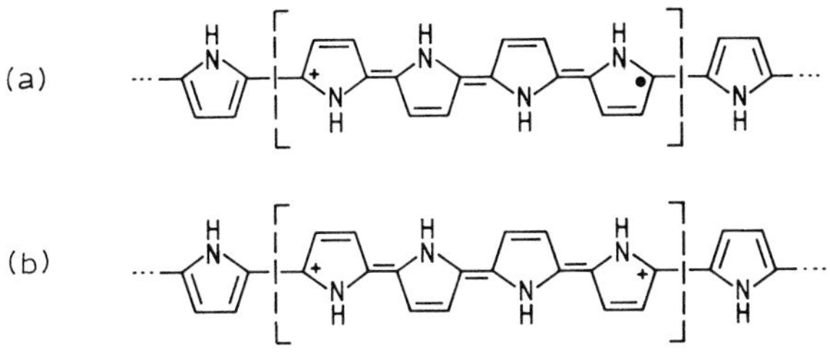

Fig. 39. The (a) polaron and (b) bipolaron in polypyrrole.

observed that, with increasing potential, the number of spins increased, reached a maximum (at a potential V_{max}), and then decreased. This behaviour was reversible and was regarded as consistent with the idea of the formation of spin $= \frac{1}{2}$ polarons upon electrochemical doping and their recombination into spin $= 0$ bipolarons. Transient experiments were those in which the potential was switched between that corresponding to neutral polypyrrole and those (V') corresponding to oxidised polypyrrole. When $V' > V_{max}$, the ESR signal was observed to rise to a maximum before decreasing to a limiting value. Nechtstein and co-workers concluded that the change from neutral to oxidised polypyrrole involved the passage through the polaron state in the formation of bipolarons, and that previous assertions that the conducting species within polypyrrole is the bipolaron are more likely incorrect and a more complex form actually exists involving both polarons and bipolarons.

The work of Nechtstein and co-workers with in-situ cells demonstrates the further information available from such techniques and, alongside more recent conductivity data on the polypyrrole system [93], suggests that further examination of the bipolaron view of polymer conductivity is necessary.

5.5.3 Molecular motion within polymer-coated electrodes

The molecular motion of redox couples within polymer-coated electrodes has recently been investigated by making use of both nitroxide spin probes and various cationic spin probes [94–97]. Spin probes, such as the nitroxide probe TEMPO (see Sect. 2.1.2) and its derivatives, have well-defined electrochemistry and their ESR spectra in viscous media exhibit effects due to incomplete rotational averaging of the g and hyperfine coupling constant tensors. Analysis of the spectra [98] allows deductions to be made concerning the molecular rotation. Such analysis has been performed for spin probes incorporated into various polymer films.

(a) Nafion-modified electrodes

Nafion, a perfluorinated ion exchange polymer in which cationic species may electrostatically bind, has found a wide applicability in the design of various polymer-modified electrodes for diverse uses. The behaviour of various cationic spin probes incorporated into Nafion has been investigated by Kaifer and Bard [94]. Films of 1 μm thickness were prepared on platinum electrodes, then ESR active cations, e.g. the protonated form of 4-amino-2,2,6,6-tetramethylpiperidine-1-oxyl (Tempamine), were incorporated into the film by soaking the electrode in aqueous solutions of the appropriate salt. The electrochemistry of protonated Tempamine was shown to be nearly reversible, but ESR spectra would suggest that the spin probe is strongly anchored by the Nafion film. This contrasts with results for the cation radical of methyl viologen, electrochemically generated within the Nafion film [96], for which the ESR spectra were shown to be consistent with the radical cation tumbling freely as in solution. Intermediate behaviour was observed for the radical cation of tetramethylphenylenediamine [94] electrochemically generated within the Nafion film. The behaviour of organic cations within the Nafion film, which ranges from strongly anchored to free tumbling, was explained by the extent of delocalisation in each cation, the more delocalised the charge over the cation, the less strongly anchored to the Nafion and the cation resides in an aqueous-like environment. In contrast, the results for simple inorganic cations [94, 97] suggested that they retain most of their solution mobility, as observed by the fast-tumbling ESR spectra characteristic of their behaviour in aqueous solution.

(*b*) *Poly(vinylchloride)-modified electrodes*

The modification of electrodes with PVC membranes has found applicability in ion selective electrode work [99] (so-called "coated wire electrode"). The molecular motion of species within such electrodes has been investigated by Compton and Waller [100]. Using a range of derivatives of the nitroxide spin probe TEMPO, they were able to show how the rotational correlational time was dependent upon the molecular volume of the probe and, by use of variable-temperature apparatus, how this varied with temperature. The effect of various plasticizers upon the molecular motion within the PVC membrane was investigated, rotational correlational times being dependent upon the nature of the plasticizer and the loading level. The effect of loading level upon the correlation time was shown to correlate with data obtained by Compton Maxwell [101] for the response times of K^+ ion selective electrodes based upon PVC modified electrodes.

5.6 A SPIN-LABELLED ELECTRODE

Analogous to the use of spin probes in the investigation of polymer coated electrodes is the use of a spin label to investigate the electrode–solution interface. A "spin-labelled electrode" has been produced by Hill and co-workers [102] by the adsorption of a spin label on to the surface of a gold electrode. The spin label was based upon compounds used for the promotion of protein electrochemistry [103], when adsorbed on electrode surfaces, and is shown in Fig. 40(a). In-situ ESR experiments on the spin-labelled electrode

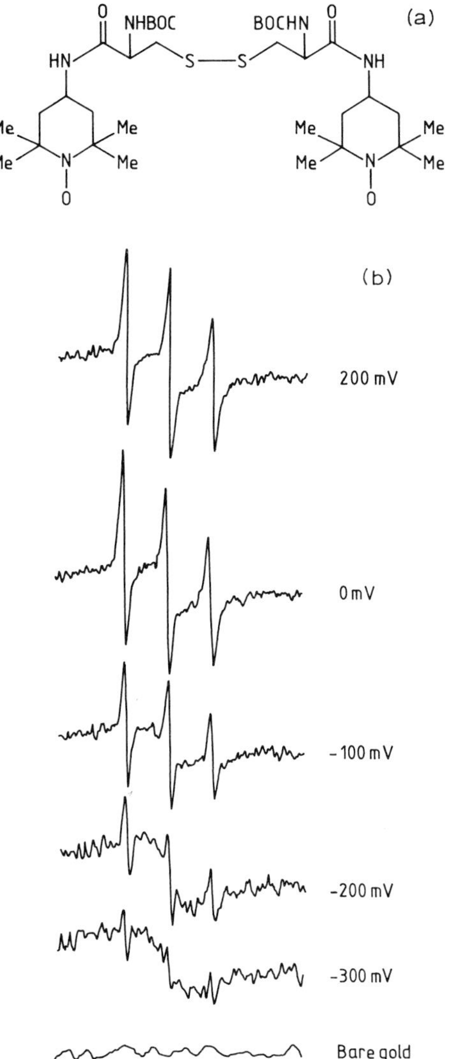

Fig. 40. (a) The spin label used by Hill and co-workers in the preparation of the "spin-labelled electrode". (b) The potential-dependent ESR spectra of the spin-labelled electrode.

trode revealed two distinct environments, one corresponding to a solution-phase radical characterised by the hyperfine splitting, superimposed on a broad spectrum due to the adsorbed nitroxide. The effect of applying a potential to the electrode is observed in Fig. 40(b). When a reducing potential was applied, the amount of solution-phase radical decreased until, a −300 mV (reference to saturated calomel), only the broad feature corresponding to the adsorbed spin label was observed. This effect was shown to be reversible; adsorbed spin label may be freed from the electrode by application of positive potentials. Within the potential range studied, the spin label and buffer solution exhibited no electrochemical activity, hence it was concluded that a potential-dependent equilibrium existed between adsorbed and free spin label.

This work was enlarged to study the behaviour of iron(III) cytochrome *c* upon electrodes modified by promoters of protein electrochemistry. For this work, an electrode was modified by the presence of the spin label and the modifier 4,4′-dithiodipyridine. Cyclic voltammetry at such an electrode was shown to be similar to that obtained with an electrode modified only by 4,4′-dithiopyridine. But in-situ ESR experiments in the presence of cytochrome *c* produced different results from those obtained with the spin-labelled electrode alone. Holding the potential at 0 mV (vs. saturated calomel) for 30 min produced a five-fold increase in the concentration of free spin label. Also, the potential dependence of the ESR spectrum was markedly different; changing the potential to −300 mV did not result in a decrease in the amount of free spin label. These results were explained by the binding of cytochrome *c* to the electrode causing a rearrangement of the 4,4′-dithiodipyridine promoter in order to maximise electrostatic interaction between positively charged cytochrome *c* lysine groups and lone pairs upon the pyridyl nitrogens. The rearrangement of the promoter caused displacement of the surface-bound groups, e.g. the spin label. The change in potential dependence was explained by steric blockage of binding sites by cytochrome *c* decreasing the amount of spin label able to bind to the electrode.

The use of in-situ ESR in the investigation of the spin-labelled electrode clearly demonstrates the applicability of the technique to the study of small amounts of radicals or electrodes and suggests the possibility of the study by in-situ ESR of adsorption upon the electrode surface.

5.7 SPIN TRAPPING

The techniques and applications described in this chapter so far are restricted by the lifetime of the radical being investigated. If the radical has a particularly short lifetime, it may not be possible to observe it by direct in-situ methods. Instead, the technique of spin trapping may have to be used. The technique of spin trapping was first introduced by Janzen and Blackburn [104, 105]; short-lived free radicals react with a diamagnetic compound, e.g. a nitrone, to produce a relatively stable paramagnetic species. Where the spin trap is a nitrone, the corresponding nitroxide is found

$$R\cdot + C_6H_5CH{=}\overset{\overset{\displaystyle O^-}{|}}{N^+}C(CH_3)_3 \rightarrow \underset{\underset{\displaystyle R}{|}}{C_6H_5CH}-\overset{\overset{\displaystyle O^\cdot}{|}}{N}C(CH_3)_3$$

Spin trap Spin adduct

From the ESR parameters of the spin adduct, the structure of the original radical may be deduced.

The potential for the use of spin trapping in the study of electrochemical reactions was demonstrated early on by Bard et al. [106]. They showed that spin trapping was possible in the environment of electrochemical experiments by trapping thermally produced phenyl radicals with the spin trap α-phenyl-*N*-*t*.-butylnitrone (PBN) in 0.1 M tetrabutylammonium perchlorate–acetonitrile solution. The spin trap PBN was shown to be electro-inactive between 1.5 and −2.5 V (vs. saturated calomel), but the resulting spin adduct from trapping of phenyl was inactive in the range 0.7 to −2.0 V. The applicability of this technique to the study of electrochemical reactions was then demonstrated by the reduction of phenyldiazonium tetrafluoroborate in the presence of PBN using ex-situ methods. It was shown that the resulting solution gave a strong ESR signal consistent with trapping of phenyl radical by PBN and concluded that the reduction of aryl diazonium salts proceeds via radical intermediates.

The use of spin trapping has been reported for both electrode reductions and oxidations [107–110]. A good example of the technique using an in-situ method is the work of Volke and co-workers [110] on the electro-oxidation of substituted 1,4-dihydropyridines

H R4
R3 R3
R2 N R2
R1

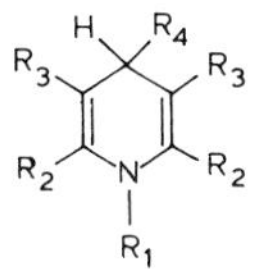

It was previously shown that oxidation produced a radical cation which could be detected by ESR [111, 112] provided the 4 position was fully substituted. If the 4 position contained a hydrogen atom, then the initial oxidation was followed by deprotonation with the formation of a neutral radical which could undergo further oxidation or dimerisation. In-situ spin trapping using PBN demonstrated the presence of radical intermediates in the electro-oxidation of substituted 1,4-dihydropyridines. That the trapped radicals were the deprotonated neutral radical, and not the primary radical cation, was demonstrated by comparison of the ESR parameters with those of the spin adduct produced by electroreduction of *N*-methylpyridine ion cations, where only the neutral dihydropyridyl radical would be produced. The work of Volke and co-workers clearly demonstrates how spin trapping may be applied to the study of more complex organic electrode reactions and how comparison of ESR spectra generated from different precursors may be used to reveal the nature of the trapped radical.

The spin traps used most often usually come from one of two classes of compounds, namely the nitrones and the nitroso compounds, PBN belonging to the former of these groups. Nitrones offer considerable advantage in that they are electrochemically inactive over a wide potential range, both in aqueous and organic solvents. Furthermore, its potential range may be extended by substitution, e.g. of the phenyl group in PBN [113]. The disadvantage of nitrones is that the trapped radical is distant from the spin of

References pp. 349–352

the resultant adduct, and consequently the nature of the trapped radical does not have a significant effect upon the spectral parameters of the spin adduct. As a result, positive identification of the trapped species needs production of the spin adduct via alternate methods.

The nitroso compounds do not suffer from this disadvantage; the nature of the trapped radical is readily identified as it is close to the unpaired spin

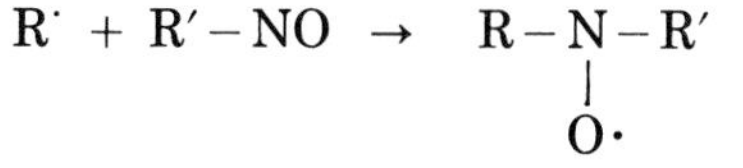

$$\mathrm{R^{\cdot} + R'-NO \rightarrow R-\underset{\displaystyle \underset{|}{} \atop \displaystyle O\cdot}{N}-R'}$$

Spin trap Spin adduct

However, the potential range within which the spin trap is electro-inactive is considerably reduced. This, the commonly used trap nitroso-*t.*-butane is reduced at a potential −0.98 V (vs. Ag/AgI) at mercury in dimethylformamide. The range may be extended by substitution of appropriate aryl groups for butane [114]. Nitroso compounds also present a problem in that some compounds have been shown to undergo a monomer/dimer equilibrium in solution in which only the monomeric form acts as a radical trap. Hence, for nitroso compounds, it is necessary to understand this equilibrium, as well as the electrochemical properties of the trap, before it may be used in the investigation of electrode reactions.

The problems of the limited potential range available to both nitrones and nitroso compounds are a major disadvantage in their application in electrochemistry, but it has been shown by Martigny et al. [115] that this can be surmounted using mediated electrode reactions. The reduction of sulphonium cations was studied using the spin traps PBN and nitrosodurene, but the first reduction step of these spin traps was shown to be at a less cathodic potential than those of the sulphonium cations under investigation. The use of mediated reduction was demonstrated for several sulphonium cations for both spin traps. A good example of the technique was the cation

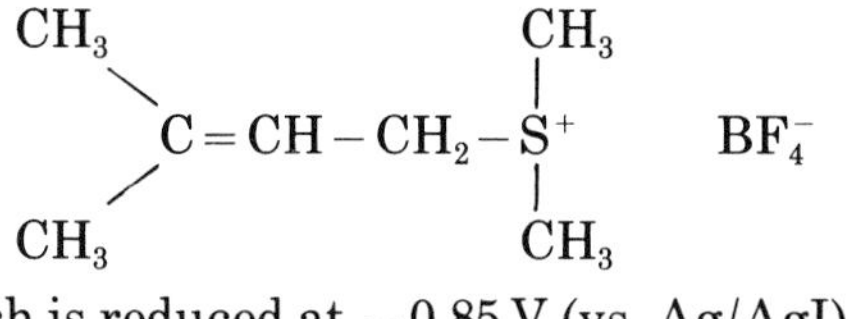

$$\mathrm{(CH_3)_2C=CH-CH_2-\overset{+}{S}(CH_3)_2 \quad BF_4^-}$$

which is reduced at −0.85 V (vs. Ag/AgI). This would preclude the use of the spin trap nitrosodurene which is reduced at −0.02 V. It was shown that mediated reduction of the sulphonium cation by dinitro-1,8-naphthalene at 0.45 V in the presence of nitrosodurene revealed ESR spectra attributed to the spin adduct

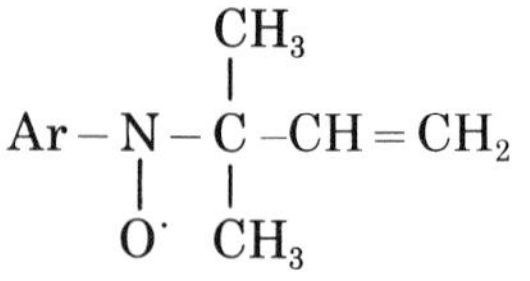

$$\mathrm{Ar-N(O^{\cdot})-C(CH_3)_2-CH=CH_2}$$

The use of mediated oxidation or reduction demonstrates how spin trapping may be applied to the study of electrode reactions and how the careful choice of conditions makes this technique applicable to a wide range of compounds over a large potential window.

6. Concluding remarks

The representative applications of in-situ ESR described in Sect. 5 demonstrate the amount of information it is possible to obtain concerning electrode reactions with this method. Since the first experiments were performed over 25 years ago, the method has been used to identify and investigate radicals produced from a vast range of organic electrode reactions. Considerable study of intermediates in inorganic electrode reactions has also taken place. The information obtained from these studies, together with the development of the improved techniques outlined above, indicate that in-situ ESR will continue to make a major contribution to the study of electrode reactions.

Acknowledgements

The authors wish to thank Dr. B.A. Coles, Mr. M.B.G. Pilkington, and Mr. G.M. Stearn for stimulating and helpful discussions.

References

1 A. Carrington and A.D. McLachlan, Introduction to Magnetic Resonance, Chapman and Hall, London, 1979.
2 D.E.G. Austen, P.H. Given, D.J.E. Ingram and M.E. Peover, Nature (London), 182 (1958) 1784.
3 A.H. Maki and D.H. Geske, J. Chem. Phys., 30 (1959) 1356.
4 A.H. Maki and D.H. Geske, J. Am. Chem. Soc., 82 (1960) 2671.
5 P. Ludwig and R.N. Adams, J. Chem. Phys., 37 (1962) 828.
6 N.M. Atherton, Electron Spin Resonance, Ellis Horwood, Chichester, 1973.
7 P.B. Ayscough, Electron Spin Resonance in Chemistry, Methuen, London, 1967.
8 M.C.R. Symons, Electron Spin Resonance Spectroscopy, Van Nostrand Reinhold, London, 1978.
9 K.A. McLauchlan, Magnetic Resonance, Oxford University Press, Oxford, 1972.
10 J.E. Wertz and J.R. Bolton, Electron Spin Resonance, McGraw-Hill, New York, 1972.
11 C.P. Poole, Electron Spin Resonance: A Comprehensive Treatise on Experimental Techniques, Wiley, New York, 2nd edn., 1983.
12 A.H. Maki and D.H. Geske, J. Chem. Phys., 33 (1960) 825.
13 A.H. Maki and D.H. Geske, J. Am. Chem. Soc., 83 (1961) 1852.
14 D.H. Geske and J.L. Ragle, J. Am. Chem. Soc., 83 (1961) 3532.
15 M.C.R. Symons and M.M. Maguire, J. Chem. Res. (S), (1981) 330.
16 L.H. Piette, R. Ludwig and R.N. Adams, Anal. Chem., 34 (1962) 916.
17 L.H. Piette, R. Ludwig and R.N. Adams, Anal. Chem., 34 (1962) 1587.

18 D. Levy and R.J. Myers, J. Chem. Phys., 41 (1962) 1062.
19 G.D. Luer and D.E. Bartak, J. Org. Chem., 47 (1972) 1238.
20 V.E. Norvell, K. Teremati, G. Mamanta and L.N. Klatt, J. Electrochem. Soc., 128 (1981) 1254.
21 R.N. Adams, J. Electroanal. Chem., 8 (1964) 151.
22 G.K. Fraenkel and P.H. Rieger, J. Chem. Phys., 39 (1963) 609.
23 G.K. Fraenkel, P.H. Rieger, I. Bernal and W.H. Reinmuth, J. Am. Chem. Soc., 85 (1963) 683.
24 G.K. Fraenkel and J. Bolton, J. Chem. Phys., 40 (1964) 3307.
25 B. Kastening, Z. Anal. Chem., 224 (1967) 196.
26 B. Kastening, Ber. Bunsenges. Phys. Chem., 72 (1968) 20.
27 B. Kastening and S. Vavricka, Ber. Bunsenges. Phys. Chem., 72 (1968) 27.
28 H. Gerischer, D. Kolb and W. Wirths, Ber. Bunsenges. Phys. Chem., 73 (1969) 148.
29 W.J. Albery, A.M. Couper, B.A. Coles and A.M. Garnett, J. Chem. Soc. Chem. Commun., (1974) 198.
30 W.J. Albery, A.M. Couper and B.A. Coles, J. Electroanal. Chem., 65 (1975) 901.
31 W.J. Albery, R.G. Compton, B.A. Coles, A.T. Chadwick and J.A. Lenkait, J. Chem. Soc. Faraday Trans. 1, 76 (1980) 139.
32 W.J. Albery, A.T. Chadwick, B.A. Coles and N.A. Hampson, J. Electroanal. Chem., 75 (1977) 229.
33 W.J. Albery and R.G. Compton, J. Chem. Soc. Faraday Trans. 1, 78 (1982) 1561.
34 W.J. Albery, R.G. Compton and C.C. Jones, J. Am. Chem. Soc., 106 (1984) 469.
35 G. Cauquis, J.P. Billon and J. Cambrisson, Bull. Soc. Chim. Fr., (1960) 2062.
36 G. Cauquis and M. Genies, Bull. Soc. Chim. Fr., (1967) 3220.
37 G. Cauquis, J.P. Billon and J. Raisson, Bull. Soc. Chim. Fr., (1967) 199.
38 G. Cauquis, M. Genies, H. Lemaire, A. Rassat and J. Ravet, J. Chem. Phys., 47 (1967) 4642.
39 G. Cauquis, C. Barry and M. Maivey, Bull. Soc. Chim. Fr., (1968) 2510.
40 B. Kastening, J. Divisek and B. Costisa-Mihelcic, Faraday Discuss. Chem. Soc., 56 (1973) 341.
41 B. Kastening, J. Divisek, B. Costisa-Mihelcic and H.G. Müller, Z. Phys. Chem. (Frankfurt), 87 (1973) 125.
42 J.K. Dohrmann, F. Galluser and H. Wittchen, Faraday Discuss. Chem. Soc., 56 (1973) 350.
43 J.K. Dohrmann and K.J. Vetter, J. Electroanal. Chem., 20 (1969) 23.
44 J.K. Dohrmann and F. Galluser, Ber. Bunsenges. Phys. Chem., 75 (1971) 432.
45 J.K. Dohrmann, Ber. Bunsenges. Phys. Chem., 74 (1970) 575.
46 I.B. Goldberg and A.J. Bard, J. Phys. Chem., 75 (1971) 3281.
47 I.B. Goldberg and A.J. Bard, J. Phys. Chem., 78 (1974) 290.
48 I.B. Goldberg, D. Boyd, R. Hirasawa and A.J. Bard, J. Phys. Chem., 78 (1974) 295.
49 V.J. Puglisi and A.J. Bard, J. Electrochem. Soc., 119 (1972) 829.
50 R.N. Bagchi, A.M. Bond and R. Colton, J. Electroanal. Chem., 199 (1986) 297.
51 T.B. Jarbowi, W.R. Heineman and G.J. Parriarche, Anal. Chim. Acta, 126 (1981) 57.
52 I. Fujita and C.K. Chang, J. Chem. Educ., 61 (1984) 913.
53 R.D. Allendoerfer, G.A. Martinchek and S. Bruckenstein, Anal. Chem., 47 (1973) 890.
54 R.D. Allendoerfer and J.B. Carroll, J. Magn. Reson., 37 (1980) 497.
55 J.B. Carroll, Ph.D. Thesis, State University of New York at Buffalo, 1983.
56 H. Ohya-Nishiguchi, Bull. Chem. Soc. Jpn., 52 (1979) 2064.
57 F. Gerson, H. Ohya-Nishiguchi and G. Wydler, Angew. Chem. Int. Ed. Engl., 15 (1976) 552.
58 J. Bruken, F. Gerson and H. Ohya-Nishiguchi, Helv. Chim. Acta, 60 (1977) 1220.
59 R.G. Compton and P.R. Unwin, J. Electroanal. Chem., 205 (1986) 1.
60 R.G. Compton and B.A. Coles, J. Electroanal. Chem., 144 (1983) 87.
61 V.G. Levich, Physiochemical Hydrodynamics, Prentice-Hall, Englewood Cliffs, 1982.
62 B.A. Coles and R.G. Compton, J. Electroanal. Chem., 127 (1981) 37.
63 R.G. Compton and P.J. Daly, J. Electroanal. Chem., 178 (1984) 45.
64 R.G. Compton and P.R. Unwin, J. Electroanal. Chem., 206 (1986) 57.

65 R.G. Compton, D.J. Page and G.R. Sealy, J. Electroanal. Chem., 161 (1984) 129.
66 R.G. Compton, D.J. Page and G.R. Sealy, J. Electroanal. Chem., 163 (1984) 65.
67 R.G. Compton, P.J. Daly, P.R. Unwin and A.M. Waller, J. Electroanal. Chem., 191 (1985) 15.
68 R.G. Compton and A.M. Waller, J. Electroanal. Chem., 195 (1985) 289.
69 P.J. Daly, D.J. Page and R.G. Compton, Anal. Chem., 55 (1983) 1191.
70 J.L. Anderson and S. Moldoveanu, J. Electroanal. Chem., 179 (1984) 107.
71 S. Moldoveanu, G. Handler and J.L. Anderson, J. Electroanal. Chem., 179 (1984) 119.
72 S.G. Weber and W.C. Purdy, Anal. Chim. Acta, 100 (1976) 531.
73 B.H. Vassos and G.W. Ewing, Electroanalytical Chemistry, Wiley, New York, 1983.
74 A. Gennaro, F. Maron, A. Maye and E. Viannello, J. Electroanal. Chem., 185 (1985) 353.
75 R.G. Compton and A.M. Waller, unpublished work.
76 B. Kraeutler and A.J. Bard, J. Am. Chem. Soc., 99 (1977) 7729.
77 N.B. Kondrikov, V.V. Orlov, V.I. Ermakov and M.Ya. Fioshin, Elektrokhimiya, 8 (1972) 920.
78 N.B. Kondrikov, V.V. Orlov, V.I. Ermakov and M.Ya. Fioshin, Russ. J. Phys. Chem., 47 (1973) 368.
79 B.C.L. Weedon, Adv. Org. Chem., 1 (1960) 3.
80 R.D. Goodwin, J.C. Gilbert and A.J. Bard, J. Electroanal. Chem., 59 (1975). 163.
81 R.G. Compton, B.A. Coles and M.J. Day, J. Electroanal. Chem., 200 (1986) 205.
82 R.G. Compton, B.A. Coles and M.J. Day, J. Chem. Res. (S), (1986) 260.
83 C. Amatore, M. Gareil and J.M. Savéant, J. Electroanal. Chem., 147 (1983) 1.
84 M. Fleischmann, F. Lasserre and J. Robinson, J. Electroanal. Chem., 177 (1984) 115.
85 R.G. Compton, F.J. Davis and S.C. Grant, J. Appl. Electrochem., 16 (1986) 239.
86 A. Dall'Olio, G. Dascola, V. Varacca and V. Bocchi, C.R. Acad. Sci., 433 (1968) 267c.
87 A.F. Diaz and K.K. Kanazawa, J. Chem. Soc. Chem. Commun., (1979) 635.
88 A.F. Diaz, J.I. Castillo, J.A. Logan and W.-Y. Lee, J. Electroanal. Chem., 129 (1981) 115.
89 W.J. Albery and C.C. Jones, Faraday Discuss. Chem. Soc., 78 (1984) 193.
90 J.C. Scott, P. Pfluger, M.T. Krounbi and G.B. Street, Phys. Rev. B, 28 (1983) 2140.
91 F. Genoud, M. Guglielmi, M. Nechtstein, M. Genies and M. Salmon, Phys. Rev. Lett., 55 (1985) 118.
92 R.G. Compton and A.M. Waller, unpublished results.
93 B.J. Feldman, P. Burgmayer and R.W. Murray, J. Am. Chem. Soc., 107 (1985) 872.
94 A.E. Kaifer and A.J. Bard, J. Phys. Chem., 90 (1986) 868.
95 R.G. Compton, A.N. Imal and A.M. Waller, unpublished work.
96 J.G. Gaudiello, P.K. Ghosh and A.J. Bard, J. Am. Chem. Soc., 107 (1985) 3027.
97 F.R. Fan, H.Y. Liu and A.J. Bard, J. Phys. Chem., 89 (1985) 4418.
98 J.H. Freed, in L.J. Berliner (Ed.), Spin Labelling Theory and Applications, Academic Press, New York, 1976.
99 J.D.R. Thomas, J. Chem. Soc. Faraday Trans. 1, 82 (1986) 1135.
100 R.G. Compton and A.M. Waller, unpublished work.
101 R.G. Compton and V.M. Maxwell, unpublished work.
102 K. di Gleria, H.A.O. Hill, D.J. Page and D.G. Tew, J. Chem. Soc. Chem. Commun., (1986) 460.
103 P.M. Allen, H.A.O. Hill and N.J. Walton, J. Electroanal. Chem., 178 (1984) 69.
104 E.G. Janzen and B.J. Blackburn, J. Am. Chem. Soc., 90 (1968) 5909.
105 E.G. Janzen, Acc. Chem. Res., 4 (1977) 31.
106 A.J. Bard, J.C. Gilbert and R.D. Goodin, J. Am. Chem. Soc., 96 (1974) 620.
107 H.N. Blount, E.E. Bancroft and E.G. Janzen, J. Am. Chem. Soc., 101 (1979) 3692.
108 B.W. Gara and B.P. Roberts, J. Chem. Soc. Perkin Trans. 1, (1978) 150.
109 P. Martigny, G. Mahon, J. Simonet and G.J. Mousset, J. Electroanal. Chem., 121 (1981) 349.
110 J. Klima, J. Ludvik, J. Volke, M. Krikava, V. Skala and J. Kuthan, J. Electroanal. Chem., 161 (1984) 205.

111 J. Klima, A. Kurfürst, J. Kuthan and J. Volke, Tetrahedron Lett., 31 (1977) 2725.
112 J. Ludvik, J. Klima, J. Volke, A. Kurfürst and J. Kuthan, J. Electroanal. Chem., 138 (1982) 131.
113 G.L. McIntire, H.N. Blount, H.J. Stronks, R.V. Shetty and E.G. Janzen, J. Phys. Chem., 84 (1980) 916.
114 G. Granchi, P. Courbis, P. Tordo, G. Mousset and J. Simonet, J. Phys. Chem., 87 (1983) 1343.
115 P. Martigny, J. Simonet, G. Mousset and J. Vigneron, Nouv. J. Chim., 7 (1983) 299.

Chapter 8

Photocurrent Spectroscopy

L.M. PETER

1. Introduction

The introduction of a range of in-situ spectroscopic techniques to electrochemistry has given access to a wealth of information to supplement the data provided by electrochemical methods. Many spectroscopic techniques rely on measurement of intensity changes that arise from the absorption of light by components of the electrolyte or by the electrode itself, and the subsequent fate of the excited state is of little interest. Light absorption in solids, however, can lead to the formation of electronically excited states that are delocalised as free charge-carrier pairs, giving rise to a photocurrent or a photovoltage, and consequently such systems can be characterised by measuring the electrical response to the optical stimulus. This is the basis of photocurrent spectroscopy. Since the photoexcitation of charge carriers is possible, at least in principle, for most solids, the technique is potentially applicable to metal as well as semiconductor or insulator electrodes. In practice, however, the quantum efficiencies of carrier generation and collection are highest for semiconducting or insulating solids, where the existence of a forbidden energy gap guarantees that a significant fraction of geminate carrier pairs escape recombination, and it is to this class of materials that most attention has been given, although photoemission from metals has also been studied extensively.

The application of photocurrent spectroscopy is not restricted to bulk semiconductors and insulator electrodes. The anodic oxidation of many metal electrodes produces surface films that are insulators or semiconductors, and in spite of the fact that these surface films are often very thin, their characterisation by photocurrent spectroscopy poses few experimental difficulties since photocurrents as small as 10^{-10} A can be measured by conventional lock-in methods. The instrumentation required for photocurrent spectroscopy is relatively modest and the technique is undemanding in terms of the degree of optical perfection of the electrode surface. Consequently, there seems to be considerable scope for the application of this type of spectroscopy to electrochemical problems such as corrosion, for example, where surface roughening may rule out methods that require an optically flat surface.

It is worth noting here that photocurrents may be generated when electrodes are illuminated during other spectroscopic measurements; in extreme

References pp. 382–383

cases, the electrochemistry of the system may be perturbed to such an extent that the light can no longer be considered as a passive probe. The most marked effects are to be expected in the case of spectroscopic techniques, such as laser Raman, that employ high-intensity visible or UV radiation. Unfortunately, it has not been common practice to check for photocurrents in such measurements, in spite of the fact that, in many cases, the incident power is sufficient to cause a surface film to be formed or destroyed in a matter of seconds. Even in the absence of semiconducting or insulating surface films, photoemission from metal electrodes may be important. These processes are all examples of photoelectrochemistry, and photocurrent spectroscopy, in addition to being a valuable method in its own right, is therefore able to provide information that is relevant to other in-situ spectroscopic techniques.

The qualitative application of photocurrent spectroscopy to bulk semiconductor electrodes is well established. This chapter is therefore largely concerned with quantitative aspects of the technique and particular emphasis is placed on the investigation of thin surface films on metal electrodes.

2. Light absorption and charge carrier generation in solids

The absorption of light in a regular crystalline lattice can be described by a number of different approaches. At one limit, in the case of molecular solids, the interaction between neighbouring molecules may be sufficiently weak for the electronic transitions in the solid to be closely related to those of the individual molecule, although the translational equivalence of the crystal lattice gives rise to a delocalised excited state known as a Frenkel exciton. Although this excited state does not, in general, involve charge separation, photocurrents may result from exciton dissociation at the crystal interface if suitable contacts are applied to the solid [1]. In the case of covalent or ionic solids, optical excitation may give rise to a bound state, known as a Wannier exciton [2], that consists essentially of an electron–hole pair interacting through the dielectric medium of the crystal. Thermal ionisation of this exciton state leads to the formation of free carriers in bands. For the purposes of this chapter, however, the most important mechanism of charge-carrier generation is that involving interband excitation of electrons from the valence band to the conduction band.

The electronic structure of a solid determines, to a large extent, the subsequent fate of the excited state produced when a photon is absorbed. It is convenient to distinguish at the outset between the excitation of electrons in a simple metal on the one hand and in a semiconductor or insulator on the other. Figure 1 contrasts the two cases. Excitation of an electron from a filled state in a metal leaves an electron vacancy below the Fermi level that can be filled rapidly by electrons from higher occupied energy levels. The

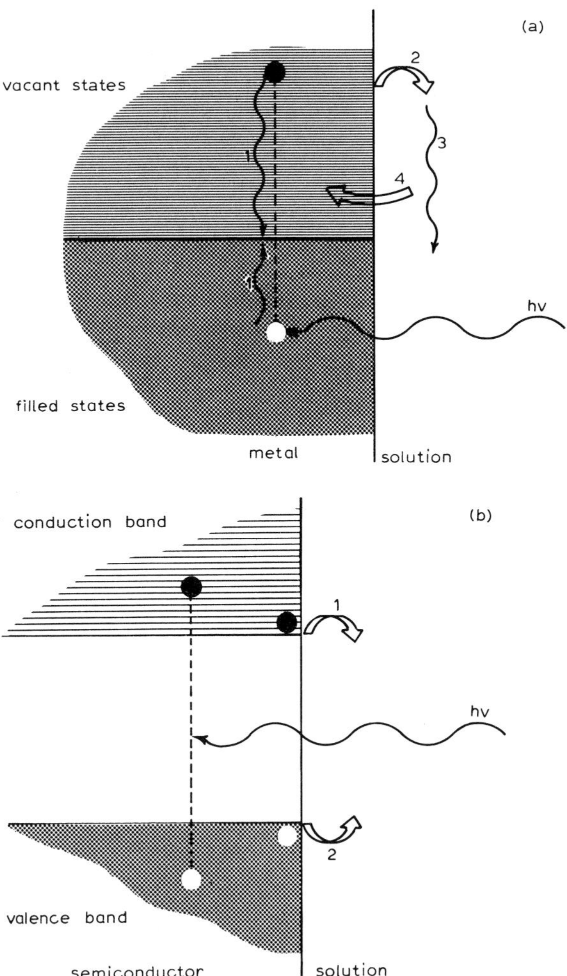

Fig. 1. (a) Optical excitation in a simple metal. The excited electron–hole pair can relax rapidly (path 1) so that the quantum efficiency of the photoemission process (2) is small. In addition, the back reaction (4) involving electron injection from the solution may occur before the photoemitted electron relaxes in the solvent (3). (b) Optical excitation in an insulator or semiconductor. Here, geminate recombination is slower due to the forbidden gap so that the quantum efficiency of photoelectrochemical reactions can approach unity under optimum conditions. In principle, either conduction band (1) or valence band (2) reactions can occur, but in practice the reaction of the minority carrier is most important for extrinsic semiconductors.

excited electron can cascade through the vacant states above the Fermi level and, as a result, the probability of electron–hole recombination is high. Alternative processes involving the excited electron–hole pair have to compete with efficient recombination and, consequently, the quantum yields for photoelectrochemical reactions at illuminated metal electrodes are usually very small. Even if the excited electron is able to escape into a neighbouring

References pp. 382–383

phase by photoemission, it is still threatened with recombination by the back reaction shown in Fig. 1(a) and electron scavengers are therefore necessary components of electrolyte solutions used for photoemission studies [3]. Internal photoemission can also occur from a metal electrode into a solid surface phase and, in this case, the quantum efficiencies may be larger if the film contains a high density of electron traps that are able to capture the photoemitted electrons and prevent their recombination.

The optical excitation and subsequent recombination of an electron–hole pair in a semiconductor or insulator differ fundamentally from the corresponding processes in a metal because they involve initial and final states that are separated by a forbidden energy gap. Since the probability of a direct radiative transition is small, the lifetime of the photoexcited minority carrier depends critically on the purity and perfection of the semiconductor crystal. The carriers are free to move in the energy bands and, in the absence of an electrical field in the solid, the electron and hole may diffuse apart, generating a photovoltage. If an electric field exists in the solid, the carriers may be separated by migration before recombination can occur and a corresponding photocurrent can be detected if suitable contacts are made. Under optimum conditions, the collection efficiency for photoexcited carriers approaches unity for single-crystal semiconductors and, for this reason, there is considerable interest in semiconductor devices for the efficient conversion of light into electrical power. In practice, sites in the bulk or at the surface of the semiconductor provide routes for recombination that can reduce the overall conversion efficiency.

The selection rules governing photon absorption in solids determine the oscillator strength of the optical transition and its energy dependence. The expressions obtained for the imaginary component of the optical dielectric constant depend on whether the transition is allowed in the dipole approximation and on whether the simultaneous absorption or emission of a phonon is involved. In pure single-crystal materials, the absorption coefficient can be described conveniently by relationships that take the general form [4]

$$\alpha h\nu = A(h\nu - E_0)^n \qquad (1)$$

where A is a constant that depends on the type of transition and E_0 is the band gap energy. The exponent n takes the value 1/2 for direct allowed transitions, 3/2 for direct forbidden transitions, and 2 for indirect allowed transitions. Graphical methods are widely used to evaluate the indirect and direct band gaps from absorption spectra; in the simplest form of analysis, plots of $(\alpha h\nu)^{1/n}$ vs. photon energy are constructed and the band gaps evaluated from the intercepts. Since indirect and direct transitions generally overlap in energy, it is desirable in an exact analysis to deconvolute their contributions to the absorption coefficient in order to obtain more reliable values of the band gaps. Polycrystalline or amorphous materials are more difficult to treat since they contain an appreciable density of states in the band gap but, in many cases, reasonably satisfactory extrapolations from the high-energy data can be made.

The oscillator strengths of indirect transitions are generally low and the corresponding absorption coefficients are usually below $10^3\,cm^{-1}$. Direct transitions, on the other hand, are more intense, with absorption coefficients as high as $10^5\,cm^{-1}$, so that even very thin films absorb an appreciable fraction of incident light. The optical properties of amorphous and polycrystalline solids are particularly relevant to the spectroscopy of films formed on metals by electrochemical reactions, and the effects of disorder are usually evident at the low-energy end of the absorption spectrum. Defects in polycrystalline materials can give rise to states located in the band gap and, as a consequence, the absorption is broadened towards the red. It has been suggested that the energy distribution of these states is responsible for the exponential dependence of the absorption coefficient on photon energy that is often observed near the band edge. The absorption coefficient of amorphous solids often exhibits a similar energy dependence, although earlier interpretations in terms of band tailing are now discounted*. At higher photon energies, the absorption coefficient of many glassy materials appears to follow eqn. (1) with $n = 2$.

3. Collection of photogenerated carriers

The photoelectrical behaviour of an illuminated solid depends not only on its optical properties but also on the nature of the contacts made to it. If a resistive material is provided with ohmic contacts, there is no barrier to the transfer of electrons to and from the solid and the current that flows when a potential difference is applied between the contacts depends only on the density and mobility of charge carriers. The current will increase if illumination raises the free carrier density considerably and the material is then termed a photoconductor. In practice, illumination may serve either to promote electrons from the valence band to the conduction band or to release carriers trapped at impurity states in the band gap. In both cases, the light gives rise to a volume photoeffect [6].

By contrast, if one of the contacts to a semiconductor or insulator presents a barrier to the flow of charge carriers, the material can be reverse biased so that only small currents flow in the dark. A space-charge region is formed beneath the contact under these conditions and a corresponding electrical field penetrates into the solid to a depth that depends on the doping density. Such a Schottky barrier, as it is called, provides a driving force for the separation of photogenerated carriers and, consequently, illumination produces a photocurrent. Here, we are dealing with a surface

* For a discussion of the optical properties of amorphous semiconductors, see ref. 5.

References pp. 382–383

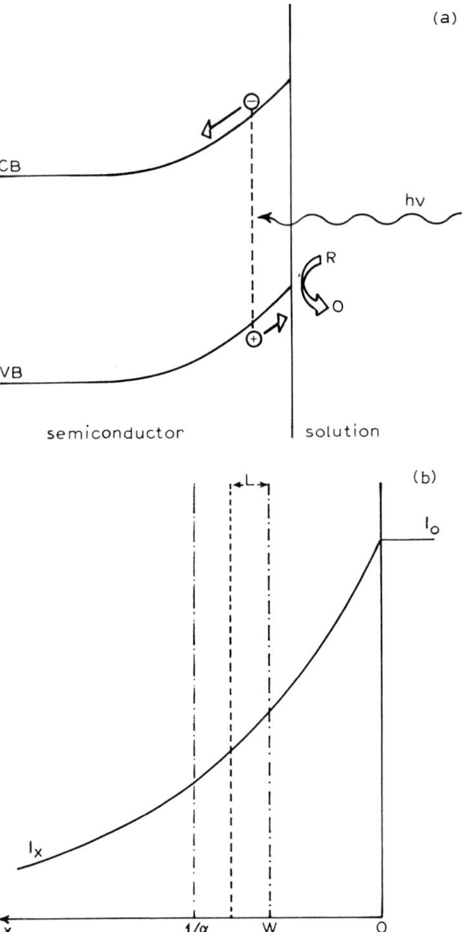

Fig. 2. (a) Band bending at the semiconductor/electrolyte interface under depletion conditions (n-type). Photogenerated carriers are separated rapidly by the electrical field in the space-charge region and the minority carriers (in this case holes) are involved in reactions at the interface. (b) The characteristic lengths involved in the description of photocurrent generation at the illuminated Schottky barrier junction. W is the width of the space-charge region and L is the diffusion length of minority carriers. The exponential decay of the light intensity is shown in order to illustrate the significance of the penetration depth $1/\alpha$.

photoeffect. Most of the photocurrents generated at the semiconductor/electrolyte junction arise from this surface mechanism.

The properties of the semiconductor/electrolyte contact are illustrated in Fig. 2(a) for an n-type semiconductor. The space-charge region develops as the potential is made more positive than the flat-band potential and its width, W, is given by the expression

$$W = \left(\frac{2\varepsilon\varepsilon_0 \Delta\phi}{eN}\right)^{1/2} \tag{2}$$

where N is the density of donor or acceptor states, ε is the relative permittivity of the solid, and $\Delta\phi$ is the potential difference between the bulk and the surface of the semiconductor.

As Fig. 2(a) indicates, photoexcited carriers created within the space-charge region are separated by the electric field and the minority carriers (in the present case, holes) move towards the interface where they may react with components of the electrolyte or with the semiconductor lattice. Minority carriers created deeper in the semiconductor can still escape recombination if, during their lifetime, they diffuse to the edge of the space-charge region. Carrier collection from the charge-free region of the semiconductor depends on the diffusion coefficient, D, and the lifetime, τ, of minority carriers and it is conveniently described in terms of the diffusion length, $L = (\pi D\tau)^{1/2}$ since carriers created within this distance from the edge of the space-charge region contribute to the photocurrent. The boundary value problem which must be solved involves consideration of the rate of photogeneration of minority carriers as a function of position in the crystal and the rates of diffusion and recombination. The inverse of the absorption coefficient is referred to as the penetration depth and it provides a useful characteristic length for comparison with the width of the space-charge region and the diffusion length. These characteristic dimensions are illustrated in Fig. 2(b). If recombination in the space-charge region and at the surface are neglected, the solution known as the Gärtner equation* is obtained. It can be written in the convenient normalised form

$$\Phi = \frac{j_{\text{photo}}}{I_0} = 1 - \frac{\exp(-\alpha W)}{1 + \alpha L} \tag{3}$$

Here, the ratio of the photocurrent, j_{photo} to the incident photon flux, I_0, (corrected for reflection) is the photocurrent conversion efficiency, Φ. Since eqn. (3) takes no account of recombination in the space-charge region or at the interface, it represents an upper limit to the photocurrent conversion efficiency. Several treatments of recombination have been discussed in the literature [7], but they lie outside the scope of the present chapter.

The Gärtner equation provides a useful theoretical basis for the analysis of the photocurrent response of illuminated semiconductor electrodes. It can be rewritten in the form

$$-\ln(1 - \Phi) = \ln(1 + \alpha L) + \alpha W \tag{4}$$

to show that a plot of $-\ln(1 - \Phi)$ against W is a straight line with a slope proportional to the absorption coefficient and an intercept that depends on the product αL. The value of the flat band potential, E_{fb}, and of the doping density needed to calculate W can be obtained from capacitance data by plotting the inverse square of the space charge capacitance, C_{sc}, against

---

* For a discussion of the Gärtner equation, see ref. 7.

electrode potential according to the Mott–Schottky relationship

$$C_{sc}^{-2} = \frac{2}{eN\varepsilon\varepsilon_0}\left[(E - E_{fb}) - \frac{kT}{e}\right] \qquad (5)$$

where E_{fb} is the flatband potential.

Quantitative measurements of the photocurrent conversion efficiency form the basis of a method to determine the minority carrier diffusion length and the absorption spectrum and Fig. 3 illustrates the application of the analysis to the characterisation of p-GaP. Photocurrent–voltage curves are recorded for different wavelengths of the exciting light [Fig. 3(a)] and plotted according to eqn. (4) as shown in Fig. 3(b). It should be noted here that precise determination of the photocurrent conversion efficiency requires independent measurement of the incident photon flux and of the reflection at normal incidence. At potentials sufficiently far from E_{fb}, the expected linear plots are obtained but the experimental points fall below the linear relationship as the potential approaches E_{fb}; this effect is due to surface recombination which becomes significant close to the flatband potential. The values of the absorption coefficient derived from the slopes and intercepts of the linear part of the plots obtained at different wavelengths are compared in Fig. 3(c) with the published absorption spectrum of GaP [9], and it is evident that the analysis is satisfactory. The diffusion length for electrons was found to be 7×10^{-6} cm. The same method can be applied to solid-state Schottky barriers, but the electrolyte "contact" has the advantage that is transparent and easily removed. Although there appears to be scope for the exploitation of this method for the non-destructive evaulation of semiconductor materials, it has not found widespread use so far.

4. Photocurrent generation in semiconducting films of finite thickness

The treatment outlined in the preceding section is appropriate provided that the thickness of the sample is greater than all of the characteristic lengths. This will nearly always be the case for bulk semiconductors, but films on metal electrodes are often sufficiently thin that a different description is necessary. We can distinguish several limiting causes on the basis of the magnitudes of W, $1/\alpha$, and L compared with the film thickness, L_f. If the film thickness is comparable with the penetration depth of the light, $1/\alpha$, then some light will be reflected at the film/substrate boundary and pass through the film for a second time. Since typical oxide film thicknesses lie in the range 10^{-7} to 10^{-5} cm, whereas the penetration depth is usually of the order of 10^{-5} to 10^{-4} cm, reflection can make an important contribution to the photocurrents observed in these systems. In addition, the absorption of light in the near-surface region of the metal substrate can excite electrons across the boundary into the surface film, giving rise to a photoemission current.

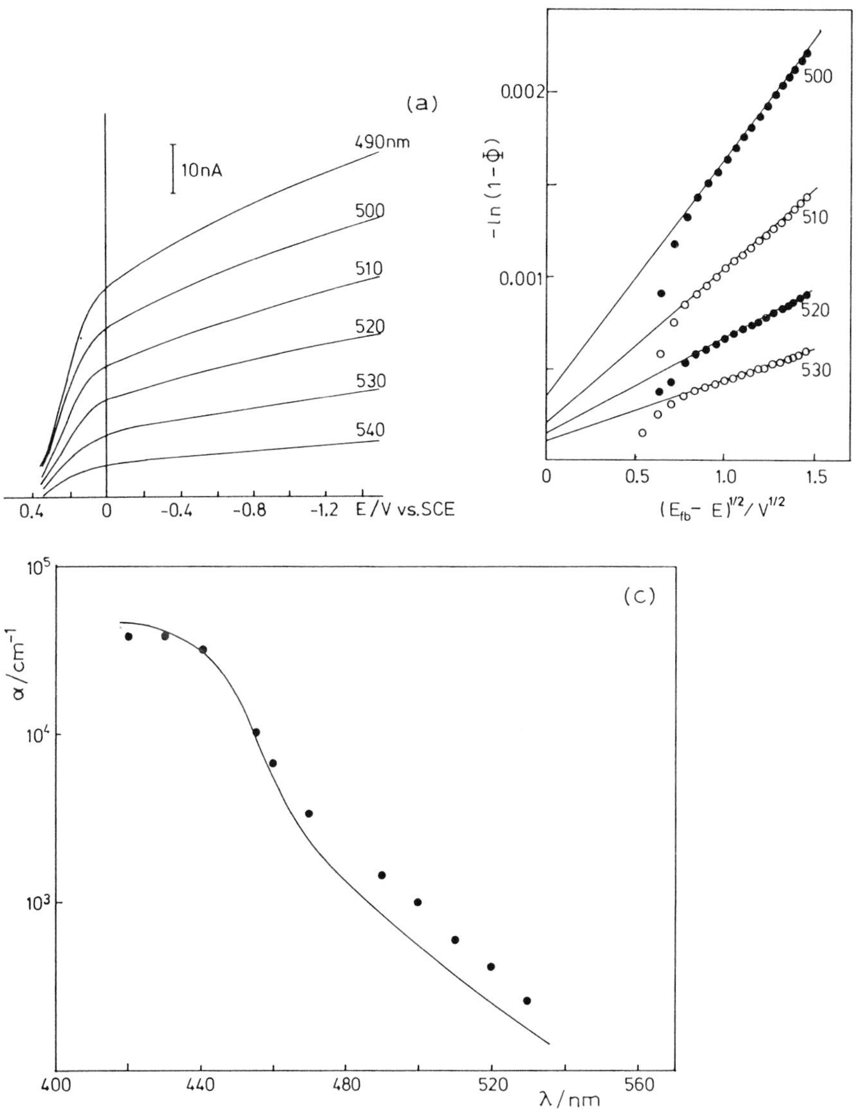

Fig. 3. (a) Photocurrent–potential curves for p-GaP in 0.5 mol dm^{-3} H_2SO_4 at the wavelengths shown (in nm). (b) Examples of the plots used to obtain values of the absorption coefficient $\alpha(\lambda)$ and electron diffusion length L [eqn. (4)]. (c) Comparison of the absorption coefficients, obtained using eqn. (4) with the published absorption spectrum of GaP. (Reproduced with permission from ref. 8.)

Examples of internal photoemission at oxide-covered metal electrodes are considered later in this chapter.

It is clear that the film thickness is an important characteristic length that must be incorporated into a general expression for the photocurrent. Figure 4 illustrates the case where the film thickness, L_f, is less than the

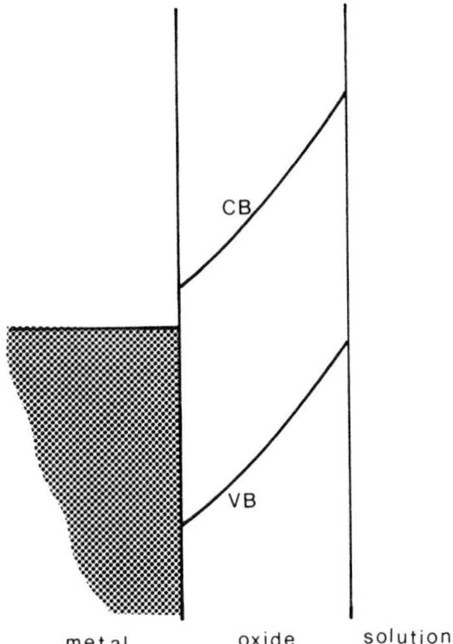

Fig. 4. Band bending in a thin film on a metal substrate where the calculated value of the space-charge layer width, W, exceeds the film thickness, L_f, so that excess charge appears on the metal. In this situation, the field assists the separation of photoexcited carriers throughout the entire film.

space-charge layer thickness, W. The condition $W > L_f$ implies that excess charge appears at the metal surface so that the electrical field penetrates throughout the film and, as a consequence, the term associated with the diffusion of minority carriers disappears from the Gärtner equation. If recombination in the space-charge region and at the surface is negligible, the photocurrent depends simply on the flux of photons absorbed in the film and, for normal incidence, the photocurrent conversion efficiency is given by the expression

$$\Phi = 1 - R_{12} + (R_{12} + R_{23} - R_{12}R_{23} - 1)\,y - (R_{23} - R_{12}R_{23})\,y^2 \quad (6a)$$

where $y = \exp(-\alpha(\lambda)L_f)$. Here, I_0 is the incident photon flux, R_{12} and R_{23} are the reflection coefficients at the solution/film and film/metal interfaces, respectively, and $\alpha(\lambda)$ is the absorption coefficient of the film at the wavelength λ.

Equation 6(a) reduces to several useful limiting forms.

$$\lim R_{12}, R_{23} \to 0; \qquad \Phi \to 1 - \exp(-L_f) \quad (6b)$$

$$\lim R_{12} \to 0, R_{23} \to 1; \qquad \Phi \to 1 - \exp(-2\alpha L_f) \quad (6c)$$

$$\lim \alpha L_f < 1; \qquad \text{eqn. (6b)} \to \Phi = \alpha L_f \quad (6d)$$

$$\lim \alpha L_f < 1; \qquad \text{eqn. (6c)} \to \Phi = 2\alpha L_f \quad (6e)$$

Although space-charge recombination may not be important for single crystal semiconductors, high densities of defects and impurities are expected to be present in anodic films on metal electrodes and recombination may therefore be significant. Equations (6a)–(6e) therefore represents the upper limit to the photocurrent generated by illumination of a metal electrode covered by a thin film of insulating or semiconducting material.

If the film contains a high density of impurity states that act as electron donors or acceptors, the space-charge region may become thinner than the film itself. A more general expression for the photocurrent coversion efficiency that takes this possibility into account can be derived if the simplifying assumption is made that the diffusion length of minority carriers is likely to be small in polycrystalline or amorphous films that contain high densities of recombination centres

$$\Phi = (1 - R_{12})(1 - w) + (R_{23} - R_{12}R_{23})\left(\frac{1}{w} - 1\right)y^2 \tag{7}$$

where $w = \exp(-\alpha W)$ and $y = \exp(-\alpha L_f)$.

Equation (7) is valid if the field in the space-charge region is sufficient to prevent recombination but, in disordered materials, we might expect some recombination effects even when migration is rapid. Space-charge recombination will also be important if slow interfacial electron transfer leads to a build-up of minority carriers.

5. Demonstration of photosensitivity of anodic films on metals

Photocurrent spectroscopy can be defined in the context of this chapter as the precise determination of the photocurrent conversion efficiency as a function of photon energy and electrode potential. The main objective of the analysis of photocurrent data is to obtain quantitative information about the system under study, but first of all some indication is needed that a photosensitive film is indeed formed under electrochemical conditions. A convenient way to establish whether a film is photosensitive is to illuminate the electrode with a beam of white light interrupted by a mechanical chopper during linear sweep voltammetry or galvanostatic anodisation. An example of the particularly striking effects than can be observed is seen when a cadmium electrode is anodised in sodium sulphide [10]. Figure 5(a) shows the clear evidence for the formation of a photosensitive CdS films provided by the appearance of an anodic photocurrent superimposed on the linear sweep voltammogram. A large effect is also observed when the galvanostatic anodisation of cadmium in Na_2S is carried out under interrupted illumination and Fig. 5(b) shows that a substantial fall in the overpotential occurs during the on period. The origin of the photoeffect is rather different here. The increase of potential during anodisation is a consequence of the need to maintain a sufficiently high electrical field across the surface layer to assist the migration of the ions involved in film growth. Illumination opens an alternative route for current flow in which electronic rather than ionic

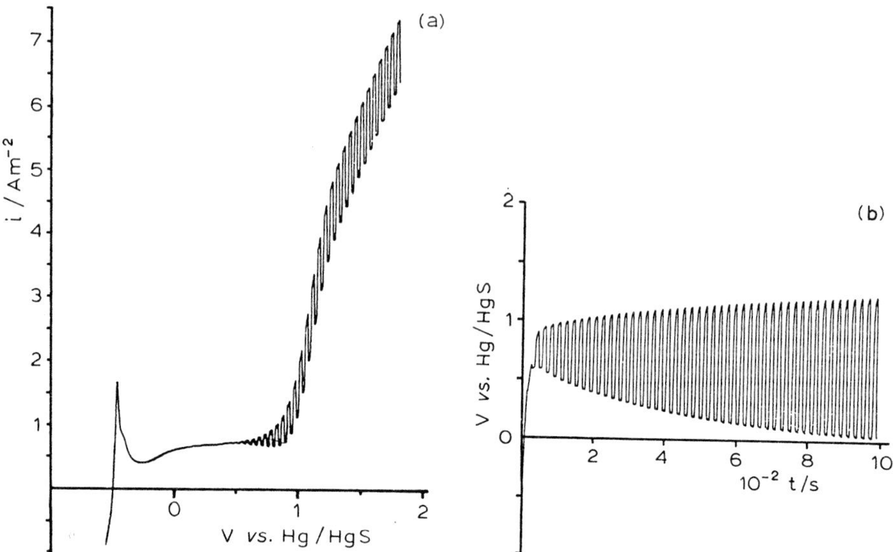

Fig. 5. (a) Linear sweep voltammogram for cadmium in 0.1 mol dm^{-3} Na_2S + 1.0 mol dm^{-3} $NaHCO_3$. Sweep rate 30 mV s^{-1}. The electrode was illuminated by a 100 W tungsten lamp via a slowly rotating chopper. Note the appearance of pulses of anodic photocurrent as film growth proceeds. (b) Galvanostatic transient for the growth of CdS on Cd in the same solution with chopped illumination. Note that the overpotential falls when the electrode is illuminated. Current density, 50 μA cm^{-2}. (Reproduced with permission from ref. 10.)

conductivity is involved and which does not require the high electric field. Under suitable conditions, illumination can cause a complete switch from ionic to electronic currents in the CdS film and, as Fig. 4 shows, the overpotential collapses entirely under illumination. This is a particularly extreme example of the effect of light on an electrochemical process.

6. Experimental aspects of photocurrent spectroscopy

Quantitative measurements are best carried out at much lower light intensities than those responsible for the large effects illustrated in Fig. 5. It is desirable to avoid, as far as possible, the photoelectrochemical oxidation or reduction of the film on the time scale of the measurements and this generally restricts incident power densities to less than 10^{-4} W cm^{-2}. Since the photocurrents generated by such low levels of illumination are too small to be measured directly, it is necessary to use a lock-in amplifier in conjunction with a mechanical chopper in the experimental arrangement shown in Fig. 6. The sensitivity of the photocurrent spectrometer is usually determined by the noise current arising from the electrode capacitance and the noise voltage in the system. Under favourable conditions, it is possible to measure photocurrents as small as 10^{-10} A. For typical illumination inten-

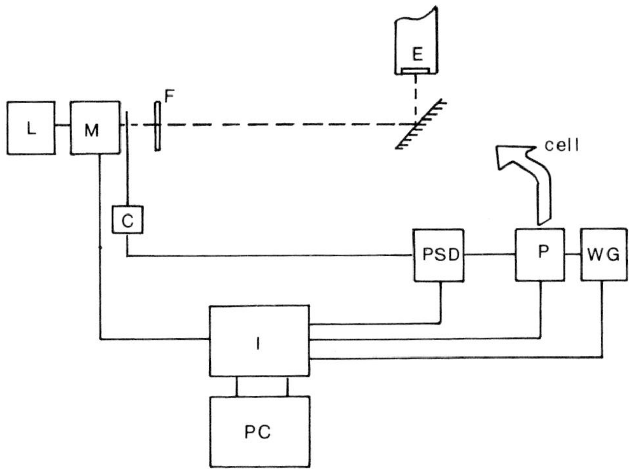

Fig. 6. Experimental arrangement for photocurrent spectroscopy. C, chopper; E, electrode; F, optical filters; I, interface; L, tungsten or xenon lamp; M, grating monochromator; PC, microcomputer with peripheral hardware; PSD, phase sensitive detector; P, potentiostat; WG, waveform generator.

sities, this limit corresponds to photocurrent conversion efficiencies smaller than 10^{-5} and, in principle at least, it should be possible to observe photocurrents generated in films as thin as a monolayer.

The determination of the photocurrent conversion efficiency requires calibration of the incident photon flux. A silicon photodiode calibrated to NBS standard is a convenient choice for the wavelength range 350–1000 nm, whereas a calibrated vacuum photodiode is required for higher photon energies. A thermopile can also be used as a calibration standard, although the slow response of most thermopiles is inconvenient when they are used to monitor light fluxes during measurements. Xenon and tungsten lamps provide adequate coverage of the spectral range of interest and a high throughput single grating monochromator offers sufficient spectral resolution. Care is needed to eliminate stray and harmonic light during measurement and calibration; appropriate cut-off and band filters form an essential part of the optical arrangement.

The spectrometer can also incorporate facilities for single beam transmission and reflectance measurements since it is often necessary to determine the amount of light reflected or absorbed in order to calculate the quantum efficiency of photocurrent generation. Optically transparent electrodes (OTEs) are particularly useful substrates since they allow simultaneous measurements of photocurrents and transmission to be made. Metal films can be prepared by vacuum deposition on the OTEs or, in some cases, the material of interest can be electroplated directly. If SnO_2-coated quartz electrodes are used, it is essential to check them periodically to make sure that they do not give rise to background photocurrents since aged electrodes

References pp. 382–383

can become photosensitive as a result of the leaching of donor species during electrochemical experiments.

The detection of the photocurrent by a lock-in amplifier is straightforward provided that the photocurrent response to chopped illumination is square. The most reliable way to set the phase is to replace the photoelectrochemical cell by a fast-response photodiode controlled by the potentiostat since the photocurrent response of electrodes is often distorted by relaxation effects such as recombination that give rise to a phase shift in the a.c. response. It is therefore always desirable to characterise the photocurrent response by a signal-averaging technique. This is particularly important if it is suspected that thermal effects may be involved in the photocurrent response since a phase lag is often the result of transient heating of the electrode during the on period of the illumination. The slow rise time associated with the heating effect can usually be distinguished in the averaged transient response.

The signal-to-noise ratio can be improved if a low-noise battery-operated potentiostat is used and it is usually not necessary to resort to a two-electrode configuration unless the photocurrents are very small or the electrode capacitance unusually large. It is often useful to reduce the sample area to match the illumination spot so as to eliminate the noise contribution from the part of the electrode that is not illuminated. Particular care is necessary to eliminate earth loops and high-frequency pick-up and a screened Faraday cage is essential. Commercial reference electrodes can be replaced by low-resistance electrodes if they cause problems or bridging capacitors can be used to bypass high-resistance liquid junctions.

7. Derivation of absorption spectra from photocurrent spectra

7.1 Bi_2S_3 FILMS

Photocurrent spectroscopy provides a powerful approach to the determination of the absorption spectra of thin semiconducting or insulating films. Equation (6c) can be rearranged to show explicitly the relationship between photocurrent conversion efficiency and absorption coefficient

$$\alpha(\lambda) = \frac{-\ln[1 - \Phi(\alpha)]}{2L_f} \tag{8}$$

However, this equation can only be used to obtain the film thickness or absorption spectrum if the assumptions made in its derivation are valid; if recombination is important, the photocurrent conversion efficiency will be reduced and, as a consequence, values of α or L_f derived by analysis of the photocurrent spectrum will be too low. Similarly, if the assumption that $W > L_f$ is incorrect, the analysis will underestimate the absorption coefficient. If, on the other hand, recombination is negligible and $W > L_f$, the

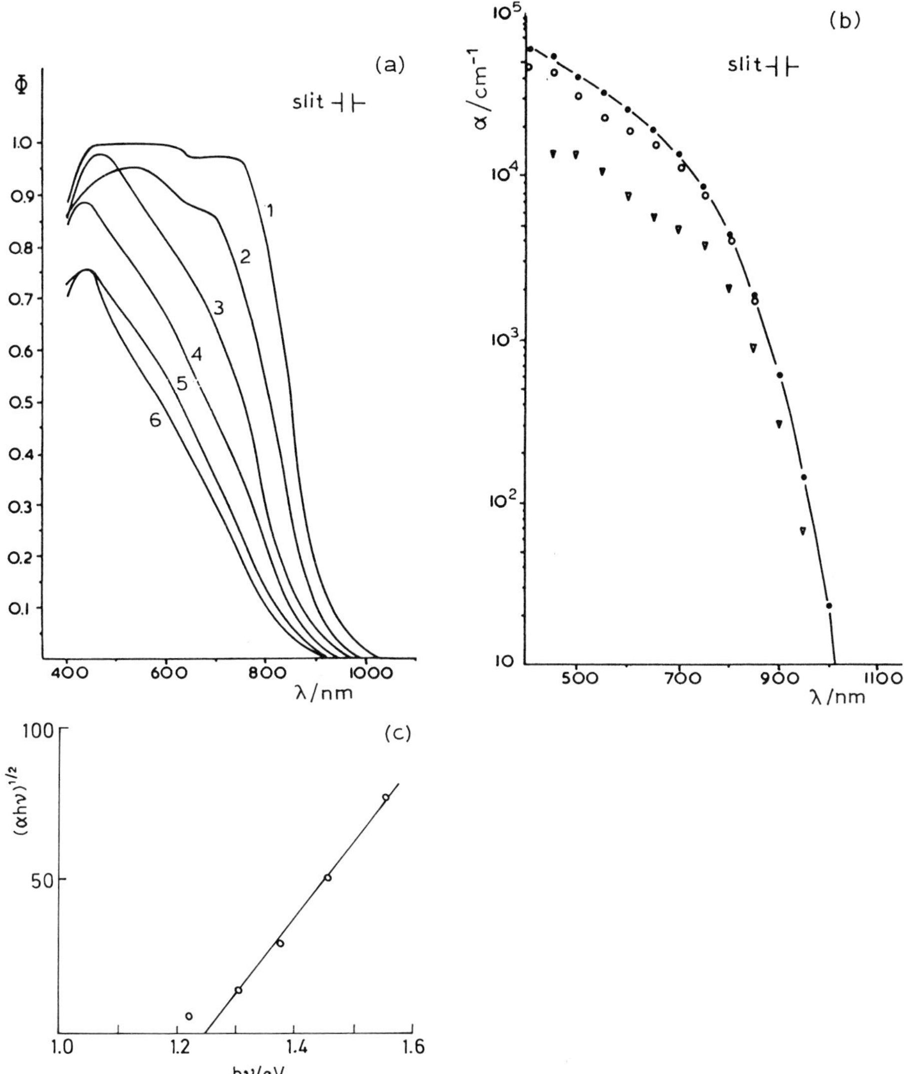

Fig. 7. (a) Set of photocurrent spectra for Bi_2S_3 films of different thicknesses grown in 1 mol dm^{-3} Na_2S. Film thicknesses (μm) (1) 3.9; (2) 2.6; (3) 1.0; (4) 0.5; (5) 0.25; (6) 0.13. (b) Comparison of α values for Bi_2S_3 obtained from photocurrent data, using eqn. (8) with the results of transmission measurements. ●, 0.13 μm film using eqn. (8); ▽, 1.3 μm film using eqn. (8); ○, 0.28 μm film by transmission. (c) Plot of absorption data used to obtain the band gap of Bi_2S_3 [see eqn. (1)]. [(a) and (b) reproduced with permission from ref. 11.]

photocurrent conversion efficiency should approach unity provided that the film is sufficiently thick to absorb the incident light. This limiting condition can easily be tested in some systems. A good example is provided by Bi_2S_3 films grown by the anodisation of bismuth in Na_2S [11]. Bismuth sulphide is an n-type semiconductor with a band gap of about 1.2 eV which is potentially suitable for solar energy conversion and films up to several microns in thickness can be grown by the anodisation of bismuth in Na_2S. The Bi_2S_3 films formed by anodisation in 1.0 mol dm^{-3} Na_2S exhibit well-developed anodic saturation photocurrents under illumination and Fig. 7(a) is a set of photocurrent spectra that shows the characteristic dependence of the response on film thickness. The photocurrent collection efficiency at short wavelengths evidently approaches unity for the thicker films, whereas the response at longer wavelengths is clearly dependent on film thickness, which suggests that light absorption is incomplete. The high limiting value of the photocurrent conversion efficiency observed at short wavelengths indicates that recombination in the space-charge region is insignificant and analysis of the photocurrent spectra with the help of eqn. (8) should therefore lead to the energy dependence of the absorption coefficient of Bi_2S_3. In the absence of reliable data for the optical constants of bismuth, it was assumed that the reflection coefficient at the Bi/Bi_2S_3 interface was unity, whereas reflection at the Bi_2S_3/solution boundary was neglected. The errors introduced by these assumptions are relatively small. It was also assumed that the space charge extended throughout the film so that the condition $W > L_f$ was fulfilled.

In order to provide an independent test of whether eqn. (8) can be used with confidence to determine absorption spectra, bismuth-coated OTEs were prepared with a range of thicknesses by vacuum deposition and were subsequently anodised in Na_2S. The evaporated bismuth film reacted completely during the anodisation, leaving a loosely adherent brown Bi_2S_3 film on the OTE, and L_f was calculated from the thickness of the initial bismuth layer. The transmission of the anodised films was measured in situ and, although the determination of the low-energy part of the spectrum was restricted by light scattering, reliable data were derived for values of the absorption coefficient greater than about 10^3 cm^{-1}.

Figure 7(b) contrasts the absorption data obtained by analysis of the photocurrent spectra with the results of the transmission measurements. It is apparent from the figure that the two methods give spectra that almost coincide at short wavelengths. At the same time, the comparison underlines the advantages of photocurrent spectroscopy which were used to derive absorption coefficients as low as 20, two orders of magnitude below the lowest reliable values obtained from the transmission measurements. The identical results obtained by absorption and photocurrent spectroscopy provide a direct proof that the assumptions made in the analysis of the photocurrent spectra were satisfactory.

The photocurrent data extend to sufficiently low photon energies to allow the indirect band gap of Bi_2S_3 to be derived by the method outlined in Sect. 2 and Fig. 7(c) shows the appropriate plot of the absorption coefficient data which gives a value of 1.25 eV for the indirect allowed band gap, close to a previous estimate of the band gap based on the cut-off wavelength of bulk crystals of Bi_2S_3 [12].

The success of this direct comparison of absorption and photocurrent spectra provides a convincing demonstration of the potential of photocurrent spectroscopy as a quantitative method. The results show that the collection efficiency of photogenerated charge carriers can approach unity, even in polycrystalline anodic layers, which might be expected to contain a high density of defects that can act as recombination centres.

7.2 THE ANODIC OXIDE ON IRON

In the previous section, the absorption coefficients derived by photocurrent spectroscopy were compared with the results obtained from transmission measurements. It is also interesting to compare the absorption coefficient data derived from photocurrent spectra with those obtained by a different optical technique, namely ellipsometry. One of the systems that has been studied extensively by ellipsometry is the anodic oxide film on iron. Remarkably, no concensus of opinion exists concerning the structure of the oxide, although there is evidence from other techniques such as Mössbauer spectroscopy [13] that it is an amorphous, possibly polymeric form rather than a crystalline oxide. Photocurrent measurements show that optical excitation leads to holes that can move through the film, probably by a thermal or field-assisted hopping process, and the excitation spectrum of the photocurrent has been analysed in order to derive the absorption coefficient as a function of photon energy [14]. Figure 8(a) compares the results with the absorption spectrum calculated from the ellipsometric data of Chen and Cahan [15]. It is clear that the shapes of the absorption spectra obtained by the two methods are very similar over most of the energy range, although they differ by about a factor of two in absolute magnitude, probably as the result of uncertainties in the reflection coefficient at the metal/oxide boundary and in the film thickness. Although the photocurrent spectrum falls rather more rapidly then the absorption spectrum at low photon energies, the overall agreement between the spectra derived by the two techniques is sufficiently striking to show that photocurrent spectroscopy can be used to give quantitative values that are in good agreement with those derived by other methods.

The absorption spectrum has been analysed in order to obtain an estimate of the band gap of the oxide and Fig. 8(b) shows that a reasonable fit is obtained for a plot of $(\alpha h\nu)^{1/2}$ against photon energy. Although this analysis is appropriate for an indirect allowed band gap, the same energy dependence of α has been reported for amorphous semiconductors and, in the present case, it seems likely that the oxide film has no regular crystal structure. It is therefore not surprising that the absorption spectrum of the anodic oxide film differs substantially from that reported for crystalline Fe_2O_3 [16]. Further support for the view that the oxide film is amorphous comes from the fact that the photocurrents show a pronounced relaxation effect consistent with charge transfer via localised trapping states [14].

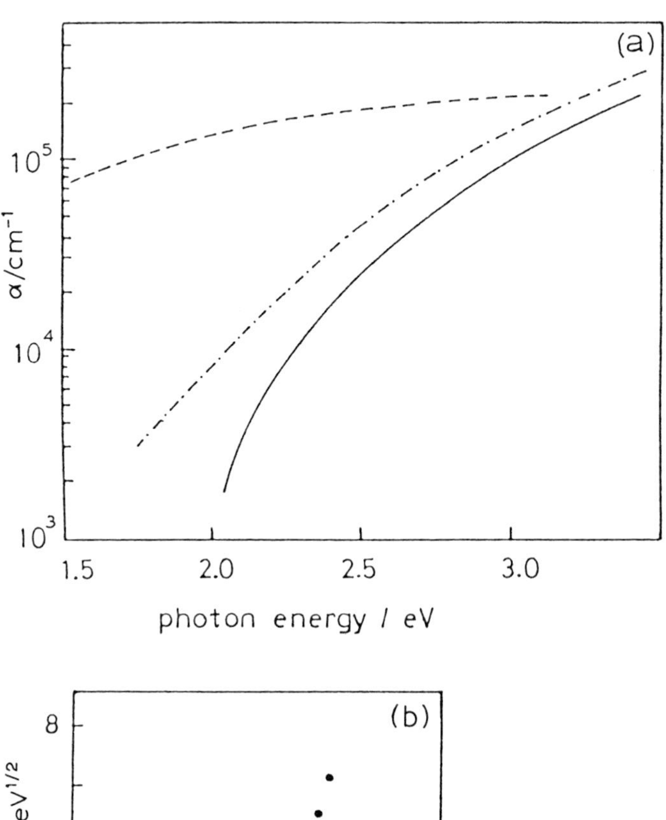

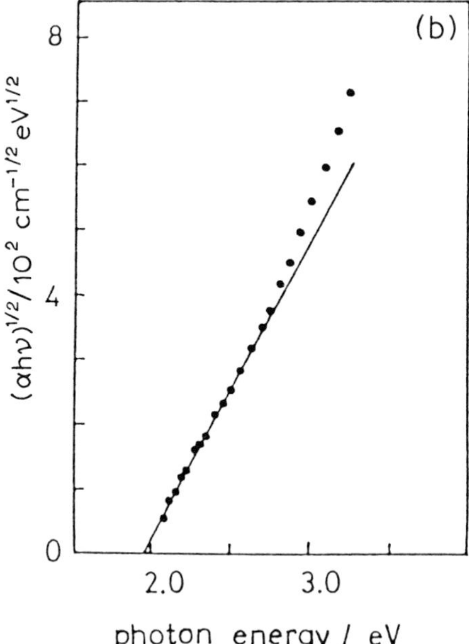

Fig. 8. (a) Comparison of the absorption spectrum of the anodic oxide on iron (——) derived by photocurrent spectroscopy with the spectrum (–·–·–) calculated from the ellipsometric data of Chen and Cahan [15]. The absorption spectrum of α-Fe_2O_3 calculated from the data of Lewis and Westwood [16] is also shown (– – – –). (b) Plot used to determine the band gap of the anodic oxide film on iron (1.95 eV). (Reproduced with permission from ref. 14.)

7.3 PbO FILMS ON LEAD ELECTRODES

The derivation of absorption spectra by photocurrent spectroscopy has proved particularly effective in the case of the PbO films that are formed on lead under different electrochemical conditions [17, 18]. The corrosion of lead in sulphuric acid is of practical interest because of its relevance to the lead–acid battery. The existence of a PbO layer under a $PbSO_4$ film has been inferred from ex-situ X-ray diffraction studies [19] and it appears that the sulphate layer is able to sustain a sufficiently large pH gradient to ensure that PbO is thermodynamically stable underneath it. Although it has been known for some time that a photosensitive oxide layer is formed on lead in H_2SO_4 [20–23], its identification as tetragonal PbO has been achieved only by a quantitative analysis of the photocurrent spectrum [17, 18]. Figure 9(a) shows than an anodic photocurrent appears in a linear sweep experiment soon after the passivation of lead in H_2SO_4 and continues to increase as the potential becomes more positive before collapsing in the region where PbO_2 is formed. The photocurrent reappears on the reverse sweep and changes sign before disappearing when the sulphate layer is reduced to lead.

Conclusive proof of the identity of the photoactive film on lead follows from the analysis of the photocurrent spectra. Figure 9(b) compares the spectrum of the anodic photocurrent with the absorption spectrum of the tetragonal form of PbO. The oxide film is sufficiently thin for the photocurrent to be a linear function of the absorption coefficient so that direct comparison of the photocurrent and absorption spectra is possible and it is clear from the coincidence of the two spectra that the film consists of tetragonal PbO. The results illustrate the sensitivity of photocurrent spectroscopy as a method for the identification of thin surface phases. In-situ Raman spectroscopy [24] and in-situ X-ray measurements [25] of the same system show no evidence for the formation of PbO unless the electrode is held at a constant potential for a time sufficient for a much thicker layer of oxide to be formed. By contrast, photocurrent spectroscopy is sufficiently sensitive to detect the formation of PbO on the much shorter timescale of a linear sweep measurement and quantitative estimates of the film thickness are feasible. Similar results have been obtained for the reduction of α-PbO_2 in alkaline solution, where the existence of the tetragonal form has also been established from the photocurrent spectra [26].

8. Determination of film thickness by photocurrent spectroscopy

8.1 TiO_2 FILMS ON Ti

The primary purpose of the studies described in the previous section was the characterisation of the optical properties of the photoactive layer. Often, however, a reliable in-situ method of following film growth is required. As

References pp. 382–383

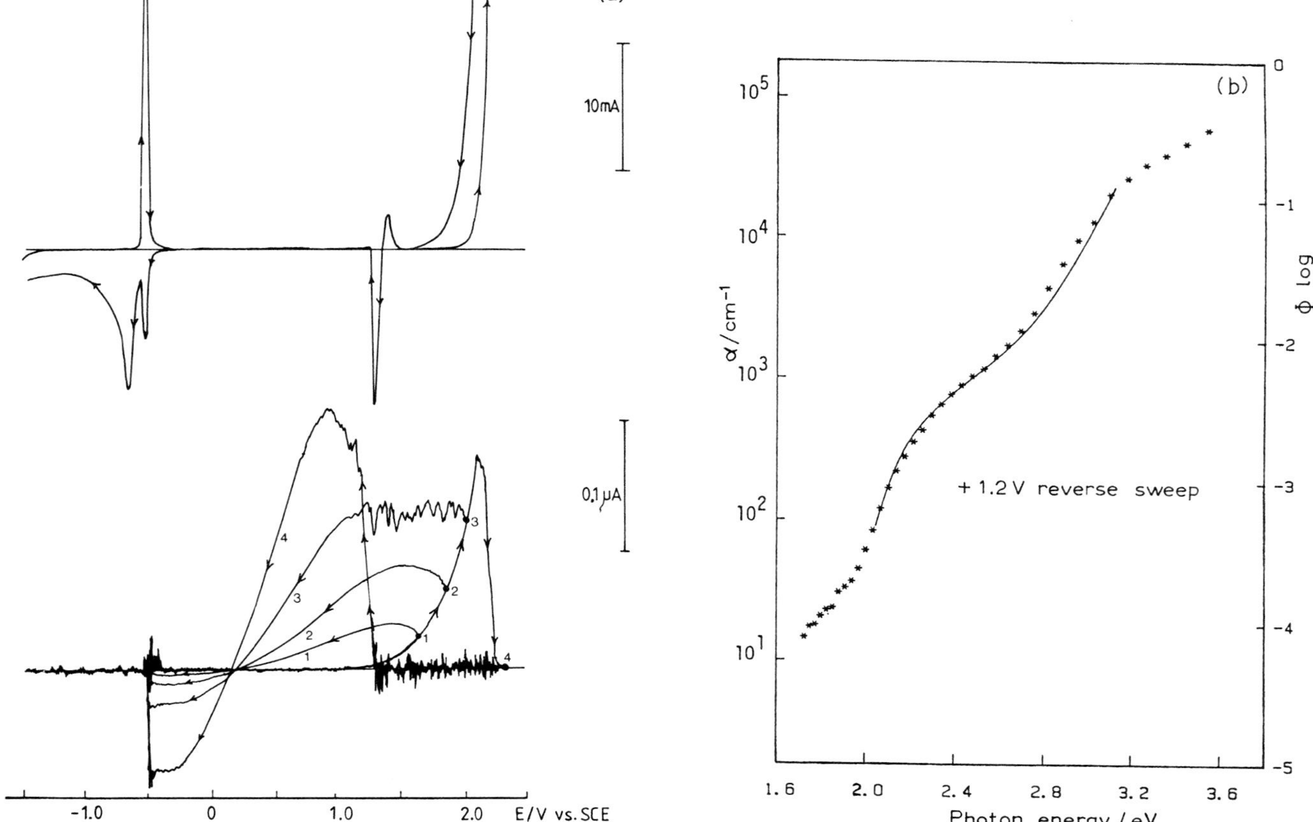

Fig. 9. (a) Simultaneous measurement of the cyclic voltammogram (upper trace) and the photocurrent (lower trace) of a lead electrode in 0.5 mol dm^{-3} H_2SO_4. Wavelength, 340 nm; sweep rate, 20 mV s^{-1}. The photocurrent curves correspond to sweep reversal at different anodic limits. Note, in particular, the change in sign of the photocurrent on the reverse sweep. (b) Comparison of the anodic photocurrent spectrum (*) with the absorption

an illustration, we consider the growth of the anodic film on titanium, which, although not a true valve metal like tantalum, has many features in common with this class of metals. The growth of oxide films on the valve metals involves the field-assisted transport of ions or ion vacancies and the film thickness usually depends linearly on potential; the ratio of film thickness to growth voltage is commonly referred to as the anodising ratio. The growth rate depends exponentially on the electrical field in the so-called high-field limit and consequently the rate of thickening becomes essentially negligible after a short period. Typical values of the anodising ratio lie in the range 2–4 nm V^{-1} and it is evident, therefore, that the residual electric field in the films exceeds 10^6 V cm^{-1}. Such extreme electrical fields greatly assist the separation of photoexcited carrier pairs and, as a result, collection efficiencies can approach unity in spite of the polycrystalline or even amorphous nature of the oxide films. The in-situ determination of oxide film thickness is a challenging objective for photocurrent spectroscopy, but the task is simplified in the case of metals like titanium because the high electric field eliminates recombination losses. L_f can therefore be obtained from the photocurrent spectrum provided that the optical constants of the oxide and metal are known. Equation 6(c) can be rearranged to show this explicitly.

$$L_f = \frac{-\ln(1 - \Phi)}{2\alpha} \tag{9}$$

Figure 10(a) is a set of photocurrent spectra obtained for titanium in H_2SO_4 at different potentials [27] and, as expected, the photocurrent conversion efficiency is seen to increase with electrode potential. The absorption spectrum of thin CVD TiO_2 films has been reported by Memming and co-workers [28] and eqn. (9) has therefore been used to calculate the film thickness. A sensitive way to test the validity of the method is to carry out the calculation of film thickness at different wavelengths; if the optical constants are correct, the same value of L_f will be obtained. Figures 10(b) and (c) show that this is indeed the case for TiO_2 films on Ti since the data taken at different wavelengths condense on to a common line when the absorption coefficient data are used to calculate the film thickness. The results are interesting since they show a break in the plot of L_f vs. potential that indicates a change in growth mechanism. The main errors in this analysis probably arise from the uncertainties in the optical constants of the substrate, but often relative rather than absolute values of film thickness are required and one of the limiting forms of eqn. 6(a) may be adequate.

8.2 CORROSION OF Pb IN H_2SO_4

The growth of PbO films on lead in H_2SO_4 has been discussed in the previous section, where it was shown that the tetragonal modification of the oxide is formed. The corrosion of lead in sulphuric acid is more rapid at elevated temperatures and a systematic study of the rate of oxide growth as a function of temperature has been made by photocurrent spectroscopy [17].

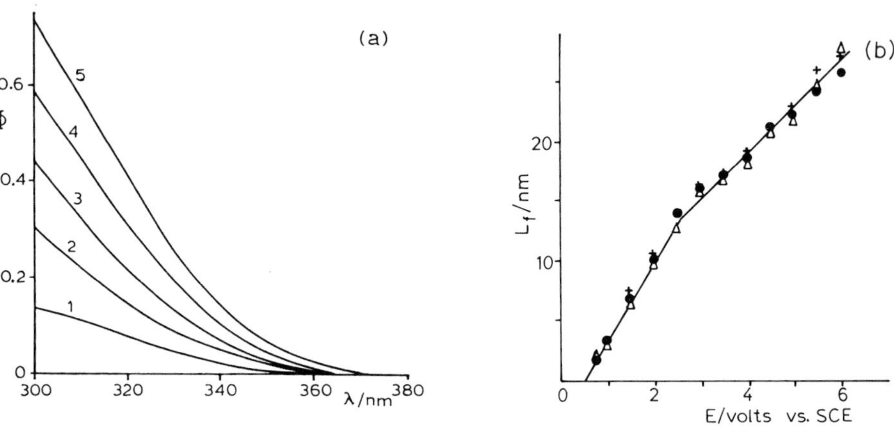

Fig. 10. (a) Wavelength dependence of the photocurrent conversion efficiency for a titanium electrode in 1 mol dm^{-3} H_2SO_4. The electrode potentials (vs. SCE) were (1) 1.0 V; (2) 1.5 V; (3) 2 V; (4) 3 V; (5) 6 V. Slit resolution 9 nm. (Reproduced with permission from ref. 27.) (b) Film thickness as a function of potential calculated from eqn. (9). Wavelengths: ●, 330 nm; +, 340 nm; △, 350 nm.

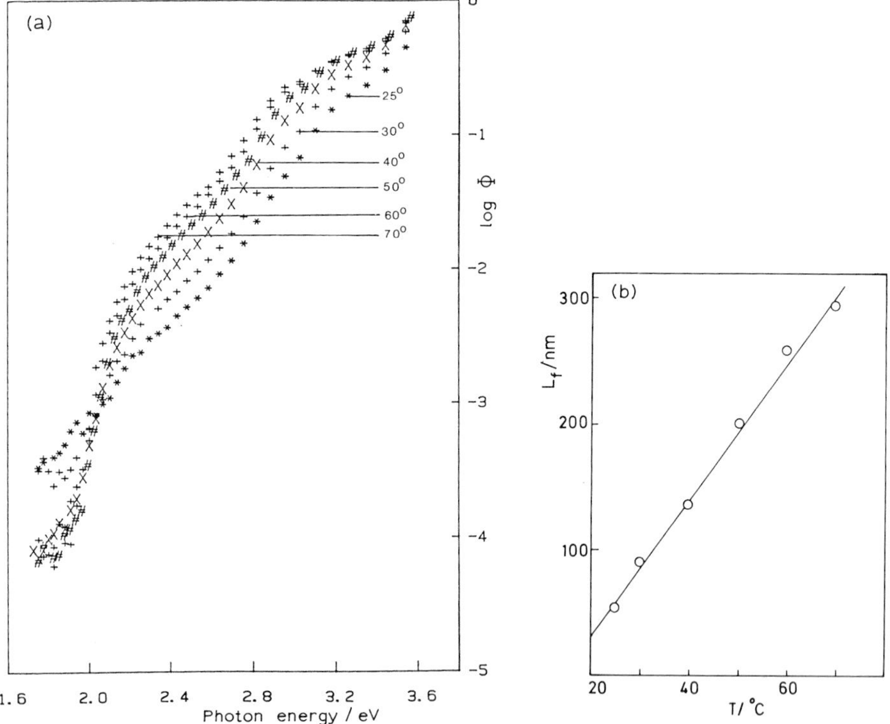

Fig. 11. (a) Set of photocurrent spectra for lead in 5 mol dm^{-3} H_2SO_4 at 1.4 V vs. SCE showing the temperature dependence (°C) due to the increasing thickness of the PbO layer. Spectra recorded after 50 min. (b) Plot of PbO film thickness as a function of temperature, obtained by fitting the photocurrent spectra shown in (a). (Data taken from ref. 18.)

Figure 11(a) shows a set of photocurrent spectra obtained under at the same potential at different temperatures and the increase of thickness is evident. Figure 11(b) shows that the film thickness increases approximately linearly with temperature under the conditions of these experiments. Photocurrent spectroscopy is probably the only in-situ spectroscopic technique that can give this information because ellipsometric measurements are hampered by the crystalline layer of $PbSO_4$ that covers the PbO film. The oxide film is evidently not a continuous insulating layer since impedance measurements suggest that only a thin dielectric film of the order of a few monolayers is present, whereas photocurrent spectroscopy reveals the existence of a much thicker layer of PbO. The continued growth of the corrosion layer over long periods of time is consistent with a mechanism in which a polycrystalline deposit is formed over a thin basal layer and the combination of impedance and photocurrent measurements gives results that are consistent with this model of the film structure.

9. Internal photoemission

It is often assumed that the majority carrier type can be deduced from the sign of the photocurrent observed at electrodes covered by surface layers and changes of sign of the photocurrent have been attributed to variations in film stoichiometry. In fact, such an interpretation of photocurrent data is not necessarily correct, since insulating surface layers can give both anodic and cathodic photocurrents; the determining factor is simply the direction of the electric field vector in the film. It is therefore important to obtain independent evidence concerning the conductivity mechanism in the film. Mott–Schottky plots of the inverse square of the space-charge capacitance against potential can be used to determine the doping density of bulk semiconductor electrodes, but the capacitance of thin films will be constant and equal to the geometric capacitance as long as the condition $W > L_f$ is satisfied. An alternative method of estimating the conductivity type and doping density in thin films is necessary. The Fermi level in the surface film phase can be determined directly if internal photoemission is observed at the metal/film junction since the barrier height corresponds to the separation between the Fermi level and the conduction band.

Results obtained for the oxide film on bismuth illustrate this approach [29]. The oxide grows by the field-assisted migration of ions and Fig. 12 shows that the forward sweep of a cyclic voltammogram exhibits the plateau currrent characteristic of this mechanism. On the reverse sweep, where the field is insufficient to sustain further film growth, the current falls to zero until the potential is reached at which the film is reduced. The figure also shows the photocurrent observed under the same conditions when the electrode is illuminated. On the forward sweep, the photocurrent rises as the film thickens, but on the reverse sweep it falls and then changes sign before the reduction potential is reached.

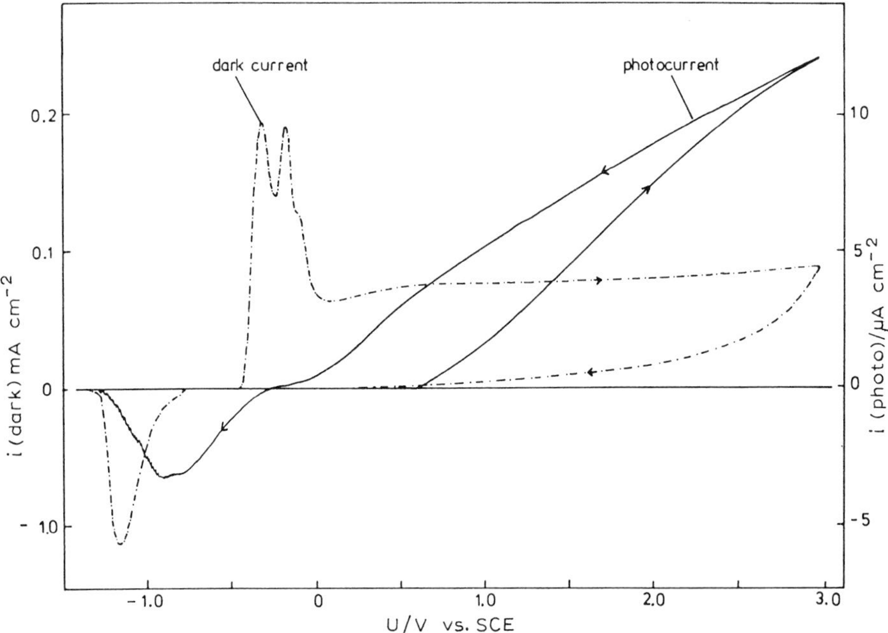

Fig. 12. Simultaneous recording of the cyclic voltammogram (- - - -) and the photocurrent (——) for a bismuth electrode in 0.1 mol dm^{-3} $NaHCO_3$. Wavelength, 380 nm; sweep rate, 30 mV s^{-1}. (Reproduced with permission from ref. 29.)

The change of sign of the photocurrent has been interpreted as evidence for a transition from n- to p-type behaviour [30], but impedance measurements have shown that the capacitance of the oxide film on bismuth is completely independent of potential on the reverse sweep and this implies that the doping density is so low that the oxide behaves as an insulator. Such a conclusion is at variance with the suggestion that the film displays either n- or p-type conductivity depending on the electrode potential. Photocurrent spectroscopy provides the solution to this problem. The anodic and cathodic photocurrent spectra are contrasted in Fig. 13(a) and (b) and it is immediately clear that they are not identical. The anodic photocurrent spectrum is typical of an amorphous oxide, showing an exponential edge below 3 eV. The apparent indirect band gap has been estimated from the absorption coefficient data derived from the photocurrent spectrum and Fig. 14(a) shows that a value of 2.8 eV is obtained from a plot of $(\alpha h\nu)^{1/2}$ against photon energy. The cathodic photocurrent spectrum [Fig. 13(b)], on the other hand, exhibits a well-defined additional low-energy response that extends well below 2 eV. This type of spectrum is characteristic of internal photoemission and the threshold energy obtained from the Fowler plot of $\Phi^{1/2}$ against photon energy shown in Fig. 14(b) is 1.4 eV, exactly half the band gap energy.

The photoemission process involves the excitation of electrons from the metal electrode into the oxide overlayer and the threshold energy therefore

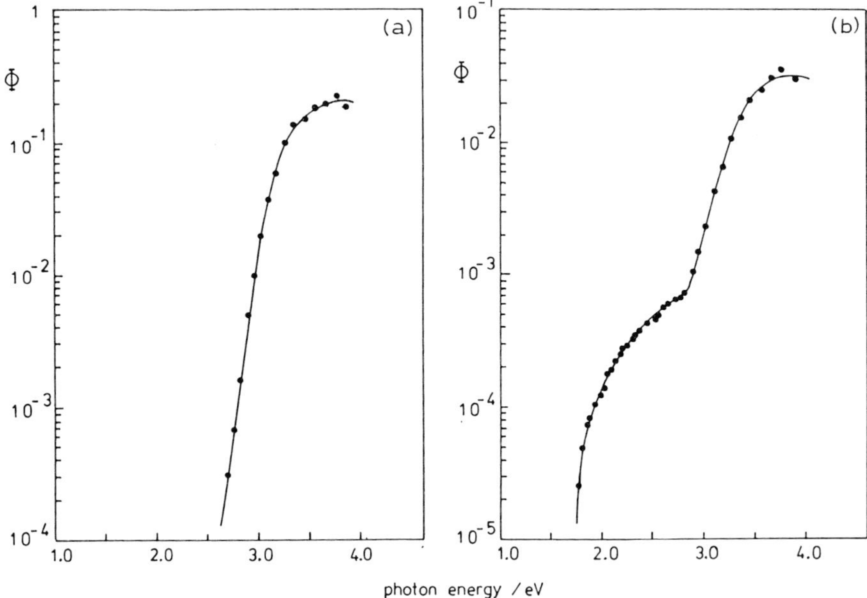

Fig. 13. (a) Spectrum of anodic photocurrent conversion efficiency of anodised bismuth at 3 V vs. SCE. (b) Spectrum of cathodic photocurrent conversion efficiency of the same film at −0.55 V vs. SCE. The additional low-energy response is due to internal photoemission. (Reproduced with permission from ref. 29.)

corresponds closely to the energy difference between the Fermi level in the metal and the conduction band in the oxide. In the present case, then, we can conclude that the Fermi level of the metal is situated at the centre of the band gap at the intrinsic level of the oxide providing conclusive proof that the oxide is either an intrinsic or a compensated insulator. This conclusion is important since it shows that the metal/oxide junction is not ohmic, but corresponds instead to the situation illustrated in Fig. 14(c).

Internal photoemission has also been observed at the Pb/PbO interface [16, 17] and Fig. 15 is an example of the Fowler plots obtained for anodised lead in H_2SO_4. Here, again, the results show clearly that the Fermi level coincides closely with the intrinsic level at the centre of the band gap and it is evident that the PbO layer behaves as an insulator. The change in sign of the photocurrent in both these systems is therefore a consequence of field reversal and not of changes in stoichiometry. The insulating behaviour exhibited by these anodic oxide layers is most probably a consequence of a high density of states that pin the Fermi level close to the centre of the band gap. The transition from monocrystalline to polycrystalline oxide structures has been discussed by van den Broek [31] who has argued that surface states will pin the Fermi level throughout the film when the average crystallite size becomes smaller than the space-charge layer thickness. This condition is met not only by the anodic oxide films discussed here, but also by the evaporated PbO films studied by van den Broek.

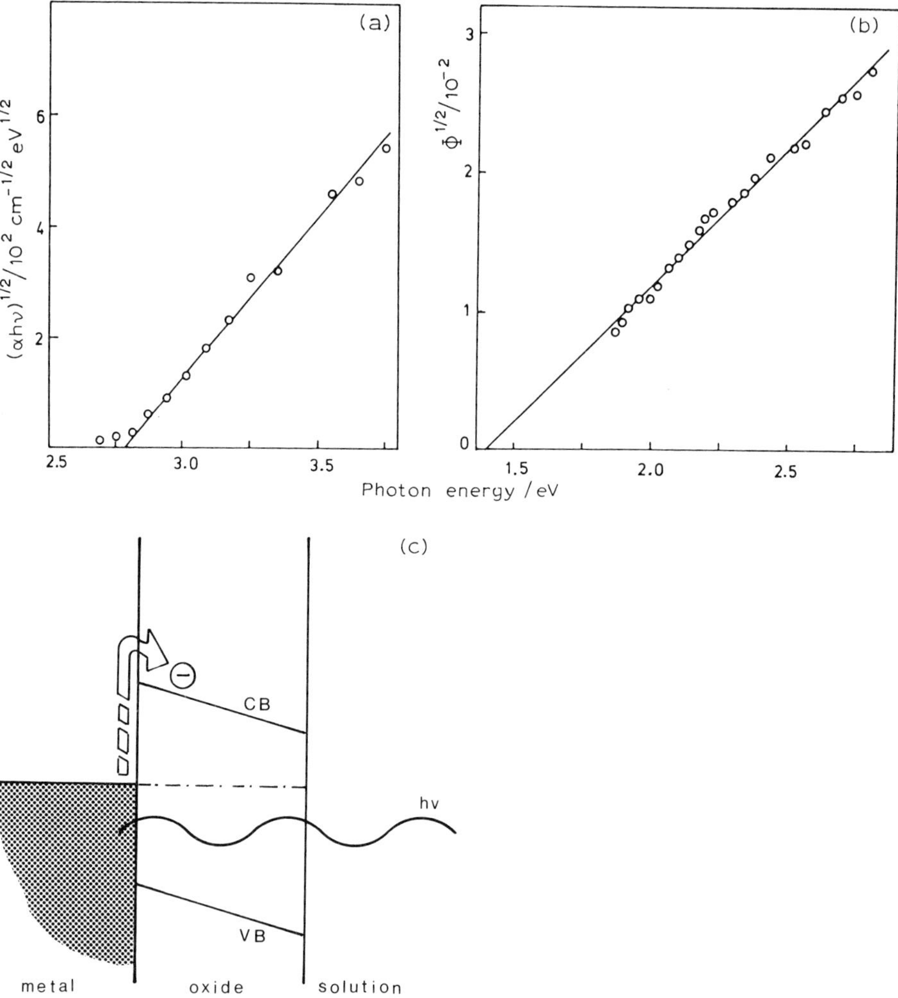

Fig. 14. (a) Plot of absorption data for the anodic oxide film on bismuth illustrating the determination of the band gap of the oxide. (b) Fowler plot of the cathodic photocurrent data below 2.8 eV showing the threshold intercept at 1.4 eV. (Reproduced with permission from ref. 29.) (c) Origin of the low-energy cathodic photocurrent response at oxide-coated electrodes. If the oxide is effectively insulating, the barrier height for photoemission is half the band gap.

10. An example of a p-type surface film: the oxide film on Cu

The cathodic photocurrents observed on Bi and Pb electrodes are clearly not evidence of p-type behaviour, but are instead a consequence of field reversal in the insulating surface film. By contrast, the anodic oxidation of copper in alkaline solution does appear to give a layer of Cu_2O with p-type conductivity [32]. Figure 16 relates the cyclic voltammogram of Cu in alkaline solution to the photocurrent. Here, no anodic photocurrent is observed;

instead, a small cathodic photocurrent appears on the forward sweep and develops on the reverse sweep before showing a discontinuous increase at a potential coincident with the first reduction peak on the cyclic voltammogram. Detailed study has shown that a Cu_2O layer grows during the initial stage of copper oxidation, but it is partially destroyed at more positive potentials, where a layer of $Cu(OH)_2$ is formed. The appearance of a cathodic photocurrent on the reverse sweep proves that a Cu_2O layer still persists on the electrode surface, whereas the abrupt increase in photocurrent at the potential of the first cathodic peak shows that the Cu_2O layer thickens by the reduction of the overlying Cu(II) film. This interpretation was confirmed in the same study by impedance measurements, which showed a corresponding decrease in film capacitance coincident with the increase in photocurrent as the film thickens by reduction of the overlayer.

The formation of a p-type layer during the anodic oxidation of a metal is particularly interesting. If the oxide grows by the field-assisted migration of ions, the electric field in the oxide is expected to correspond to the transfer of metal ions from the metal/oxide to the oxide/solution interface. If the material is p-type, however, the field must have the opposite sign if a cathodic photocurrent is to be observed. The potential dependence of the photocurrent observed at copper electrodes indicates that the Cu_2O layer is formed at potentials negative of the flatband potential, E_{fb}, and it therefore follows that the direction of the electric field actually opposes film growth by ion migration. At the same time, the stability of the Cu_2O layer is restricted because holes will accumulate at the oxide solution interface as the potential approaches E_{fb} and oxidation of Cu_2O to Cu(II) species becomes possible. The oxidation of copper in alkaline solution can therefore be described in terms of the solid-state properties of the surface phase.

The photocurrent spectra for Cu_2O films grown in different alkaline solutions have been analysed in order to derive the absorption spectrum and Fig. 17 shows that the film grown in at pH 9 appears to differ from the oxides

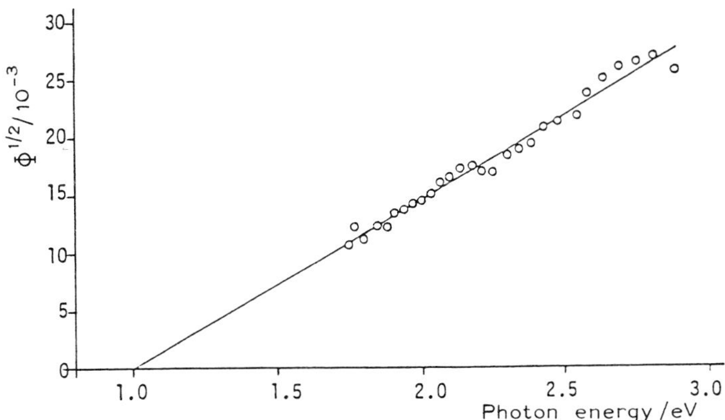

Fig. 15. Fowler plot of the cathodic photocurrent for lead anodised in 5.0 mol dm^{-3} H_2SO_4 showing the threshold intercept at 1 eV. (Data taken from ref. 18.)

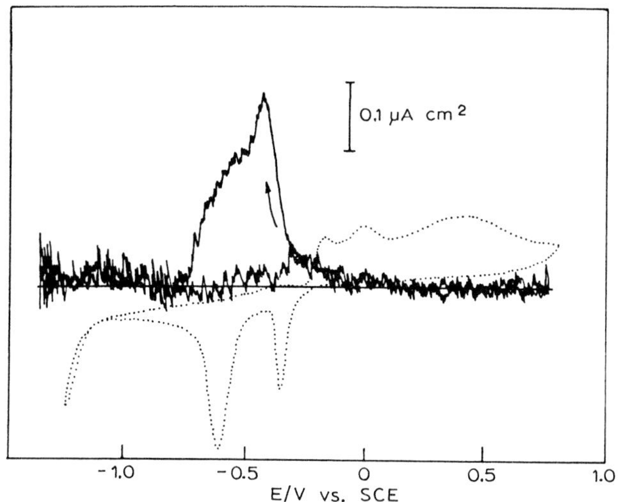

Fig. 16. Simultaneous measurement of the cyclic voltammogram (· · · · ·) and photocurrent response (———) of a copper electrode in 0.1 mol dm^{-3} $Na_2B_4O_7$. Wavelength, 350 nm; sweep rate, 10 mV s^{-1}. (Reproduced with permission from ref. 32.)

formed in solutions of higher pH. The optical absorption of Cu_2O depends critically on stoichiometry and the results suggest that the oxide films formed at higher pH values contain excess oxygen. The film thicknesses derived from the photocurrent spectra are of the order of a few monolayers, underlining the sensitivity of photocurrent spectroscopy. Ex-situ methods are likely to be unreliable when such thin corrosion layers are involved unless rigorous high-vacuum transfer techniques are available.

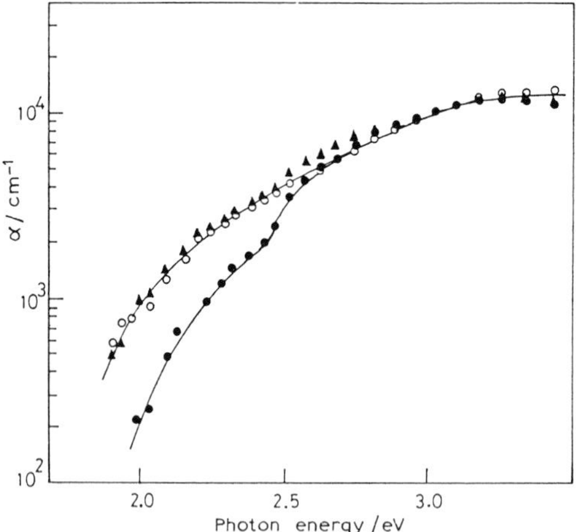

Fig. 17. Absorption spectra for Cu_2O films on copper calculated using eqn. (6a). Electrolytes: ▲, 0.1 mol dm^{-3} LiOH; ○, 0.1 mol dm^{-3} NaOH; ●, 0.1 mol dm^{-3} $Na_2B_4O_7$. (Reproduced with permission from ref. 32.)

11. Organometallic films on metal electrodes

The formation of photoactive films on metal electrodes is not restricted to inorganic materials. Copper, for example, can be anodised to form polymeric phenylacetylide [33] and acetylide [34] layers that appear to behave as p-type organic semiconductors. The photoconducting properties of the arylethynyl polymers have been known for some time, although the mechanism of photoconductivity is not well understood. It seems probable that charge carriers are created by the annihilation of mobile Frenkel excitons at electron traps such as adsorbed oxygen rather than by direct interband excitation.

Photoeffects at copper electrodes coated with copper phenylacetylide (CuPA) films were discovered during an in-situ laser Raman investigation [35], where substantial cathodic photocurrents were observed. A subsequent detailed study by photocurrent spectroscopy showed that the photocurrent conversion efficiency was of the order of a few percent, even for thicker films which absorbed a substantial fraction of the incident light. Figure 18 contrasts the photocurrent excitation spectrum with the absorption spectrum measured on an OTE coated with the CuPA polymer. The coincidence in the vibrational structure in the two spectra is striking, suggesting that the absorption gives rise to a state with considerable molecular character.

The existence of exciton states in polymers such as CuPA has not been considered in detail in the literature, but it seems reasonable to suppose that efficient energy transfer can occur along the polymer chains. The absorption transition has been attributed to the formation of a charge transfer state and it is therefore possible that exciton dissociation is enhanced by the local electric field. Alternatively, exciton dissociation may occur at the polymer surface, with the electron being transferred to an acceptor such as molecular oxygen. The reaction scheme

$$(\mathrm{M}) + h\nu \rightarrow (\mathrm{M}^*) \qquad \text{optical excitation}$$

$$(\mathrm{M}^*) \rightarrow (\mathrm{M}^*_{\mathrm{surf}}) \qquad \text{energy transfer}$$

$$(\mathrm{M}^*_{\mathrm{surf}}) + \mathrm{O}_2 \rightarrow (\mathrm{M}^+_{\mathrm{surf}}) + \mathrm{O}_2^- \qquad \text{exciton dissociation}$$

$$(\mathrm{M}^+_{\mathrm{surf}}) \rightarrow (\mathrm{M}^+_{\mathrm{bulk}}) \qquad \text{hole migration}$$

where (M) represents a unit in the polymer chain, describes this process. The M^+ species in this scheme represents a hole which can migrate along and between the polymer chains until, eventually, it recombines with an electron at the metal/polymer junction. The importance of oxygen as an electron acceptor is demonstrated by the fact that the cathodic photocurrent is increased substantially in oxygenated solutions. The low photocurrent conversion efficiencies are not surprising, since the diffusion lengths of excitons in molecular solids and polymers are generally rather small and radiative or non-radiative quenching of the excited state compete with electron transfer. In fact, the photocurrent conversion efficiencies observed at CuPA and related polymers are much higher than those for other organic solid state

References pp. 382–383

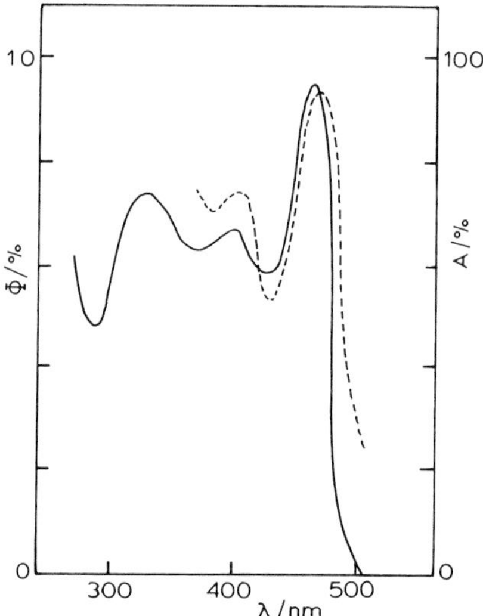

Fig. 18. Comparison of photocurrent spectrum (——) of CuPA-coated copper electrode with the absorption spectrum (- - - -) of the film. (Data taken from ref. 33.)

systems; this may be due, in part, to the high internal surface area of the polymer coating, which appears to consist of interwoven filaments. Further studies of these interesting polymer systems should prove rewarding.

Acknowledgements

Much of the work described in this chapter was carried out with the financial support of the Science and Engineering Research Council. The author also thanks his co-workers who collaborated in these investigations.

References

1 See, for example, F. Willig, Adv. Electrochem. Electrochem. Eng., 12 (1981).
2 R.S. Knox, Solid State Phys., Suppl. 5 (1963).
3 Yu.Ya. Gurevich, Yu.V. Pleskov and Z.A. Rotenburg, Photoelectrochemistry, Consultants Bureau, New York, 1980.
4 See, for example, G. Harbecke, in F. Abeles (Ed.), Optical Properties of Solids, North-Holland, Amsterdam, 1972, p. 23.
5 J. Tauc, in F. Abeles (Ed.), Optical Properties of Solids, North-Holland, Amsterdam, 1972, p. 277.
6 R.H. Bube, Photoconductivity of Solids, Wiley, New York, 1960.
7 L.M. Peter, Electrochemistry, Specialist Periodical Report, Vol. 9, Royal Society of Chemistry, London, 1984.
8 J. Li and L.M. Peter, J. Electroanal. Chem. Interfacial Electrochem., 165 (1984) 41.
9 W.K. Subashiev and S.A. Abagyan, in M. Hulin (Ed.), Proceedings of the 7th International Conference on the Physics of Semiconductors, Academic Press, New York, 1964, p. 226.

10 L.M. Peter, Electrochim. Acta, 23 (1978) 1073.
11 L.M. Peter, J. Electroanal. Chem. Interfacial Electrochem., 98 (1979) 49.
12 L. Gildart, J.M. Kline and D.M. Mattox, J. Phys. Chem. Solids, 18 (1961) 286.
13 W.E. O'Grady, J. Electrochem. Soc., 127 (1980) 555.
14 L.M. Abrantes and L.M. Peter, J. Electroanal. Chem. Interfacial Electrochem., 150 (1983) 593.
15 C.T. Chen and B.D. Cahan, J. Electrochem. Soc., 129 (1982) 17.
16 D.C. Lewis and W.D. Westwood, Can. J. Phys., 42 (1964) 2367.
17 J.S. Buchanan, N.P. Freestone and L.M. Peter, J. Electroanal. Chem. Interfacial Electrochem., 182 (1985) 383.
18 J.S. Buchanan, Ph.D. thesis, University of Southampton, 1985. J.S. Buchanan and L.M. Peter, Electrochim. Acta, to be submitted.
19 D. Pavlov, J. Electrochem. Soc., 110 (1969) 316.
20 D. Pavlov, S. Zanova and G. Papzov, J. Electrochem. Soc., 124 (1977) 1522.
21 D. Pavlov, Electrochim. Acta, 23 (1978) 845.
22 S. Fletcher and D.B. Matthews, J. Electroanal. Chem. Interfacial Electrochem., 126 (1981) 131.
23 R.G. Barradas, D.S. Nadeshdin, J.R. Webb, A.P. Roth and D.F. Williams, J. Electroanal. Chem. Interfacial Electrochem., 126 (1981) 273.
24 R.J. Thibeau, C.W. Brown, A.Z. Goldfarb and R.H. Heidersbach, J. Electrochem. Soc., 127 (1980) 37.
25 L.G. Li and L.M. Peter, unpublished data.
26 L.M. Peter, J. Electroanal. Chem. Interfacial Electrochem., 144 (1983) 315.
27 J.F. McAleer and L.M. Peter, Faraday Discuss. Chem. Soc., 70 (1980) 67.
28 F. Mollers, H.J. Tolle and R. Memming, J. Electrochem. Soc., 121 (1974) 1160.
29 L.M. Castillo and L.M. Peter, J. Electroanal. Chem. Interfacial Electrochem., 146 (1983) 377.
30 M. Metikos-Hukovic, Electrochim. Acta, 26 (1981) 989.
31 J. van den Broek, Philips Res. Rep., 22 (1967) 367.
32 L.M. Abrantes, L.M. Peter and L.M. Castillo, J. Electroanal. Chem. Interfacial Electrochem., 163 (1984) 209.
33 L.M. Abrantes, L.M. Castillo, M. Fleischmann, I.R. Hill, L.M. Peter, G. Mengoli and G. Zotti, J. Electroanal. Chem. Interfacial Electrochem., 177 (1984) 129.
34 G. Zotti, S. Catterin, G. Mengoli, M. Fleischmann and L.M. Peter, J. Electroanal. Chem. Interfacial Electrochem., 200 (1986) 341.
35 L.M. Abrantes, L.M. Castillo, M. Fleischmann, I.R. Hill, L.M. Peter, G. Mengoli and G. Zotti, J. Electroanal. Chem. Interfacial Electrochem., 164 (1984) 177.

Chapter 9

Electroreflectance at Semiconductors

A. HAMNETT, R.L. LANE, P.R. TREVELLICK and S. DENNISON

1. Introduction

Considerable controversy surrounds the description of the potential distribution at the surface of a semiconductor electrode in contact with a concentrated electrolyte. As compared with a metal [Fig. 1(a)], the classical semiconductor model invokes a depletion layer, as shown in Fig. 1(b), that extends an appreciable distance, W, into the semiconductor [1]. Within this layer, the potential varies quadratically with distance as discussed elsewhere [1] and the magnitude of W depends inversely on the square root of the donor density; for typical dopant density levels of 10^{16}–10^{18} cm^{-3}, W takes values from several hundred to several thousand Ångstroms, depending on the potential. The effect of varying the potential within the classical model is shown in Fig. 1(c)–(e) and it can be seen that all the potential change is accommodated inside the semiconductor. In effect, the band edges are fixed in energy at the surface.

Although this model is a natural extension of that derived for metal/semiconductor or p–n junctions, it has proved remarkably difficult to verify it for semiconductors in contact with those electrolytes normally employed by electrochemists. As an example, the electrochemistry of germanium initially proved very difficult to understand in aqueous solution [2] and it was only with DeWald's studies of n-ZnO [3] that a paradigmatic example of the classical model was discovered. The data found by DeWald in his study of the ZnO electrolyte interface confirmed quantitatively the behaviour of the a.c. response of the semiconductor/electrolyte as predicted by the classical model. In particular, DeWald confirmed that the series capacitance of the interface obeyed the Mott–Schottky relationship [1]

$$C_{sc}^{-2} = (V - V_{fb}) \quad (1)$$

where V is the applied potential, V_{fb} the flat-band potential (i.e. the applied potential at which there is no potential drop in the semiconductor), and the dopant density is small compared with the concentration of the electrolyte. However, it was rapidly discovered that even n-ZnO does not behave entirely predictably. If the pH of the electrolyte is altered, then the flat-band potential changes in a regular way; in fact, for a large number of semiconducting oxides, the relationship [4]

References pp. 425–426

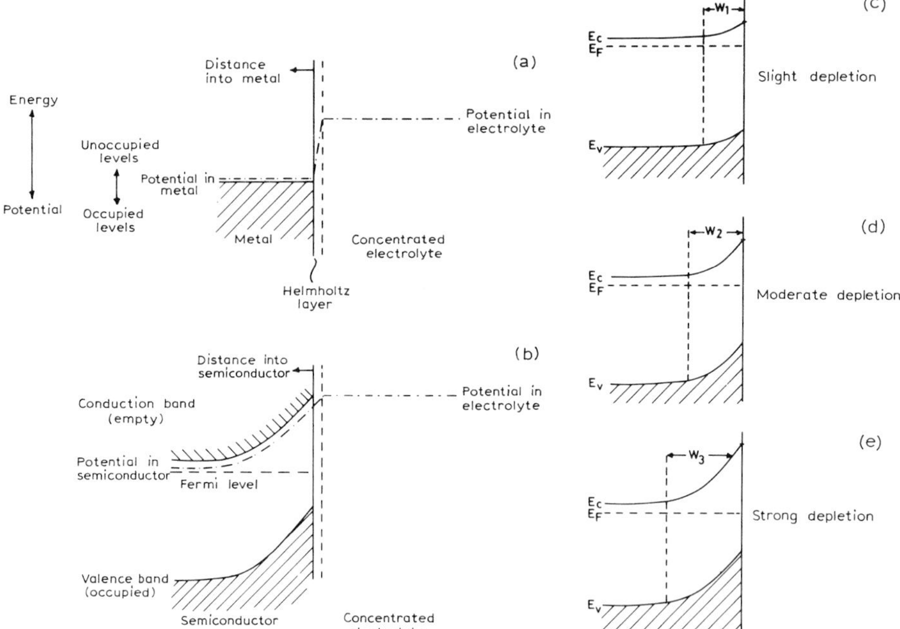

Fig. 1. Potential distribution at the interface between a concentrated electrolyte and (a) a metal (b)–(e) a semiconductor. In (c)–(e), the (positive) potential applied to the interface increases; this potential is accomomodated entirely within the semiconductor and is manifested in a steadily increasing depletion-layer width.

$$V_{fb} = \frac{V_{fb}^0 - 2.303kT}{e_0 \cdot pH} \tag{2}$$

is satisfied. This is most naturally explained by ionosorption of protons, which alters the potential dropped in the Helmholtz layer. However, to good accuracy, the surface coverage by protons does *not* alter as the applied potential is altered, thus, the *change* in potential is accommodated entirely within the depletion layer and eqn. (1) is recovered. This effect is illustrated in Fig. 2.

A much more serious problem was identified by Green in 1959 [5]. Quantitative analysis of the extent of coverage by protons required to give rise to eqn. (2) reveals that only a fractional change in coverage could account for the entire 0.83 V implied by eqn. (2) on going from acid to base and Green pointed out that a similar fractional coverage by electronic surface states, whose population would be a strong function of the Fermi level in the semiconductor, would profoundly modify the potential distribution shown in Fig. 1. Two cases can be distinguished: if the electronic surface states are in equilibrium with the bulk of the semiconductor, then the situation shown in Fig. 3 will result; the potential is now accommodated both within the depletion layer and aross the Helmholtz layer, with a partition coefficient that depends critically on the surface state density. As a rule of thumb, it is

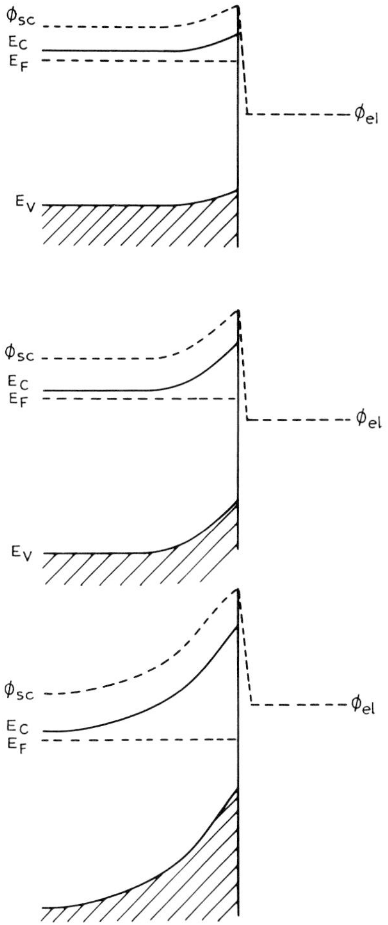

Fig. 2. Potential distribution at the semiconductor/electrolyte interface in the presence of a (fixed) positive charge at the semiconductor surface. Curves (a)–(c) show the effect of a steadily increasing positive bias which, as for Fig. 1, is accommodated entirely within the semiconductor depletion region.

usually suggested that surface state densities in the range $\geqslant 10^{13}\,\text{cm}^{-2}$ will have a significant effect on the potential distribution. The second case occurs if the surface states are in sluggish equilibrium with the bulk of the semiconductor but in rapid equilibrium with redox couples in solution. The effect is that the occupancy of these states alters until the steady state is reached. Energetically, the states fill up until the topmost occupied level has the same energy as the redox energy $-e_0 U_R$. The effect is shown in Fig. 4 and, since U_R depends, through the Nernst equation, on the relative concentrations of oxidised and reduced forms, substantial swings in V_{fb} can occur when these concentrations are altered with, for example, illumination. However, for a given redox system, eqn. (1) will still be recovered.

References pp. 425–426

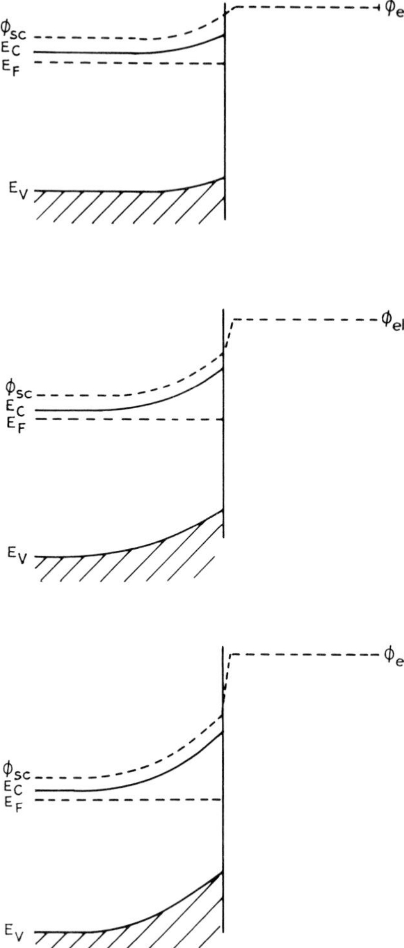

Fig. 3. Potential distribution at the semiconductor/electrolyte interface in the presence of surface electronic levels whose occupany varies with potential. As the potential bias is increased, some of the potential drop is accommodated within the depletion layer, but some is accommodated across the Helmholtz layer.

It is obviously highly desirable that the case applicable to a given experimental configuration should be known, and known with some precision, since our interpretation of, for example, any photocurrent observed is likely to be sensitively dependent on the potential distribution. For this reason, a variety of techniques have been developed to study the internal field and potential.

The earliest technique to be used extensively was the measurement of the a.c. current response to an applied a.c. potential. Provided very small-amplitude a.c. signals are used, the resultant equations can be linearised and solved directly. If there are no surface electronic levels, eqn. (1) is recovered immediately from this theory [6], but very substantial theoretical problems

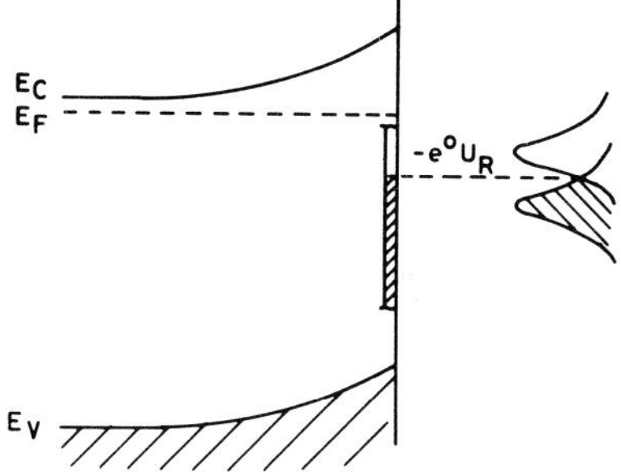

Fig. 4. Potential distribution at the semiconductor/electrolyte interface in the presence of surface electronic levels whose occupancy is determined by equilibration with a redox couple in solution.

arise if surface states are present. Provided that these states have relaxation frequencies within the frequency range employed, they can be detected, but is now accepted that attempts to model such states in terms of simple components, such as frequency-independent resistors and capacitors, is unlikely to be very accurate. As an example, the a.c. response of p-GaP in aqueous acid electrolyte is uninterpretable in terms of the simple series circuit

An attempt to model the behaviour using the more complex circuit

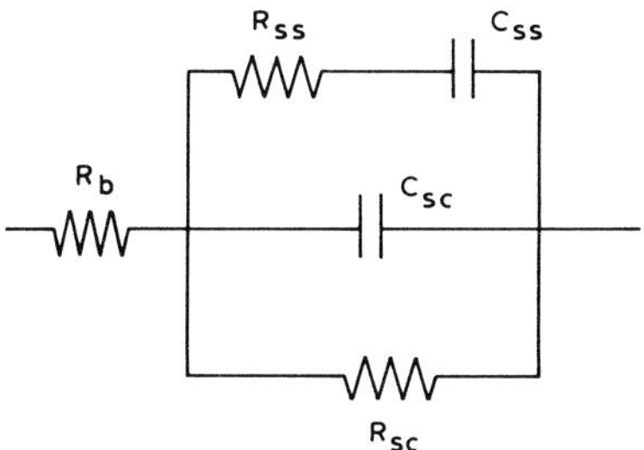

was made by Dare-Edwards et al. [6] on the basis of a theoretical analysis that made the simplifying assumption that the major contributions to the a.c. response of the electronic surface states could be modelled by a simple series resistor–capacitor network $R_{ss}C_{ss}$. This is equivalent to the assumption that, at any particular potential, one particular set of surface states dominates the a.c. response in the range studied. Under this circumstance, C_{sc} can be extracted by measuring the a.c. response over a frequency range of several decades. The results are shown in Fig. 5(a); it can be seen that C_{sc}^{-2} is linear with potential V over the range 0.1 to -1.0 V vs. SSE. Its slope is consistent with eqn. (1), giving a donor density close to that quoted by the manufacturer. Even above 0.1 V, there is some variation of C_{sc}, suggestive of further changes in the depletion layer. However, for p-GaAs, the situation is

References pp. 425–426

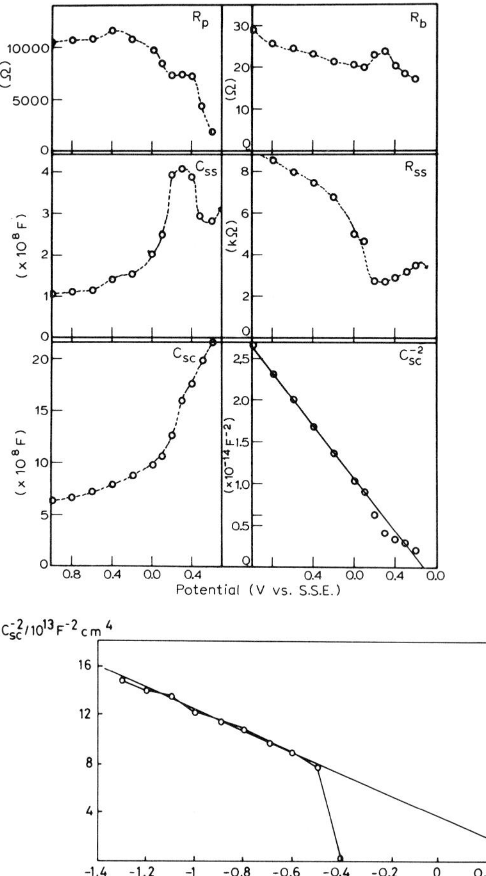

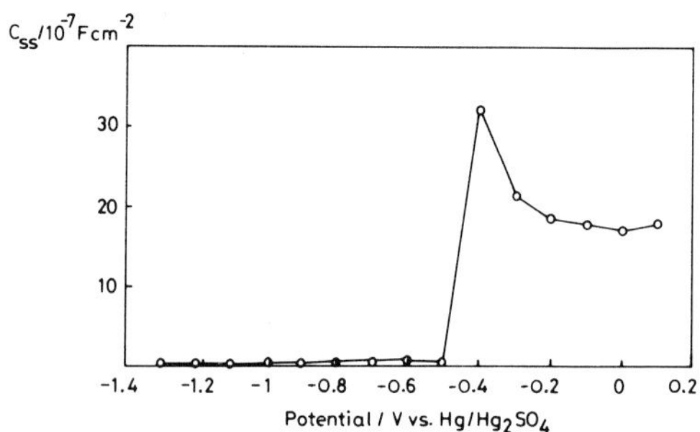

C_ss/10^-7 F cm^-2
30
20
10
-1.4 -1.2 -1 -0.8 -0.6 -0.4 -0.2 0 0.2
Potential / V vs. Hg/Hg2SO4

Fig. 5. (a) Analysis of the a.c. impedance data for p-GaP in 0.5 M H_2SO_4 according to the five-component model circuit given in the text. The final graph is a plot of C_{sc}^{-2} vs. potential and is clearly linear below 0.1 V vs. SSE. (b) Analysis of the a.c. impedance data for p-GaAs in 0.5 M H_2SO_4 according to the five-component model given in the text. Potential vs. the Hg/Hg_2SO_4 electrode.

rather different, as can be seen in Fig. 5(b). Once again, below a certain potential (here -0.5 V vs. Hg/Hg_2SO_4), C_{sc}^{-2} is linear with potential; above this, the values of both C_{ss} and C_{sc} increase dramatically, provided the semiconductor is allowed several minutes to equilibriate at each potential. However, the sudden increase in C_{sc} almost certainly does not reflect the fact that the potential drop in the depletion layer has fallen to zero, but rather that cross-talk between C_{sc} and C_{ss} arising from the inadequacy of the equivalent circuit, is responsible. In other words, a.c. techniques depend critically upon the interpretative framework and the commonly used frameworks are clearly not appropriate in the case cited above.

A second technique that has been used is the measurement of photovoltage. The basis of this technique is that, on illumination under open-circuit conditions, the potential distribution will be modified so as to eliminate the potential drop within the depletion layer. In fact, as has been demonstrated by Kautek and Gerischer [7], the theory of the photovoltage effect is far from straightforward, especially in the presence of surface states. The effect is a steady-state rather than equilbrium phenomenon; the potential distribution will change until the flux of holes to the surface is equal to the flux of electrons and the potential at which this occurs will depend on the recombination kinetics at the surface. Only when these kinetics are slow, i.e. when the surface states are slow and the main surface state equilbrium is with the redox couple in solution, is the technique likely to give results that can be interpreted within a consistent framework.

The serious problems with the techniques described above have led to a revival of interest in electroreflectance. The essence of this technique can be illustrated in Fig. 6, which shows the experimental configuration necessary to perform the measurement. Light from a suitable source, such as a high-pressure xenon lamp, is incident on a high-radiance monochromator and, after passage through this and an appropriate polariser, it is collimated and reflected from the surface of a semiconductor and detected by a photomultiplier. The basis of the effect is that the dielectric constant of the semiconductor is a function of the electric field: the reflectivity of the surface is, in turn, dependent on the dielectric constant through the well-known Fresnel relations [8] and so, in principle, we have a direct probe of the potential distribution. In spite of this apparently attractive feature, electroreflectance has been little used by electrochemists for two reasons. The first problem is experimental; the effect is extremely small and phase-sensitive techniques must be used: the d.c. potential must therefore have, superimposed on it, an a.c. component usually of ca. 20 mV rms or more. The photomultiplier signal thus contains both an a.c. and a d.c. component that can be separated electronically; the former is passed to a lock-in amplifier from which a measure of $\Delta R/\Delta V$ can be obtained and the latter gives a measure of R. To obtain a reasonable number of readings, say one every 1 nm in the UV/visible region, is, furthermore, extremely tedious unless micro-processor control is exercised over the entire experiment.

References pp. 425–426

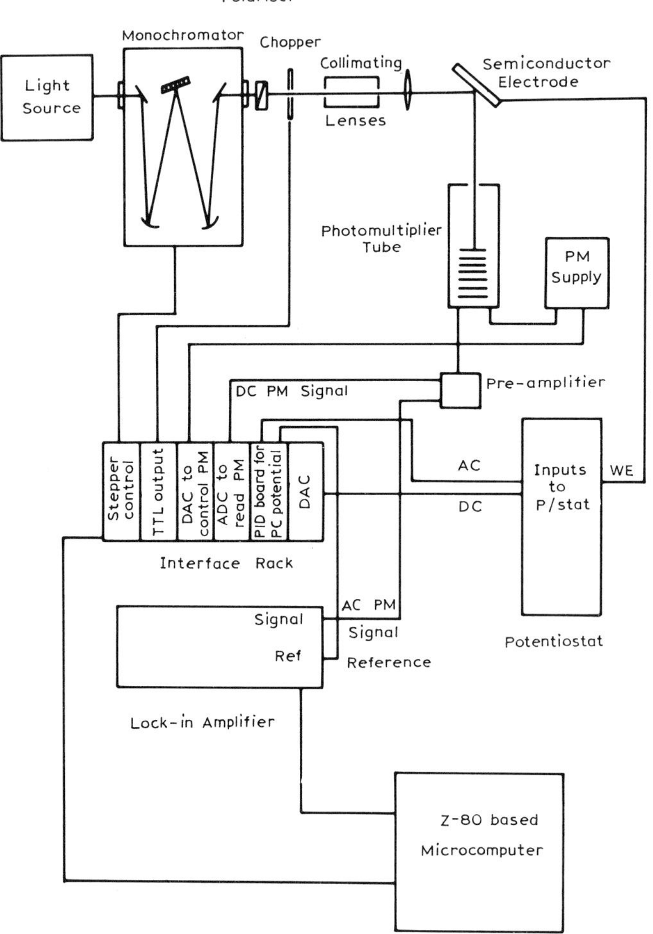

Fig. 6. Experimental configuration for electroreflectance spectroscopy.

The second reason for the lack of exploitation of electroreflectance, at least as a probe of electric field distribution, has been the rather limited development of the theory. Whilst the basic equations describing the effect have been known for some time [9], they are of considerable complexity and the simplifications that have been made, such as the Aspnes "third derivative modulation spectroscopy" [10] and the extended lineshape theories of Raccah et al. [11] have regions of applicability that may not include all commonly found experimental conditions. There are two difficulties with these theories. The first is that the electric field strengths found in practice in the semiconductors commonly used in electrochemical research may be too high for simple lineshape theories to be applicable; the essential requirement of such theories is that the lineshape should be independent of applied d.c. potential, a result not always found in practice, as discussed below. The second is that the presence of a depletion layer implies that the electric field is not constant within the semiconductor but decreases linearly with distance. This problem of inhomogeneous electric fields has been treated within the WKB approximation by Aspnes and Frova [12], but this approximation is again not applicable to many of the cases of electrochemical interest and the treatment of inhomogeneous fields in the general case is of considerable complexity.

In this chapter, the basic theory is first reviewed and the nature of the fundamental equations for the direct allowed transition given; extension to the direct forbidden case is reviewed elsewhere [13]. Following this, the types of behaviour encountered in practice for the III/V semiconductors will be given and placed in context.

2. Basic theoretical development

The basic theoretical difficulty of electroreflectance arises from the fact that the perturbation itself, the electric field $\mathscr{E}(x)$, does not display the periodicity of the lattice. Instead, it creates an overall force that accelerates the electron and destroys the translational symmetry of the Hamiltonian in the direction of the applied field. Other forms of modulation spectroscopy, such as thermoreflectance (in which the temperature of the sample is modulated) cause an isotropic differential dilatation of the lattice and lead to much simpler lineshapes. This is illustrated in Fig. 7. Figure 7(a) shows the valence and conduction bands for a simple direct transition; changing the temperature causes changes in the energies of the two bands, which may be represented as vertical displacements. The effect is to displace horizontally the values of ε_2, the imaginary part of the dielectric constant, and the difference then appears as a simple first-order spectrum as shown. Figure 7(b) shows the corresponding situation for low-field electroreflectance. In this case, the main effect is not on the band energies, but upon the selection rules governing the transition. The possibility of accelerating the electron by the electric field allows it to change its momentum during the transition; non-vertical transitions are, therefore, allowed and the type of complex lineshape that results from the difference spectrum is also shown.

In order to calculate the effect quantitatively, explicit expressions are needed for the complex dielectric constant, ε, in the presence and absence of an electric field; the effects of thermal broadening and inhomogeneities in the electric field may be included at a later point in the theoretical development.

In the absence of any electric field, the value of the imaginary part of the dielectric constant, $\varepsilon_2(\omega)$, for a transition from bands n to n' is

$$\varepsilon_2(\omega) = \frac{\pi e_0^2}{\varepsilon_0 m^2 \omega^2}\left(\frac{2}{(2\pi)^3}\right)\int_S \mathrm{d}S_{n'n} \int_{E_{n'n}} \mathrm{d}E \, \frac{|\boldsymbol{\varepsilon}\cdot\mathbf{P}_{n'n}(\mathbf{k})|^2}{|\boldsymbol{\nabla}_k E_{n'n}(\mathbf{k})|}\,\delta(E_{n'n} - \hbar\omega) \qquad (3)$$

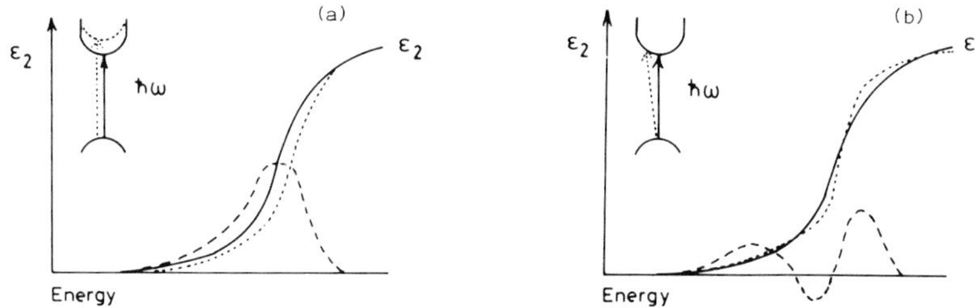

Fig. 7. Changes in ε_2 induced by (a) a first-order effect such as a temperature modulation and (b) a higher-order effect (in this case that of a change in the electric field). ———, Unperturbed; ····, perturbed; ----, difference.

Here, the integration is over the surface in **k**-space at which $E_{n'n} = \hbar\omega$. By evaluating the second integral explicitly, we obtain

$$\varepsilon_2(\omega) = \frac{\pi e_0^2}{\varepsilon_0 m^2 \omega^2}\left(\frac{2}{(2\pi)^3}\right)|\boldsymbol{\varepsilon}\cdot\mathbf{P}_{n'n}(\mathbf{k})|^2 \int_S \frac{\mathrm{d}S_{n'n}}{|\nabla_k[E_{n'}(\mathbf{k}) - E_n(\mathbf{k})]|}\bigg|_{E_{n'n}=\hbar\omega} \tag{4}$$

In these formulae, $\mathbf{P}_{n'n}(\mathbf{k})$ is the momentum matrix element between states in bands n,n' at wave-vector **k**, $\boldsymbol{\varepsilon}$ is a unit vector in the direction of the electric field vector of the incident electromagnetic radiation, $E_{n'n}(\mathbf{k}) \equiv E_{n'}(\mathbf{k}) - E_n(\mathbf{k})$ is the transition energy $\hbar\omega$, m is the effective mass of the electron and the other symbols have their ususal meanings. The integral in eqn. (4) is now explicitly over that surface in k-space at which $E_{nn'} = \hbar\omega$.

If the bands are parabolic (i.e. $E \simeq k^2$) and we are within the effective mass approximation (for which it is assumed that all electron–electron and electron–nuclear interactions can be absorbed into an effective electron mass), explicit expressions for $\varepsilon_2(\omega)$ can be obtained. These expressions will be most accurate near the so-called critical points, which are points in the energy spectrum at which a new transition has its onset or disappearance, or at which there is a change in the type of transition observed. The nature of a critical point is illustrated in Fig. 8, which shows a valence band and a wider conduction band. For this very simple system, there are four critical points, labelled A–D in the figure. The valence band (VB) and conduction band (CB) are characterised, within the effective-mass approximation, by effective masses m^*_{vi} and m^*_{ci}, where

$$\frac{1}{m^*_{vi}} = \frac{1}{\hbar^2}\frac{\partial^2 E_v}{\partial k_i^2} \quad i = x,y,z \tag{5}$$

and the joint effective mass μ_i is defined as

$$\frac{1}{\mu_i} = \frac{1}{m^*_{ci}} - \frac{1}{m^*_{vi}} \tag{6}$$

For the diagram above, consider transition A; for this, $m^*_{vi} < 0$, $m^*_{ci} > 0$ and so $\mu_i > 0$. If we look at Table 1, we see that A corresponds to an M_0 transition

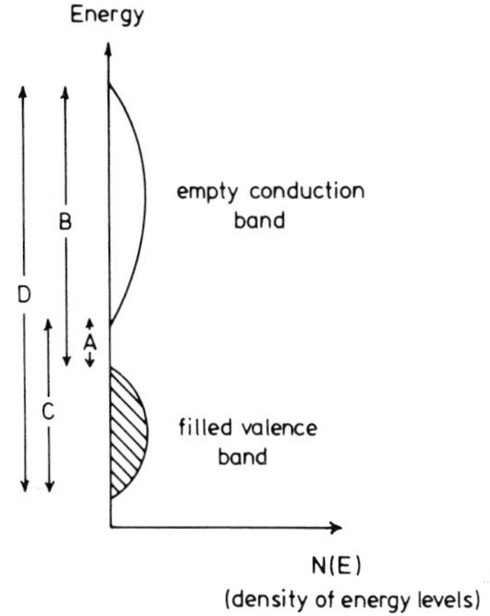

Fig. 8. Filled and empty energy levels in the solid and the corresponding critical points.

and the dielectric constant is obtained by explicit calculation in the region of M_0 assuming the parabolic approximation

$$E_{n'n}(\mathbf{k}) = E_{n'n}(\mathbf{k}_{cr}) + \frac{1}{2}\sum_{i,j} \left.\frac{\partial^2 E_{n'n}}{\partial k_i \partial k_j}\right|_{\mathbf{k}=\mathbf{k}_{cr}} (k_i - k_{cri})(k_j - k_{crj}) \tag{7}$$

where $\mathbf{k}_{cr}$ is the critical $\mathbf{k}$ value at which the transition occurs and $i, j \equiv x,y,z$.

The value of ε_1, the real part of the dielectric constant, may be obtained from the Kramers–Kronig transformation. The basis of this approach is that it can be shown that $\tilde{\varepsilon}(\omega) - 1 \equiv \varepsilon_1(\omega) - 1 + i\varepsilon_2(\omega)$ is analytic in the upper half of the complex plane; from Cauchy's theorem, we have, therefore

$$\tilde{\varepsilon}(\omega_0) - 1 = \frac{1}{2\pi i} \oint_{C_2} \frac{[\tilde{\varepsilon}(\omega) - 1]}{\omega - \omega_0} \mathrm{d}\omega \tag{8}$$

where C_2 is a circular contour which includes the point ω_0. We can show immediately from this that

$$\varepsilon_1(\omega) - 1 = \frac{1}{\pi} \mathscr{P} \int_{-\infty}^{\infty} \frac{\varepsilon_2(\chi)\mathrm{d}\chi}{\chi - \omega} \tag{9}$$

where $\mathscr{P}$ is the principal value of the integral.

In the presence of an electric field, $\mathscr{E} \equiv (\mathscr{E}_x, \mathscr{E}_y, \mathscr{E}_z)$, the imaginary part of the dielectric constant, must be recalculated. The electric field introduces a term linear in $\mathbf{r}$ into the Schroedinger equation. On resolving the equation, we find

$$\varepsilon_2(\omega, \mathscr{E}) = \frac{\pi e_0^2}{\varepsilon_0 m^2 \omega^2} |\boldsymbol{\varepsilon}\cdot\mathbf{P}|^2 \left(\frac{2}{(2\pi)^3}\right) \int \mathrm{d}^3 k \frac{1}{\hbar\Omega} \mathrm{Ai}\left(\frac{\hbar\omega_g + \sum_i (\hbar^2 k_i^2/2\mu_i) - \hbar\omega}{\hbar\Omega}\right) \tag{10}$$

TABLE 1

Behaviour of ε_1 and ε_2 at the critical points M_0–M_3

Critical point	Condition on effective mass	$\varepsilon_1(\omega)$	$\varepsilon_2(\omega)$
M_0	$\mu_x, \mu_y, \mu_z > 0$	$1 + (B/\omega^2)\,[2(\omega_g)^{1/2} - (\omega_g + \omega)^{1/2} - (\omega_g - \omega)^{1/2}\,H(\omega_g - \omega)]$	$(B/\omega^2)\,(\omega - \omega_g)^{1/2}\,H(\omega - \omega_g)$
M_1	$\mu_x, \mu_y > 0$ $\mu_z < 0$	$1 - (B/\omega^2)\,(\omega - \omega_g)^{1/2}\,H(\omega - \omega_g)$	$C_1 - (B/\omega^2)\,(\omega - \omega_g)^{1/2}\,H(\omega_g - \omega)$
M_2	$\mu_x, \mu_y < 0$ $\mu_z > 0$	$1 - (B/\omega^2)\,[2(\omega_g)^{1/2} - (\omega_g + \omega)^{1/2} - (\omega_g - \omega)^{1/2}\,H(\omega_g - \omega)]$	$C_1 - (B/\omega^2)\,(\omega - \omega_g)^{1/2}\,H(\omega - \omega_g)$
M_3	$\mu_x, \mu_y, \mu_z < 0$	$1 + (B/\omega^2)\,(\omega - \omega_g)^{1/2}\,H(\omega - \omega_g)$	$(B/\omega^2)\,(\omega_g - \omega)^{1/2}\,H(\omega_g - \omega)$

TABLE 2

The analytical form of the electroreflectance response at the critical points

Critical point	Sign of $\hbar\theta$	$\Delta\varepsilon_1(\omega_1\ \mathscr{E}) = (B/\omega^2)\lvert\theta\rvert^{1/2}\ \times$	$\Delta\varepsilon_2(\omega,\ \mathscr{E}) = (B/\omega^2)\lvert\theta\rvert^{1/2}\ \times$
M_0	$\hbar\theta > 0$	G(x)	F(x)
M_1	$\hbar\theta_{\parallel} < 0$	G(x)	− F(x)
	$\hbar\theta_{\perp} > 0$	− F(x)	G(x)
M_2	$\hbar\theta_{\parallel} > 0$	− G(x)	− F(x)
	$\hbar\theta_{\perp} < 0$	F(x)	G(x)
M_3	$\hbar\theta < 0$	− G(x)	F(x)

ωhere $x = (\hbar\omega_g - \hbar\omega)/\hbar\theta$; $\theta = [\theta_x^3 + \theta_y^3 + \theta_z^3]^{1/3} = [e_0^2\lvert\mathscr{E}\rvert^2/2\mu_{\parallel}\hbar]^{1/3}$

$\Delta\varepsilon_1 = \varepsilon_1(\omega,\mathscr{E}) - \varepsilon_1(\omega,0)$

$\Delta\varepsilon_2 = \varepsilon_2(\omega,\mathscr{E}) - \varepsilon_2(\omega,0)$

$F(x) = \pi[\mathrm{Ai}'^2(x) - x\mathrm{Ai}^2(x)] - (-x)^{1/2}H(-x)$

$G(x) = \pi[\mathrm{Ai}'(x)\mathrm{Bi}'(x) - x\mathrm{Ai}(x)\,\mathrm{Bi}(x)] + (x)^{1/2}H(x)$

where Ai (x) is the Airy function regular at the origin and at $+\infty$ and the important parameter Ω is a measure of the electric field. In fact, $\hbar\Omega = 2^{-2/3}\hbar\theta$ and

$$\hbar\theta = \left\{\left(\frac{\hbar^2 e_0^2}{2}\right)\left(\frac{\mathscr{E}_x^2}{\mu_x} + \frac{\mathscr{E}_y^2}{\mu_y} + \frac{\mathscr{E}_z^2}{\mu_z}\right)\right\}^{1/3} \tag{11}$$

In the limit, as $\mathscr{E} \to 0$

$$\frac{1}{\hbar\Omega}\,\mathrm{Ai}\left[\frac{E - \hbar\omega}{\hbar\Omega}\right] \to \delta(E - \hbar\omega) \tag{12}$$

and eqn. (4) is recovered. For parabolic bands, eqn. (5) may, once again, be explicitly evaluated to give the results in Table 2.

These are the key results of the theory, but they mask two difficulties: the first is that no thermal broadening has been included in the basic equations and the second is that the equations hide formidable numerical problems that can only be surmounted with some difficulty. Incorporation of thermal broadening into the equations is straightforward and relies on finding suitable functions that are analytic in the upper complex plane. The reason for this is that thermal broadening may be expressed in integral form as

$$\varepsilon_1(\omega,\Gamma) - 1 + i\varepsilon_2(\omega,\Gamma) = \frac{1}{\pi}\int_{-\infty}^{\infty}\frac{\varepsilon_2(\omega')\mathrm{d}\omega'}{(\omega' - \omega) - i\Gamma/\hbar} \tag{13}$$

where Γ is a thermal broadening energy. This integral can be transformed by recalling the Kramers–Kronig equations; in particular, we recall that $\tilde{\varepsilon}(\omega) - 1$ is analytic in the upper complex plans and we find

$$\varepsilon_1(\omega,\Gamma) - 1 = \frac{1}{\pi}\,\mathrm{Re}\int_{-\infty}^{\infty}\frac{\mathrm{d}\omega'\cdot\varepsilon_2(\omega')}{(\omega' - \omega) - i\Gamma/\hbar} = \frac{\Gamma}{\pi\hbar}\int_{-\infty}^{\infty}\frac{[\varepsilon_1(\chi) - 1]\,\mathrm{d}\chi}{(\chi - \omega)^2 + \Gamma^2/\hbar^2} \tag{14}$$

and

$$\varepsilon_2(\omega, \Gamma) = \frac{1}{\pi} \mathrm{Im} \int_{-\infty}^{\infty} \frac{\mathrm{d}\omega' \cdot \varepsilon_2(\omega')}{(\omega' - \omega) - i\Gamma/\hbar} = \frac{\Gamma}{\pi\hbar} \int_{-\infty}^{\infty} \frac{\varepsilon_2(\chi)\mathrm{d}\chi}{(\chi - \omega)^2 + \Gamma^2/\hbar} \tag{15}$$

whence

$$\tilde{\varepsilon}(\omega, \Gamma) - 1 = \frac{\Gamma}{\pi\hbar} \int_{-\infty}^{\infty} \frac{[\tilde{\varepsilon}(\chi) - 1]\mathrm{d}\chi}{(\chi - \omega)^2 + \Gamma^2/\hbar^2} \tag{16}$$

If we extend the integration to the upper complex plane and integrate about the infinite semicircle, the analyticity of $\tilde{\varepsilon}(\chi) - 1$ ensures that this contribution to the total integral must vanish and we have, finally

$$\varepsilon(\omega, \Gamma) - 1 = \frac{\Gamma}{\pi\hbar} \oint \frac{[\varepsilon(\chi) - 1]\mathrm{d}\chi}{(\chi - \omega)^2 + \Gamma^2/\hbar} \tag{17}$$

This contour can be deformed to a residue about $\chi = \omega + i\Gamma/\hbar$ since the analyticity of $\tilde{\varepsilon}(\chi) - 1$ guarantees that this is the only residue. Whence

$$\tilde{\varepsilon}(\omega, \Gamma) - 1 = \tilde{\varepsilon}(\omega + i\Gamma) - 1 \tag{18}$$

i.e.

$$\begin{aligned} \varepsilon_1(\omega, \Gamma) &= \mathrm{Re}\{\tilde{\varepsilon}(\omega + i\Gamma)\} \\ \varepsilon_2(\omega, \Gamma) &= \mathrm{Im}\{\tilde{\varepsilon}(\omega + i\Gamma)\} \end{aligned} \tag{19}$$

Consider, explicitly, the case of an M_0 transition. It is clear that the function

$$\varepsilon_1 - 1 + i\varepsilon_2 = \frac{B[2(\omega_g)^{1/2} - (\omega + \omega_g)^{1/2} + i(\omega - \omega_g)^{1/2}]}{\omega^2} \tag{20}$$

is well behaved in the upper complex plane. Putting $\omega \to \omega_c + i\Gamma$, we find

$$\varepsilon_1(\omega, \Gamma) =$$

$$1 + \mathrm{Re}\left\{\frac{B\{2(\omega_g)^{1/2} - (\omega_c + i\Gamma + \omega_g)^{1/2} + (\omega_c + i\Gamma - \omega_g)^{1/2}\}}{(\omega_c + i\Gamma)^2}\right\} \tag{21}$$

$$\varepsilon_2(\omega, \Gamma) =$$

$$\mathrm{Im}\left\{\frac{B\{2(\omega_g)^{1/2} - (\omega_c + i\Gamma + \omega_g)^{1/2} + (\omega_c + i\Gamma - \omega_g)^{1/2}\}}{(\omega_c + i\Gamma)^2}\right\} \tag{22}$$

and the others can be evaluated similarly.

A similar analysis may be carried through in the case where an electric field is present. The analytic function in this case takes the form $\mathrm{F} + i\mathrm{G}$ where F and G are defined in Table 2. Explicitly

$$\mathrm{F} + i\mathrm{G} = \pi[\mathrm{Ai}'^2(x) - x\mathrm{Ai}^2(x) + i\mathrm{Ai}'(x)\mathrm{Bi}'(x) - ix\mathrm{Ai}(x)\mathrm{Bi}(x) + ix^{1/2}] \tag{23}$$

If now $x \rightarrow [(\omega_g - \omega)/\theta] + i\Gamma/\theta$, we recover thermally broadened F and G values to obtain

$$\varepsilon_2(\omega, \mathscr{E}, \Gamma) = \varepsilon_2(\omega, \Gamma) + \text{Im}\left\{\frac{B|\theta|^{1/2}}{(\omega + i\Gamma)^2}(\text{F} + i\text{G})\right\} \tag{24}$$

with a similar formula for $\varepsilon_1(\omega, \mathscr{E}, \Gamma)$.

The numerical problems arise because, individually, either Ai, Ai′, Bi, or Bi′ may become very large in the upper complex plane. It therefore becomes essential to work with F + iG if a well-behaved function is to be evaluated. The asymptotic forms of the Airy functions permit us to evaluate F + iG for various ranges of the argument x, though the complexity of the resultant formulae inevitably impels us to use computational techniques.

We have now established the dependence of the dielectric function $\tilde{\varepsilon}$, and hence of the complex refractive index $\tilde{n} \equiv n + ik \equiv (\tilde{\varepsilon})^{1/2} \equiv (\varepsilon_1 + i\varepsilon_2)^{1/2}$ on the electric field and there remains the problem of the variation of the electric field within the semiconductor. The plane wave, propagating through a medium of varying refractive index, is a problem considered in electromagnetic theory and analytical solutions are far from easy to obtain, even when the variation in ε or n is a simple function of x. The complex formulae given above rule out any hope of an analytical solution and so a different approach must be adopted. The earlier theoretical work in this area was presented by Aspnes and Frova [12] who derived results in closed integral from for the case of normal incidence. Extension to non-normal incidence is not trivial since propagation now occurs both parallel and perpendicular to the surface. To avoid this difficulty, another approach has been used. The region of inhomogeneity may be divided into layers, within which the electric field may be imagined to be a constant. Reflection may then be described using a simple matrix method [14]. To show how this works, consider first a pure substrate within which ε is constant. If electromagnetic radiation is incident through medium 1 on this substrate, which we label medium 2, we may write

$$\begin{pmatrix} E_1^+ \\ E_1^- \end{pmatrix} = |\mathbf{N}|_1 \begin{pmatrix} E_2^+ \\ E_2^- \end{pmatrix} \tag{25}$$

where E_1^+, E_1^- are the incident and reflected magnitudes of the electric vector in medium 1, and E_2^+, E_2^- are the same for medium 2. The matrix $|\mathbf{N}|_1$ is given by

$$|\mathbf{N}|_1 = \frac{1}{t_{12}} \begin{pmatrix} 1 & r_{12} \\ r_{12} & 1 \end{pmatrix} \tag{26}$$

where t_{12} is the transmission coefficient and r_{12} the reflection coefficient. Without loss of generality at this point, we can consider just plane polarised incident light: the theory of optics shows that, for an arbitrary angle between the electric vector of the incident light and the plane of reflection, the reflected light will be elliptically polarised, but if the electric vector is *parallel* to the plane of reflection (denoted by p) or *perpendicular* (denoted by s), then the reflected light is also plane polarised in the same direction. For these two polarisation directions, the reflection and transmission coefficients take particularly simple forms

$$r_{12}^{\mathrm{p}} = \frac{n_1 \cos\phi_2 - \tilde{n}_2 \cos\phi_1}{\tilde{n}_2 \cos\phi_1 + n_1 \cos\phi_2} \qquad r_{12}^{\mathrm{s}} = \frac{\tilde{n}_2 \cos\phi_2 - n_1 \cos\phi_1}{n_1 \cos\phi_1 + \tilde{n}_2 \cos\phi_2} \tag{27}$$

$$t_{12}^{\mathrm{p}} = \frac{2n_1 \cos\phi_1}{n_1 \cos\phi_2 + \tilde{n}_2 \cos\phi_1} \qquad t_{12}^{\mathrm{s}} = \frac{2n_1 \cos\phi_1}{n_1 \cos\phi_1 + \tilde{n}_2 \cos\phi_2} \tag{28}$$

where ϕ_1 is the angle of incidence (i.e. the angle between the Poynting vector of the incident light and the perpendicular to the surface) and ϕ_2 is the angle of refraction.

Evidently, for an infinite slab of medium 2, $E_2^- \equiv 0$ so

$$\frac{E_1^-}{E_1^+} = r_{12} \tag{29}$$

If a film of thickness d and complex refractive index $\tilde{n}_2$ is now placed on a substrate of refractive index $\tilde{n}_3$, we find

$$\begin{pmatrix} E_1^+ \\ E_1^- \end{pmatrix} = |\mathbf{N}|_1 \, |\mathbf{M}| \, |\mathbf{N}|_2 \begin{pmatrix} E_3^+ \\ E_3^- \end{pmatrix} \tag{30}$$

where

$$|\mathbf{M}| = \begin{pmatrix} e^{iD_2} & 0 \\ 0 & e^{-iD_2} \end{pmatrix} \qquad D_2 = \frac{2\pi \tilde{n}_2 d \cos\phi_2}{\lambda} \tag{31}$$

and, in general, if there are n films

$$\begin{pmatrix} E_1^+ \\ E_1^- \end{pmatrix} = |\mathbf{N}|_1 \, |\mathbf{M}|_1 \, |\mathbf{N}|_2 \, |\mathbf{M}|_2 \, . \, . \, |\mathbf{N}|_j \, |\mathbf{M}|_j \, . \, . \, |\mathbf{N}|_n \begin{pmatrix} E_{n+1}^+ \\ E_{n+1}^- \end{pmatrix} \tag{32}$$

where

$$|\mathbf{N}|_j = \frac{1}{t_{j,j+1}} \begin{pmatrix} 1 & r_{j,j+1} \\ r_{j,j+1} & 1 \end{pmatrix}$$

$$|\mathbf{M}|_j = \begin{pmatrix} \exp(iD_{j+1}) & 0 \\ 0 & \exp(-iD_{j+1}) \end{pmatrix} \tag{33}$$

$$D_{j+1} = \frac{2\pi \tilde{n}_{j+1} d_{j+1} \cos\phi_{j+1}}{\lambda}$$

Given that E_{n+1}^{-} is set equal to zero (i.e. the substrate is again assumed to be semi-infinite), it is possible to establish the *overall* reflection coefficients

$$r_T^p = \frac{E_{1p}^-}{E_{1p}^+}$$
$$r_T^s = \frac{E_{1s}^-}{E_{1s}^+} \quad (34)$$

Working at 45°, the total reflected intensity is then given by

$$\frac{I^-}{I^+} = \tfrac{1}{2}(|r_T^s|^2 + |r_T^p|^2) \quad (35)$$

Evidently, this theory allows us to incorporate an arbitrarily large number of films, the number effectively being limited by the computing power available. As we shall see below, a description of the classical depletion layer must be given in terms of at least ten such films.

With at least two strands to the theoretical development, the dependence of $\tilde{\varepsilon}$ on $\mathscr{E}$ and thence on x (the distance into the semiconductor), it is possible to derive theoretical electroreflectance spectra which can be compared with the experimental spectra in an effort to understand the latter. Clearly, some parameters must be chosen in order that such computations may be made and it is instructive to consider these in some detail. These parameters fall into two categories: those that are invariant and well-defined for a given semiconductor and experiment and those that are, to some extent, arbitrary. In the first category come the dielectric constant of the electrolyte, the nature of the transition, and, for the materials considered here, the critical energy for the interband transition being studied (since this is known with some accuracy and only rather weakly dependent on such factors as donor density, majority carrier charge polarity, and the joint effective mass (again well-known for the III/V materials). In addition, we can fix such experimental variables as the polarisation of the incident beam and its angle of incidence, the spectrometer wavelength resolution, and the d.c. electrode potential.

Parameters that are, to some degree, arbitrary include the Debye length, L_D, and the thermal broadening function $\Gamma(\omega)$. The Debye length can be obtained, rather approximately, from the Mott–Schottky slope provided the semiconductor exhibits any form of classical behaviour. However, surface roughness factors as high as two have been suggested, even for well-polished and etched surfaces and the derived L_D values are, therefore, rather inaccurate. The donor density N_D can also be derived from conductivity and Hall measurements; again, these suffer from some uncertainty in the r factor, though this technique is inherently more reliable than the electrochemical one since fewer assumptions are made. Values of N_D from the Hall experiment do allow us to estimate L_D fairly well and it follows that the range over which L_D can be allowed to vary is quite limited. The main variables remain-

References pp. 425–426

ing are $\Gamma(\omega)$ and the matrix element B defined above. The former is normally assumed to be a weak function of ω and is discussed further below, whereas the latter can be obtained from the known optical spectrum for the first direct transition. For subsequent transitions, it is more difficult to strip out other absorption processes and B must be estimated. Fortunately, within a fairly wide range, the theoretical electroreflectance spectra are not especially sensitive to B since $\Delta R/R$ is being measured and the influence of B tends to cancel to first order.

The computational difficulties encountered in the theory above have led many workers to inquire whether or not simplifications may be found that would allow us to a simple physical insight into the spectra. Aspnes and later Raccah and co-workers were able to develop "low field" theories of electroreflectance that have proved to be of great value, particularly in spectroscopic studies, and before considering the results of the more complete theory outlined above, we will turn to a consideration of these theories and discuss some applications.

3. Low-field theories

The expression in eqn. (24) for $\varepsilon(\omega, \mathscr{E}, \Gamma)$ may be further transformed by making use of integral representations of the various Airy functions. After a considerable degree of manipulation, it is found that, if $\hbar\Gamma \geqslant 3|\hbar\Omega|$, then to a good approximation:

$$\Delta\varepsilon(\omega, \mathscr{E}, \Gamma) \simeq \frac{(\Omega)^3}{3(\omega^2)} \frac{\partial^3}{\partial\omega^3} (\omega^2\tilde{\varepsilon}(\omega, \Gamma)) \tag{36}$$

and

$$\frac{\Delta R}{R} \simeq \mathrm{Re}(Ce^{i\theta}\Delta\tilde{\varepsilon}) \tag{37}$$

where C is a real constant that depends on the joint effective masses of the carriers, the optical absorption matrix elements in the absence of the field and, through Ω^3, on the square of the electric field strength $\mathscr{E}$; θ is a phase-factor that depends in a complex way on the inhomogeneity of the electric field and the optical properties of the crystal. The condition $\hbar\Gamma \geqslant 3|\hbar\Omega|$ can be tested by considering the case of p-GaAs referred to below. For a donor density of $10^{17}\,\mathrm{cm}^{-3}$ (which is fairly typical of those encountered in electrochemical studies) the electric field at the surface, which is given by

$$\mathscr{E}_s = \frac{kT/e_0, \sqrt{2}}{L_D} \left(\frac{|V - V_{fb}|e_0}{kT}\right)^{1/2}$$

where $L_D = (\varepsilon_0\varepsilon_{sc}kT/e_0^2N_D)^{1/2}$ will be of the order of $2 \times 10^5\,\mathrm{V\,cm}^{-1}$ for a bias of 1 V from flat band. The joint effective mass in this material is of the order of $0.05\,m_e$ [15], so $\hbar\Omega$ takes a value of ca. 40 meV. This is comparable with the

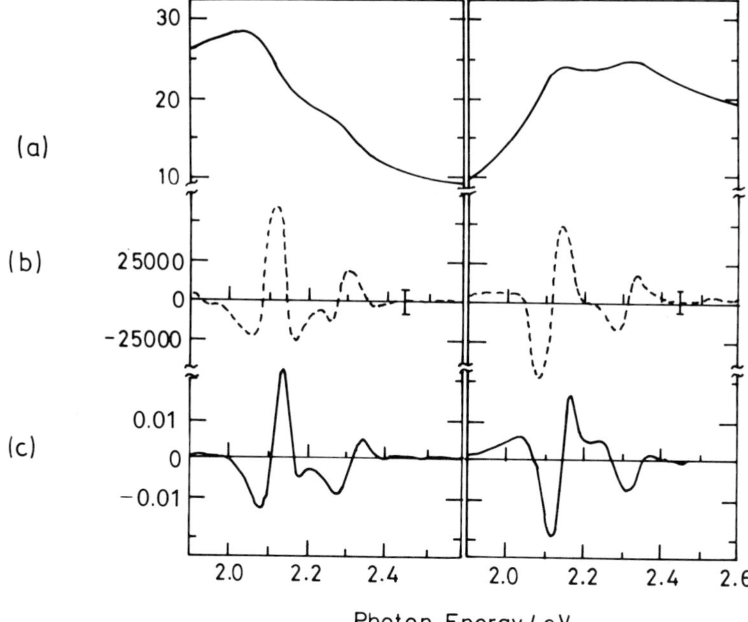

Fig. 9. (a) Real and imaginary parts of ε for germanium as measured by spectroscopic ellipsometry. (b) $E^{-2}(\partial^3[E^2\varepsilon]/\partial E^3)$, where E is the photon energy $\hbar\omega$, calculated from curve (a) numerically. (c) Electrotreflectance spectrum obtained at $\mathscr{E} = 38\,\mathrm{kV\,cm^{-1}}$, analysed to yield $\Delta\varepsilon_1$ and $\Delta\varepsilon_2$ (after ref. 10).

values of $\hbar\Gamma$ encountered by direct use of the full Franz–Keldysh theory and, in this case, the low-field theories will not hold. Only for donor densities below $10^{16}\,\mathrm{cm^{-3}}$ are low-field theories likely to be strictly applicable for this material. By contrast, for TiO_2, with a donor density of $10^{18}\,\mathrm{cm^{-3}}$ but an effective mass of ca. $10m_e$, $\hbar\Omega$ will have a value of ca. 6 meV, sufficiently far below the likely value of kT for the third derivative theory to hold. It follows that great care must be taken with most of the materials reported in the literature, which are likely to hover close to the borderline of applicability of the low-field limit.

If eqn. (37) is valid, two predictions can be made immediately: the first is that the lineshape should depend solely on the third derivative of the dielectric function of the semiconductor. This has been verified for i-Ge as shown in Fig. 9: here, the dielectric function determined from spectroscopic ellipsometry is differentiated numerically three times and the results compared with the electroreflectance spectrum. The second consequence of eqn. (37) is that the lineshape depends quadratically on $\mathscr{E}$ and, for a classical depletion layer, this means, in turn, that the electroreflectance spectrum should be *independent* of the d.c. potential provided θ does not alter.

To summarise, in the low-field limit [16], we expect that

$$\frac{\Delta R}{R} = -\frac{2e_0 N_D \Delta V}{\varepsilon_0 \varepsilon_{sc}} \mathscr{L}_n(\hbar\omega)$$

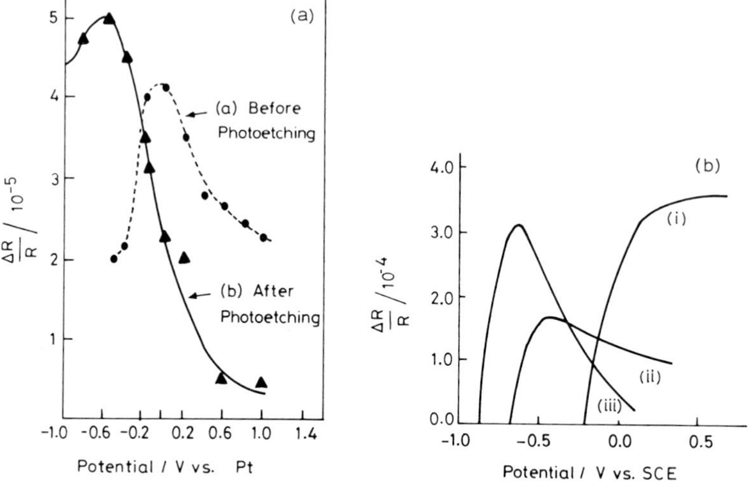

Fig. 10. Variation of the experimental ER signal with potential at ●, 1.8 eV (before photoetching) and ▲, 2.0 eV (after photoetching) for $CdIn_2Se_4$ in polysulphide solution. Modulation voltage is 0.2 V p–p. (b) Dependence of the ER signal at 4.0 eV on applied potential for n-TiO_2 ($N_D \simeq 8 \times 10^{19}$) in contact with (i) 0.23 M Na_2SO_4/H_2SO_4 at pH 3.4 ($V_{fb} = -0.22$ V vs. SCE from Mott–Schottky analysis), (ii) 0.23 M Na_2SO_4/NaOH at pH 11.3 ($V_{fb} = -0.70$ V vs SCE from Mott–Schottky analysis), and (iii) 0.23 M Na_2SO_4/NaOH + 0.1 M H_2O_2 (where V_{fb} shifts, from MS analysis, to −0.9 V vs. SCE owing to a specific adsorption of HO_2^- : frequency = 65 Hz, V_{ac} = 0.2 V rms).

where $\mathscr{L}_n(\hbar\omega)$ takes the form

$$\mathscr{L}_n(\hbar\omega) = \mathrm{Re}[Ce^{i\theta}(\hbar\omega - \hbar\omega_g + i\Gamma)^{-n}]$$

and we have assumed parabolic bands. The value of n depends on the dimensionality of the critical point and, for a three-dimensional point, has the value 5/2.

This approach has been used by Tomkiewicz et al. [17] to rationalise the electroreflectance spectrum of $CdIn_2Se_4$; unfortunately, the crystal studied in this report disintegrated before measurement of the donor intensity could be carried out to verify that the experimental conditions were such that the low-field theories could reasonably be expected to hold. Nevertheless, the *intensity* of the electroreflectance peak showed a marked dependence on potential, decreasing by a factor of ten over a 1 V range as shown in Fig. 10(a).

It has also been suggested by Salvador and co-workers [18] who have reported the variation of the $\Gamma_{5v}^+ \rightarrow \Gamma_{1,4c}^+$ was used to track the variation of $\Delta R/R$ with potential. The donor density of the TiO_2 sample is high, judging by the Mott-Schottky slope, and can be estimated from the data in the paper to be ca. $8 \times 10^{19}\,cm^{-3}$ so that $\hbar\Omega$ will have a value in the order of 25 meV. However, it appears that the shape of the ER spectrum is rather insensitive to applied potential, suggesting a large value of $\hbar\Gamma$ at RT ($\geqslant$ 100 meV). The variation of $\Delta R/R$ at 4.0 eV with potential is shown in Fig. 10(b) and two factors are of interest. Firstly, the intensity of $\Delta R/R$ becomes very small at a potential corresponding closely to the flat-band potential of TiO_2 under the conditions specified. This would be expected provided that the potential could vary classically either side of V_{fb}, since the dependence of $\Delta R/R$ on $\mathscr{E}^2$ causes a complete cancellation in the signal at flat band [20], as the phase of the ER signal will change by 180° on traversing the flat-band condition. More interesting is the fact that, in basic solutions, instead of reaching a saturation value rapidly (as expected in the exhaustion region for $|V_{applied} - V_{fb}| > \sqrt{2}\,V_{rms}$ and as is, indeed, seen in acid solution), the ER signal goes through a maximum and then decreases in magnitude. It is clear that, as for the results on $CdIn_2Se_4$, there is a fraction of the potential that is dropped within the depletion layer and a fraction droppped in the Helmholtz layer. Furthermore, this fraction is clearly sensitive to electrolyte and must, therefore, be associated with surface states that are *fast* at the frequency employed (65 Hz). The fraction of charge dropped in the Helmholtz layer for a potential change ΔV is clearly $(e_0 \Delta N_{ss}/C_H \Delta V)\Delta V$ and so

$$\frac{\Delta R}{R} = -\frac{2e_0 N_D}{\varepsilon_0 \varepsilon_{sc}} \mathscr{L}_n(\hbar\omega)\left[1 - \frac{e_0 \Delta N_{ss}}{C_H \Delta V}\right]\Delta V$$

a result obtained by Tomkiewicz et al. [17].

4. Applications of the Franz–Keldysh theory to more heavily doped III/V semiconductors

In this section, we will present results mainly on the III/V materials since theoretical interpretation is at its most advanced for these semiconductors. We shall concentrate primarily on the first and second direct allowed M_0 transitions of p-GaP and of p- and n-GaAs whose energies are ca. 2.78 and 3.87 eV for GaP and 1.43 and 3.04 eV for GaAs. To place these transitions in electronic perspective, the band diagrams of GaP and GaAs are presented in Figs. 11 and 12, respectively.

4.1 THE EFFECT OF INHOMOGENEOUS ELECTRIC FIELD

The effect of allowing the electric field to vary inside the semiconductor is shown most graphically in Fig. 13. This shows a theoretical calculation of the E_1 and $E_1 + \Delta_1$ transitions of p-GaP at ca. 320 nm for the case (i) that the field constant in the semiconductor at a level equal to that calculated using the classical Mott–Schottky model at the surface, and extending a distance W into the semiconductor, and (ii) that the electric field varies linearly in the semiconductor over a distance W in the classical model, using twelve layers of differing electric field as discussed above to represent the depletion layer. This sample was very heavily doped ($N_A \simeq 2 \times 10^{18}\,cm^{-3}$) and this leads, with the known joint effective mass, to a large value of $\hbar\Omega$ and hence

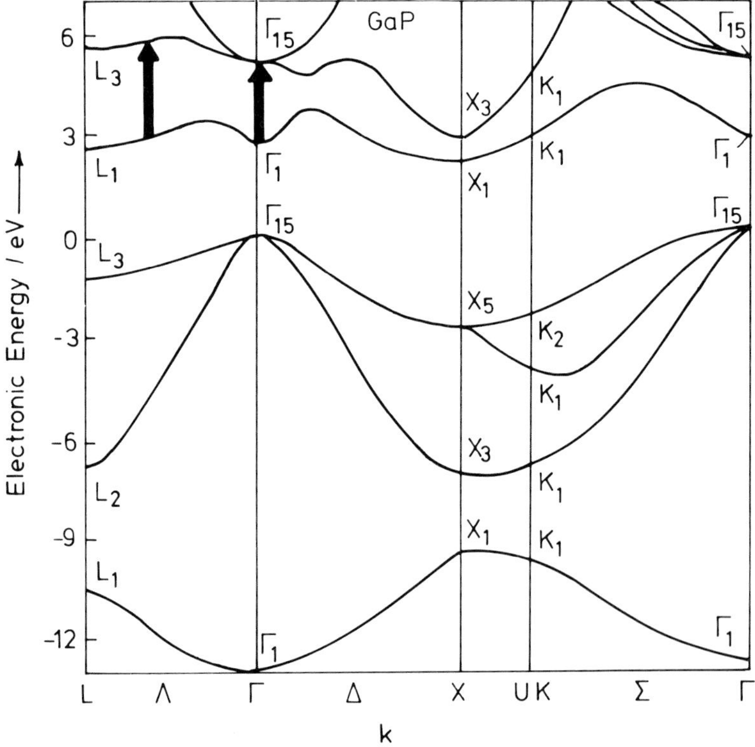

Fig. 11. Band structure of GaP showing the first ($\Gamma_{8v} \rightarrow \Gamma_{6c}$) and second ($\Lambda_{4,5v}$, $\Lambda_{6v} \rightarrow \Lambda_{6c}$) sets of transitions at $\simeq 2.78$ and 3.87 eV, respectively. These transitions are denoted as (E_0, E_0 and Δ_0) and (E_1, E_1 and Δ_1).

to the expectation of large-ampltitude Franz–Keldysh oscillations within the intermediate-field approximation. These large oscillations can clearly be seen in the calculated spectrum of Fig. 13(a), but they are strongly damped in Fig. 13(b) owing to interference effects from the different layers. In addition, there is an appreciable broadening of the spectrum as the number of layers increases, again traceable to interference effects. In this particular case, the calculated spectrum is essentially independent of the number of layers, but below this number significant changes in the spectrum do appear. This is illustrated more clearly in Fig. 14, which shows the changes in the calculated E_1 and $E_1 + \Delta_1$ spectra of p-GaAs as the number of layers is increased from one to twelve. Gross changes are evident, especially as the number of layers becomes very small.

4.2 THE EFFECT OF d.c. POTENTIAL

The effect of varying the potential on the electroreflectance spectra of typically doped semiconductors is shown in the Figs. 15–18. In all four cases, marked changes are predicted, especially near flat-band, sufficient by itself

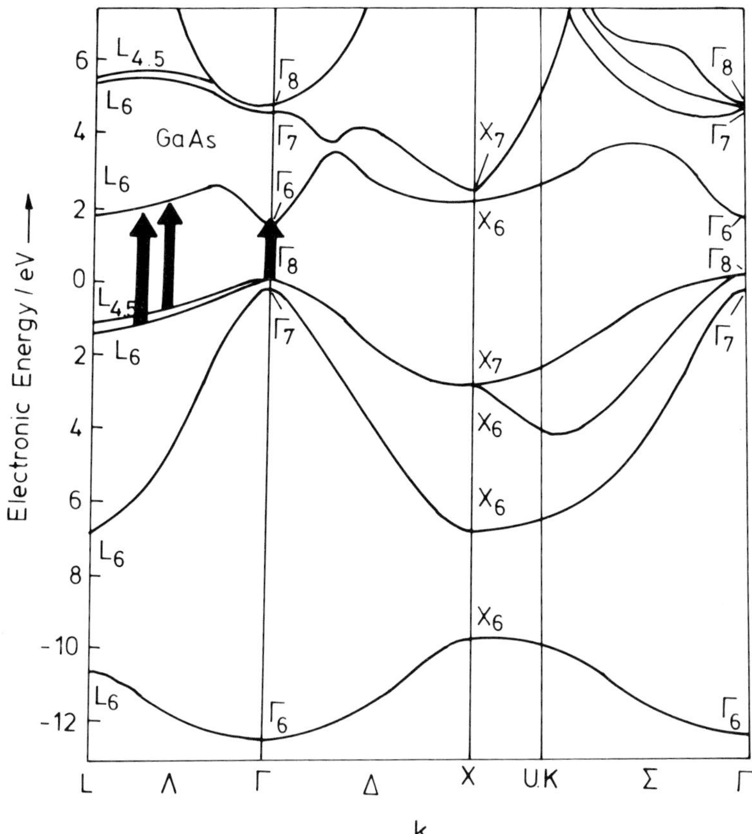

Fig. 12. Band structure of GaAS showing the first and second sets of transitions as for Fig. 11, at ≃ 1.43 and at 2.90, 3.17 eV, respectively. Note the increase in the spin-orbit coupling for GaAs as compared with GaP.

to raise serious doubts about the reasonableness of using a low-field limit approximation. Very generally, the following observations can be made of the variation in shape with potential.

(1) Far from flat-band, the spectra are comparatively broad and show increasing Franz–Keldysh structure with increasing depletion. In addition to increased structure at more remote energies, some spectra, such as the E_0 and $E_0 + \Delta$ structure of p-GaP, actually show, at higher depletion, the development of an increasingly well-resolved shoulder.

(2) As the potential approaches flat-band for the p-type semiconductors, the positive-going structure close to the band-edge increases in amplitude and decreases in width, whereas the negative-going structure at immediately higher energies decreases in amplitude.

4.3 THE EFFECT OF DONOR DENSITY

The effect of increasing the donor density, and decreasing L_D, is shown in

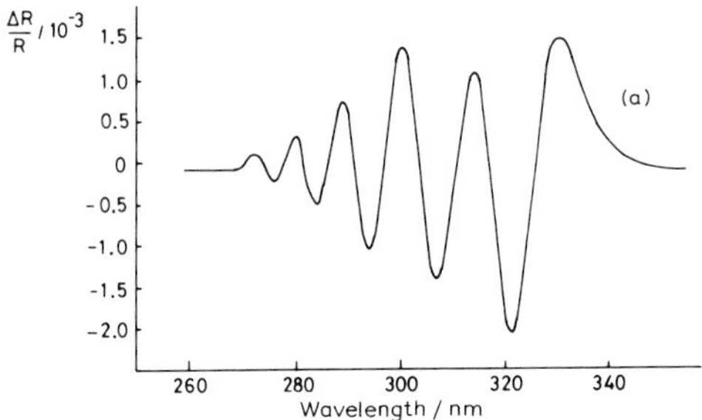

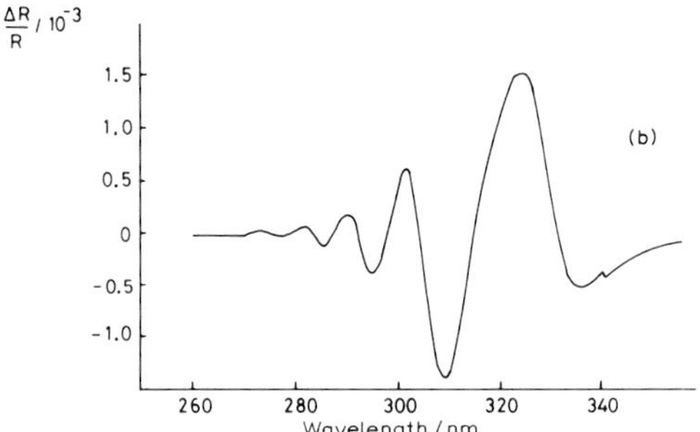

Fig. 13. ER spectra generated theoretically for the E_1 and $E_1 + \Delta_1$ transitions of p-GaP ($N_A \simeq 10^{18}\,\mathrm{cm}^{-3}$) assuming (a) that the depletion layer can be modelled as a single slab of thickness W in which the electric field is fixed at a constant value corresponding to half that calculated at the surface using the classical Schottky model and (b) that the depletion layer can be modelled by dividing the depletion layer into twelve slabs within each of which the electric field is fixed at the value obtained from the Schottky model for the mid-point of the slab.

Fig. 19 for p-GaAs and Fig. 20 for p-GaP. It can be seen that, for both cases, as L_D is decreased there is a marked increase in the overall amplitude of the ER spectrum as well as a loss in resolution associated with a general broadening of all the spectral features. In addition to these features, the short-wavelength Franz-Keldysh structure shows a considerable degree of enhancement at lower L_D values, though the fact that such structure originates from the overlap of a large number of layers militates against any simple interpretation of it in terms of asymptotic expansions of the intermediate-field equations.

4.4 THE EFFECT OF THERMAL BROADENING

The effect of thermal broadening can be seen in Fig. 21. As might be expected, the main effect is, in fact, to reduce both the resolution of the spectrum and the overall amplitude as Γ is increased.

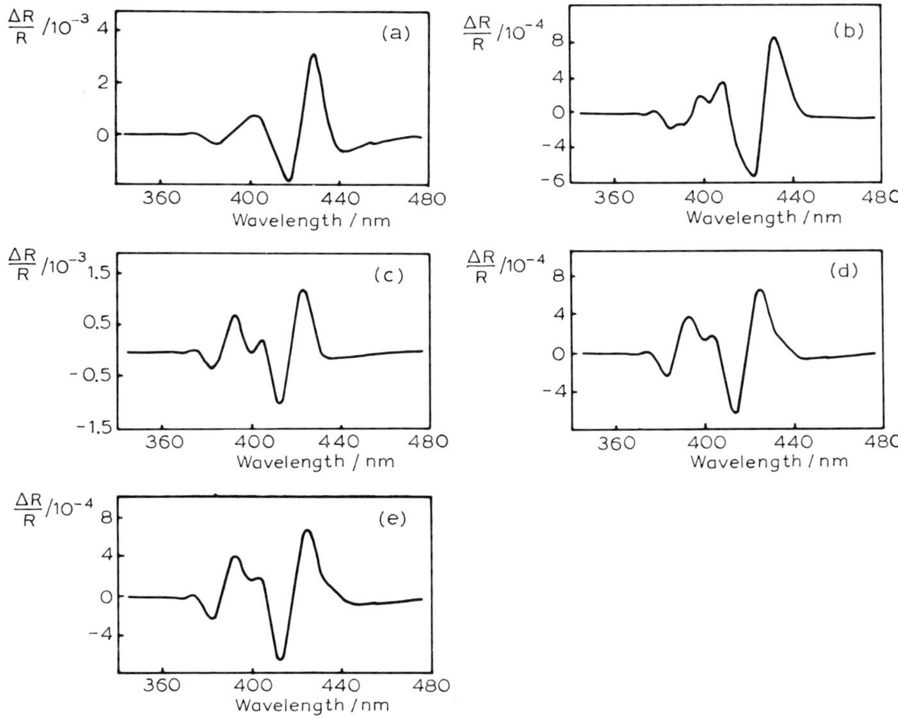

Fig. 14. ER spectra for the E_1, E_1 and Δ_1 transitions of p-GaAs (thermal broadening parameter $\hbar\Gamma \simeq 40\,\text{meV}$, $N_A \simeq 10^{17}\,\text{cm}^{-3}$) calculated assuming an increasing number of slabs taken to model the depletion layer: (a) one slab, (b) two slabs, (c) three slabs, (d) six slabs, and (e) nine slabs. Increasing the number of slabs above nine gave little further change in the shape of the ER spectrum. $V_{sc} = 0.2\,\text{V rms}$.

4.5 THE EFFECT OF a.c. AMPLITUDE

The effect of increasing the amplitude of the a.c. potential is demonstrated in Fig. 22, which shows calculations on the E_0 structure of p-GaAs for a donor density of $10^{17}\,\text{cm}^{-3}$. It has already been established that simple low-field theories are not applicable to this material, as shown in Fig. 15, but a plot of the amplitude of the first negative-going peak vs. the amplitude of the applied a.c. potential is remarkably linear, as shown in Fig. 23. The important point here is that this linearity is often used as a test for the applicability of low-field theories and the effect of inhomogeneities in the electric field seem to wash out the non-linearity expected from an uncritical application of intermediate-field theories.

4.6 THE EFFECT OF OPTICAL CROSS-SECTION

The optical absorption cross-section will depend sensitively on the surface from which reflection takes place and the relative orientation of the surface and the electric vector of the incoming radiation. The effect of

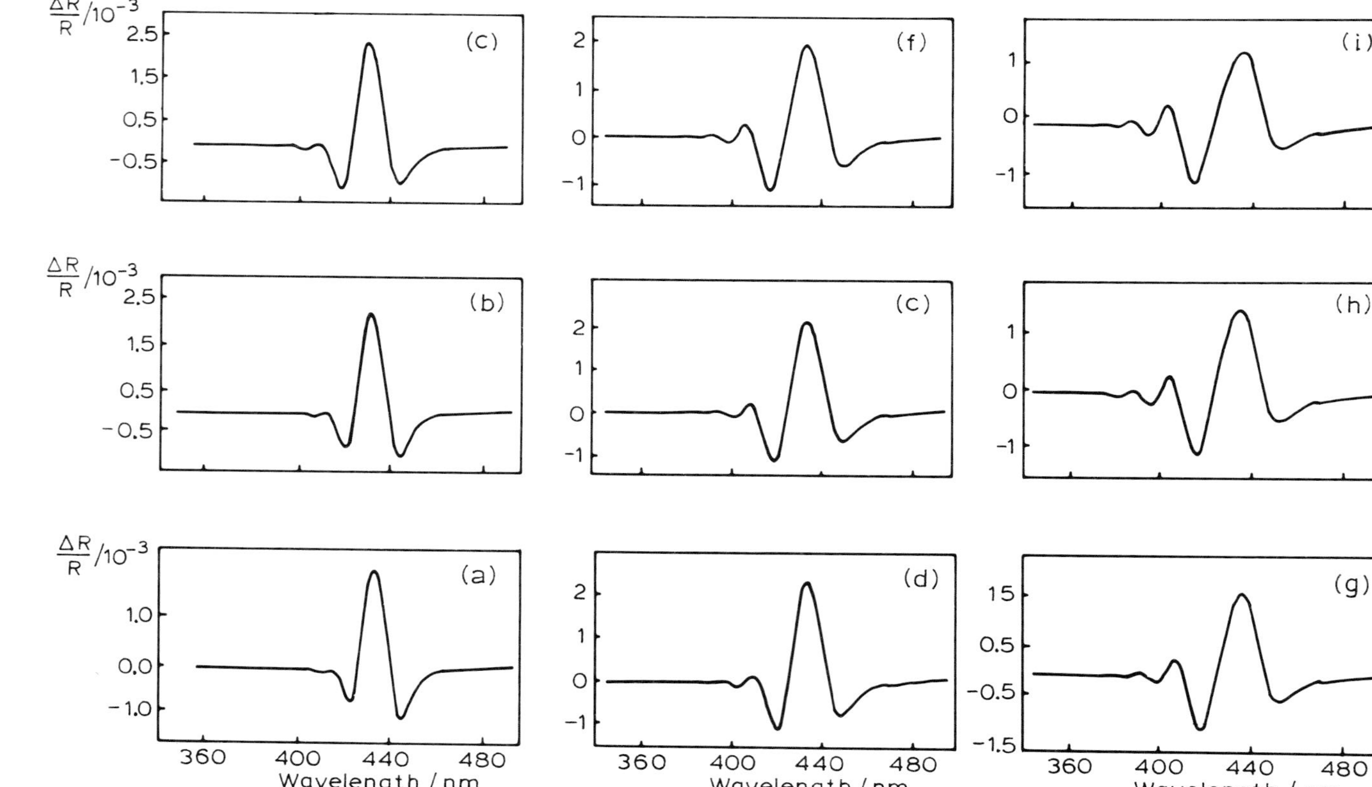

Fig. 15. Calculated ER spectra for the E_0, $E_0 + \Delta_0$ transitions of p-GaP ($N_A \simeq 10^{18}\,cm^{-3}$, $\hbar\Gamma \simeq 40\,meV$) as a function of d.c. potential for $V_{ac} = 0.2\,V$ rms. The potential bias away from flat-band potential is (a) 0.2 V, (b) 0.3 V, (c) 0.5 V, (d) 0.7 V, (e) 0.9 V, (f) 1.1 V, (g) 1.3 V, (h) 1.5 V, and (i) 1.7 V.

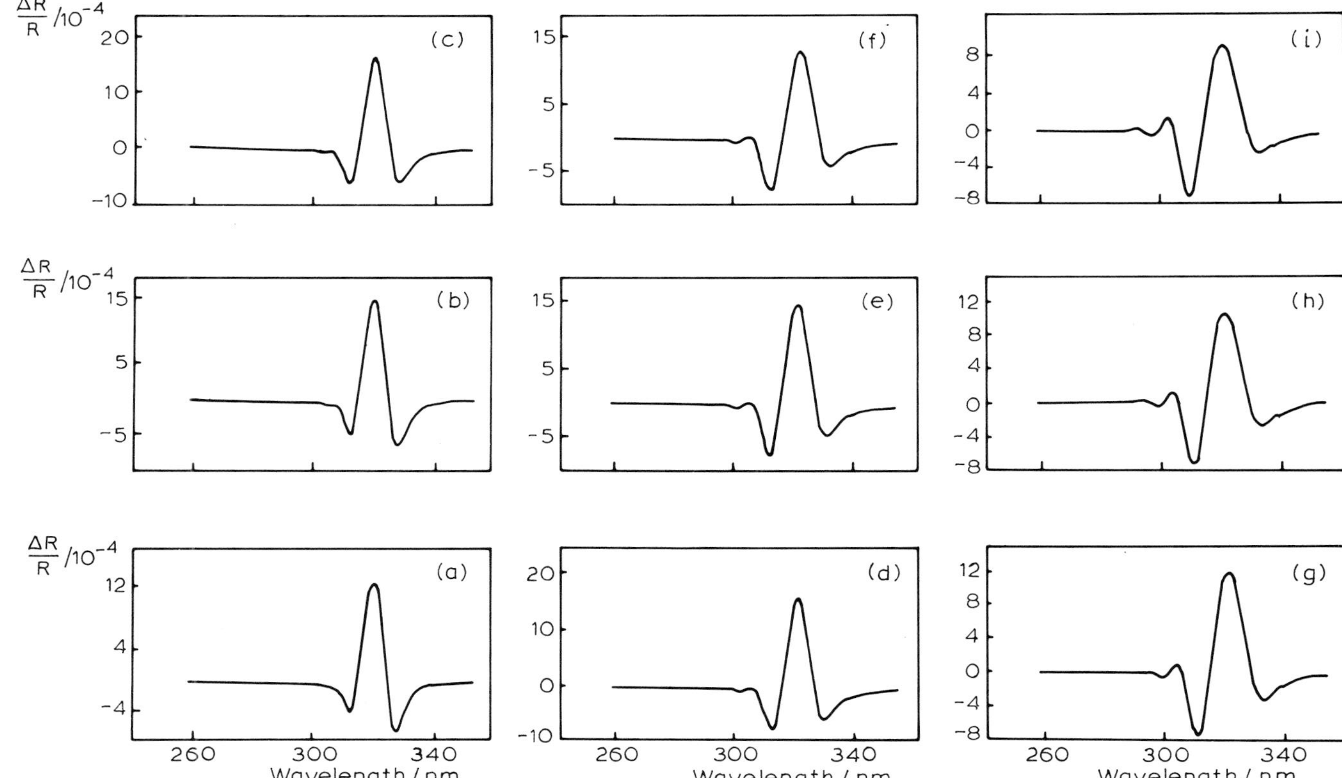

Fig. 16. Calculated ER spectra for the E_1, $E_1 + \Delta_1$ transitions of p-GaAs ($N_A \simeq 10^{17}\,\mathrm{cm}^{-3}$, $\hbar\Gamma \simeq 40\,\mathrm{meV}$) as a function of the d.c. potential for $V_{ac} = 0.2\,\mathrm{V}$ rms. The potential bias away from flat-band potential is as for Fig. 15.

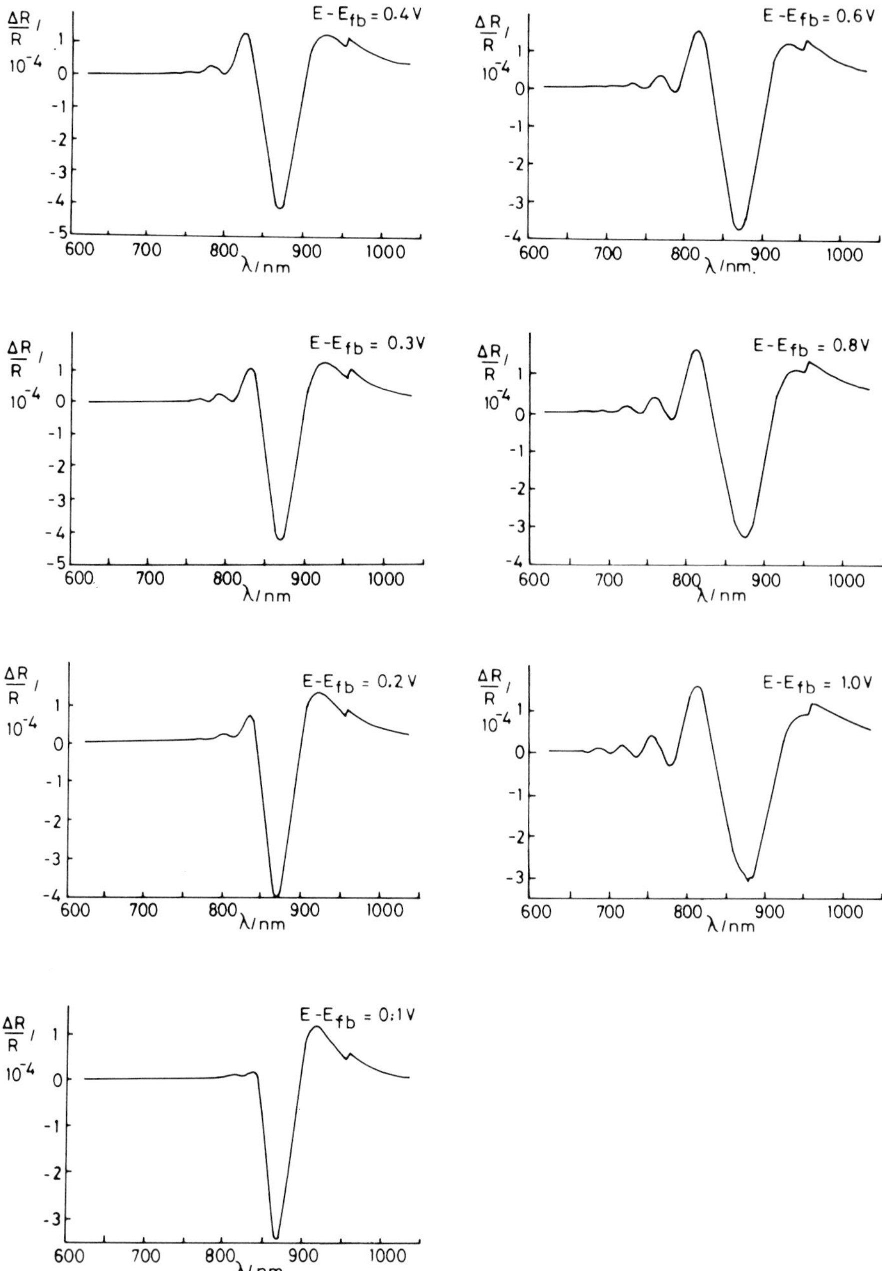

Fig. 17. Calculated ER spectra for the E_0 transition of n-GaAs ($N_D \simeq 10^{17}\,cm^{-3}$) as a function of the d.c. potential for $V_{ac} = 0.016$ V rms. The discontinuity at $\simeq 970$ nm is an artifact of the computer program arising from the folding in of the instrumental resolution and should be ignored.

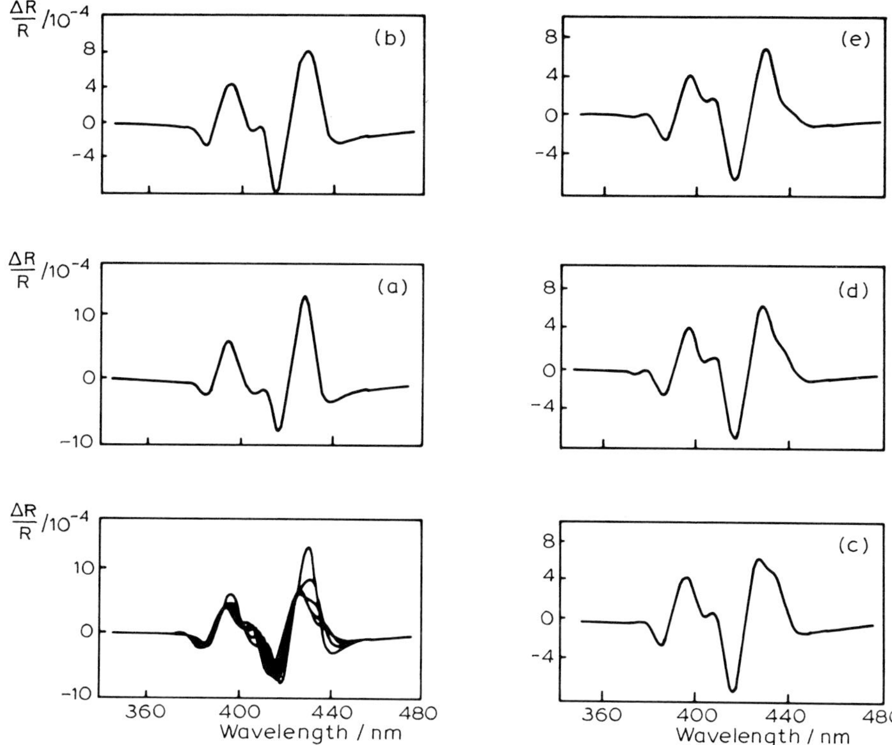

Fig. 18. Calculated ER spectra for the E_1, $E_1 + \Delta_1$ transitions of p-GaAs ($N_A \simeq 10^{17}\,cm^{-3}$) as a function of the d.c. potential for $V_{ac} = 0.2\,V$ rms. The potential bias away from flat-band potential is (a) 0.3 V, (b) 0.5 V, (c) 0.7 V, (d) 0.9 V, and (e) 1.1 V.

varying the optical matrix elements for absorption are shown for the E_0 structure of p-GaAs in Fig. 24. It can be seen that the major effect is on the structure of the first positive-going peak.

5. Experimental results

We consider first the results on p-GaP. The impedance data for p-GaP has been a fruitful source of controversy, though not of comprehension. If a sample of p-GaP is held at a negative potential for a considerable period and then slowly ramped towards positive potentials, the a.c. impedance data cannot be analysed within the framework of the two-component model. Attempts to do so lead to Mott–Schottky plots whose slopes and intercepts are both frequency-dependent as shown in Fig. 25. If the data are analysed according to the more complex five-component equivalent circuit shown above, then a much better fit is obtained for the potential region more than about 0.6 V negative of the predicted flat-band potential. In this region, the Mott–Schottky plot is linear with a slope that corresponds reasonably well

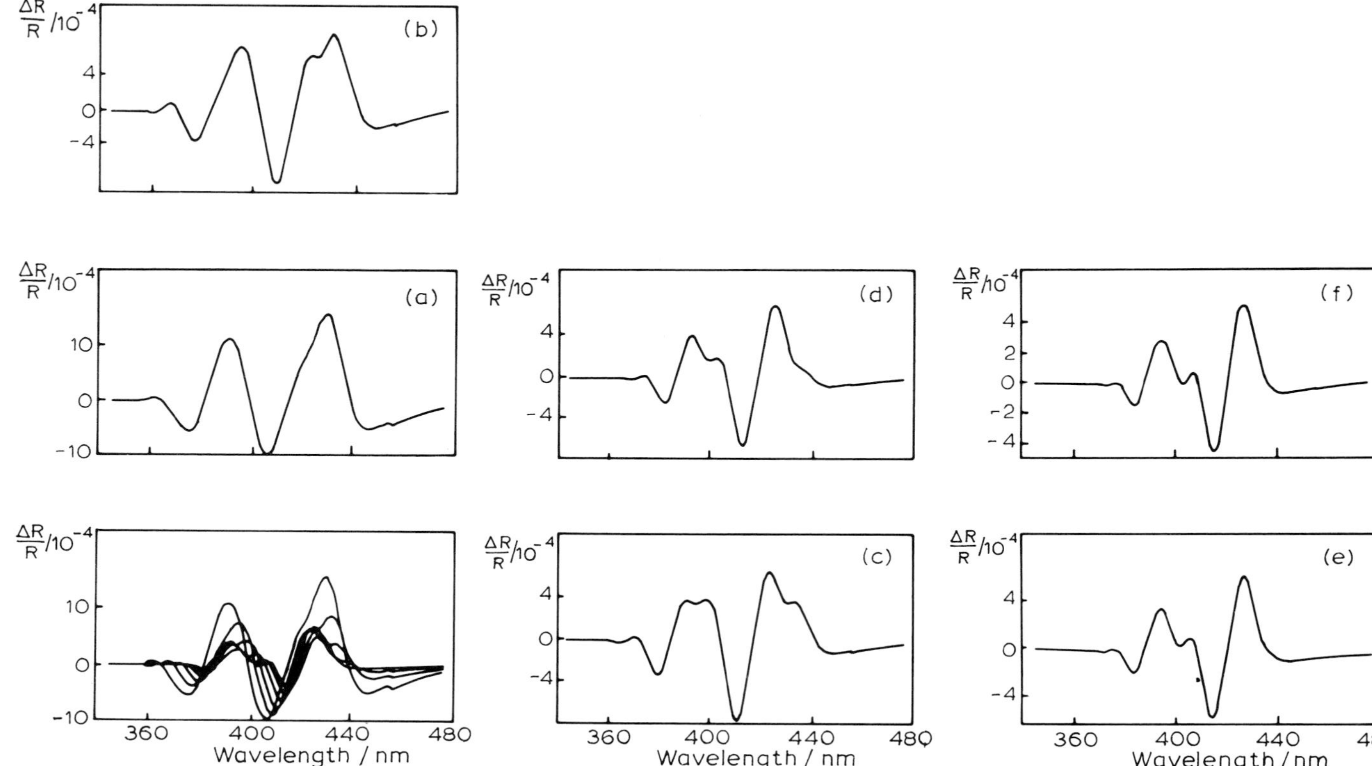

Fig. 19. Calculated ER spectra for the E_1, $E_1 + \Delta_1$ transitions of p-GaAs as a function of Debye length, L_D, as defined in the text, for $V_{ac} = 0.2$ V rms, $|V_{dc} - V_{fb}| = 1.0$ V. L_D takes the values (a) 60 Å, (b) 80 Å, (c) 100 Å, (d) 120 Å, (e) 140 Å, and (f) 160 Å.

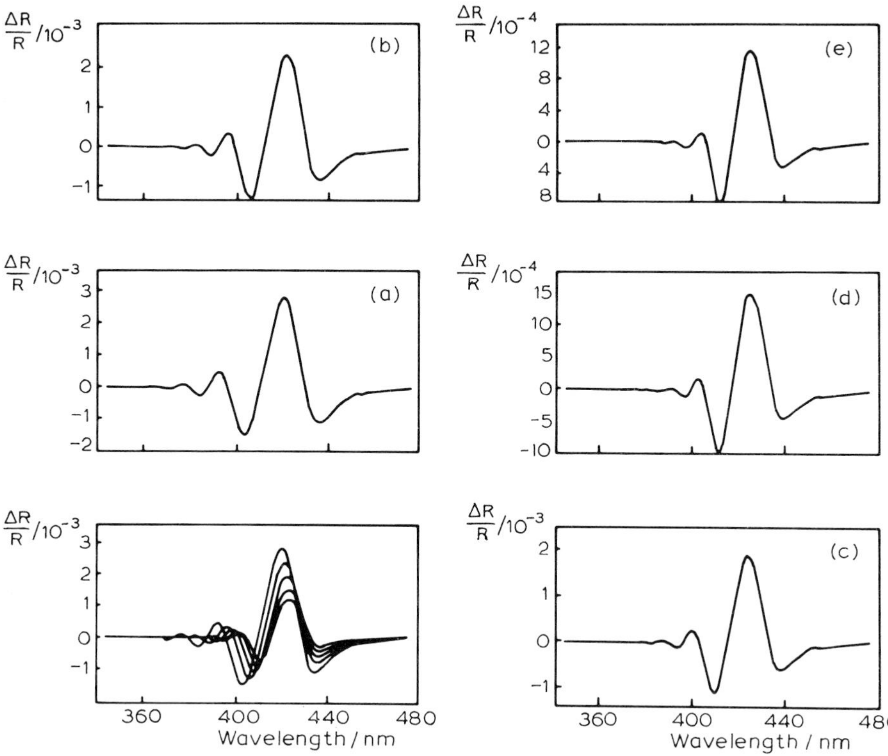

Fig. 20. Calculated ER spectra for the E_0, $E_0 + \Delta_0$ transitions of p-GaP as a function of Debye length, L_D, and with $V_{ac} = 0.2$ V rms, $|V_{sc} - V_{fb}| = 1.0$ V. L_D has the values (a) 40 Å, (b) 50 Å, (c) 60 Å, and (e) 80 Å.

to that expected from Hall data. Nearer to flat band, this linearity is lost and, indeed, the data becomes much less reproducible with results dependent on the time elapsed between each measurement. In this region close to flat band, the surface-state capacitance rises dramatically as indicated in Fig. 5(a); as the potential rises to within ca. 0.5 V of flat band, a dark current commences and ellipsometric data indicate the onset of surface oxide formation.

The electroreflectance spectra of a p-GaP sample is shown as a function of amplitude in Fig. 26 and a plot of peak height vs. amplitude is linear. This, as was indicated above, has been suggested as a test for the applicability of the low-field limit theories, but it will be evident from the data below that we are not in the low-field region for this sample and the linearity is a feature of the inhomogeneous electric field. The potential dependence of the electroreflectance signal for p-GaP is shown in Fig. 27 and a careful analysis of the data for Figs. 27(a) and (b) shows the following features occur in the experimental data that correspond to effects seen in the theoretical modellings of Fig. 16.

(a) The small negative peak on the high-energy side of the peak at 440 nm becomes accentuated as we approach flat-band potential.

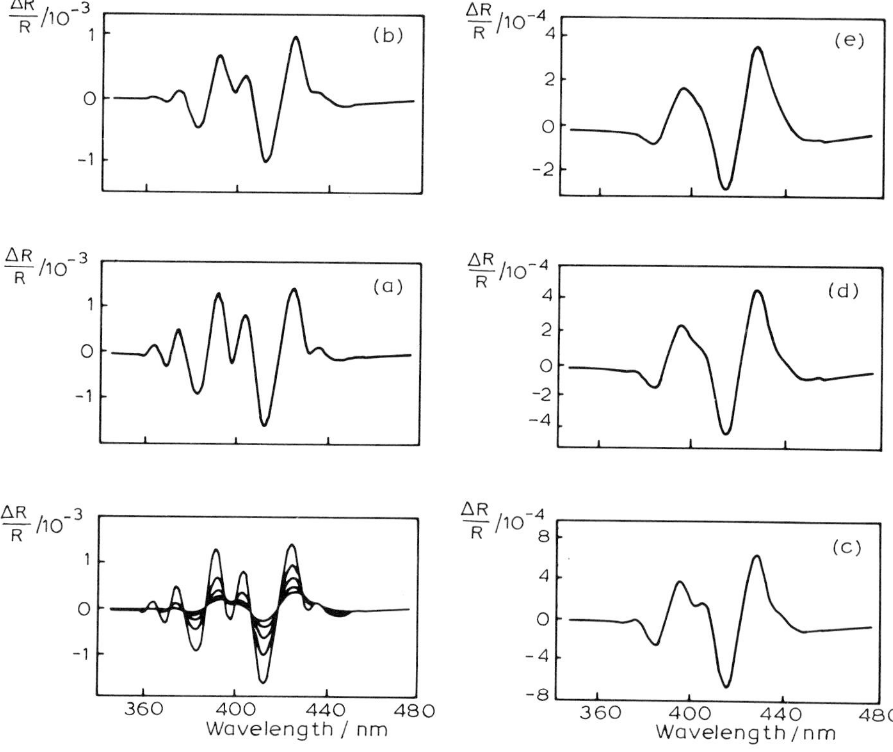

Fig. 21. Calculated ER spectra for the E_1, $E_1 + \Delta_1$ transitions of p-GaAs ($N_A \simeq 10^{17}\,\mathrm{cm}^{-3}$) for various values of the thermal broadening parameter $\hbar\Gamma$ with $V_{ac} = 0.2\,\mathrm{V}$ rms, $|V_{dc} - V_{fb}| = 1.0\,\mathrm{V}$. The values of $\hbar\Gamma$ are (a) 0.012 eV, (b) 0.025 eV, (c) 0.037 eV, (d) 0.050 eV, and (e) 0.063 eV.

(b) The negative peak to the low-wavelength side of the same peak is reduced as flat band is approached.

(c) The small shoulder on the low-wavelength side of the same peak, at about 430 nm, decreases and finally disappears.

(d) The overall amplitude of the positive peaks at both 440 and 320 nm increases.

(e) The peaks, especially those of the E_1 and $E_1 + \Delta_1$ transitions at 320 nm, become appreciably sharper.

(f) The negative peak at 330 nm increases in amplitude near flat band.

Such a detailed qualitative fit is quite remarkable and, taken with the a.c. data of Fig. 5(a), does strongly support the thesis that most, if not all, the potential change in the region more negative than ca. 0.7 V from flat band is accommodated across the depletion layer. The qualification in the last sentence arises from the fact that detailed quantitative fits have not proved easy and it seems increasingly likely that a constant proportion rather than all the potential change may be accommodated in this way: the point is returned to below in the consideration of the ER spectra of p-GaAs.

Further careful examination of Fig. 27(c) reveals that, however, for poten-

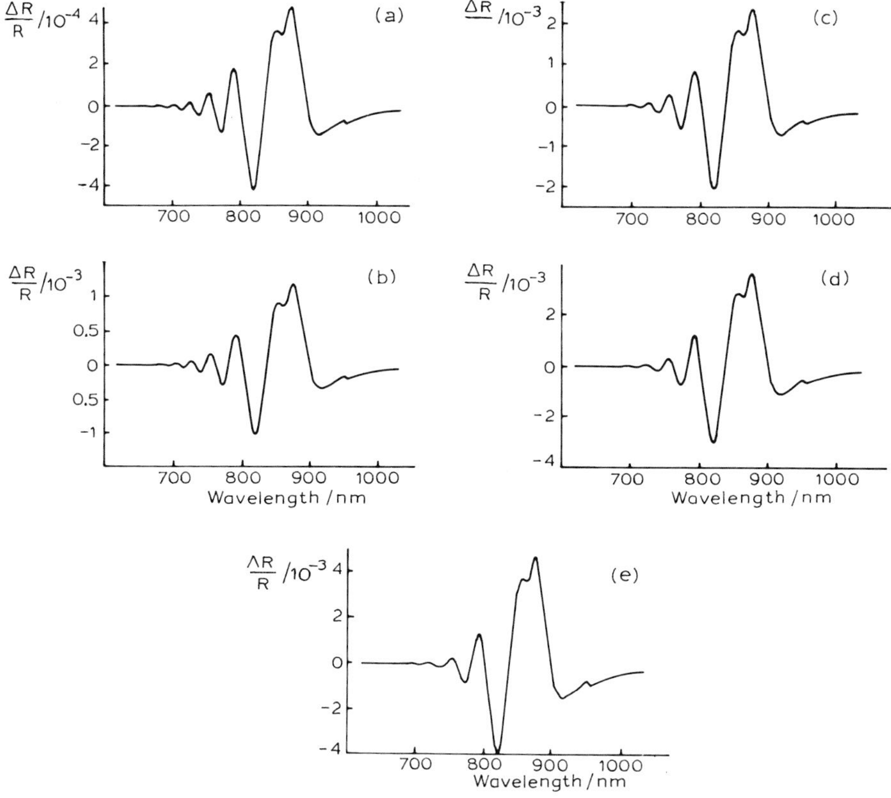

Fig. 22. Calculated ER spectra for the E_0 transition of p-GaAs ($N_A \simeq 10^{17}\,cm^{-3}$, $\hbar\Gamma = 20\,meV$) for various values of the amplitude of the a.c. modulation potential V_{ac} (rms). (a) 0.04 V, (b) 0.10 V, (c) 0.20 V, (d) 0.30 V, and (e) 0.40 V.

tials within ca. 0.7 V of flat band, the alteration in the line shape becomes appreciably less marked than the theory would suggest and that, above 0.3 V, within 0.5V of flat-band potential, the magnitude of the ER signal for the E_1 signal decreases rapidly. It does not seem reasonable to suggest that

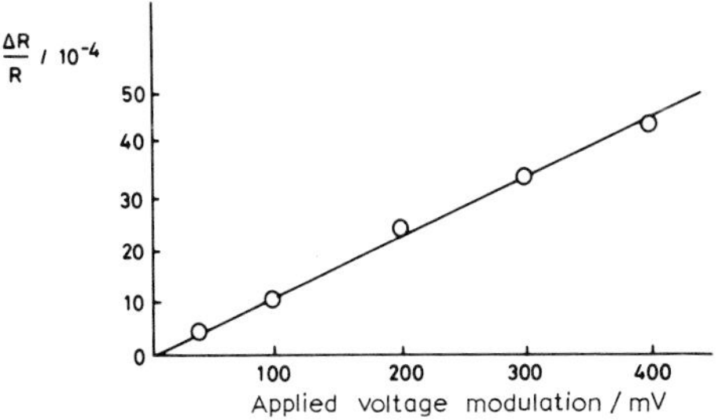

Fig. 23. Magnitude of the calculated ER signal at 870 nm vs. the applied a.c. modulation amplitude from the data of Fig. 22.

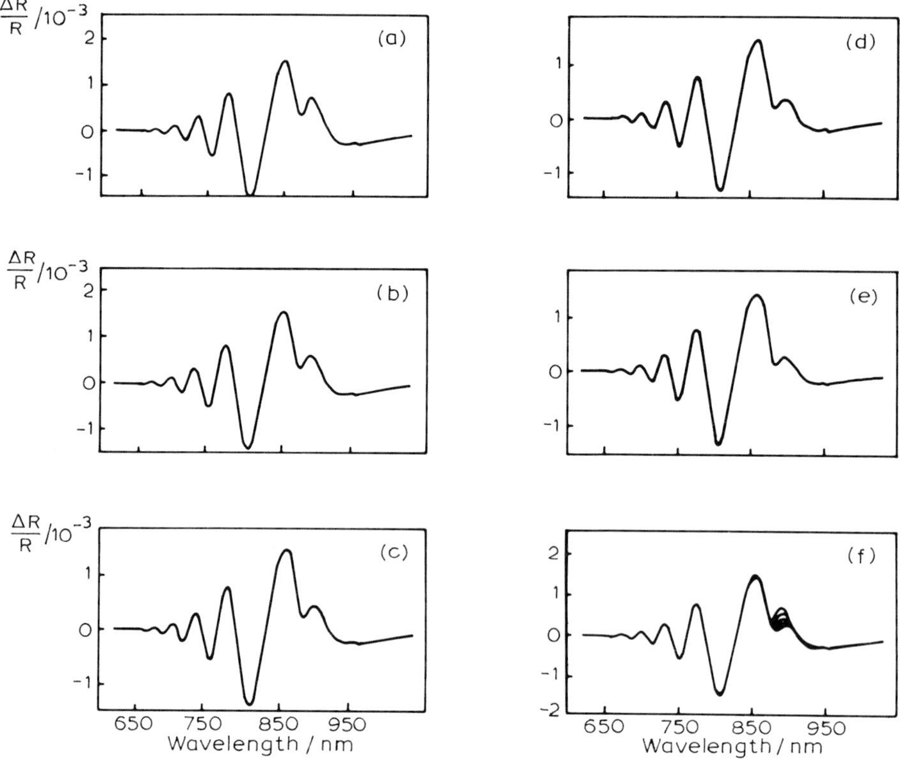

Fig. 24. Calculated ER spectra for the E_0 transition of p-GaAs as a function of the optical absorption coefficient, which increases by a factor of 50% from (a) to (e).

the theory suddenly ceases to be valid at rather *lower* electric fields and we are then faced with the fact that the system must have become partially pinned by surface states somewhat similar to those identified by Tomkiewicz for $CdIn_2Se_4$, where the amplitude of the electroreflectance signal fell to zero as the flat-band potential was approached and where there was clear evidence for the presence of *fast* surface states. There is, however, a further subtlety: the theoretical spectra of Fig. 16, whilst they do not fit the data precisely, nevertheless reveal that a continued evolution in the *shape* of the ER spectrum is expected as the d.c. potential flat band. In fact, whilst the magnitude of the ER spectrum *decreases*, the shape remains remarkably constant. This suggests that, when the d.c. potential across the interface is altered in this region, the d.c. potential dropped across the depletion layer is changing much more slowly than the total d.c. potential. It follows that, in the case of p-GaP, we are also dealing with slow states that cannot respond to the 240 Hz a.c. potential imposed on the sample and an increasing proportion of the d.c. potential is accommodated primarily across the Helmholtz layer as we approach flat band. In a sense, it is rather as if the goalposts are being moved in the course of the game: as the potential is ramped more

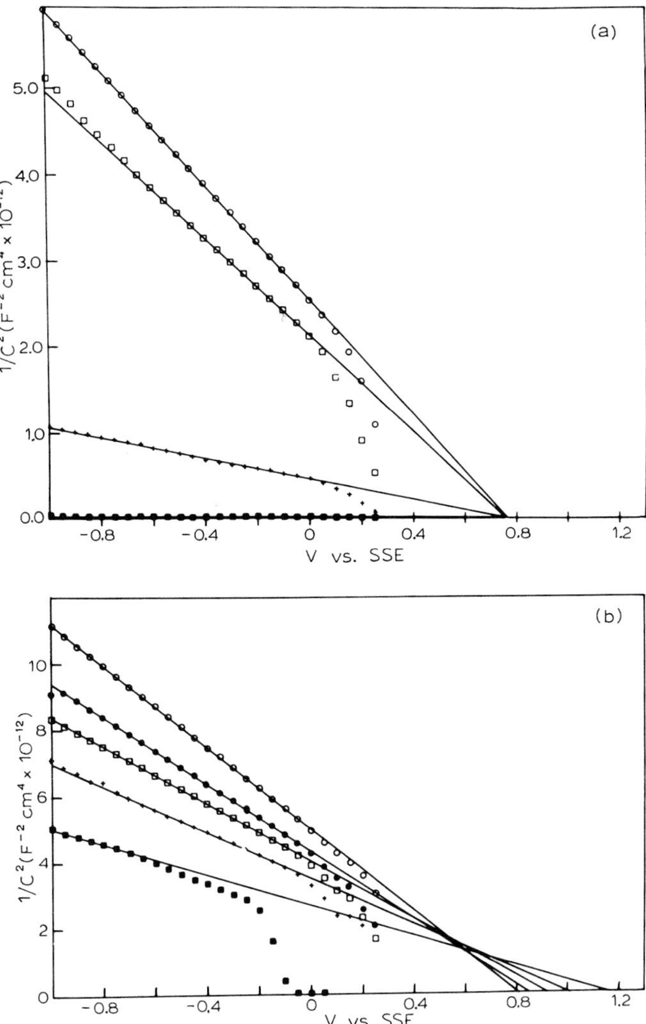

Fig. 25. Variation of the inverse square of the series capacitance for two faces of p-GaP ($N_A \simeq 10^{18}$ cm^{-3}) in 0.5 M H_2SO_4 for various frequencies. (a) GaP(100):○, 10 kHz; □, 1 kHz; +, 330 Hz; ■, 87 Hz. (b) GaP($\bar{1},\bar{1},\bar{1}$):○, 10 kHz; ●, 3 kHz; □, 1 kHz; +, 330 Hz; ■, 87 Hz.

positive, the flat-band potential itself moves anodically and instead of attaining flat-band at the potential derived from the Mott-Schottky analysis (which is clearly valid at potentials corresponding to high depletion), the band edges become unpinned in this region. A very similar result is found for p-GaAs, though the quality of the data here is less good as the material is sustaining quite a brisk dark current in this region and cross-talk leads to serious difficulties in obtaining reasonably noise-free spectra.

It might reasonably be asked what the origin of this behaviour pattern is and Fig. 28 shows the variation in ellipticity of a sample of p-GaP measured

References pp. 425–426

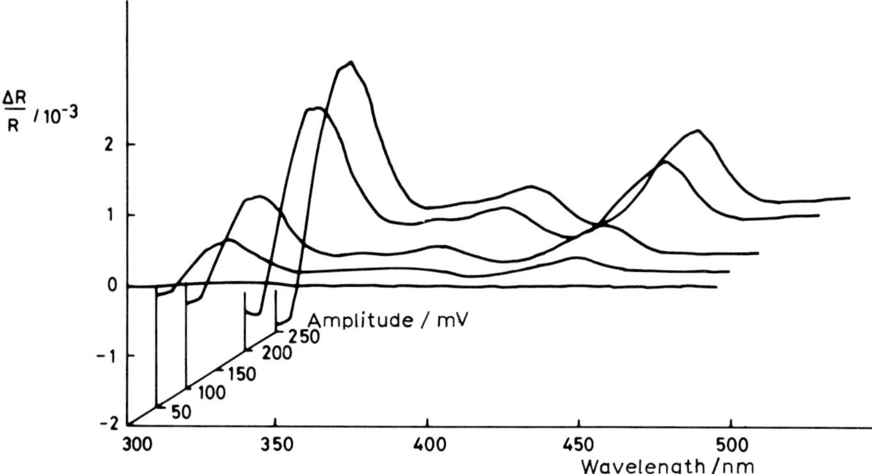

Fig. 26. Experimental ER spectra for p-GaP ($N_A \simeq 10^{18}\,\mathrm{cm}^{-3}$) in 0.5 M H_2SO_4 as a function of the amplitude of the a.c. potential modulation.

in an automatic ellipsometer as the potential is ramped near to flat band. It can be seen that, near 0 V vs. Hg/Hg_2SO_4 (about 0.4–0.5 V from flat-band potential vs. the reference electrode used), a significant increase in ellipticity takes place. This corresponds to the formation of a sub-monolayer of oxide on the surface of the GaP and it is clear that this is the main cause of the change in potential distribution.

The E_0 transition in GaAs is the simplest single transition that can be investigated within the III/V materials and a detailed quantitative fit has been attempted to the data of Fig. 29. A careful analysis of these data lead to the fit shown in Fig. 17 and it is clear that all the features of the experimental spectrum can be reproduced with some precision provided that the manufacturer's acceptor density be taken as the basis for the analysis. By comparing the changes in the Franz–Keldysh oscillation near 820 nm with those calculated using the intermediate field model with a presumed parabolic decay of potential inside the semiconductor depletion layer, it is found that some 70 ± 10% of the potential is dropped inside the depletion layer of this n-type material, as can be seen in Fig. 30, and there is no evidence for the phenomenon described in the previous paragraph whereby the band

---

Fig. 27. Experimental ER spectra of the E_0 and E_1 structure of p-GaP ($N_A \simeq 10^{18}\,\mathrm{cm}^{-3}$) in 0.5 M H_2SO_4 as a function of the applied d.c. potential. The flat-band potential is estimated from Mott–Schottky analysis to be 0.8 V vs. SCE. For (a), the a.c. modulation is 0.25 V rms and the applied potentials are, from top to bottom: 0.1 V, 0.0 V, −0.1 V, −0.2 V, −0.3 V, −0.4V, −0.5 V, and −0.6 V. For (b), which shows only the E_1 structure, and for which the applied a.c. modulation is 0.1 V rms, the applied potentials are, from top to bottom, 0.0 to −0.1 V in 0.1 V steps and (c), which also has an a.c. modulation of 0.1 V, the applied potentials are, from top to bottom, 0.5 to 0.1 V in 0.1 V steps.

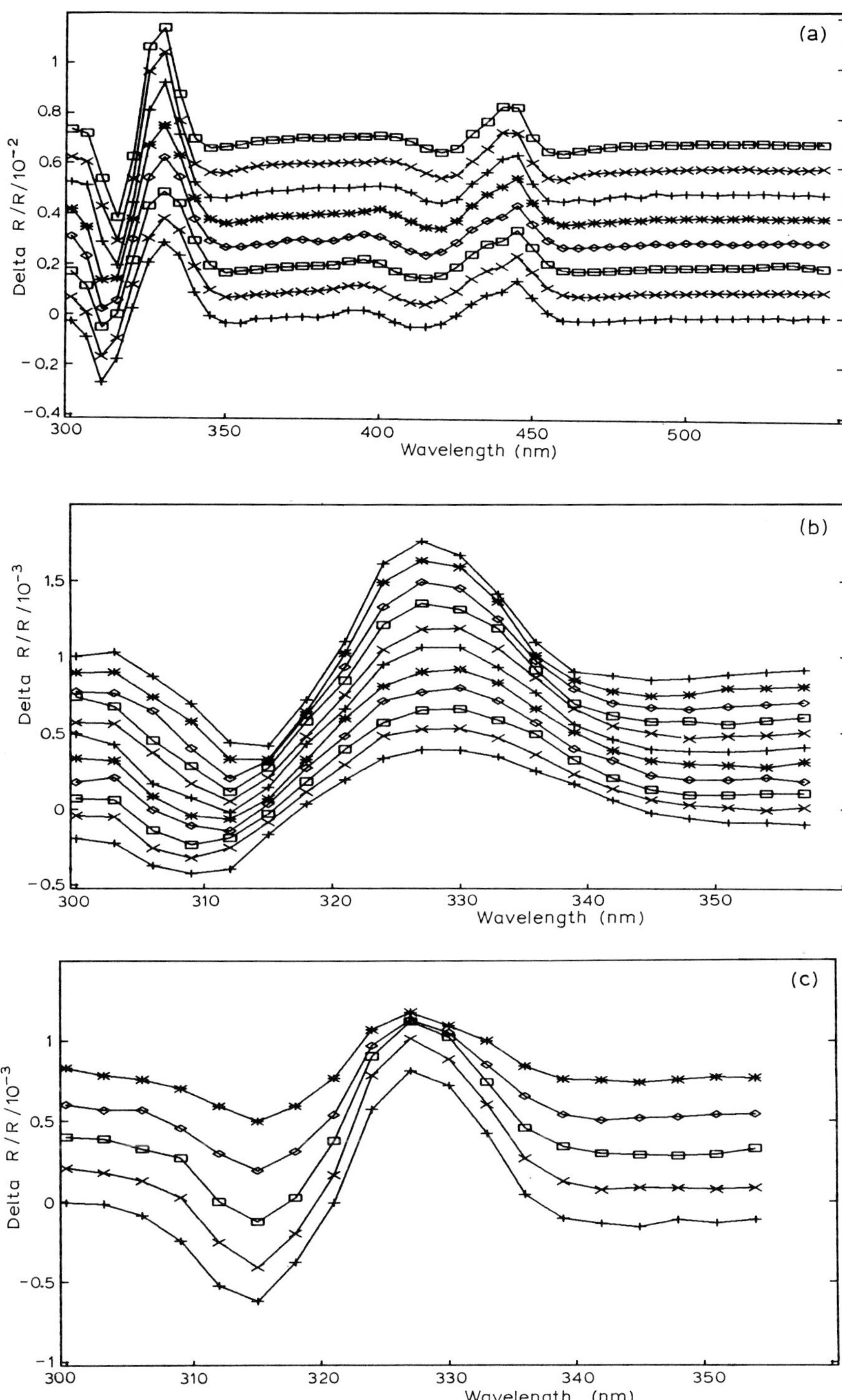

(a)
Delta R/R/10^-2
Wavelength (nm)
(b)
Delta R/R/10^-3
Wavelength (nm)
(c)
Delta R/R/10^-3
Wavelength (nm)

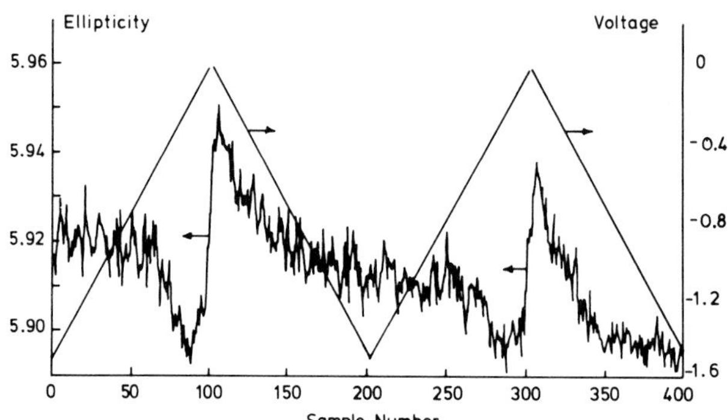

Fig. 28. Ellipsometric measurements of the ellipticity changes at a p-GaP ($N_A \simeq 10^{18}\,cm^{-3}$)/ 0.5 M H_2SO_4 interface as a function of time during two potential cycles from -1.5 to 0.0 V vs. Hg/Hg_2SO_4. The flat-band potential vs. this reference is close to 0.4 V.

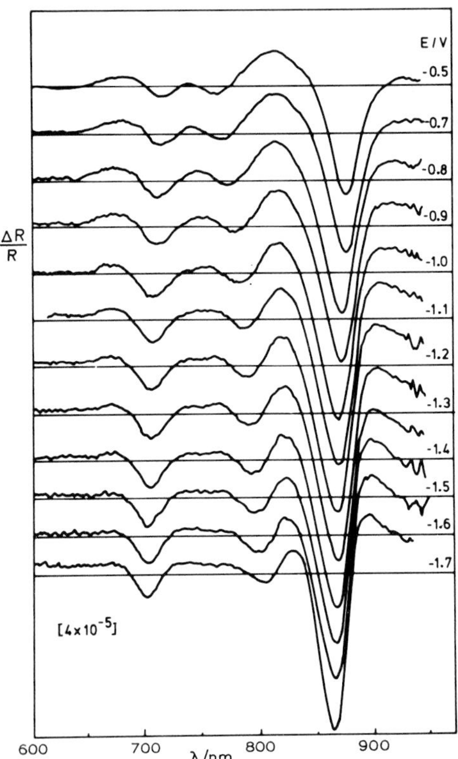

Fig. 29. Experimental ER spectra for n-GaAs ($N_D \simeq 10^{17}\,cm^{-3}$) showing E_0 structure as a function of potential (vs. SCE). The flat-band potential is estimated to be near -1.8 V in the 0.1 M KOH solution used.

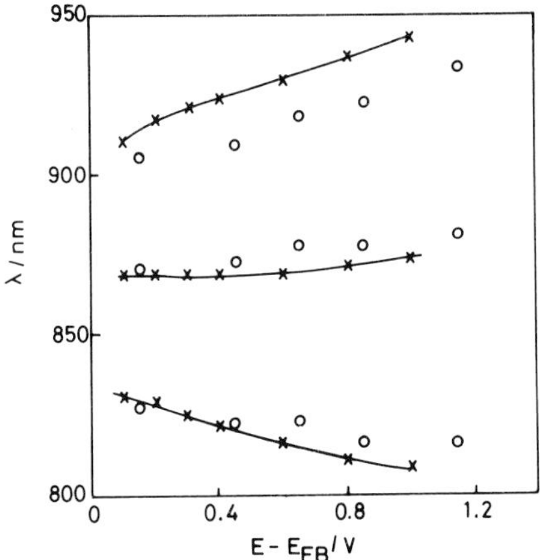

Fig. 30. Variation in the position of the maxima or minima in the ER spectra of Fig. 29 (○) and variation calculated from the extended theory discussed in the text assuming a classical Schottky barrier (×).

edges become unpinned near flat band. The detailed fit between experiment and theory for this single transition is very powerful evidence that the model is fundamentally correct and that, at least at these donor densities, the additional effects described by Raccah and co-workers are less important.

5.1 NON-AQUEOUS SOLVENTS

The use of non-aqueous solvents in "wet" photovoltaic devices has been carefully investigated by a number of workers in recent years. The major impetus for this work has come from the observation by Lewis and co-workers that highly efficient photovoltaic devices can be obtained by the use of reversible one-electron couples such as ferrocene or cobalticinium carboxylic acid in anhydrous aprotic solvents such as acetonitrile (ACN) or propylene carbonate (PC) [21, 22].

There are some experimental difficulties encountered with these solvents: they are difficult to purify and are often highly hygroscopic, so that special precautions are necessary. In our own work, PC was found to be sufficiently involatile that transfer of pre-dried PC to the cell could be effected within a high-integrity dry-box. The cell was then placed in a welded-steel box with a viton O-ring seal, within which it could be transported to the spectrometer and studied for several days without degradation. Studies with ACN are appreciably more difficult since its voltatility effectively precludes the use of a high-integrity dry box. The techniques adopted involved essentially standard Schlenk tube manipulations with carefully predried ACN in baked-out, equipment, but the presence of small ($\simeq$ 1 ppm) quantities of water cannot be eliminated by the techniques used.

The main difference between aqueous and non-aqueous solutions in the case of p-GaAs is that there is no sign, in the latter case, of the large density

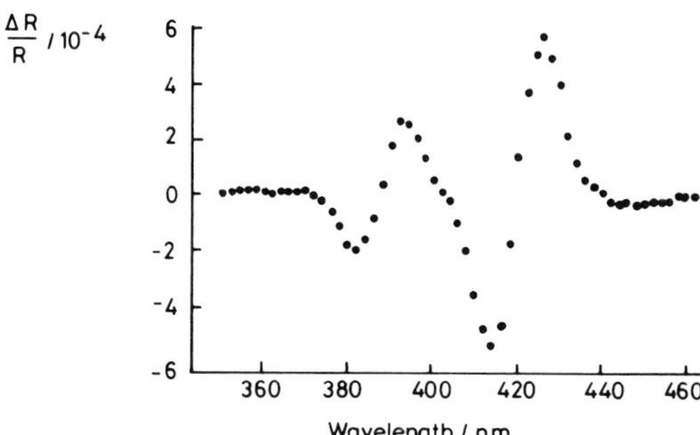

Fig. 31. Experimental ER spectrum for p-GaAs ($N_A \simeq 10^{17}\,\mathrm{cm}^{-3}$) showing E_1, $E_1 + \Delta_1$ structure taken in propylene carbonate solution ($V_{ac} = 0.25$ V rms) with 0.5 M $(\mathrm{Bu_4N})\mathrm{PF_6}$ as electrolyte.

of surface states detected by standard a.c. techniques in aqueous electrolytes. Given that we believe that these states are associated with oxide formation, which should be suppressed in anhydrous solvents, this is added confirmation of the chemical origin of these states. The electroreflectance spectrum of the E_1, $E_1 + \Delta_1$ transitions of p-GaAs in 0.5 M $(n\text{-}\mathrm{Bu_4N})^+\,\mathrm{PF_6^-}$ / PC is shown in Fig. 31 and is very similar to that found in aqueous electrolytes. A study of the variation of the electroreflectance signal with potential in non-aqueous electrolyte is currently under way.

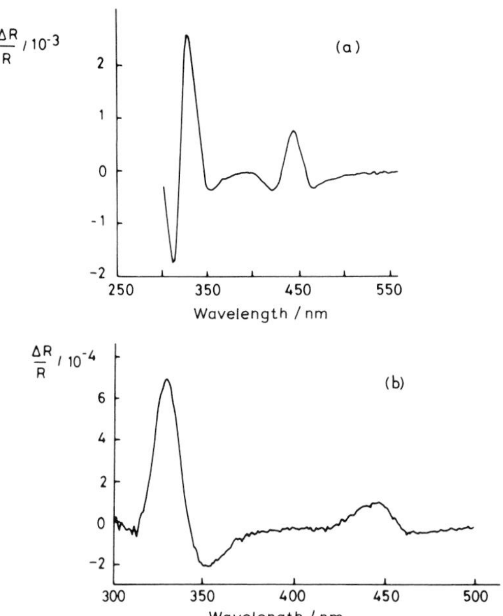

Fig. 32. Experimental ER spectra for p-GaP ($N_A \simeq 10^{18}\,\mathrm{cm}^{-3}$) in the E_0 and E_1 regions in (a) acetonitrile (b) propylene carbonate. The electrolyte in both cases is 0.5 M $(\mathrm{Bu_4N})\mathrm{PF_6}$ and $V_{ac} =$ 0.25 V rms. Note the difference in magnitude of the spectra, particularly in the E_0 peak at 440 nm.

For p-GaP, the situation is rather different. In ACN, p-GaP again appears to behave classically, but in PC solutions, the behaviour is much more difficult to analyse. The a.c. data, analysed with the 5-component circuit described above, suggests that C_{sc} is nearly constant and very large indeed (1–2 $\mu F\,cm^{-2}$). The constancy of C_{sc} can be interpreted as deriving from near total Fermi-level pinning by slow surface states and the large value suggests that very little potential is dropped across the depletion layer: a value of 2 $\mu F\,cm^{-2}$ suggests less than 20 mV actually appears across this layer.

The electroreflectance spectrum of p-GaP in PC is shown in Fig. 32; it clearly differs both in size and magnitude from that found in aqueous solution. A detailed analysis along the lines developed above suggests that a spectrum of this shape could only arise from a sample with very little potential dropped across the depletion layer, in agreement with the a.c. data, and the magnitude of the spectrum is consistent with the suggestion that the magnitude of the potential drop in the depletion layer is only a few tens of mV.

6. Conclusions

The detailed interpretation of electroreflectance spectra is still in its infancy, but enough has already been learnt to indicate that the technique will form a most valuable adjunct to other methods that have recently been developed to study the semiconductor/electrolyte interface. The next few years should see this technique become a standard weapon in the armoury of the semiconductor electrochemist.

Acknowledgements

We are grateful to Drs. L.M. Peter and R. Peat for providing their data on n-GaAs. We would like to thank Prof. P.M. Raccah for a most helpful conversation in Oxford that clarified, to a large extent, the nature of the differences between his approach and our own, and last, but not least, we would like to thank Prof. J.B. Goodenough and Dr. M.P. Dare-Edwards.

References

1 V.A. Myamlin and Yu. V. Pleskov, Electrochemistry of Semiconductors, Plenum Press, New York, 1982. A. Hamnett, in R.G. Compton (Ed.), Comprehensive Chemical Kinetics, Vol. 27, Elsevier, Amsterdam, 1987.
2 H. Gerischer, A. Mauerer and W. Mindt, Surf. Sci., 4 (1966) 431.
3 J.F. DeWald, in N.B. Hannay (Ed.), Semiconductors, Reinhold, New York, 1959.
4 F. Lohmann, Ber. Bunsenges. Phys. Chem., 70 (1966) 428.
5 M. Green, J. Chem. Phys., 31 (1959) 200.

3 J.F. DeWald, in N.B. Hannay (Ed.), Semiconductors, Reinhold, New York, 1959.
4 F. Lohmann, Ber. Bunsenges. Phys. Chem., 70 (1966) 428.
5 M. Green, J. Chem. Phys., 31 (1959) 200.
6 M.P. Dare-Edwards, A. Hamnett and P.R. Trevellick, J. Chem. Soc. Faraday Trans. 1, 79 (1983) 2111.
7 W. Kautek and H. Gerischer, Electrochim. Acta, 27 (1982) 355.
8 B.I. Bleaney and B. Bleaney, Electricity and Magnetism, Oxford University Press, Oxford. 3rd edn., 1976.
9 M. Cardona, Modulation Spectroscopy, Academic Press, New York, 1969.
10 D.E. Aspnes, Surf. Sci., 37 (1973) 418.
11 P.M. Raccah, J.W. Garland, Z. Zhang, U. Lee, D.Z. Xue, L.L. Abels, S. Ugar and W. Wilmsky, Phys. Rev. Lett., 55 (1984) 1958.
12 D.E. Aspnes and A. Frova, Solid State Commun., 7 (1969) 155.
13 A. Hamnett, R.A. Batchelor, P.R. Trevellick and S. Dennison, J. Chem. Soc. Faraday Trans. 1, to be submitted for publication.
14 W.-K. Paik, in J. O'M. Bockris (Ed.), MTP International Review of Science, Series One, Vol. 6, 1973, p. 239.
15 O. Madelung (Ed.), Landolt-Börnstein, Neue Serie, Vol. 17c, Springer-Verlag, Berlin, 1979.
16 B.O. Seraphin (Ed.), Optical Properties of Solids, New Developments, North-Holland, Amsterdam, 1976.
17 M. Tomkiewicz, W. Siripala and R. Tenne, J. Electrochem. Soc., 131 (1984) 736.
18 I.J. Ferrer, H. Muraki and P. Salvador, J. Phys. Chem., 90 (1986) 2805.
19 K. Vos and H.J. Krusemayer, J. Phys. C, 10 (1977) 3893.
20 Y. Hamakawa, P. Handler and F.A. Germano, Phys. Rev., 167 (1968) 709.
21 A. Hamnett and S. Dennison, Nature (London), 300 (1982) 687.
22 C.M. Gronet and N.S. Lewis, Nature (London), 300 (1982) 733.

Chapter 10

Ellipsometry

R. GREEF

1. Introduction. What is Ellipsometry?

Ellipsometry is a branch of specular reflection spectroscopy which is particularly useful for characterising thin films on surfaces in situ, e.g. in the presence of supernatant liquids. It is applicable to many problems in corrosion and passivation, lubrication, the physics and chemistry of new materials, and biological chemistry. The principles behind it have been known for a long time, but it remains a rather under-used technique. This situation is now changing because of a number of factors, such as the advent of cheap computing power, and the impetus, brought about by the great economic importance of the semiconductor industry, to produce more powerful theories to interpret the results.

Another reason for its relative neglect may be that it deals with properties that are rather unfamiliar, such as complex refractive index and the nature of polarised light, and has a mathematical basis that can seem obscure. To some extent this is the fault of ellipsometrists who have sometimes been content to talk only to other specialists in the field and have not devoted much effort to popularising the technique. Understanding what ellipsometry is and what it can do does require some acquaintance with the basic concepts, and, in particular, with three important parameters that are often ignored.

2. Theory. Three important parameters

The detailed theory has been well and comprehensively described in a book that has become the "bible" of practitioners [1]. This is not primarily a tutorial text for beginners, but a more readable introductory article is available [2]. Therefore only the basic ideas and essential practicalities are presented here.

In most applications of reflectance spectroscopy, there is only one important parameter, the intensity of the reflected light at the wavelength of interest, and this leads to just one result, the relative absorbance of the film at each wavelength. In ellipsometry, three other parameters are of equal importance: (a) the phase change of the light upon reflection, (b) the refractive index of the film material, and (c) the thickness of the film.

References p. 452

2.1 PHASE

Light can be described as an oscillatory electric and magnetic field which is represented as two orthogonal sinusoidally varying vectors travelling through space in the direction of the beam. If we rule magnetic materials out of the discussion, we need only consider the electric vector and can define polarisation of the light in terms of this vector in relation to spatial coordimates. Therefore, in the rest of this discussion, the magnetic component is not considered further.

For light obliquely incident on a surface, the natural coordinates are based on the plane of incidence and we define the direction parallel to this plane the p direction. The perpendicular to this plane is called the s direction (from the German senkrecht). Any electric vector can be resolved into components in these two orthogonal directions.

If the two component vectors are in phase, the light is linearly polarised and this is what is meant in popular parlance when light is said to be simply "polarised". However, there is also the possibility that the two component vectors may not be in phase. The resultant of these will be an electric vector whose tip traces out a spiral in space, the projection of which on a plane perpendicular to the ray path is an ellipse. This is the origin of the description "elliptically polarised" for such a state of the light beam. If the phase difference between the components is 90°, the light is circularly polarised, and because this phase difference may be either plus or minus 90°, the polarisation has an associated direction (clockwise or anticlockwise). The most general state of polarised light is therefore elliptical, with linear and circular polarisation being particular cases.

Ellipticity can be induced in a linearly polarised state either by trans-

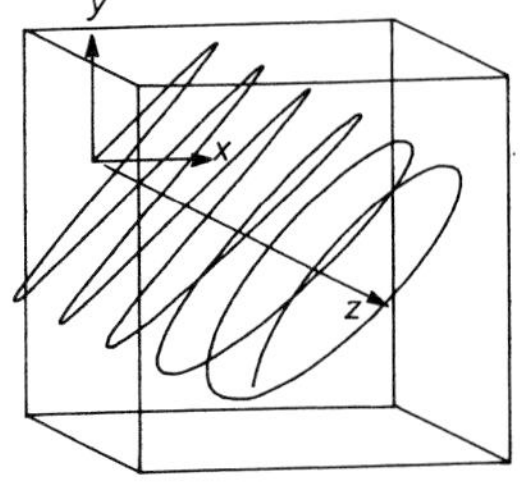

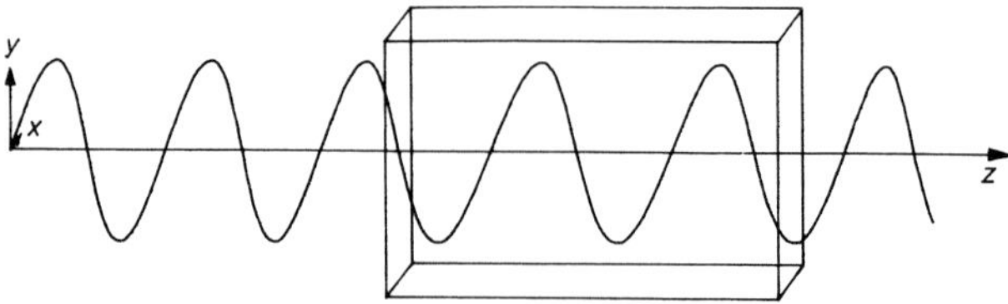

Fig. 1. The electric vector of light passing through a bi-refringent material, with $n_x = 1.10$ and $n_y = 1.15$. The light is initially polarised at 45° to the x,y directions. The two perspective views show the progressive induction of ellipticity within the material.

mission or reflection, of which the former process may be easier to appreciate. Figure 1 shows the passage of a linearly polarised light beam through a slab of birefringent material, i.e. a substance, usually a crystal, having two different refractive indices designated here as being in the x and y directions. Light polarised linearly at 45° to the x direction therefore has equal x and y components, which in the crystal have different velocities in these two directions and therefore become more and more out of phase in their passage through the material. If the thickness of the slab is chosen to be such that the phase difference induced is exactly 90°, the wave is said to be a quarter wave retarder or quarter wave plate.

If such a retarder is placed with its optic axis aligned exactly with the x or y directions, however, no change in the polarisation is induced, i.e. linearly polarised light stays linearly polarised. A combination of linear polariser and quarter wave plate can therefore be used to create any state of polarisation by proper orientation of their relative azimuths.

Ellipticity can be induced by reflection because of the fact that the reflection coefficients for the two components in the p and s directions are different. A reflection coefficient comprises two parts, an amplitude term and a phase term, and for this reason complex number representation is used to describe it. The "phasor" notation is a convention that provides a compact way of representing reflection coefficients

$$r_v = |r_v| \exp \delta_v$$

where v can be p or s.

The r_v terms are *amplitude* reflection coefficients, which define the ratio of the electric field amplitudes of the reflected and incident rays. They can have values between plus and minus unity. The observable quantities are the *intensity* reflection coefficients, R_v, which are the square of the amplitude reflection coefficients.

There are differences in behaviour between r_p and r_s which are greatest for reflection at the surface of transparent materials (dielectrics). At a certain angle of incidence, the Brewster angle ϕ_B, the p component of the reflected light goes to zero, i.e. this component of the incident light is totally *refracted*, whereas the s component increases smoothly from $\phi = 0$ to $\phi = 90°$, at which angle both components are totally reflected ($r_p = r_s = 1$). This difference is smoothed out for reflection from an absorbing material.

By contrast with the transmission case, where the phase difference induced by birefringent materials is smooth and progressive, the phase and amplitude changes of the p and s components induced by reflection at a boundary between two materials are discontinuous. Figure 2 shows the effect of these stepwise changes on the polarization of light obliquely incident on a metallic mirror.

The differences in the optical properties of dielectrics, semiconductors, absorbing materials, and metals can be understood by reference to their optical constants, i.e. their complex refractive indices.

References p. 452

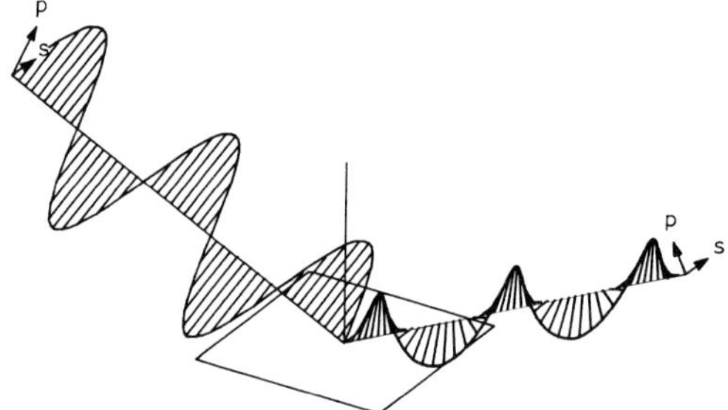

Fig. 2. Light initially linearly polarised at 45° to the plane of incidence is elliptically polarised after reflection due to the stepwise changes in the amplitude and phase of both the p and s components of the electric vector.

2.2 COMPLEX REFRACTIVE INDEX AND THE FRESNEL EQUATIONS

The concept of refractive index and its variation with wavelength is well known when applied to transparent materials. So is the idea of absorbance and the way it varies with wavelength for coloured materials and metals. It is less widely understood that dispersion and absorption of light are properly considered together and can be represented as the real and imaginary parts of a single quantity, the complex refractive index, N.

$$N = n - ik$$

For a transparent material, the imaginary part (the extinction coefficient k) is zero; it is simply related to the absorption coefficient α, measured by transmission, by

$$\alpha = \frac{4\pi k}{\lambda}$$

The reflection and transmission (refraction) of light obliquely incident on the interface between two isotropic media is entirely controlled by the angle of incidence and the complex refractive indices of the media, being described by the Fresnel reflection and refraction equations (see Appendix). Originally worked out for transparent materials, these equations apply with complete generality when the refractive indices are complex rather than simple numbers. If the refractive indices are complex numbers, the angles of refraction must also be complex. For a description of the meaning of such quantities, see ref. 3.

The form of the Fresnel equations is deceptively simple. They encapsulate the different reflectivity in the p and s planes referred to, but also describe such phenomena as internal reflection, where the refractive index of the incident medium is higher than that of the substrate, and total and frustrated total internal reflection.

While it is comparatively easy to measure the modulus of the reflection coefficient (or rather its square), it is difficult to measure the phase change of the separate p and s components. It is, however, feasible to measure both

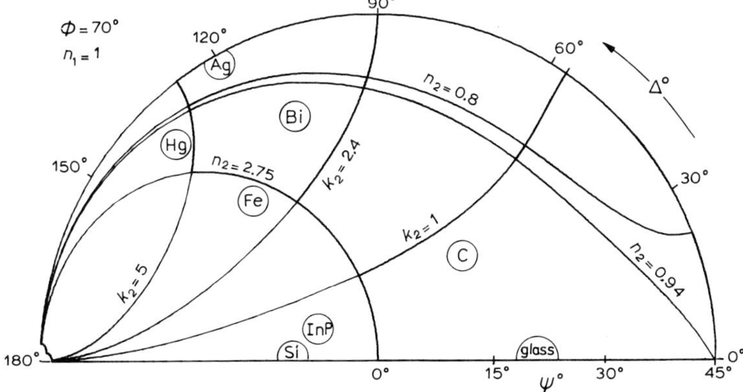

Fig. 3. The one-to-one mapping between Δ and ψ and n and k. Dielectrics, having $k = 0$, are located on the basal diameter. "Perfect" (high conductivity) metals are located near the periphery. This diagram is for $\phi = 70°$; refractive index of the immersion medium = 1.

parts of the *ratio* of the p and s components and this is what ellipsometry does. The complex ratio is defined as ρ

$$\rho = \frac{r_p}{r_s}$$

ρ could be split up into real and imaginary parts, but it is more useful to use the modulus and argument representation previously used for reflection coefficients because of the clearer physical meaning of these quantities. Tan ψ is the ratio $|r_p|/|r_s|$ and Δ is the *difference* between the phase changes for the p and s components.

$$\rho = \tan\psi \exp i\Delta$$

Ellipsometry is therefore a unique kind of spectroscopy in which each datum point contains two pieces of information. Moreover, since these two quantities are ratiometric and derived from the same beam of light, ellipsometry has the important operational advantage of high immunity to noise in the light source.

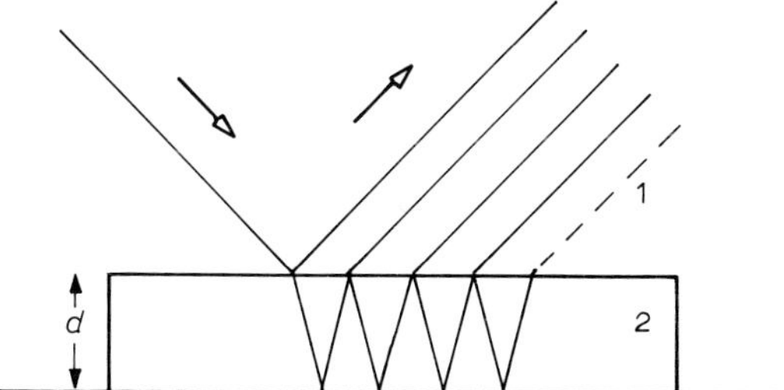

Fig. 4. The "ideal" three-phase model. Medium 1 is the immersion medium, medium 2 is a parallel-sided film of homogeneous, isotropic properties, and medium 3 is the perfectly plane, smooth substrate.

References p. 452

Of course there would be little point in measuring Δ and ψ if they could not be interpreted in terms of the properties of surfaces. Obviously they can be and to round out this discussion of bare surfaces, and the first two of the "forgotten parameters", a pictorial representation of an important property of Δ and ψ, the one-to-one mapping between them and the two parts of the complex refractive index of a reflecting surface, is given in Figure 3. There is quite a lot of information in this figure (see ref. 4), which is a representation of the Fresnel equations at one angle of incidence.

As well as being a graphic illustration of how ellipsometry can be used to measure the optical properties of clean surfaces, itself an important practical application, Fig. 3 also gives an insight into the meaning of n and k in terms of how materials approach the "ideal" properties of metals and dielectrics.

2.3 FILM THICKNESS

The third parameter comes into play when considering film-covered surfaces. The reason for calling it a "forgotten" parameter is that it cannot be measured directly by intensity–reflection methods and so it is too often ignored completely.

The model to be used as a starting point in discussing the ellipsometry of filmed surfaces is the "ideal" one-film model, sometimes called the three-phase or three-layer model (Fig. 4) in which a parallel-sided slab, the film, is sandwiched between a transparent immersion medium and a perfectly smooth substrate. All three phases are isotropic. The essential point to note is that multiple reflections occur, so that the detected beam, which is a vector sum of all these reflections, is always influenced by the presence of the film, no matter how thin it is, and whether or not it is absorbing. It is also clear that, as the film grows in thickness, d, the resultant beam will change in a complicated way. The reflection from the front surface is unchanging, and for a transparent film there is no atttenuation within the film, but the reflection coefficients are low and at each reflection there is a change in Δ and ψ. For an absorbing film, both p and s components are attenuated equally in their passage through the film but, again, there are multiple changes in Δ and ψ at each reflection that contribute in a complicated way to the polarisation state of the detected beam.

The problem of predicting the outcome of this kind of interaction was solved by Drude, whose equations (see Appendix) form the basis of the interpretation of ellipsometric results from film-covered surfaces.

3. Instrumentation

There are two basic kinds of ellipsometer: nulling and intensity modulating. The nulling ellipsometer is historically the first to be invented and was the only kind available until the late sixties. The working principle is to

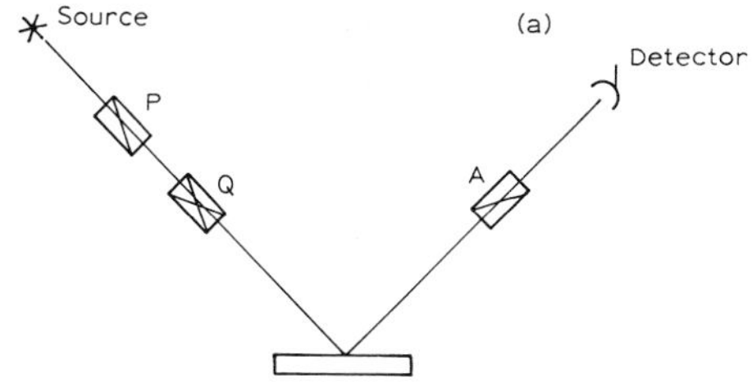

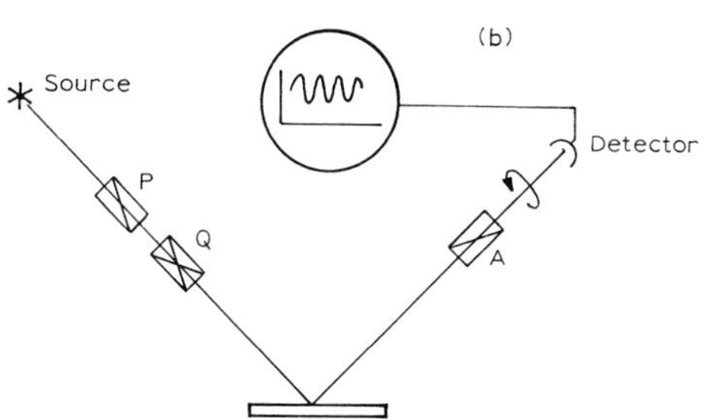

Fig. 5. (a) Shows the essential elements of a null-seeking ellipsometer. P is a linear polariser, which in combination with quarter wave retarder Q creates any required degree of ellipticity in the incident beam. At null, this ellipticity is exactly reversed by reflection and reflected light is extinguished by crossing the linear polariser A (the analyser) with its plane of polarization. The azimuthal settings of P, Q, and A relative to the plane of incidence lead directly to Δ and ψ. (b) One of the many possible configurations for a modulation ellipsometer. In this case, the modulating element is a continuously rotating analyser. The intensity–time signal at the detector has then to be analysed in a more complicated way to give Δ and ψ. Other configurations without a quarter-wave plate, or with more than one rotating element, or with continuous rotation replaced with oscillation about a mean position, are possible.

combine the creation of a known degree of ellipticity by transmission, using a combination of linear polariser (P) and quarter wave plate (Q), with reversal of this ellipticity by reflection from the specimen surface [Fig. 5(a)]. In this way, linear polarisation is restored and by crossing the plane of polarisation with another polariser (the analyser A), the null condition of light at the detector can be achieved. The quantities Δ and ψ can then be calculated from the azimuths of P, Q and A.

With a strong light source, the null can be detected by eye. Sensitivity is increased by using a photomultiplier detector and the sensitivity is then limited by that of the divided circles in which the optics are mounted. However, finding repeated nulls is a laborious process, because it involves iterative approach to the null by successive adjustments to two divided circles. It is also very time-consuming, so that dynamic measurements, except on a very long time scale, are impossible. The process can now be automated, however, making it feasible to record several points per second.

This method has the usual advantages of null techniques, i.e. immunity to noise and potentially very high precision. The presence of a quarter-wave plate in the optical train, however, makes wavelength scanning more difficult. The retardance of transmission wave plates varies with wavelength, so compensating corrections have to be made, or more expensive reflection-type quarter-wave retarders, which are essentially achromatic, have to be used. Non-idealities such as residual birefringence limit absolute accuracy in nulling instruments.

The second kind of ellipsometer relies upon measuring the intensity changes produced by continuously rotating polarising elements or modulation elements based on cyclic variations of the plane of polarisation or of retardance [Fig. 5(b)]. A great many designs have been published, and a comprehensive review of the theory and practice of such designs is available [5].

One quantity which is obtainable from some of the rotating element ellipsometers is the relative total intensity change of the light upon reflection. This is an additional piece of information which is obviously not available from nulling instruments and which can be useful in interpreting results from multilayer and rough interfaces.

4. The ideal three-phase model. SiO_2 on Si

Most ellipsometers in existence today are in semiconductor processing plant, being used as a quick, accurate, and non-destructive method of measuring the refractive index of oxide and other films on silicon wafers. As well as being economically the most important use of ellipsometry, this is also an example of a well-nigh ideal example of a three-phase model, involving as it does materials that are stable and as perfect as high technology can make them. A lot can be learnt of the theory and practice of ellipsometry from this one system.

Figure 6 shows the results to be expected from a system of this kind, comprising immersion medium (usually air or a low pressure gas, in the semiconductor world), a transparent film, and an absorbing substrate. The figure is in the form of a delta–psi "signature", i.e. it shows the evolution of a pattern in the Δ–ψ plane as the film thickness is smoothly increased from zero, the wavelength of the illumination and angle of incidence being held constant. Azzam and Bashara [1] have called these curves constant angle of incidence contours (CAICs).

The first point to notice is that the contours are closed curves, a separate one for each refractive index. The Δ–ψ values for the difference refractive indices necessarily all pass through the zero-thickness point which is characteristic of the n and k of the bare substrate. The thickness at which the curves cycle through this point, d_r, (i.e. the thickness period) is characteristic of the film refractive index, the angle of incidence, and the wavelength.

$$d_r = \frac{\lambda}{2}(N_2^2 - n_1^2 \sin^2\phi)^{-1/2}$$

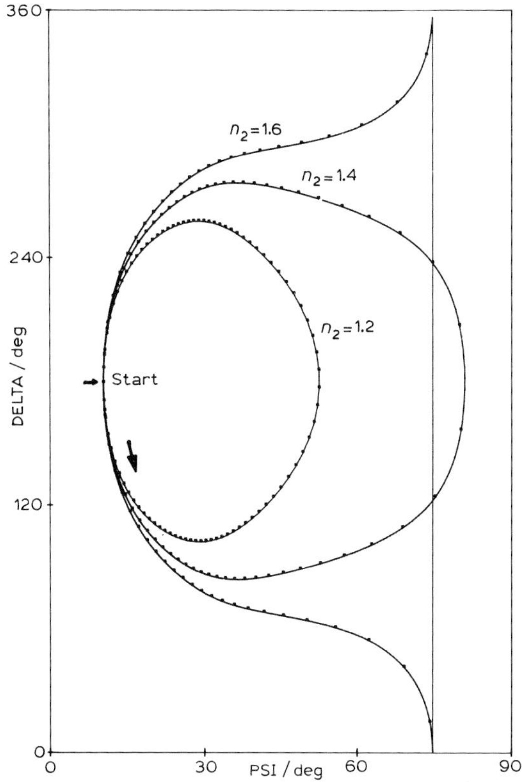

Fig. 6. Signatures in the Δ,ψ plane for the growth of SiO_2 films on Si, with air as the immersion medium. Wavelength = 633 nm, ϕ = 70°, n_{Si} = 3.85 − 0.02*i*.

This periodicity is obviously akin to the phenomenon of interference fringes. The d_r term is important in ellipsometry because it is the only one that contains the wavelength of light.

Notice also that the change in Δ and ψ for a given thickness step varies with the thickness, being greatest on the apex of the egg-shaped curves. For higher refractive indices, the ψ value at the apex approaches 90° and, for further increases of refractive index, the egg "bursts", with Δ decreasing suddenly through the 0–360° boundary.

Graphs containing many of these 0–360° crossings have an untidy appearance and it is here that plots of Δ and ψ in polar form are more useful. Figure 7 shows a family of CAICs for a range of film indices. The lower-index curves lie on the kidney-shaped figures, while the higher-index curves lie on the "eyeball" shapes which intersect the horizontal axis forming the 0–360 deg boundary.

These graphs demonstrate the vitally important property that any point in the Δ–ψ plane is interesected by only *one* contour, so a single Δ–ψ measurement for a substrate of known optical constants is able to give *both* the

References p. 452

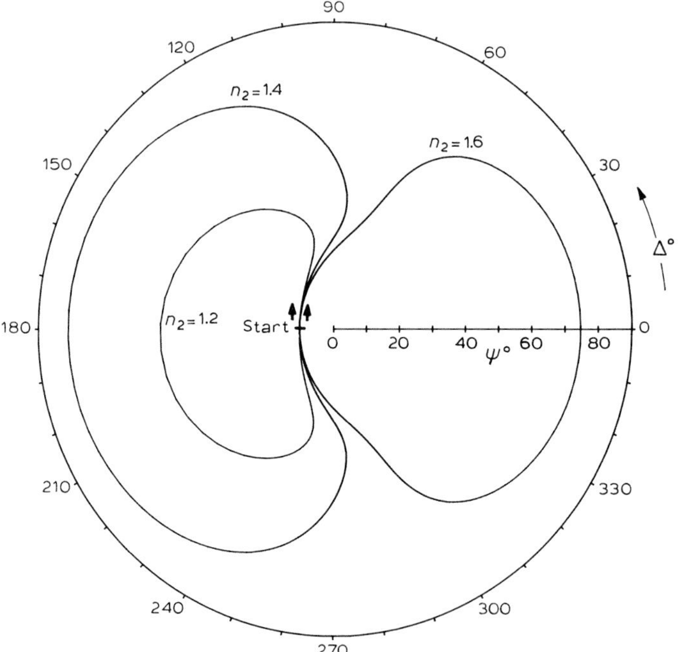

Fig. 7. Polar plot of data in Fig. 6. The radial coordinate is ψ; the angular coordinate is Δ.

thickness and the refractive index of a transparent film. There is an obvious snag, due to the cyclic nature of the contours, in that multiples of the thickness period give the same Δ–ψ value. For films of the kind used in the semiconductor industry, this is usually not an important restriction as the films are usually below the thickness period (around 3000 Å for oxide at 70° and wavelength 6328 Å) or the thickness is known to a resolution much higher than this. For films that are very much thicker than d_r, the thickness cannot be determined by ellipsometry alone; some supplementary information is needed. This can sometimes be provided if the film can be grown in steps, with intermediate stages being masked so that measurements can be made on thinner parts of the film, or if the film can be chamfered at the edge. Best of all, of course, is to measure the film growth in situ from the start. An electrochemical example of unambiguous thick film measurement (PbO_2) by this technique is given later. Ellipsometry is not necessarily the correct tool for measuring very thick films and some other technique such as stylus profilometry may be better.

The latter technique does not, of course, give the refractive index (RI) of the material, which is an important parameter in semiconductor applications that can be used diagnostically as a quality control parameter.

Another area worth attention is the part of the Δ–ψ signature near the starting point, which marks a confluence point for all the RIs, where discrimination of this property is low (Fig. 8). In this case, further information is

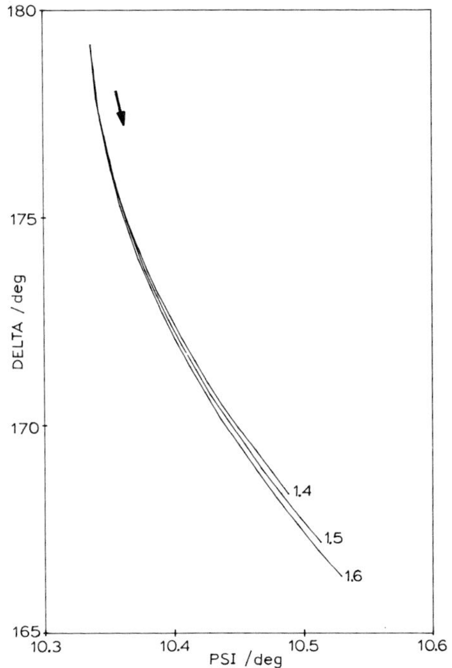

Fig. 8. Early part of growth signatures on an enlarged scale. The final thickness in each case is 4 nm.

necessary to give RI and thickness resolution and it is easily provided by varying either (or both) of the two parameters held constant in CAIC diagrams, i.e. the angle of incidence and the wavelength. Figure 9 gives an overview of the kind of results obtained at different angles of incidence. Generally speaking, ϕ variations will allow resolution of n and d at near-period thickness. Figure 8 shows that the parameter that gives refractive index discrimination at the other end of the thickness range is the slope of the Δ–ψ signature.

Here, for d values less than 10 nm or so, the discrimination available by altering ϕ is far less and wavelength variation becomes a necessity [6].

In the case of very thin films, below about 5 nm, unambiguous determination of both thickness and RI is more difficult and it is essential to use spectroscopic ellipsometry over as wide a wavelength range as possible [7]. This type of analysis works best when there is structure in the RI spectrum of the substrate or in both substrate and film. However, it has been shown that it is possible to gain information on thin transparent layers, such as spread monolayer films on water, by using spectroscopic ellipsometry [8].

Of course, the "ideal" three-phase model begins to break down for good chemical reasons for very thin compound films such as SiO_2, which are formed from the substrate, as there is, by definition, always an interphasial region of non-stoichiometric composition. Nevertheless, progress has been

References p. 452

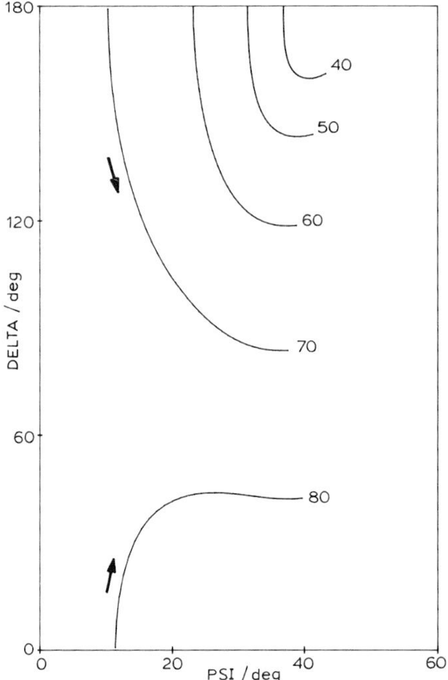

Fig. 9. Signatures for the growth of 100 nm of SiO_2, n = 1.4, on Si at various angles of incidence marked in deg, showing the variation in sensitivity of Δ and ψ to thickness. The pseudo-Brewster angle for Si is at about 75 deg, and this marks the division between the two branches of the signatures.

made in analysing the composition of such "connecting layers" as SiO_x between bulk-property SiO_2 and Si by using spectroscopic ellipsometrry [9].

Some further points about ellipsometer performance are implicit in Fig. 8, which shows that resolution in the Δ and ψ terms may be necessary down to a few hundredths of a degree. This is not a severe challenge to commercially available ellipsometers when the surface is highly reflecting. Special purpose ellipsometers can resolve 0.001° in favorable circumstances.

5. Calculations for the three-phase model

Estimation of the thickness and RI of perfect films can be done by inspection from a Δ–ψ graph of suitable scale or by consulting published tables of data [10]. With the ready availability of powerful desk-top computers, however, a computer search is the most usual way of solving the problem. Many search algorithms are available and, for most regions of the diagram, the problem of matching thickness and refractive index is not a very demanding one; most algorithms will work. Near the zero-thickness point, of course, where the RI sensitivity is low, the program should report this, for example as a high correlation between the thickness and RI parameter. A general purpose program for doing this kind of search, as well as many other calculations for ellipsometry, is available and widely used [11].

6. A "nearly perfect" three-phase system

The electrochemical deposition of lead dioxide provides an example of a three-phase system which is so nearly perfect that it can be analysed by the foregoing techniques, but which has some instructive non-idealities. Divalent lead ions from a wide variety of salts can be oxidised electrochemically in acid or alkaline solution to give a smooth adherent deposit on inert surfaces such as platinum. Either of the two crystalline forms α or β, or possibly a mixture of them, make up the layer, according to the ionic environment. If a mixture of sodium acetate, acetic acid, and plumbous acetate is aqueous solution is used, the α form is thought to be produced exclusively [12] and ellipsometry has been used to learn more about the growth details [13].

Figure 10, taken from unpublished work in the author's laboratory, shows the Δ–ψ signature in polar form (the Cartesian plot of the data is confused by a plethora of zero crossings). This plot is strongly reminiscent of the Si–SiO_2 family (Fig. 6) but, in this case, the cross-over from the "kidney"

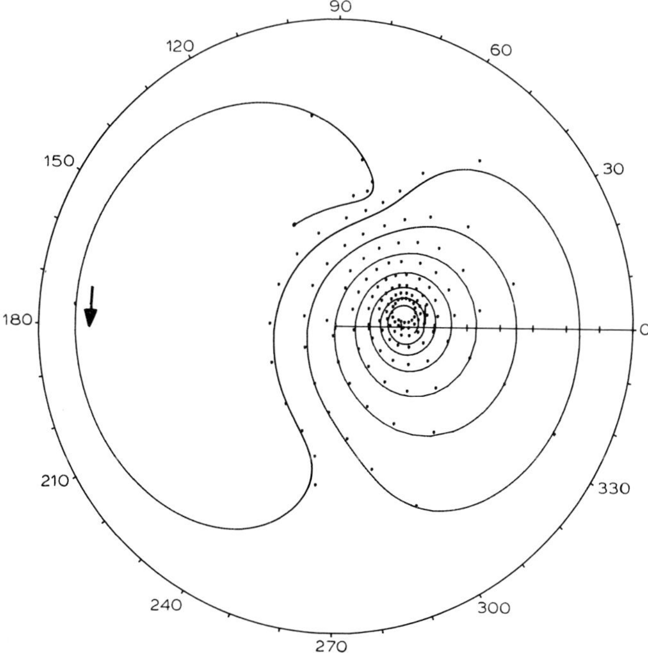

Fig. 10. Polar signature for the growth of PbO_2 from acid–acetate solution at 2 mA cm^{-2} (points). The smooth curve is calculated for a film of optical constants 2.05 − 0.068i growing at a rate of 3.5 nm s^{-1}. The final film thickness is about 1600 nm.

References p. 452

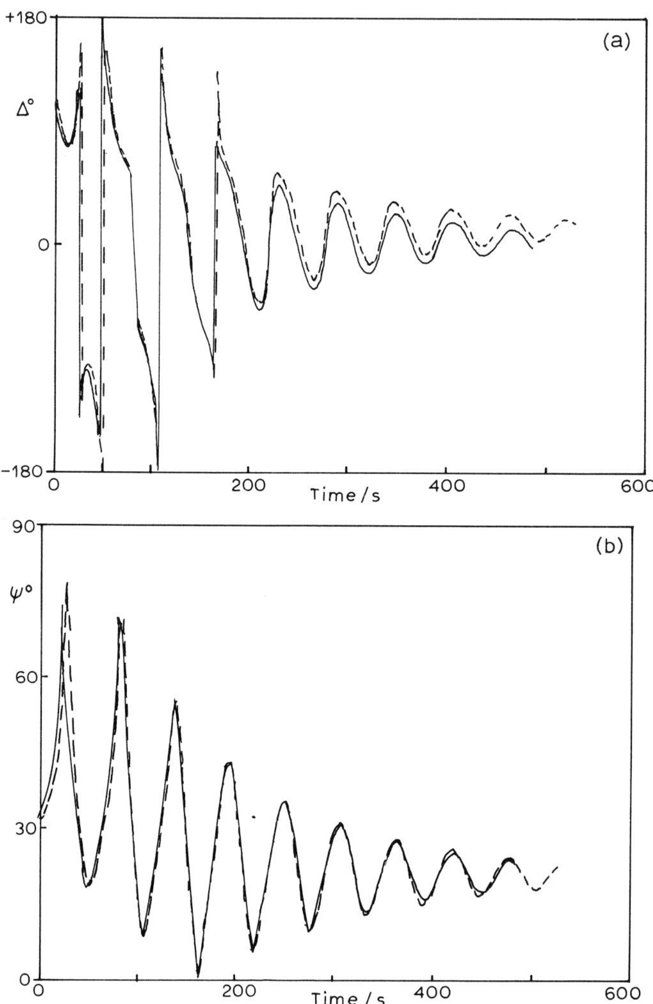

Fig. 11. (a) and (b) Data of Fig. 10 plotted on a time axis. The optical constants of the film can be estimated from the Δ,ϕ values to which the curves are converging.

curve to the "eyeball" curve with increasing thickness is a characteristic of a weakly absorbing film. The experiment was conducted at constant deposition current and previous experiments had suggested that the oxide production proceeded with very high current efficiency at this current density ($2\,\mathrm{mA\,cm^{-2}}$). With this assumption, a theoretical fit to the "ideal" three-phase model entails finding three parameters: n and k of the film and the thickening constant Tc ($\mathrm{nm\,s^{-1}}$) for the film growth. This fit was carried out and showed that, certainly in its early stages, the growth constant and the optical constants are indeed almost invariant. The theoretical and experimental curves are plotted together on a time axis (Fig. 11) to show the excellent agreement. The three parameters so obtained are independent, i.e.

the effect of a small change in one cannot be compensated by changes in either, or both, of the others, and so this set of parameters is unique. With this unambiguous film thickness, the density of the film was worked out assuming 100% current efficiency for its production and it was found that the oxide had a density of 70% of the value for dry PbO_2 as measured by X-ray crystallography. This density deficit is one of a number of intriguing features of this system, which is currently still under investigation.

There is one other purely ellipsometric point to be made from Fig. 10 concerning the fact that the end-point to which the measured points are converging is slightly different from that for a material having the optical constants of the earlier part of the film growth. At the last experimental point, the layer is about 1600 nm thick and its outer surface is no longer completely smooth. This slight degree of roughness causes some light scattering, which is responsible for the discrepancy.

7. Non-uniform and rough films. Aluminium corrosion

The corrosion of aluminium can be divided broadly into four regimes. In near neutral solutions, anodisation of aluminium gives dense, stable oxide films called barrier layers whose thickness is linearly proportional to the formation potential over a wide range. Anodisation in acids gives layers with a complex structure, having a dense part in contact with the metal and a honeycomb layer of voids and oxide above. The oxide is strong and can be grown thickly, so is very widely used as a protective layer that will also absorb dyes for a decorative finish. In alkaline solutions, aluminium corrodes freely, but at high potentials a hydrated oxide layer is formed which gives some protection to the underlying metal but which is itself slowly soluble.

These three regimes are examples of uniform corrosion and can be contrasted with the fourth, i.e. solutions containing aggressive ions such as chloride, where protective layers are not formed, and localised, pitting corrosion takes place.

Ellipsometry can be used to gain information about the films formed in the three uniform corrosion regimes. Details are given here about two of those cases; firstly the barrier or valve metal oxide layers, which optically are nearly ideal films, and secondly alkaline corrosion, which gives some insight into the applications of ellipsometry to non-uniform films and roughened metallic surfaces.

Figure 12 shows the Δ–ψ signature for the growth of a barrier layer on a super-pure aluminium electrode which had been metallographically polished. The points represent steady values which were achieved after polarisation with a constant potential across a two-electrode cell. The potential was advanced manually after each point had been recorded, up to a maximum of 60 V. Upon disconnecting the cell, the optical signal remained constant and

References p. 452

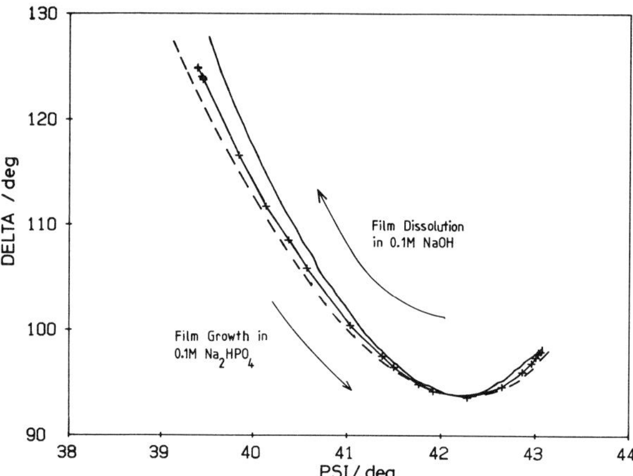

Fig. 12. Growth of a barrier layer on Al as the voltage across the film is increased to a maximum of 60 V, The broken line is a three-phase calculation for a film of $n = 1.60$.

so the dissolution of the film could be studied by sucking the original solution (Na_2HPO_4) out of the cell and replacing it with 0.1 M NaOH. Dissolution proceeds slowly, but in a regular manner, as shown by the way the Δ–ψ contour is retraced close to the growth curve.

In calculating a three-phase model signature (superimposed in Fig. 12 as a broken line) there is some doubt as to the correct starting point. It is known that Al always has an air-formed protective film whose thickness has been estimated by various techniques to be about 2 nm. The starting point was estimated by using this value and a film RI of 1.65.

The agreement between theory and experiment is very good: the barrier layer behaves as an almost ideal phase, with a refractive index very close to that of the anhydrous crystalline material. This is confirmed by the Δ–ψ trajectory for the dissolution of the layer, which, in contrast to the behaviour of the porous hydrated oxide formed in alkaline solution, dissolved uniformly from the outer surface towards the metal.

8. Aluminium in alkaline solution: non-uniform and rough films

The Δ–ψ signature for a super-pure aluminium electrode polarised anodically at a series of constant potentials is shown in Fig. 13. As the film grows, the familiar egg-shaped curve is traced out, except that, in this case, the shape is distorted from that typical of the growth of a uniform film. The film of hydrated oxide is necessarily non-uniform, because it is slowly soluble in the alkaline medium, so the outer surface, having been in contact with the electrolyte for longer than the more deeply lying part, has already been partially dissolved at any stage of layer growth. The layer, in fact, is porous, having fewer voids close to the metal where it is being formed. This also

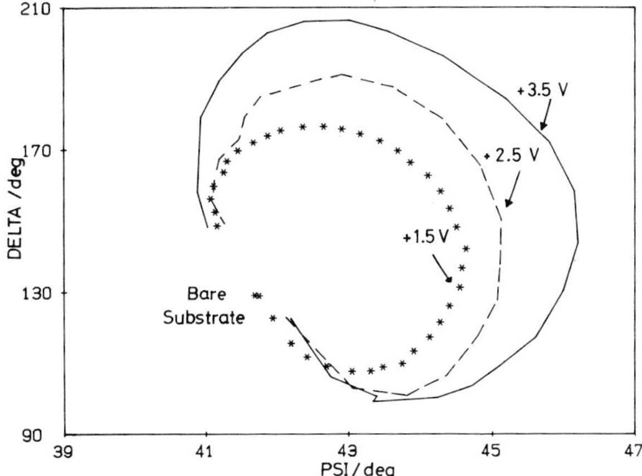

Fig. 13. Growth of a hydrated oxide on aluminium in 0.1 M NaOH at three anodisation potentials.

explains why the size of the "egg" is larger at high potentials as the rate of formation is then higher, while the dissolution rate, being non-electrochemical, is unaffected. This process has been modeled quantitatively, by replacing the single film by a "pile of plates" of graded refractive index [14].

By dropping the potential to a value where the oxide stops forming, the dissolution process can be watched in its entirety (Fig. 14). The Δ–ψ curve now follows a trajectory that lies within the original figure until almost all the oxide has been dissolved. This effect, too, has been modeled in a similar way, taking account that, as soon as the formation process stops, dissolution occurs throughout the depth of the oxide, including the denser layer in contact with the metal.

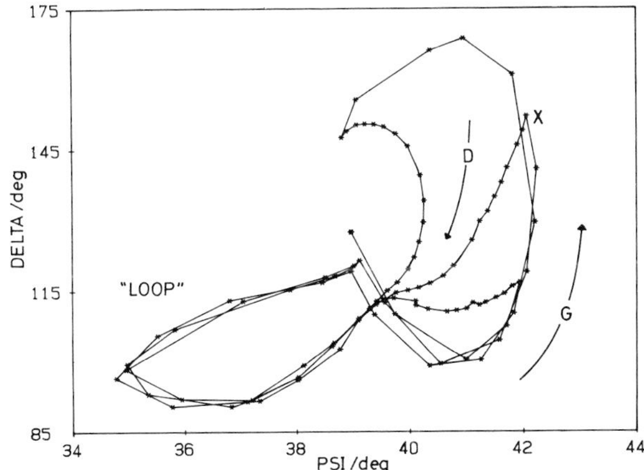

Fig. 14. Growth (in direction G) followed by dissolution (direction D) of anodic aluminium oxide film. The "loop" is an ellipsometric signature of a spontaneous substrate roughening–self-smoothing cycle.

References p. 452

The surprising feature of the dissolution curves is the large and very reproducible excursion or loop as the starting-point is approached. This traverses a region of the Δ–ψ surface that is inaccessible by an combination of transparent layers and therefore has to be due to the development, and the subsequent disappearance by a remarkable "self-healing" process, of a rough layer or pitted aluminium surface.

The modeling technique applied to the transparent oxide layers also works in this case provided a suitable depth profile for the roughened surface can be found. Trial-and-error procedures were used and a profile consisting of triangular-section furrows was found to give good agreement with experiment.

Later work was carried out on aluminium foils, removing the electrode at a suitable point on the formation–dissolution cycle, then cracking the foil after cooling in liquid nitrogen, and examining the cracked cross-section by scanning electron microscopy. The images obtained provided dramatic confirmation of the surface dynamics worked out on the basis of ellipsometry alone, showing the porous layer atop a smooth metal surface (Fig. 15) at point X in the growth cycle and a rough metal surface sparsely covered with oxide at a point on the apex of the loop.

9. Thin films. A biological example

There are many applications of ellipsometry in the measurement of monolayer and sub-monolayer films. The theory of the optical signal to be expected from an adsorbed layer less than one monolayer thick has been placed on a firm footing by Smith [15] in some elegant experiments on adsorption in a Langmuir trough. Simultaneous ellipsometric and surface potential measurements were made on various molecules spread in thin layers on mercury as the surface pressure was varied. One conclusion was the simple result that the effective thickness divided by the thickness of the island molecules in the adsorbed islands was equal to the fractional coverage of the surface area.

An application that is presently coming to the fore is the investigation of biological materials, for example antibody–antigen reactions, with the antibody immobilised by attachment to a surface. This is, in fact, one of the original applications of modern ellipsometry and an early polarisation-measuring instrument was used by Rothen and Landsteiner [16] to follow an immunological reaction some years before Rothen coined the work ellipsometry to describe his apparatus [17]. Present-day quantitative immunology relies upon radio-tracer assay but, by an interesting twist of scientific history, ellipsometry is one of the techniques vying to replace it in the search for more environmentally acceptable techniques.

Two examples follow of the kind of information that ellipsometry can give about the adsorption of biological molecules. Ma et al. [18] studied the

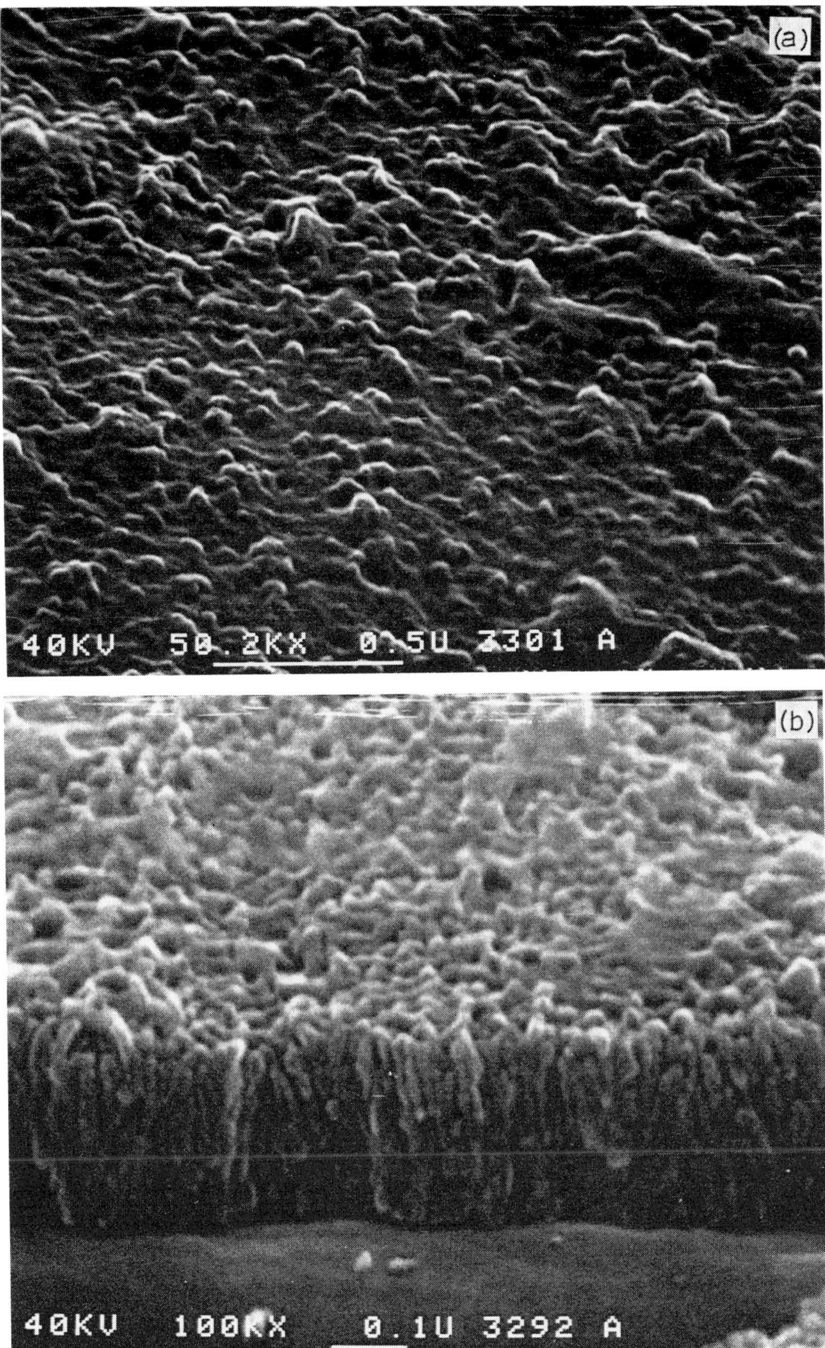

Fig. 15. (a) Scanning electron micrograph of sectioned anodic film on aluminium grown at 2.5 V in 0.1 M NaOH, showing a porous oxide on a smooth substrate. (b) SEM of the anodic film dissolved at − 0.5 V for 80 s, showing a patchy residual oxide on a roughened substrate.

References p. 452

adsorption of bovine serum albumin (BSA) on oxides of tantalum and silicon. Recognising that such films may desorb, become denatured, or participate in competitive adsorption processes, these authors then attempted to stabilise the films by treatment with glutaraldehyde and then exposed the modified films to further albumin solution. The first albumin layer had a high refractive index of 2.0 and a thickness of 2 nm, consistent with a close-packed monolayer structure. When this layer was exposed to an albumin–glutaraldehyde mixture, the thickness increased, promoted by glutaraldehyde cross-linking in random orientation, though the density of the additional layer was lower. When this assembly was further treated with albumin solution, there was further thickening of the layer, with comcomitant reduction in density, as the albumin was adsorbed only on the sites of dangling glutaraldehyde bonds.

These ideas are being taken further in recent papers by Jonsson et al. [19] where the reactions between adsorbed and solution species are of the more specific antigen–antibody kind. Great attention is paid in this work to pretreating the silicon dioxide substrate to ensure that the active adsorbate is not denatured or changed in conformation.

The molecules studied, human fibronectin and rabbit fibronectin antibodies, were taken in turn as the species first immobilised on the surface, the other being held in solution. It was found that immobilised fibronectin interacted to a much greater extent with antibodies in solution than in the inverse situation when adsorbed antibody was taken as the starting point. This effect was shown to be due to the fact that antigens have several antibody-binding sites, so that multiple attachment of antibody molecules can occur around adsorbed fibronectin molecules. When the antibody is attached to the surface, however, only a fraction may have their antigen-specific groups pointing towards the solution, where each can be occupied by only one antigen. This idea was further confirmed by using another protein (Protein A) as a mediator between the surface and the adsorbed antibody, the effect of which is to orient the molecule so that the antigen-binding part points towards the solution, giving a higher thickness build-up on exposure to the fibronectin solution.

Ellipsometry was used in situ at all stages of the preparation of the substrates with their intermediate coatings as a monitor of surface concentration, and this study provides an excellent illustration of the power of the technique as a non-destructive and highly sensitive surface probe.

10. Spectroscopic ellipsometry

Among spectroscopic techniques, ellipsometry is unique in that it is capable of giving useful information at only *one* wavelength, as the previous examples show. Most of the ellipsometry literature is, in fact, still dominated by single-wavelength applications. However, it is only in its wavelength-

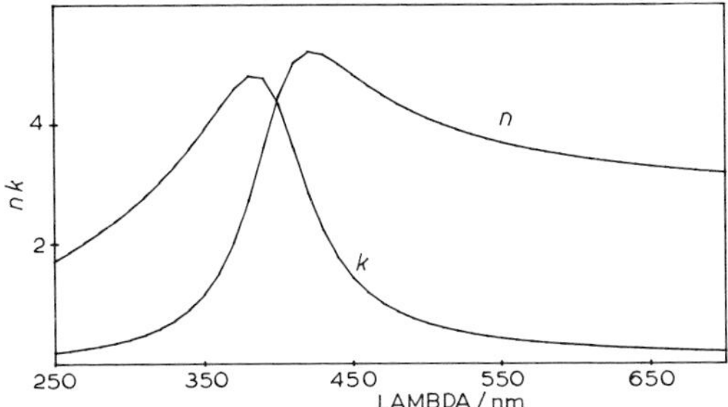

Fig. 16. Wavelength dependence of the optical constants of a typical strongly absorbing dye.

scanning mode that ellipsometry can show its full potential. It may be useful to draw a direct comparison between intensity-measuring reflectance spectroscopy and ellipsometry, and this is done in Figs. 16 and 17, showing calculated results for a simple hypothetical system. This consists of a substrate of unchanging optical properties having metallic optical constants, covered by a thin layer of a strongly absorbing material with a single

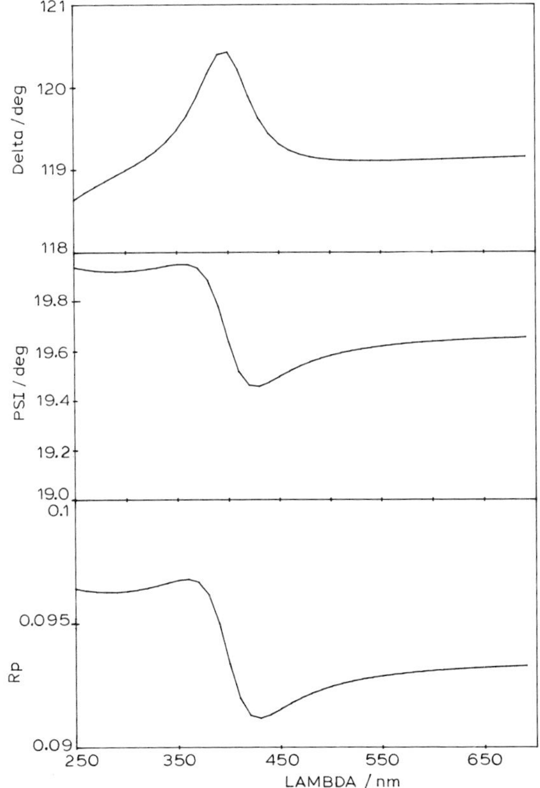

Fig. 17. Comparison between the ellipsometric and reflectance signals for a 0.1 nm thick film of the dye (Fig. 16) on a metal at angle of incidence 70°.

References p. 452

Lorentzian absorption band in the middle of the wavelength range. The optical characteristics of the film material as a bulk phase are shown in Fig. 16. Note that the absorption band is unsymmetrical and distorted by being plotted on a wavelength rather than an energy scale. Figure 17 shows the results for reflection from a thin film of this material upon the metallic substrate ($n = 3 - 2i$), from which it can be seen that there is no problem, in principle, in detecting the film ellipsometrically, while the technique also permits the measurements of the film thickness by using results from a spectral region where the film is transparent. The change in the intensity reflectance is small and would be difficult to measure in the presence of source noise without signal averaging.

Moreover, ellipsometry has the added operational advantage that, because the technique is affected only by surface layers and has a high degree of discrimination against the optical properties of the medium, measurements on such adsorbed dye layers can be made with the dye also present in the supernatant liquid, as has been shown in the author's laboratory [20]. An application of this technique to the detection of Rhodamine B as a metal-plating additive has recently been published [21].

11. Profiling of complex layers

As has been pointed out by Aspnes [22], spectroscopic ellipsometry can now rival cross-sectional transmission electron microscopy as a means of obtaining structural information about thin films. With the caveat that this really only applies at present to examples where the materials are highly purified and of well-known spectral properties, as in the world of semiconductors (although see the aluminium corrosion example above), this statement seems only slightly optimistic. More and more examples of comparison between the two techniques are appearing, but the most convincing at the present time is that published by Vedam et al. [23]. These authors studied single-crystal silicon bearing multi-layer structures formed by ion implantation. Using a model which corresponded to the known sequence of layers, but with thickness and composition for the mixed-phase layers as unknown parameters, these quantities were fitted to spectroscopic ellipsometry data by standard non-linear least-squares procedures. The results gave layer thicknesses of 270, 511, 119, and 24 Å, with error limits of a maximum of 20%. Subsequent XTEM measurements confirmed these layer thicknesses exactly, but with higher error limits on the thicknesses. The XTEM measurements could not give compositional information, whereas ellipsometry provided relative proportions of crystalline and amorphous silicon in the mixed layers with assigned uncertainties of about 3%.

12. Geometric roughness. An exact solution

A rather special case of a non-uniform or "rough" layer will be considered as a final example of the present state of development of theoretical optics allied to ellipsometry. The system consists of a smooth substrate randomly seeded with mercury spheres that are of uniform diameter. Their separation is, on average, large compared with the wavelength of light and they grow from zero size to a diameter comparable with the wavelength. This situation is created by the electrochemical deposition of mercury nuclei on a highly polished glassy carbon substrate from a dilute solution of mercury ions. These conditions allow a high degree of control of the deposition to be achieved. The population density of the nuclei can be set in the range 15^5–$10^7\,cm^{-2}$ by controlling the potential of deposition from 100 to 200 mV negative to the mercury reversible potential. Nucleation is close to being instantaneous rather than progressive, so the nuclear diameters are confined to a narrow range and the theory of nucleation allows both the nuclear density and diameter to be determined by analysis of the current–time response.

The ellipsometry of the system has been studied [24] and the results show a very characteristic Δ–ψ signature consisting of overlapping spirals, the pitch and scale of which become progressively larger for deposition at higher overpotentials (Fig. 18). An attempt was originally made to explain these results on the basis of an effective medium theory, but an exact physical solution has subsequently been found based on Mie theory for light scattering for small particles, taking account of optical interaction of the spheres with the substrate [25]. Work is now in progress to see whether the theory can be extended to non-spherical particles.

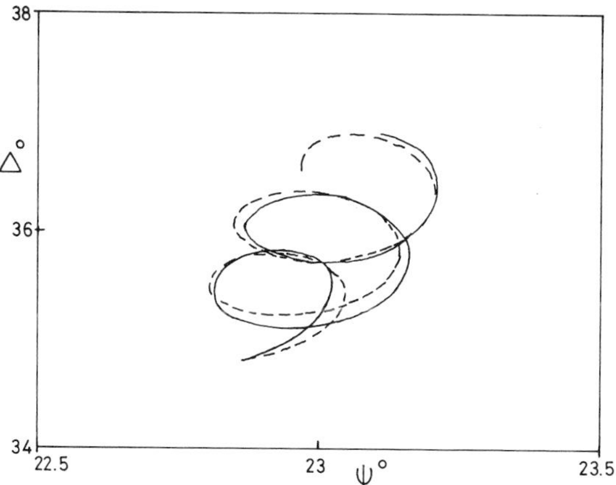

Fig. 18. Comparison between the experimental and theoretical optical signals from a glassy carbon surface randomly seeded with growing mercury nuclei, electrochemically deposited at an overpotential of 258 mV. The theoretical curve (broken line) is for a nuclear density of $7 \times 10^5\,cm^{-2}$, with a density spreading coefficient (see ref. 25) of 0.08.

13. Reviews of ellipsometry applications

No attempt has been made to survey comprehensively the wide and diverse field of ellipsometry applications, as several such reviews are available. A most important area which has hardly been mentioned, for example, is that of corrosion and a review of the field up to 1972 is provided by Kruger [26]. Another useful review covers a wider range of the literature up to about 1978 [27]. For semiconductor applications, with emphasis on later developments in the theory of multilayer structures, two reviews by Aspnes can be recommended [22, 28]. The proceedings of the international conferences on ellipsometry are an excellent source of information on developments in instrumentation, theory, and applications in all areas of surface physics and chemistry [29–33].

14. Outlook

Ellipsometry is an old technique whose practical foundations were laid by Rayleigh in his experiments on reflection from liquids a hundred years ago. Its development as a routine tool in surface research received a boost in the mid-forties with the experiments of Rothen [17], but it was not until automatic recording instruments began to be available in the late sixties that its potential began to be realised. Single-wavelength instruments have been commercially available for some time, but spectroscopic instruments are only now moving out of the development laboratories on to the commercial scene.

Other developments in techniques are still taking place. For example, it has been shown that it is possible to study the corrosion of metals *beneath* the surface of a (pigmentless!) paint film [34]. Another technical achievement, which is certain to be important in the future, is the development of a microscopic ellipsometer [35] with which it is possible to study localised (pitting) corrosion of individual grains in an alloy surface.

The other area of rapid progress is in the development of theories and computational tools with which to interpret spectroscopic results for rough surfaces, discontinuous layers and multilayered heterostructures.

Compared with some of the high-vacuum techniques now being used routinely in semiconductor research and production facilities, ellipsometry is an inexpensive technique and it can be expected to take an increasing role in surface research.

15. Appendix

Ellipsometry is based fundamentally on the Fresnel equations for the reflection and refraction of light at a sharp two-phase boundary. They can

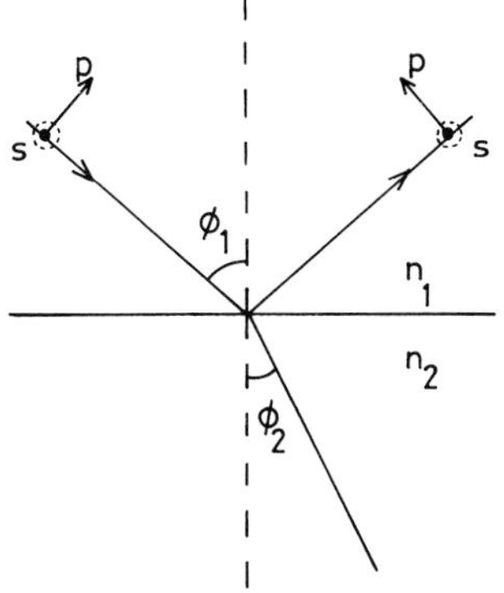

Fig. A1. Reflection at a two-phase boundary.

be stated in various forms, most compactly as (see Fig. A.1)
(a) for reflection

$$r_p = \frac{\tan(\phi_1 - \phi_2)}{\tan(\phi_1 + \phi_2)}$$

$$r_s = -\frac{\sin(\phi_1 - \phi_2)}{\sin(\phi_1 + \phi_2)}$$

(b) for refraction

$$t_p = \frac{2\sin\phi_2\cos\phi_1}{\sin(\phi_1 + \phi_2)\cos(\phi_1 - \phi_2)}$$

$$t_s = \frac{2\sin\phi_2\cos\phi_1}{\sin(\phi_1 + \phi_2)}$$

The Fresnel coefficients are complex numbers, describing the amplitude and phase changes of the p and s components.

In considering reflection by a surface bearing a film, the Drude equations, below, take into account the infinite number of reflections at substrate–film and film–ambient interfaces. Using the phase numbering of Fig. 4, the overall reflection coefficients for the assembly are designated by R

$$R_p = \frac{r_{12p} + r_{23p}\exp(-2\beta i)}{1 + r_{12p}r_{23p}\exp(-2\beta i)}$$

$$R_s = \frac{r_{12s} + r_{23s}\exp(-2\beta i)}{1 + r_{12s}r_{23s}\exp(-2\beta i)}$$

where β is a function of the film thickness and wavelength

$$\beta = 2\pi\left(\frac{d}{\lambda}\right)(N_2^2 - n_1^2\sin^2\phi_1)^{1/2}$$

and the r terms are Fresnel reflection coefficients for the two-phase boundaries.

References p. 452

Acknowledgements

Figures 13 and 14 are taken from ref. 14 and are reprinted by permission of The Electrochemical Society, Inc. Figure 15 is printed by permission of Alcan International Ltd., Banbury Laboratories.

References

1 R.M.A. Azzam and N.M. Bashara, Ellipsometry and Polarized Light, North-Holland, Amsterdam, 1977.
2 R.H. Muller, Adv. Electrochem. Electrochem. Eng., 9 (1973) 168.
3 Z. Knittl, Optics of Thin Films, Wiley-Interscience, New York, 1976, p. 182.
4 R. Greef and M.M. Wind, Appl. Opt., 25 (1986) 1627.
5 P.S. Hauge, Surf. Sci., 96 (1980) 108.
6 S.S. So and K. Vedam, J. Opt. Soc. Am., 62 (1972) 16.
7 H. Arwin and D.E. Aspnes, Thin Solid Films, 113 (1984) 101.
8 G.T. Ayoub and N.M. Bashara, J. Opt. Soc. Am., 68 (1978) 978.
9 D.E. Aspnes and J.B. Theeten, J. Electrochem. Soc., 127 (1980) 1359.
10 G. Gergely (Ed.), Ellipsometric Tables of the Si–SiO_2 System for Mercury and He–Ne Laser Spectral Lines, Akadémiai Kiadó, Budapest, 1971.
11 F.L. McCrackin and J. Colson, Natl. Bur. Stand. (U.S.) Tech. Note, 242 (1964).
12 L.M. Peter, Surf. Sci., 101 (1980) 162.
13 J.L. Ord, Z.Q. Huang and D.J. De Smet, J. Electrochem. Soc., 132 (1985) 2076.
14 R. Greef and C.F.W. Norman, J. Electrochem. Soc., 132 (1985) 2362.
15 T. Smith, J. Opt. Soc. Am., 58 (1968) 1069.
16 A. Rothen and K. Landsteiner, J. Exp. Med., 76 (1942) 437.
17 A. Rothen, Rev. Sci. Instrum., 16 (1945) 26.
18 S.M. Ma, D.L. Coleman and J.D. Andrade, Surf. Sci., 56 (1976) 117.
19 U. Jonsson, M. Malmqvist, I. Ronnberg and L. Berghem, Prog. Colloid Polym. Sci., 70 (1985) 96.
20 R. Greef and P.J. Pearson, J. Phys. C, 10 (1983) 505.
21 J.C. Farmer and R.H. Muller, J. Electrochem. Soc., 132 (1985) 313.
22 D.E. Aspnes, Surf. Sci., 101 (1980) 84.
23 K. Vedam, P.J. McMarr and J. Narayan, Appl. Phys. Lett., 47 (1985) 339.
24 R. Greef, Ber. Bunsenges. Phys. Chem., 88 (1984) 150.
25 P.A. Bobbert, J. Vlieger and R. Greef, Physica, 137A (1986) 243.
26 J. Kruger, Adv. Electrochem. Electrochem. Eng., 9 (1973) 227.
27 W.E.J. Neal, Appl. Surf. Sci., 2 (1979) 445.
28 D.E. Aspnes, Appl. Surf. Sci., 22/23 (1985) 792.
29 Ellipsometry in the Measurement of Surfaces and Thin Films, U.S. Dept. of Commerce, National Bureau of Standards Miscellaneous Publication 256, 1963.
30 Proceedings of the Symposium on Recent Developments in Ellipsometry, Surf. Sci., 16 (1969).
31 Proceedings of the Third International Conference on Ellipsometry, Surf. Sci., 56 (1976).
32 Proceedings of the Fourth International Conference on Ellipsometry, Surf. Sci., 96 (1980).
33 Ellipsometry and Other Optical Methods for Surface and Thin Film Analysis, J. Phys. C, 10 (1983).
34 J.J. Ritter and J. Kruger, J. Phys. C, 10 (1983) 225.
35 K. Sugimoto, S. Matsuda, Y. Ogiwara and K. Kitamura, J. Electrochem. Soc., 132 (1985) 1791.

Chapter 11

A Computing Strategy for the On-line Accumulation and Processing of Electrochemical Data

J.A. HARRISON

Abstract

A strategy for controlling and collecting data from experiments in electrode kinetics is described. In view of the many methods available, a consideration is given to which electrochemical methods are suitable for automation and which give the maximum information. A number of electrochemical methods have been investigated from this point of view and all can be implemented using the present equipment. The problem is that the methods which are easy to use, such as the linear potential sweep method and the polorographic methods, contain less electrical information than the methods such as the impedance-based methods, which are much more difficult to use and interpret. The use of computers completely eliminates this problem. However, in a complete system, both types of measurement have a role. The simultaneous measurement of steady state current–potential and impedance–potential data with suitable data analysis, has proved, in practice, to be the most versatile methods for investigations in electrode kinetics. In the second line of importance have been found to be measurements of current and impedance under various combinations of applied potential and time. Thirdly, measurements such as linear potential sweep and large amplitude pulses take their place. This system has been developed over the last ten years at the University of Newcastle, Gt. Britain and is completely automatic. It is run in the manner of a spectrometer, often overnight, and has been applied to rotating-disc, stationary, and other electrode systems and electrochemical devices.

1. Introduction

Electrochemistry and the investigation of electrode kinetics still seems to have an expanding role in chemical investigations. However, the group of obvious and straightforward electrochemical reactions have been investigated in great detail over the decades of this century and the emphasis must shift to more complex reactions. For investigations in this subject area, electrochemistry is only one of the techniques which will need to be employed. Exactly what the most powerful combination will be is not yet certain. The scientific areas in which electrochemistry may play a major role in the future have been reviewed by the Pimentel Committee in a report which was published in 1985 [1]. If the electrode kinetics is to enter a wider chemical appreciation, it will be necessary to carry out data collection and data reduction procedures more efficiently than at present. Using the usual methods, these steps are very time-consuming. It will also be necessary to use more powerful electrochemical methods. This will doubtless mean the more expert use of computers. It is now accepted that the more ambitious use

References p. 497

of computers can have benefits to chemistry similar to the benefits that have been gained in other disciplines, such as engineering. This is one of the points that have been emphasised in the literature (see, for example, ref. 2).

Computers are commonly used in separate areas on a small scale in applications such as

(a) control of instrumentation,
(b) data collection,
(c) data analysis,
(d) mathematical calculations, and
(e) graphics

Larger scale applications which involve one or more of these functions and have some built-in intellegence are more difficult to implement. However, this situation is changing rapidly as developments in computer science become more readily available. In this respect, it is unfortunate that few scientists are trained in, say, computing and chemistry.

Outstanding examples of the use of computers in chemistry are in the fields of mass spectrometry, nuclear magnetic resonance, and the many techniques for surface analysis. Developments in these fields have been a matter of slow, but sustained improvement. One reason for the slowness of the developments seems to have been the time needed to introduce the different areas of science which are necessary for a complete system. Foremost amongst these areas is computing science. Computing systems are notoriously difficult and slow to develop because of the complexity and number of the necessary and interlocking software and hardware modules.

However, the appreciation of the role of computers in science is developing rapidly. For example, the introduction of more powerful computers with integral graphics facilities have led to significant applications in chemistry (see a number of general articles which have appeared on the use of mini- and microcomputers as workstations in chemical laboratories [3–5]).

As can be seen from the electrochemical journals and the literature, computers have not been applied to problems in fundamental electrochemistry in any significant way, although the problems are virtually identical to those encountered in the computerisation of spectroscopic techniques. A recent review of fundamental electrode kinetics and instrumentation, for example [6], is similar in content to a review of more than ten years earlier [7].

Analytical applications of electrochemistry, where the objectives are well defined, have fared better. There is a long list of papers going back twenty years on the applications of computers and then microprocessors. Reviews of this subject appear in the Fundamental Reviews sction of *Analytical Chemistry* (see refs. 8 and 9). In general, the aim in electroanalytical methods is to avoid interfering effects, such as the ohmic loss and the double layer capacity charging, and to use the Faradaic response peak current–potential curve as an analytical tool. Identification of the electroactive species is achieved by the position of the response peak on the potential axis and "pattern recognition", and quantitative analysis by peak shape and height. A recent development is squarewave voltammetry [10].

The electrochemical aspects of corrosion research have been the subject of some computer-controlled instrumentation [11]. It seems likely, in this case, that electrochemistry alone is not the answer and other techniques, such as spectroscopy and analytical methods, need to supplement the electrochemical measurements.

Before some of the desirable features of an automatic electrochemistry system is described, it would be useful to assess the degree of difficulty involved. Analysis of the structuring of computer programs, and the effort needed to reach a certain goal, is a favourite occupation of computer scientists. A rule of thumb [12] seems to be that up to 1000 lines of source code and a structure up to 10 modules would require a programmer for 6 months and up to 10000 lines with up to 100 modules would require 5 programmers for up to 2 years. An order of magnitude rise in the number of lines of source code would require 20 programmers for up to 3 years. The last category is about the requirements for computerising a GLC laboratory and, in our experience, is about the effort necessary for a worthwhile computerisation of measurements in electrochemistry and electrode kinetics.

2. On-line accumulation of electrochemical data

Data collection itself in making measurements in electrode kinetics is relatively straightforward. There are a number of commercial systems that will carry out rudimentary data collection, such as those marketed by Solartron for impedance measurements, PAR for cyclic voltammetry and impedance measurements, BAS for cyclic voltammetry. The difficulty comes in designing a reasonably intelligent system that will deal with carrying out suitable experiments and cope with large amounts of accumulated data. The latter is not trivial and, in operational terms, is more important than the first.

The elements needed for an electrochemical computer system are common to other scientific areas, where a suitable transducer is available to convert the experimental data into electrical signals. Electrochemical measurements have a big advantage in that the electrical signals are directly available. A list of the software and hardware components might be

(a) experimental control and data collection,

(b) storage of data and database design for long term availability of data,

(c) data handling and interpretation of particular experiments,

(d) electrochemical model development for use in the experimental interpretation, and

(e) extensive graphics.

There are many ways in which the points (a)–(e) could be built into a system and the strategy described here has evolved out of many "blind alleys" over ten years. There exist many methods of investigation in electrode kinetics. However, these all involve imposing a particular electrical signal on the system under test and observing the response. Although the methods are largely equivalent, there are subtle differences between them for particular reaction types. This means that equipment must be flexible enough to encompass at least the most important methods and not be dedicated to a particular method, for example steady-state current–potential measurements at the rotating disc or ring–disc electrodes, or the impedance method at rotating or stationary electrodes, or cyclic voltammetry curves at stationary electrodes [13]. A decision must be made about how the electrochemical methods are to be ranked in importance. In the system described here, the ranking is steady-state current–potential and impedance, non-stationary linear potential sweep (staircase), and finally potential pulse methods. As will be seen below, this ranking has been arrived at as the result of experience and the importance of the stationary state in electrochemistry. However, all the methods are equally accessible if this should be needed.

The details of the hardware and software which will accomplish a reasonably intelligent electrode kinetic measuring system depend on the particular computer system used. This part of system development is getting easier. The computer must have a large enough memory and storage capacity and must be able to work with a flexible high level language such as Fortran and Pascal. Suitable computers are the IBM PC [14] and compatibles and the various DEC [15] and similar machines. Figure 1 shows in diagrammatic form the strategy that can be employed to achieve a reasonable system. It is convenient to be able to set two potentials, E_1 and E_2, which can be added together at particular times read from a clock. These can then be applied using the usual servo amplifier (potentiostat) to the system under test and the result, either a current i or an impedance $Z(\omega)$, measured.

There are two strategies that can be invoked in setting the potentials and measuring the result. One is to use A to D and D to A convertors, which are an integral part of the bus system of the computer. The second strategy is to use dedicated instruments for carrying out these functions and to communicate with the computer using serial lines. For example, it is very convenient to use a well-developed piece of equipment, such as a two-channel correlator, to measure the impedance or a signal-conditioning unit to measure and set potentials. The role of the computer is then to set the parameters, start, stop, and collect data from an instrument. The latter strategy seems to be gaining ground as it has the advantage that the peripherials and the computer and software can be optimised separately and indepently of one another. However, a large amount of "redundancy" is built into the complete system and this adds to the cost. Instruments like a correlator are very expensive and, in many ways, it would be cheaper and more efficient to carry out the impedance measurement by using the computer, via suitable peripheral devices, to generate the small amplitude sine waves required. As the programs to generate the impedance are well known, this is an attractive proposition. It will probably be the route used in the next generation of electrochemical investigation systems.

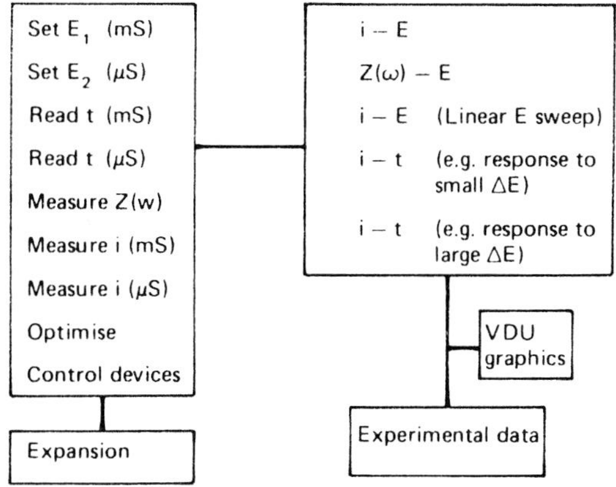

Fig. 1. Representation of how a number of hardware and software units can be used to generate the usual electrochemical methods which can be applied to a three-terminal or a two-terminal electrochemical cell or device via a suitable servo amplifier (potentiostat).

Combinations of the potential setting, time and current measurement unit operations will generate, essentially, all the methods of electrode kinetics, the ones used here being indicated in the diagram. The use of a computer to generate the signals has the advantage that the operation of the parts of the whole can be optimised and other parameters, such as liquid flow rates, can be controlled if this proves to be necessary. The whole system can also easily be made automatic. If this stage is reached, then the analogy to spectroscopic and the analytical methods, which have been automated, is complete.

Figure 2 shows the data analysis system used in our laboratory [16, 17]. The form that this takes depends on the way that the experimenter perceives the problems of electrode kinetics. In the present system, the data are analysed in a manner closely allied to the theory, as set out below, in order

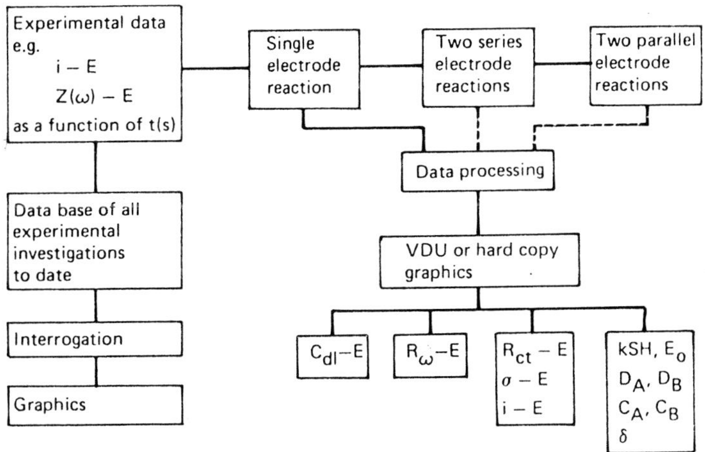

Fig. 2. Representation of a data handling system for measurements in electrode kinetics.

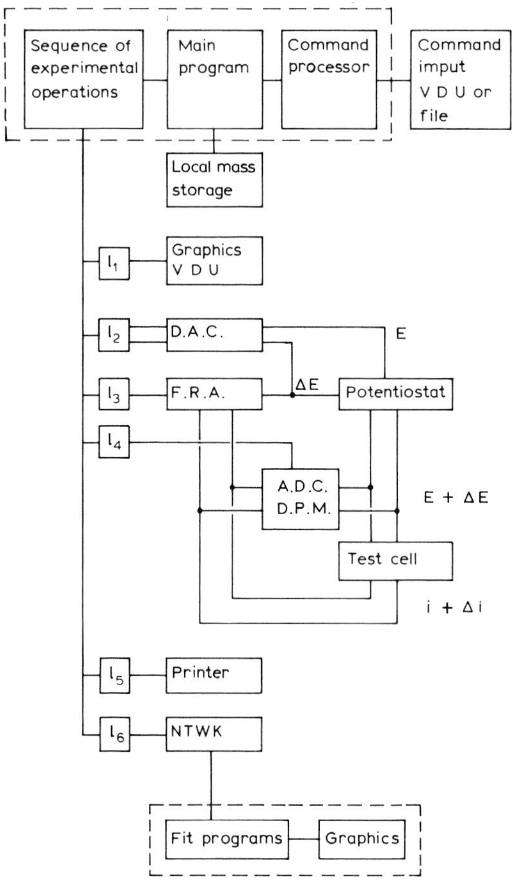

Fig. 3. Sketch of the hardware and software system used in the author's laboratory to investigate the electrode kinetics of electrochemical reactions. D.A.C are digital-to-analogue convertors, F.R.A. is a frequency response analyser, A.D.C. are analogue-to-digital convertors (used to switch devices), D.P.M is a panel meter, and NTWK is the university network serial connection to an Amdahl computer. The units I are the appropriate interfaces.

to try and make the interpretation as universal as possible. Minor modification to the theory is all that is needed, as shown below, to accommodate all the common known reaction types. An electrochemical experimentation system of this type generates a large amount of data and it is necessary to make it tractable by data reduction. The way that this is carried out, and the way in which the results are presented graphically, is very much bound up with the theory of electrode kinetics and that is the subject of the next sections. A further important problem, which is a subject in its own right, is that of database construction. This is important so that expertise does not become lost. The best solution to this problem is still under investigation.

Figure 3 gives a sketch of the implementation of the data collection and data reduction systems shown in Figs. 1 and 2.

3. Choice of electrochemical methods and the processing of electrochemical data

A review of the methods by which the kinetics of electrode reactions can be investigated is well outside the scope of the present chapter. A complete account would have to cover work carried out over a good part of this century. Methods used in practice would include steady-state current–potential measurements in polarography and the various types of pulse polarography, the rotating disc and the ring–disc, methods using a stationary electrode and potential pulses or trains of pulses or a linear potential sweep, and the a.c method. Good accounts of all these methods are available from many sources (see, for example, the review series such as the *Comprehensive Treatise of Electrochemistry* published by Plenum Press, New York and London). Much work has gone into adapting these methods to different reaction schemes. At the moment, a major preoccupation in the electrochemical journals is transferring the techniques of mathematics and numerical methods into calculating reaction schemes for particular electrochemical methods. Without experimentation, this a fairly fruitless exercise.

The strategy that will be used here is to select the methods in a general way and not be locked into a particular electrochemical technique. However, having said that, it is essential to rank the methods in order of importance so that unnecessary experiments are not carried out. Most useful electrode reactions can be observed to proceed in the steady state and there seems no doubt the most powerful general way of investigating these reactions is by the direct investigation of the steady-state current–potential curve combined with the impedance–potential curve (see, for example, refs. 18 and 19). If the impedance is measured at each potential together with the current, it provides a means of analysing the components of the steady state. Slow time effects are detected by means of following the current and potential with time or scanning the current–potential and impedance–potential curves in a some predetermined manner. The use of computers makes complicated sequences of experiments feasible for the first time.

The measurement of current and impedance are the major ways in which an electrode reaction can be investigated. Other methods such as linear potential sweep and various pulse methods can easily be implemented using the present system and may be useful under some conditions.

In the present work, the theory for single-electrode reactions (E), series reactions (E.E) and parallel reactions (E + E) only will be considered. This represents a minimum theory set that is necessary for model calculation and graphical display and for integration into the data analysis system. In the experimental section, all the data over the entire potential range has been tested against this theory to find the most appropriate theory with which to interpret the experiments.

4. Current–potential curve for a single electrode reaction with single electron transfer: description by standard rate constant

For a single electrode reaction

$$A + e \underset{k_b}{\overset{k_f}{\rightleftharpoons}} B \tag{1}$$

the following set of equations, obtained by combining Fick's first law of diffusion with the boundary conditions, describes the steady-state concentration–distance profiles of A and B in the diffusion layer

$$k_f C_A^s - k_b C_B^s = D_A \frac{dC_A^s}{dx} \tag{2}$$

$$k_f C_A^s - k_b C_B^s = -D_B \frac{dC_B^s}{dx} \tag{3}$$

$$\frac{dC_A^s}{dx} = \frac{C_A^b - C_A^s}{\delta} \tag{4}$$

$$\frac{dC_B^s}{dx} = \frac{C_B^b - C_B^s}{\delta} \tag{5}$$

These equations can be easily solved analytically to give

$$\frac{dC_A^s}{dx} = \frac{(k_f C_A^b - k_b C_B^b)}{D_A\left(1 + \dfrac{k_b\delta}{D_B} + \dfrac{k_f\delta}{D_A}\right)} \tag{6}$$

$$\frac{dC_B^s}{dx} = -\frac{(k_f C_A^b - k_b C_B^b)}{D_B\left(1 + \dfrac{k_b\delta}{D_B} + \dfrac{k_f\delta}{D_A}\right)} \tag{7}$$

and hence the surface concentrations

$$C_A^s = C_A^b - \delta\left(\frac{dC_A^s}{dx}\right) \tag{8}$$

$$C_B^s = C_B^b - \delta\left(\frac{dC_B^s}{dx}\right) \tag{9}$$

The reaction rate constants k_f and k_b can be defined in a number of ways. If the standard potential is known, or can be obtained from the experimental data, then for given Tafel slopes b_a and b_c

$$k_f = k_{SH} \exp\{-2.303(E - E^0)/b_c\} \tag{10}$$

$$k_b = k_{SH} \exp\{\ 2.303(E - E^0)/b_a\} \tag{11}$$

On the other hand, if the equilibrium potential can be measured with the species A and B present in the bulk solution then

$$k_f = (i_0/FC_A^b)\exp\{-2.303(E - E^0)/b_c\} \quad (12)$$

$$k_b = (i_0/FC_B^b)\exp\{2.303(E - E^0)/b_a\} \quad (13)$$

The Tafel slopes are not independent and are related by

$$\frac{1}{b_a} + \frac{1}{b_c} = \frac{F}{2.303RT} = \frac{1}{60} \quad (14)$$

The concentration gradient or the surface concentrations are related to the current by, say

$$i = -Ff = -D_A F\left(\frac{dC_A^s}{dx}\right) = -F(k_f C_A^s - k_b C_B^s) \quad (15)$$

It is convenient to express eqns. (2)–(5) as a matrix equation

$$\begin{bmatrix} -D_A & 0 & k_f & -k_b \\ 0 & D_B & k_f & -k_b \\ \delta & 0 & 1 & 0 \\ 0 & \delta & 0 & 1 \end{bmatrix} \begin{bmatrix} dC_A^s/dx \\ dC_B^s/dx \\ C_A^s \\ C_B^s \end{bmatrix} = \begin{bmatrix} 0 \\ 0 \\ C_A^b \\ C_B^b \end{bmatrix} \quad (16)$$

An advantage of the matrix formulation is that computer subroutines are available which will allow a direct solution of the matrix equation. The subject of matrix manipulation by computer is well advanced and reliable subroutines, which have been developed by mathematicians, are available and can be applied to problems in quantum chemistry and spectroscopy. The solution of eqns. (2)–(5) for a single electrode is fairly straightforward; however, the matrix formulation, as in eqn. (16), is a convenient shorthand for the equations and the method comes into its own when more complicated reaction schemes are considered.

5. Impedance–potential curve for a single electrode reaction with single electron transfer: description by standard rate constant

The expected current–time response of an electrode reaction of the redox type (1) to a small (< 5 mV) potential step from a particular steady state at a given potential and corresponding current can also be easily calculated. As is well known for linear systems, the result also includes the impedance of the steady state at the initial potential. The calculation can be carried out by solving Fick's second law, with the incorporation of the boundary conditions, for the change in concentration, dC, and the change in current, di, resulting from the potential change, dE. These changes can then be inserted in a suitable expansion about the steady-state value. It is convenient to carry out this calculation using Laplace transforms to transform Fick's second law equation, which is a partial differential equation; into an ordinary differential equation. This is a standard technique for dealing with linear partial differential equations and has been used to derive eqns. (18) and (19), below. It is also useful to use the changes in the flux of A, df, instead of the current change, di in the calculation, as these are directly related to the rates. The change in flux and the change in current are related the Faraday constant as given in eqn. (20). The complete set of equations for solution is

$$\frac{\overline{\mathrm{d}f}}{\overline{\mathrm{d}E}} = \frac{\partial f}{\partial E} + \left(\frac{\partial f}{\partial C_{\mathrm{A}}^{\mathrm{s}}}\right)\frac{\overline{\mathrm{d}C_{\mathrm{A}}^{\mathrm{s}}}}{\overline{\mathrm{d}E}} + \left(\frac{\partial f}{\partial C_{\mathrm{B}}^{\mathrm{s}}}\right)\left(\frac{\overline{\mathrm{d}C_{\mathrm{B}}^{\mathrm{s}}}}{\overline{\mathrm{d}E}}\right) \tag{17}$$

$$\frac{\overline{\mathrm{d}C_{\mathrm{A}}^{\mathrm{s}}}}{\overline{\mathrm{d}E}} = -\left(\frac{\overline{\mathrm{d}f}}{\overline{\mathrm{d}E}}\right)\frac{1}{\sqrt{pD_{\mathrm{A}}}} \tag{18}$$

$$\frac{\overline{\mathrm{d}C_{\mathrm{B}}^{\mathrm{s}}}}{\overline{\mathrm{d}E}} = \left(\frac{\overline{\mathrm{d}f}}{\overline{\mathrm{d}E}}\right)\frac{1}{\sqrt{pD_{\mathrm{B}}}} \tag{19}$$

$$\overline{\mathrm{d}f} = -\frac{\overline{\mathrm{d}i}}{F} \tag{20}$$

where p is the Laplace variable in the definition

$$\overline{\mathrm{d}C_{\mathrm{A}}^{\mathrm{s}}} = \int_0^\infty \mathrm{d}C_{\mathrm{A}}^{\mathrm{s}} \exp - pt \, \mathrm{d}t \tag{21}$$

The well-known solution is

$$\frac{\overline{\mathrm{d}i}}{\overline{\mathrm{d}E}} = \frac{\dfrac{\partial i}{\partial E}}{1 - \left(\dfrac{\partial i}{\partial C_{\mathrm{A}}^{\mathrm{s}}}\right)\left(\dfrac{1}{F\sqrt{j\omega D_{\mathrm{A}}}}\right) + \left(\dfrac{\partial i}{\partial C_{\mathrm{B}}^{\mathrm{s}}}\right)\left(\dfrac{1}{F\sqrt{j\omega D_{\mathrm{B}}}}\right)} \tag{22}$$

Here, the substitution $p = j\omega$ has been made to obtain the impedance. Equation (22) leads to the Randles equivalent circuit for an electrode reaction and is equivalent to eqn. (70) given below. In all electrode reactions, the impedance due to the reaction must have a differential capacity in parallel with it, to give the impedance of the interface

$$\frac{1}{Z} = \frac{\overline{\mathrm{d}i}}{\overline{\mathrm{d}E}} + j\omega C_{\mathrm{dl}} \tag{23}$$

A matrix form of the linear simultaneous equations (17)–(19) is

$$\begin{bmatrix} -k_f & k_b & 1 \\ \sqrt{pD_A} & 0 & 1 \\ 0 & \sqrt{pD_B} & -1 \end{bmatrix} \begin{bmatrix} \overline{dC_A^s/dE} \\ \overline{dC_B^s/dE} \\ \overline{df/dE} \end{bmatrix} = \begin{bmatrix} df/dE \\ 0 \\ 0 \end{bmatrix} \tag{24}$$

which is very convenient for direct solution by computer.

6. Current–potential curve for a single electrode reaction with single electron transfer and complexing of the reactant

Electron transfer reactions from or to labile complexes in solution constitute an important and common class of electrochemical reactions. The calculations for the steady-state current–potential and impedance–potential curves simply involve the coupling between the distribution of labile complexes in a particular solution with the reacting complex. If the complexes are in an equilibrium and one of the complexes is electroactive, then

$$MX_4^{2-} \rightleftharpoons MX_3^- \rightleftharpoons \ldots MX_n^{(2-n)-} \ldots M^+$$

$$k_f \uparrow\downarrow$$

$$M + nX^- - 2e \tag{25}$$

If the species M is a metal, then

$$\begin{aligned} i &= -2F(k_f[MX_n]_s - k_b[X^-]_s^n) \\ &= -2F(k_f[MX_n]_s - k_b'] \end{aligned} \tag{26}$$

where $[MX_n]_s$ and $[X^-]_s$ are surface concentrations and the superscript n means raised to the power n and where k_b' is a new rate constant. This restriction is not necessary, but it simplifies the formalism slightly and is appropriate for the experimental discussion which follows. With this equation, a set of equations, an extension of those for a single-electron transfer reaction, can easily be written which describes the complete behaviour of the electrode reaction. Diffusion is a function of the total concentration of the individual complexes and as a simplification it is advantageous to write the flux of material to the electrode using an average single diffusion coefficient, D_T. The set of equations to be solved is

$$k_f[MX_n]_s - k_b' = D_T \frac{dC_T^s}{dx} \tag{27}$$

$$\frac{dC_T^s}{dx} = \frac{C_T^b - C_T^s}{\delta} \tag{28}$$

$$K_1[M^{2+}][X^-] - [MX^+] = 0 \tag{29}$$

$$K_2[MX^+][X^-] - [MX_2] = 0 \tag{30}$$

$$K_3[MX_2][X^-] - [MX_3^-] = 0 \tag{31}$$

$$K_4[MX_3^-][X^-] - [MX_4^{2-}] = 0 \tag{32}$$

$$[M^{2+}]_s + [MX^+]_s + [MX_2]_s + [MX_3^-]_s + [MX_4^{2-}]_s = C_T^s \tag{33}$$

$$[M^{2+}]_b + [MX^+]_b + [MX_2]_b + [MX_3^-]_b + [MX_4^{2-}]_b = C_T^b \tag{34}$$

These can be solved by setting up a suitable matrix equation and iterating for the surface concentration of each complex.

7. Impedance–potential curve for a single electrode reaction with single electrode transfer and complexing of the reactant

In a similar way, the equations for the impedance can be formulated. The set of equations to be solved is

$$\frac{\overline{\mathrm{d}f}}{\overline{\mathrm{d}E}} = \frac{\partial f}{\partial E} + \left(\frac{\partial f}{\partial [MX_n]_s}\right)\frac{\overline{\mathrm{d}[MX_n]_s}}{\overline{\mathrm{d}E}} \tag{35}$$

$$\frac{\overline{\mathrm{d}[MX_n]_s}}{\overline{\mathrm{d}E}} = -\frac{\overline{\mathrm{d}f}}{\overline{\mathrm{d}E}}\frac{1}{K\sqrt{pD_T}} \tag{36}$$

$$[MX_n]_s = KC_T^s \tag{37}$$

8. Single electrode reaction with more than one electron transfer

The first step in dealing with more complex reactions is to define the relation between current and flux to accommodate more than one electron. Diffusion effects and the fluxes remain as treated above. For example, for two electrons

$$A + 2\,e \underset{k_b}{\overset{k_f}{\rightleftharpoons}} B \tag{38}$$

eqn. (15) becomes

$$i = -2F(k_f C_A^s - k_b C_B^s) \tag{39}$$

This is consistent, at equilibrium, with the Nernst equation for the two-electron reaction and requires that

$$\frac{1}{b_a} + \frac{1}{b_c} = \frac{2F}{2.303\,RT} = \frac{1}{30} \tag{40}$$

The consequences of treating the overall reaction as two consecutive steps

$$\mathrm{A} + \mathrm{e} \underset{k_{b_1}}{\overset{k_{f_1}}{\rightleftharpoons}} \mathrm{B} \tag{41}$$

$$\mathrm{B} + \mathrm{e} \underset{k_{b_2}}{\overset{k_{f_2}}{\rightleftharpoons}} \mathrm{C} \tag{42}$$

is that, in the steady state

$$i = 2i_1 = -2F\left(\frac{k_{f_1}}{k_{b_1}} C_A^s k_{f_2} - k_{b_2} C_c^s\right) \tag{43}$$

Given the necessary equality

$$2E^0 = E_1^0 + E_2^0 \tag{44}$$

eqn. (43) is consistent with eqn. (39) with the condition

$$k_{SH} = k_{SH_2} \exp\left[2.303(E_2^0 - E^0)/b_c\right] \tag{45}$$

It is also possible to separate multielectron reactions into slightly more complicated separate one-electron steps. The common gas evolving, electrocatalytic, reactions are of this type. However, this is only useful if the intermediate one-products are stable and show a definite presence in the electrochemical kinetics. As an example, consider the consequence of splitting eqn. (38) into its component one-electron steps

$$\mathrm{Cl}^- \underset{k_{f_1}}{\overset{k_{b_1}}{\rightleftharpoons}} \mathrm{Cl} + \mathrm{e} \tag{46}$$

$$\mathrm{Cl}^- + \mathrm{Cl} \underset{k_{f_2}}{\overset{k_{b_2}}{\rightleftharpoons}} \mathrm{Cl}_2 + \mathrm{e} \tag{47}$$

with separate rate constants labelled k_{f_1}, k_{f_2}, k_{b_1}, and k_{b_2}. In the steady state

$$\theta_{Cl} = \frac{k_{b_1} C_{Cl^-}^s + k_{f_2} C_{Cl_2}^s}{k_{b_2} C_{Cl^-}^s + k_{f_1}} \tag{48}$$

and the total current is given by

$$i = 2i_1 = 2i_2 = F(k_{b_1} C_{Cl^-}^s - k_{f_1}\theta_{Cl}) + F(k_{b_2}\theta_{Cl} C_{Cl^-}^s - k_{f_2} C_{Cl_2}^s) \tag{49}$$

Consider the limiting case when the first electron transfer is faster than the second. Then $k_{b_1} C_{Cl^-}^s > k_{f_2} C_{Cl_2}^s$ and $k_{f_1} > k_{b_2} C_{Cl^-}^s$ and the current is

$$i = 2F(k_{b_2}(C_{Cl^-}^s)^2 \frac{k_{f_1}}{k_{b_1}} - k_{f_2} C_{Cl_2}^s) \tag{50}$$

On the other hand, when the second electron transfer is faster than the first, $k_{f_2} C_{Cl^-}^s > k_{b_1} C_{Cl^-}^s$ and $k_{b_2} C_{Cl^-}^s > k_{f_1}$ and the current approaches

$$i = 2F\left(k_{b_1} C_{Cl^-}^s - k_{f_1} k_{f_2} \frac{C_{Cl_2}^s}{C_{Cl^-}^s}\right) \tag{51}$$

References p. 497

9. Two electrode reactions in parallel

Parallel reactions, as a class, are very common in electrode kinetics due to the proximity of many reactions to solvent breakdown. It is usually found that the reactions are additive and the individual reactions can be treated separately using the theory above. Trial calculations are easy to carry out using computers and it is necessary to include calculation of this type in a data analysis system.

10. Current–potential curve for two electrode reactions in series

For a reaction scheme

$$A + e \underset{k_{b_1}}{\overset{k_{f_1}}{\rightleftharpoons}} B \tag{52}$$

$$B + e \underset{k_{b_2}}{\overset{k_{f_2}}{\rightleftharpoons}} C \tag{53}$$

where A, B, and C are solution-soluble species, the following set of equations, obtained by combining Fick's first law of diffusion with the boundary conditions, gives

$$k_{f_1} C_A^s - k_{b_1} C_B^s = D_A \frac{dC_A^s}{dx} \tag{54}$$

$$k_{f_1} C_A^s - (k_{b_1} + k_{f_2}) C_B^s + k_{b_2} C_C^s = - D_B \frac{dC_B^s}{dx} \tag{55}$$

$$k_{f_2} C_B^s - k_{b_2} C_C^s = - D_C \left(\frac{dC_C^s}{dx} \right) \tag{56}$$

$$\frac{dC_A^s}{dx} = \frac{(C_A^b - C_A^s)}{\delta} \tag{57}$$

$$\frac{dC_B^s}{dx} = \frac{(C_B^b - C_B^s)}{\delta} \tag{58}$$

$$\frac{dC_C^s}{dx} = \frac{(C_C^b - C_B^s)}{\delta} \tag{59}$$

These equations can be easily solved analytically. It is convenient to express eqns. (54)–(59) as a matrix equation

$$\begin{bmatrix} -D_A & 0 & 0 & k_{f_1} & -k_{b_1} & 0 \\ 0 & D_B & 0 & k_{f_1} & -(k_{b_1}+k_{f_2}) & k_{b_2} \\ 0 & 0 & D_c & 0 & k_{f_2} & -k_{b_2} \\ \delta & 0 & 0 & 1 & 0 & 0 \\ 0 & \delta & 0 & 0 & 1 & 0 \\ 0 & 0 & \delta & 0 & 0 & 1 \end{bmatrix} \times \begin{bmatrix} \mathrm{d}C_A^s/\mathrm{d}x \\ \mathrm{d}C_B^s/\mathrm{d}x \\ \mathrm{d}C_C^s/\mathrm{d}x \\ C_A^s \\ C_B^s \\ C_C^s \end{bmatrix} = \begin{bmatrix} 0 \\ 0 \\ 0 \\ C_A^b \\ C_B^b \\ C_C^b \end{bmatrix} \tag{60}$$

which, on solution, gives the current by

$$i = -D_A F \frac{\mathrm{d}C_A^s}{\mathrm{d}x} - D_B F \frac{\mathrm{d}C_B^s}{\mathrm{d}x} \tag{61}$$

$$= -F(k_{f_1} C_A^s - k_{b_1} C_B^s) - F(k_{f_2} C_B^s - k_{b_2} C_C^s) \tag{62}$$

11. Impedance–potential curve for two electrode reactions in series

Using an extension of the method for single-electrode reactions, the appropriate set of equations is

$$\overline{\frac{\mathrm{d}f_1}{\mathrm{d}E}} = \frac{\partial f_1}{\partial E} + \left(\frac{\partial f_1}{\partial C_A^s}\right)\overline{\frac{\partial C_A^s}{\mathrm{d}E}} + \left(\frac{\partial f_1}{\partial C_B^s}\right)\overline{\frac{\mathrm{d}C_B^s}{\mathrm{d}E}} \tag{63}$$

$$\overline{\frac{\mathrm{d}f_2}{\mathrm{d}E}} = \frac{\partial f_2}{\partial E} + \left(\frac{\partial f_2}{\partial C_B^s}\right)\overline{\frac{\partial C_B^s}{\mathrm{d}E}} + \left(\frac{\partial f_2}{\partial C_C^s}\right)\overline{\frac{\mathrm{d}C_C^s}{\mathrm{d}E}} \tag{64}$$

$$\overline{\frac{\mathrm{d}C_A^s}{\mathrm{d}E}} = -\left(\overline{\frac{\mathrm{d}f_1}{\mathrm{d}E}}\right)\frac{1}{\sqrt{pD_A}} \tag{65}$$

$$\overline{\frac{\mathrm{d}C_B^s}{\mathrm{d}E}} = \left(\overline{\frac{\mathrm{d}f_1}{\mathrm{d}E}} - \overline{\frac{\mathrm{d}f_2}{\mathrm{d}E}}\right)\frac{1}{\sqrt{pD_B}} \tag{66}$$

$$\overline{\frac{\mathrm{d}C_C^s}{\mathrm{d}E}} = \overline{\frac{\mathrm{d}f_2}{\mathrm{d}E}}\frac{1}{\sqrt{pD_C}} \tag{67}$$

References p. 497

which can be solved analytically or by writing the set of linear simultaneous equations (63)–(67) as a matrix equation

$$\begin{bmatrix} -k_{f_1} & -k_{b_1} & 0 & 1 & 0 \\ 0 & -k_{f_2} & k_{b_2} & 0 & 1 \\ \sqrt{pD_A} & 0 & 0 & 1 & 0 \\ 0 & -\sqrt{pD_B} & 0 & 1 & -1 \\ 0 & 0 & \sqrt{pD_C} & 0 & 1 \end{bmatrix} \times \begin{bmatrix} \overline{dC_A^s}/\overline{dE} \\ \overline{dC_B^s}/\overline{dE} \\ \overline{dC_C^s}/\overline{dE} \\ \overline{df_1}/\overline{dE} \\ \overline{df_2}/\overline{dE} \end{bmatrix} = \begin{bmatrix} df_1/dE \\ df_2/dE \\ 0 \\ 0 \\ 0 \end{bmatrix} \tag{68}$$

when the substitution $p = j\omega$ is made and the result taken together with the equation for the impedance of the interface

$$\frac{1}{Z} = \frac{1}{(\overline{di_1}dE) + (\overline{di_2}/\overline{dE})} + j\omega C_{dl} \tag{69}$$

the result can be solved directly by computer.

12. Single electrode reaction: comparison of experiment and theory

In this work, the main aim has been to determine the steady-state behaviour behaviour by measuring the current–potential curve. In general, the steady state is the most important characteristic of an electrode reaction. Fortunately, most known electrochemical reactions have a steady state and are variations of the redox type of reaction. As shown above, the steady current–potential curve can be exactly interpreted for redox reactions. In order carry out a complete analysis, it is essential to measure the components of the steady state by impedance–potential measurements. In addition, impedance delivers information about the charging processes as they appear in the high-frequency double layer capacity–potential curve. This last parameter is the parameter which should connect electrochemistry and surface science. The unfortunate fact is that it is still not very well understood.

From the list of controlling parameters, a number of other possible ways

REACTION	PARAMETERS
E	δ, D_A, D_B, k_{f_1}, k_{b_1}
E + E	δ, D_A, D_B, k_{f_1}, k_{b_1}, D_C, D_D, k_{f_2}, k_{b_2}
E • E	δ, D_A, D_B, D_C, k_{f_1}, k_{b_1}, k_{f_2}, k_{b_2}

Fig. 4. List of kinetic parameters which determine the electrical response of model electrochemical reaction schemes. The single electrode reaction (E), two reactions in series (E.E), and two reactions in parallel (E + E) are common types of electrode reactions, discussed in the text, which every electrochemical investigation has to be capable of assessing as a matter of course.

of comparing experiment and theory could be devised. A table of controlling parameters is shown in Fig. 4. A convenient method, favoured here, is to interpret the current–impedance data in terms of a standard rate constant determined at each potential to generate a standard rate constant–potential curve. The advantage of the standard rate constant, for a particular reaction scheme, is that it controls the forward and backward rates, as seen in Fig. 4, and takes into account diffusion, interfacial reaction, and the order of reaction with respect to reactant and product. Most importantly for practical investigations, this includes the Tafel slopes. The standard rate constant–potential values are convenient for calculation from current or high frequency impedance data and allow a particular reaction mechanism to be tested from both sets of data.

The second method is to analyse the impedance–potential data on the basis of the equation

$$\frac{1}{Z(\omega) - R_\omega} = \frac{1}{R_{ct} + (1 - j)\sigma\omega^{-1/2}} + j\omega C_{dl} \tag{70}$$

and produce an electrochemical "spectrum" as charge transfer–potential, double layer capacity–potential, ohmic resistance–potential, and Warburg coefficient–potential plots. Together with the current–potential curve, these present a useful representation of the steady-state electrochemical behaviour.

The examples briefly discussed in this article are limited to the behaviour of simple metal and oxide electrodes. Even in this apparently simple situation, complications abound. An essential feature of all the electrochemical systems presented here is the automation and many of the runs were carried out overnight. A larger number of organic and inorganic reactions would be amenable to similar methods of investigation. The two types of data analysis mentioned in this section are used here. In the case of the reactions on titanium, the chloride and bromide oxidation, reactions on ruthenium dioxide electrodes, and the reactions of palladous ions at palladium electrodes, the data are compared with the theory given above. In more complex situations, such as the dissolution of stainless steel and the dental amalgams, the double layer potential curves and the charge transfer resistance–potential curves have been abstracted from the data and used as a "spectrum" of the behaviour. This approach is essentially the electrochemical anologue of surface science.

References p. 497

13. Dissolution and passivation of titanium in acid solution

Titanium is a possible electrode material for carrying out electrochemical synthesis reactions and for that purpose it is necessary to be able to survey the electrochemical behaviour of the metal up to high current densities. Titanium metal, which is an important structural material, is also of interest in its own right. In this section, the behaviour of the metal in simple acid solutions will be discussed. The aim is to measure the rate of the hydrogen

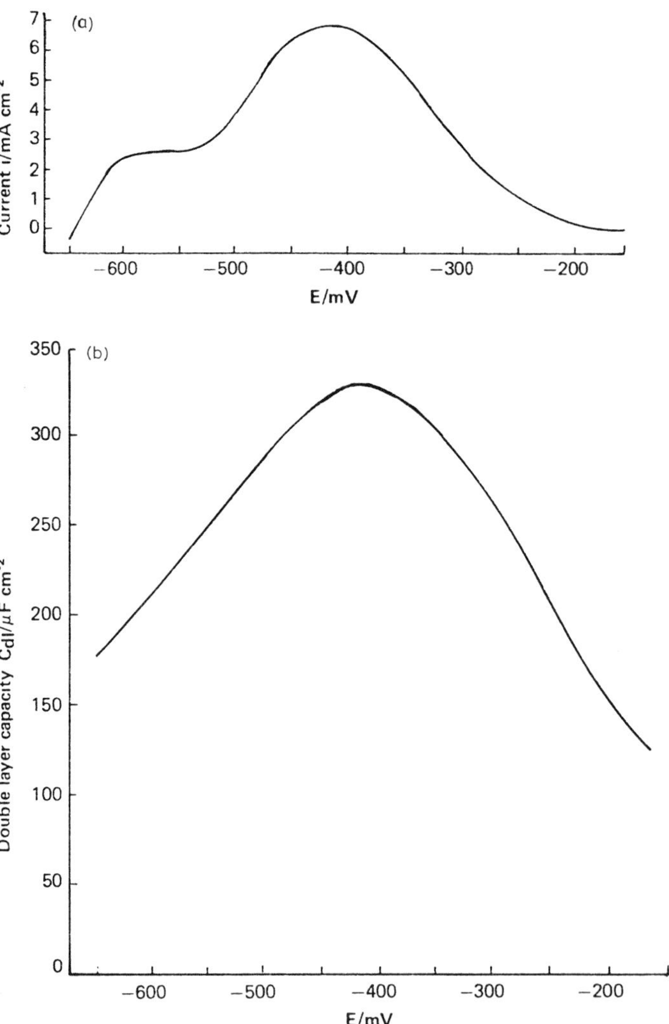

Fig. 5. Analysis of the experimental steady-state current–potential and impedance–potential data from $E = -650\,\text{mV}$ to $E = -150\,\text{mV}$ for a titanium rotating-disc electrode (45 Hz) in a solution of 3.0 M sulphuric acid at 65°C. (a) Steady state current–potential curve. The potentials are the measured potentials. (b) High-frequency double layer capacity–potential curve. The potentials are the measured potentials.

evolution reaction as a function of the anion and to detect the onset of the reactions involving the titanium metal.

The active–passive behaviour of titanium in acid solution has been investigated in a large number of papers. These have been comprehensively reviewed by Kelly [20]. The basic steady-state current–potential and double layer capacity–potential behaviour of titanium in sulphuric acid at 65°C is shown in Fig. 5. The features of the curve, going from negative to positive potentials, can be identified with the hydrogen evolution reaction, the active dissolution of the titanium to form Ti^{3+} ions, and a passivation reaction in which TiO_2 is formed. Under these temperature conditions, the three reactions are reasonably well separated from each other. At the same time, the double layer capacity–potential curve [Fig. 5(b)] rises with potential in the potential region where hydrogen evolution and titanium dissolution occur and falls with the formation of the TiO_2 layer. It is a characteristic of the passivation reaction in this system that it occurs gradually with potential.

However, the active dissolution of titanium depends markedly on temperature in acid solution. At lower temperatures, the picture is not so clear. It is necessary to have a quantitative measure of the rate of the hydrogen reaction and the titanium dissolution reaction. The complete set of current–potential and impedance–potential data has been tested against the theory given above. The best strategy seems to be to fit to a single electrode reaction and then to look for deviations from the expected behaviour for a perfect redox reaction. A convenient way of doing this is to represent the electrochemical data as a standard rate constant–potential curve in conjunction with a double layer capacity–potential curve [21].

An example is shown in Fig. 6 for 2 M perchloric acid solution. In this potential range, hydrogen evolution is practically the sole electrochemical reaction. In Fig. 6(a), the analysis assumes a Tafel slope of 120 mV. It is necessary to analyse using a 211 mV Tafel slope in order to get a potential-independent standard rate constant in the potential region up to −600 mV where hydrogen evolution is the major reaction. Figure 7 shows an analysis of some data for a 2 M HC1 solution. The Tafel slope of 211 mV is also necessary. The standard rate constant is similar in value. It is interesting also that the double layer capacity–potential curves are almost identical. In contrast, nitric acid solutions have much higher apparent rates of reaction as evidenced by Fig. 8 for a 2 M nitric acid solution and Fig. 9 for a 0.02 M nitric acid solution. Because high rates of reaction are involved, the ohmic resistance in the system strongly limits the available potential range. It is seen from Fig. 8 that the standard rate constant decreases from about −620 mV. The double layer capacity–potential curve begins to change shape about the same potential and has a maximum at about −550 mV. The inference is that, from −620 to −550 mV, the hydrogen reaction is being impeded by TiO_2 formation. In complete contrast, in Fig. 5 the standard rate constant curve would diverge from a constant value at about −650 mV, even though the capacity–potential curve carries on rising until the maximum at

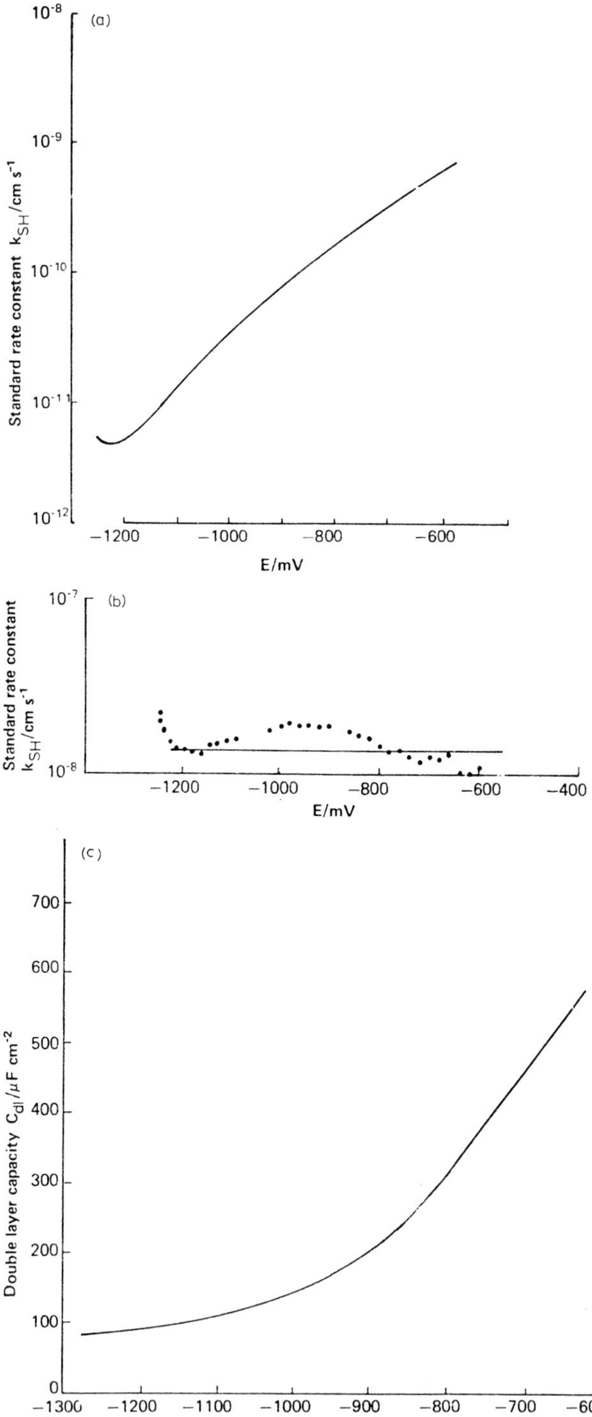
(a)
Standard rate constant k_{SH}/cm s^{-1}
10^{-8}
10^{-9}
10^{-10}
10^{-11}
10^{-12}
−1200
−1000
−800
−600
E/mV
(b)
Standard rate constant k_{SH}/cm s^{-1}
10^{-7}
10^{-8}
−1200
−1000
−800
−600
−400
E/mV
(c)
Double layer capacity C_{dl}/μF cm^{-2}
700
600
500
400
300
200
100
0
−1300
−1200
−1100
−1000
−900
−800
−700
−600
E/mV

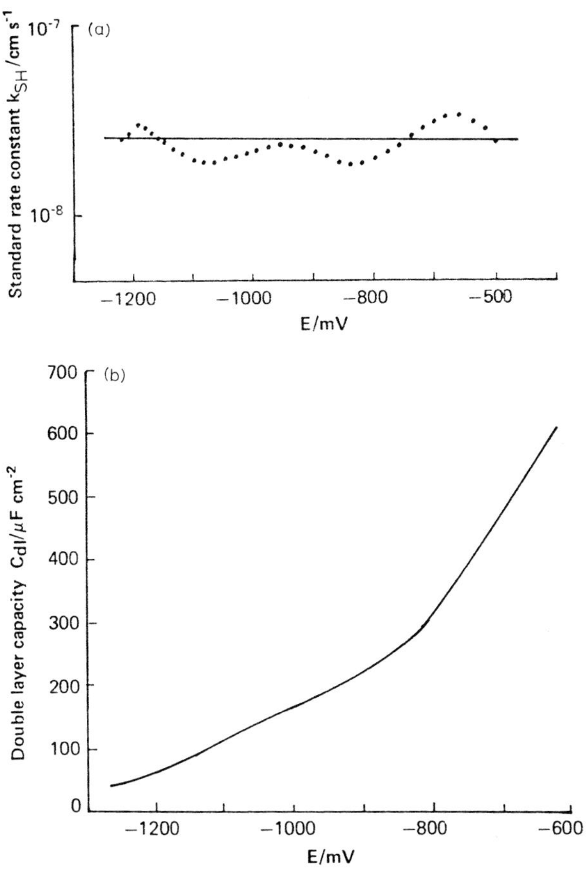

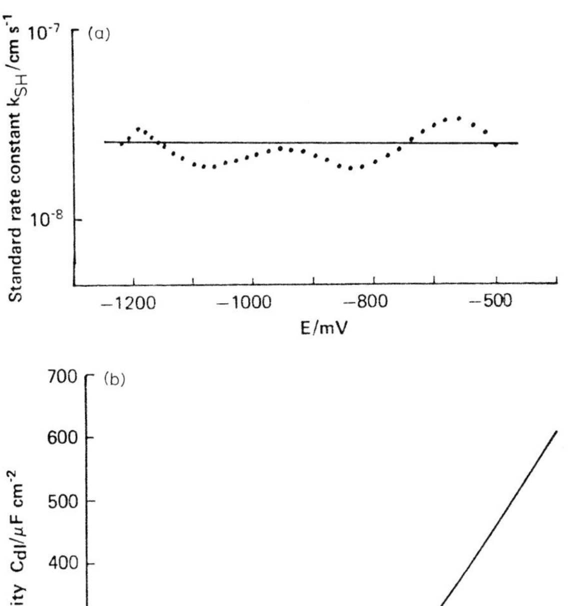

Fig. 7. Analysis of the experimental steady-state current–potential and impedance–potential data from $E = -1300$ mV to $E = -600$ mV for a titanium rotating-disc electrode (45 Hz) in a solution of 2 M hydrochloric acid. (a) Standard rate constant–potential curve calculated for the hydrogen evolution reaction on titanium assuming that $D_A = 7.5 \times 10^{-5}$ cm^{-1} s^{-1} and $E^0 = -246$ mV. The Tafel slope $b_c = 211$ mV and the measured ohmic resistance was 0.4 ohm cm^2. The potentials are the "true" potentials. (b) High-frequency double layer capacity–potential curve. The potentials are the measured potentials.

Fig. 6. Analysis of the experimental steady-state current–potential and impedance–potential data from $E = -1300$ mV to $E = -600$ mV for a titanium rotating-disc electrode (45 Hz) in a solution of 2 M perchloric acid. (a) Standard rate constant–potential curve calculated for the hydrogen evolution reaction on titanium assuming that $D_A = 7.5 \times 10^{-5}$ cm s^{-1} and $E^0 = -246$ mV. The Tafel slope $b_c = 120$ mV and the measured ohmic resistance was 0.3 ohm cm^2. The potentials are the "true" potentials. (b) Standard rate constant–potential curve calculated for the hydrogen evolution reaction on titanium assuming that $D_A = 7.5 \times 10^{-5}$ cm s^{-1} and $E^0 = -246$ mV. The Tafel slope $b_c = 211$ mV and the measured ohmic resistance was 0.3 ohm cm^2. The potentials are the "true" potentials. (c) High-frequency double layer capacity–potential curve obtained from the impedance data. The potentials are the measured potentials.

References p. 497

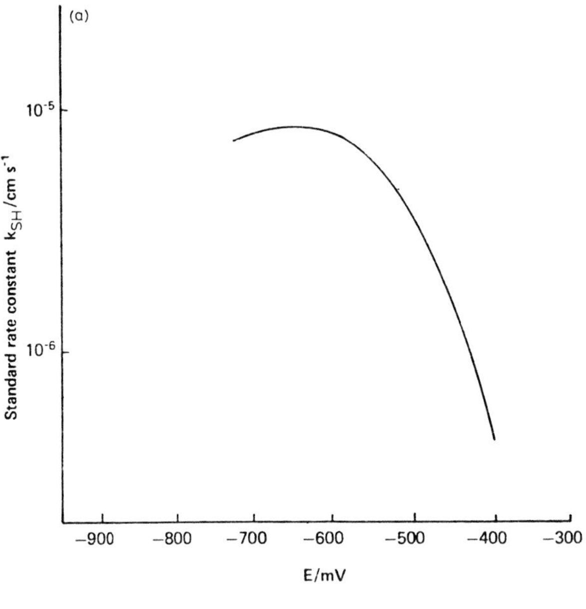

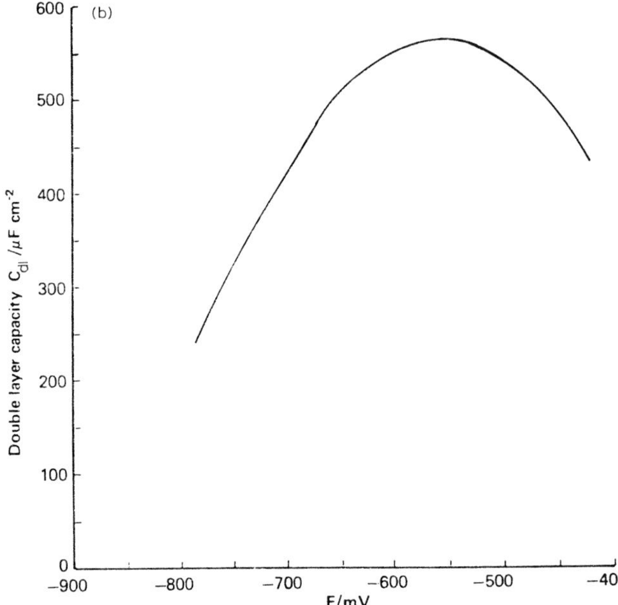

Fig. 8. Analysis of the experimental steady-state current–potential and impedance–potential data from $E = -1300$ mV to $E = -600$ mV for a titanium rotating-disc electrode (45 Hz) in a solution of 2 M nitric acid. (a) Standard rate constant–potential curve calculated for the hydrogen evolution reaction on titanium assuming that $D_A = 7.5 \times 10^{-5}$ cm s^{-1} and $E^0 = -246$ mV. The Tafel slope $b_c = 211$ mV and the measured ohmic resistance was 0.38 ohm cm^2. The potentials are the "true" potentials. (b) High-frequency double layer capacity–potential curve. The potentials are the measured potentials.

-425 mV. In this case, dissolution to form Ti^{3+} is strongly indicated. For the 0.02 M nitric acid solution, the standard rate constant seems to be roughly similar to the 2 M solution and, although in the diagram a straight line has been put through the points, the value of the standard rate constant probably decreases about -700 mV. In this system, the capacity has a maximum around -650 mV. In agreement with these results, analysis of the solution by polarography does not detect Ti^{3+} in HNO_3 solutions, although it does in sulphuric acid solution.

14. Corrosion of stainless steel and its component metals in acid solution

The electrochemical behaviour of stainless steel has not been worked out completely, although the measured data are available. However, one aspect of the behaviour, based on the measured double layer capacity data, seems to be susceptible to interpretation. The capacity–potential curves are determined by the state of the metal surface and by the ionic environment. In this work, it has been assumed that the ionic environment is a constant. This means that the double layer capacity–potential curves should reflect the nature of the metal surface just as, say, an electron energy spectrum in surface science. Stainless steel has a complicated electrochemical behaviour. In previous work [22] an attempt has been made to compare the double layer capacity curves measured during dissolution and passivation of the stainless steel with that of the pure components. It seems that all the data in the high frequency regime can be fitted to eqn. (70) with the Warburg coefficient set equal to zero.

14.1 DOUBLE LAYER CAPACITY CURVES

Nickel in perchloric acid solution [Fig. 10(a)] has three peaks at about $E = -425$ mV, $E = -200$ mV, and $E = -50$ mV. Similar peaks are observed on an electropolished nickel electrode [Fig. 10(b)]. The nature of the solution phase plays a role although, in chloride-containing solution, two peaks appear at $E = -425$ mV and $E = -275$ mV, but at more positive potentials a fall in capacity occurs, indicating probable formation of a layer on the electrode. On the other hand, chromium in perchloric acid solution has a quite different peak structure with a number of peaks [Fig. 10(d)]. Iron in the same solution has a single peak at about $E = -475$ mV and active dissolution starts at $E = -400$ mV. The stainless steel, on the other hand, seems to show features of nickel and chromium. At potentials from $E = -700$ mV to $E = 0$ mV, the curve is similar to that of nickel while at more positive potentials, it is similar to chromium.

In general, corrosion reactions are complicated electrochemical reactions which can be studied in more detail using electrochemical methods. This

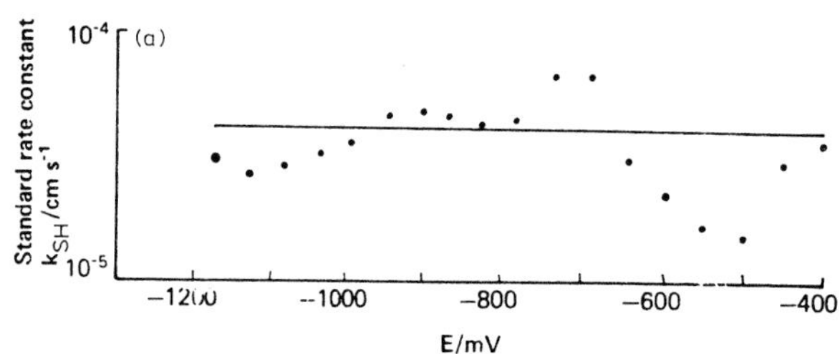
(a)
Standard rate constant
k_{SH}/cm s^{-1}
10^{-4}
10^{-5}
−1200
−1000
−800
−600
−400
E/mV

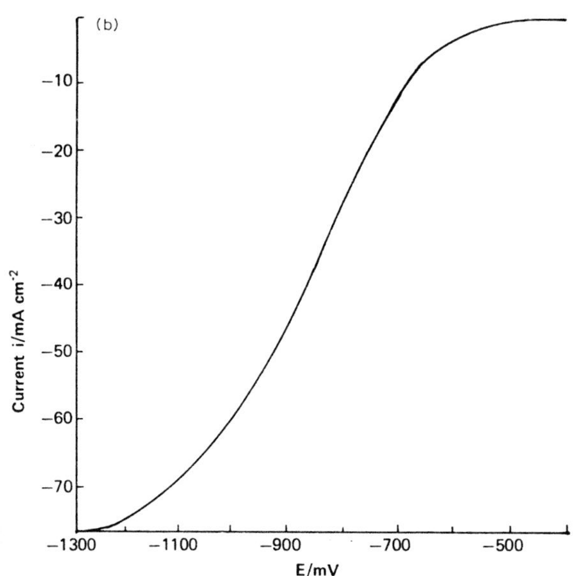
(b)
Current i/mA cm^{-2}
−10
−20
−30
−40
−50
−60
−70
−1300
−1100
−900
−700
−500
E/mV

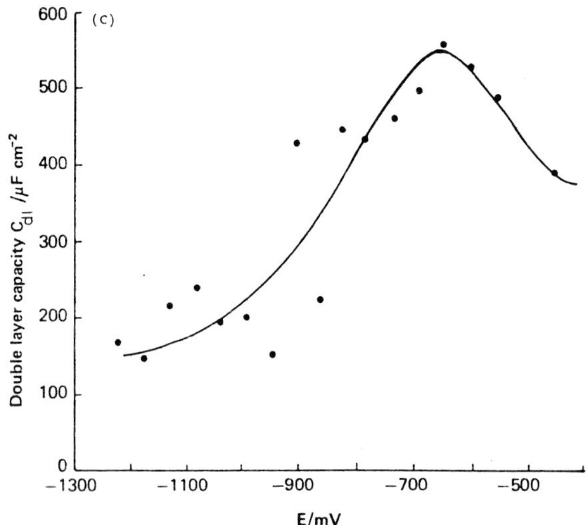
(c)
Double layer capacity C_{dl} /μF cm^{-2}
600
500
400
300
200
100
0
−1300
−1100
−900
−700
−500
E/mV

exploratory investigation has concentrated on stainless steel and its component metals. The main conclusions from this work are:

(a) The combination of "steady-state" current–potential and high-frequency (0.1–1000 Hz) impedance measurements are powerful electrochemical methods, especially when interpreted in terms of parameter curves by means of eqn. (70). The method would be suitable for the investigation of other, possibly more complex, corrosion systems.

(b) The double layer capacity–potential curves indicate that, at negative potentials, the surface of stainless steel behaves like nickel and at positive potentials like chromium.

(c) The charge transfer resistance–potential and current–potential curves, not presented here, suggest that nickel lowers the overpotential for the hydrogen evolution reaction on stainless steel and protects against the iron and chromium dissolution reactions at negative potentials and that chromium provides the protecting layer at positive potentials. An attempt to describe these features qualitatively is shown in Fig. 11, where the potential regions for the various processes which occur are plotted.

(d) In a corroding environment, the metal that corrodes will be controlled by a reactant, which is reduced. These reactions will largely control the corrosion potential. Once the behaviour of the corrosion potential is known, the corrosion rates can be calculated using the charge transfer–potential diagram and the current–potential data obtained in the absence of the other reactant (except for the hydrogen-evolution reaction).

15. Oxidation of chloride and bromide ions on ruthenium dioxide/titanium electrodes

The oxidation of chloride and bromide ions is an example of an electrocatalytic reaction and the mechanism has been the subject of much speculation [23, 24]. The oxidation of chloride ions is also the major industrial route to chlorine gas [25, 26]. Reactions of this family are termed electrocatalytic because the rate of the reaction, as measured by the current or the charge transfer resistance at a given potential, depends on the electrode material. This is a major reaction type in electrode kinetics.

Fig. 9. Analysis of the experimental steady-state current–potential and impedance–potential data from $E = -1300$ mV to $E = -600$ mV for a titanium rotating-disc electrode (45 Hz) in a solution of 0.02 M nitric acid. (a) Standard rate constant–potential curve calculated for the hydrogen evolution reaction on titanium assuming that $D_A = 7.5 \times 10^{-5}$ cm s^{-1} and $E^0 = -246$ mV. The Tafel slope $b_c = 211$ mV and the measured ohmic resistance was 1.0 ohm cm^2. The potentials are the "true" potentials. (b) Steady-state current–potential curve. The potentials are the "true" potentials. The squared symbols refer to the calculated "reversible" curve. (c) High-frequency double layer capacity–potential curve. The potentials are the "true" potentials.

References p. 497

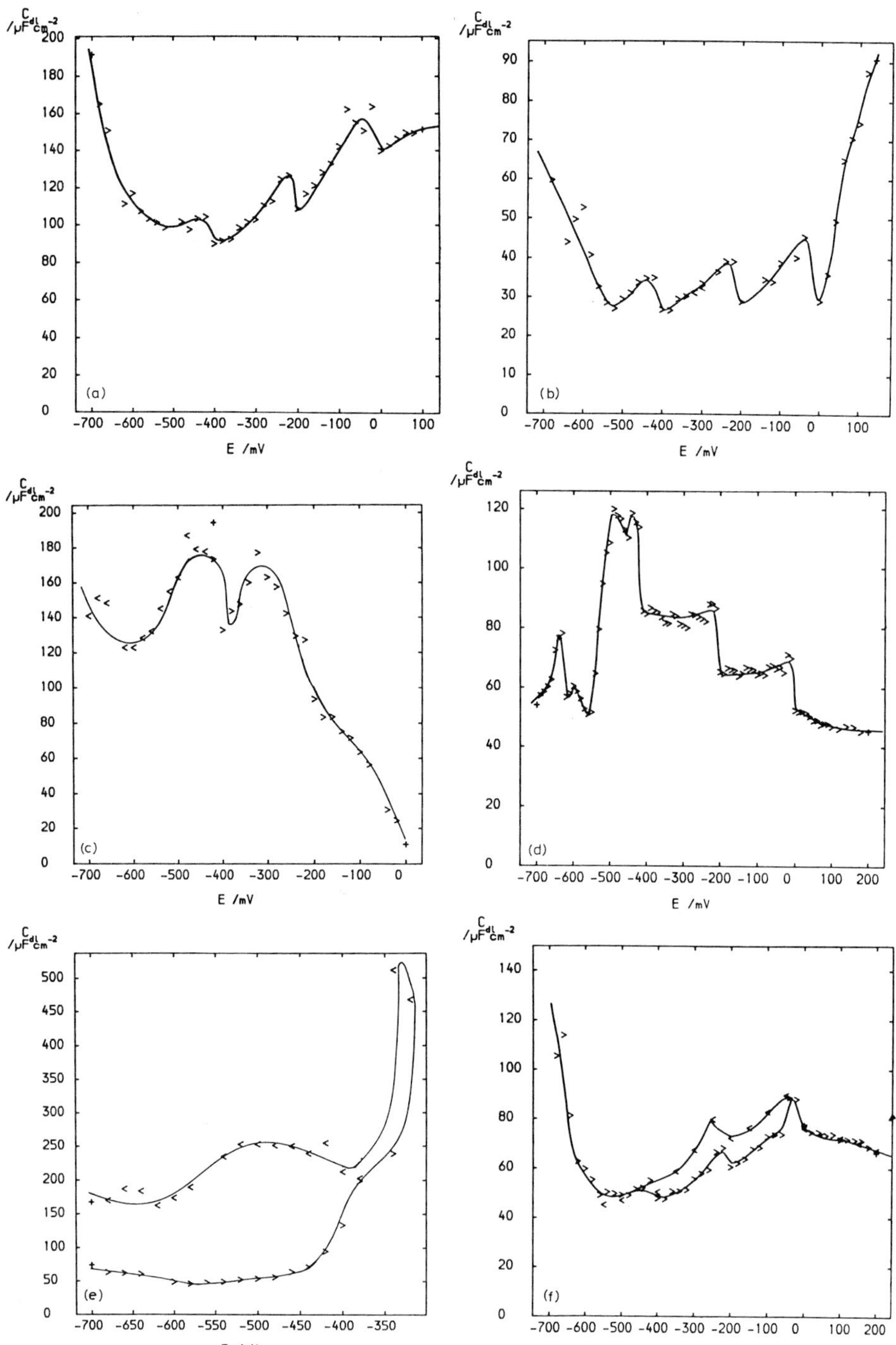

Fig. 10. Double layer capacity–potential curves for different metal rotating-disc electrodes (45 Hz rotation speed) in the given acid solutions. (a) Nickel in 1 M $HClO_4$; (b) electropolished nickel in 1 M $HClO_4$; (c) nickel in 1 M HCl; (d) chromium in 1 M $HClO_4$; (e) iron in 1 M $HClO_4$; and (f) stainless steel (304 L) in 1 M $HClO_4$.

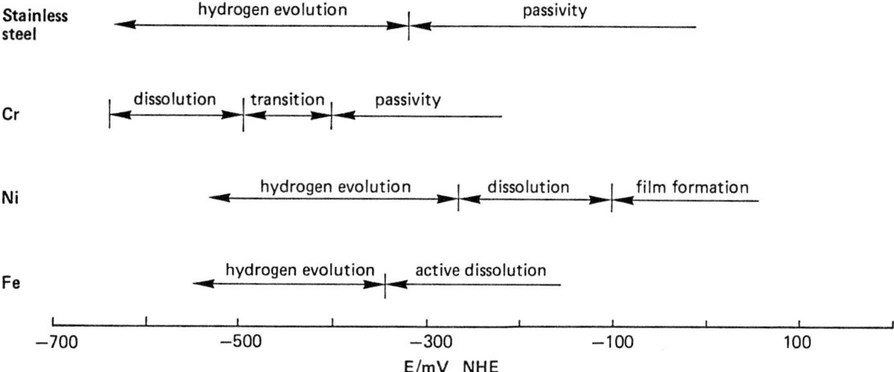

Fig. 11. Stainless steel potential regions diagram.

Attempts to analyse electrochemical current–potential and impedance potential data in terms of the electrode kinetics of a simple reaction with a single Tafel slope and standard rate constant have failed. Even so, it is useful to analyse the data as if the reaction was a perfect electrochemical reaction and then to look for divergences from expected behaviour. Comparison of experiment with the theory given above shows no sign of solution intermediates. If a single Tafel slope and standard rate constant are assumed, then the divergence from perfect behaviour has been attributed to two causes. The first is the inhibition of the reaction by the product [27] and the second is the nature of the two-electron reaction, which may involve an intermediate. An example of inhibition, which is a major feature of the reaction at high current density, is shown in Fig. 12(a)–(d) for the oxidation of 1 M NaCl at a rotating ruthenium dioxide electrode. Figure 12(a) and (b) show the sensitivity of the analysis to the Tafel slope. At low potentials and currents, the correct description of the data is via a 40 mV Tafel slope and a standard rate constant about 3.2×10^{-7} cm s^{-1}. As can be seen in Fig. 12(c), the analysis has been carried out up to very high current densities. An interesting feature of the electrochemical data is the double layer capacity–potential curve in Fig. 12(d). At potentials where the oxidation of the chloride ions is proceeding, the capacity rises and then falls. The fall in capacity at about 1220 mV roughly coincides with the potential at which the standard rate constant falls. If the capacity can be interpreted as a measure of the ability of anions and cations to move into the metal–electrolyte interfacial region [28], the rise in capacity with potential indicates where the reaction is behaving simply without inhibition.

At a stationary electrode, which is more akin to the industrial situation in chlorine manufacture, the effect is as shown in Fig. 13(a)–(d). As the concentration of chloride ions is reduced, an increasing part of the steady-state current goes into the oxygen-evolution reaction and it is necessary to correct for it [see Fig. 14(a) and (b)]. There has been a suggestion that the

References p. 497

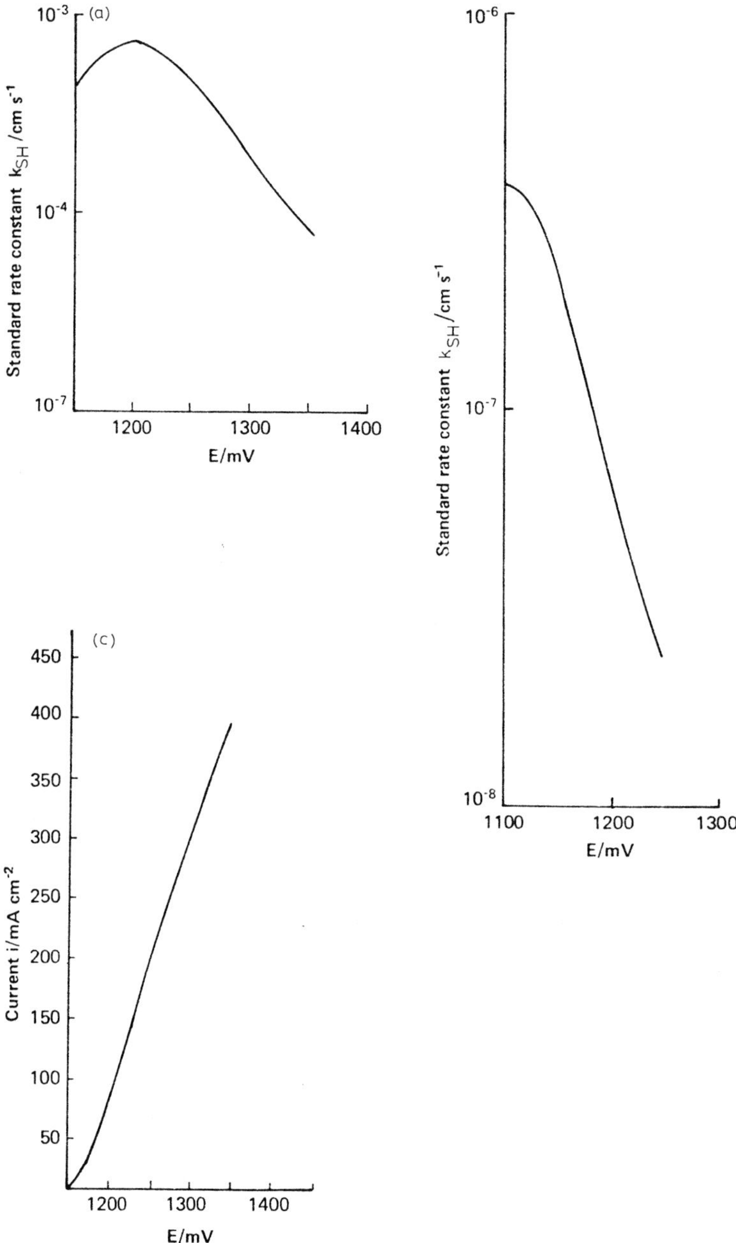

Fig. 12. Rotating (45 Hz) ruthenium dioxide/titanium dioxide electrode (35% w/w ruthenium dioxide) in 1.0 M NaCl solution. (a) Standard rate constant–potential curve assuming a constant Tafel slope of 70 mV. $D_{Cl^-} = 5 \times 10^{-6}$ cm s^{-1}, $D_{Cl_2} = 7 \times 10^{-6}$ cm s^{-1}, $E^0 = 1050$ mV SCE, and $R = 0.8$ ohm cm^2. (b) Standard rate constant–potential curve assuming a constant Tafel slope of 40 mV. $D_{Cl^-} = 5 \times 10^{-6}$ cm s^{-1}, $D_{Cl_2} = 7 \times 10^{-6}$ cm s^{-1}, $E^0 = 1050$ mV SCE, and $R = 0.8$ ohm cm^2. (c) Common experimental and calculated current–potential curve using the parameters of Fig. 12(b). (d) Double layer capacity curve.

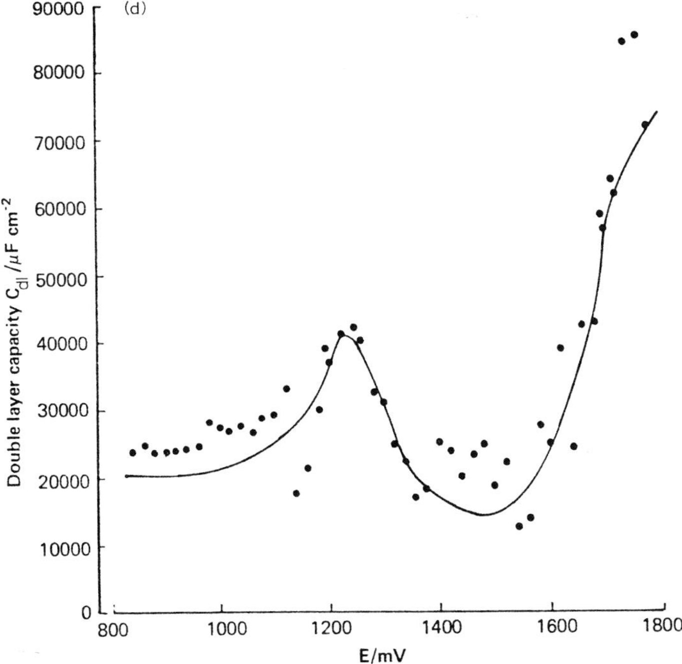

chloride-oxidation and oxygen-evolution reactions are linked by a common intermediate [27].

Similar results have been obtained for the oxidation of bromide ions on ruthenium dioxide electrodes. A typical result is shown in Fig. 15.

16. Reduction of palladium chloro complexes

The electrochemistry of palladium in chloride-containing solution has been used in this laboratory as a model system for active metal deposition–dissolution reactions which involve complexing. These represent a large and important class of electrochemical reactions which occur often in applied electrochemistry. It is impossible to detect the salient features of this and electrochemical reactions which involve surface changes without significant automation in obtaining and processing the electrode kinetic data. The full details of the electrode kinetics have not yet been worked out, so only a provisional account can be given here.

In principle, the reaction scheme could involve the reduction of any of the known palladium labile chloro and hydroxo complexes in the chain

$$PdCl_4^{2+} \underset{}{\overset{K_4}{\rightleftharpoons}} PdCl_3^- \overset{K_3}{\rightleftharpoons} PdCl_2 \overset{K_2}{\rightleftharpoons} PdCl^+ \overset{K_1}{\rightleftharpoons} Pd^{2+} \overset{K_{\text{hydrol}}}{\rightleftharpoons} PdOH^+ \tag{71}$$

References p. 497

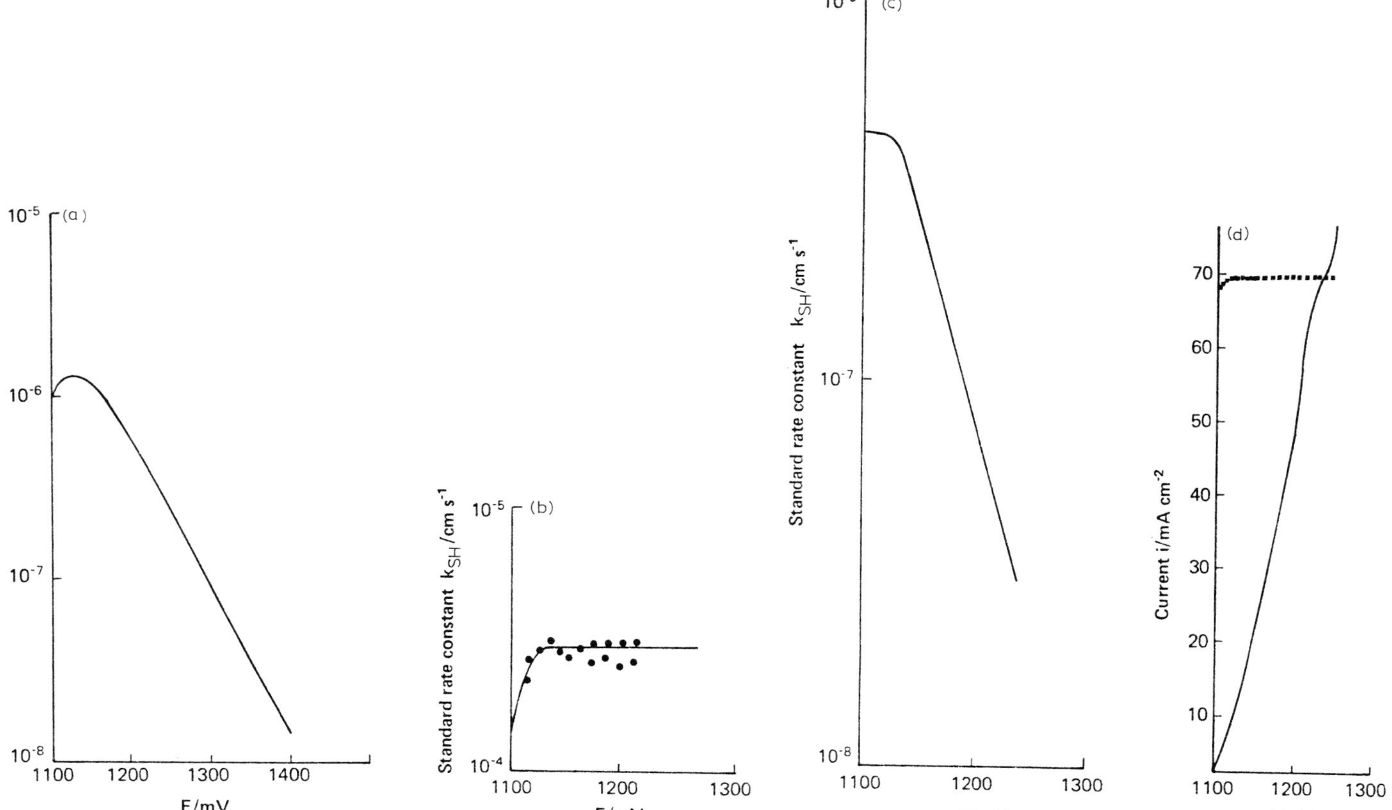

Fig. 13. Stationary ruthenium dioxide/titanium dioxide electrode (UN5) in 5.13 M NaCl solution. (a) Standard rate constant–potential curve assuming a constant Tafel slope of 70 mV. $D_{Cl^-} = 5 \times 10^{-6}$ cm s^{-1}, $D_{Cl_2} = 7 \times 10^{-6}$ cm s^{-1}, $E^0 = 1050$ mV SCE, no ohmic correction. (b) Standard rate constant–potential curve assuming a constant Tafel slope of 70 mV. $D_{Cl^-} = 5 \times 10^{-6}$ cm s^{-1}, $D_{Cl_2} = 7 \times 10^{-6}$ cm s^{-1}, $E^0 = 1050$ mV SCE, and $R = 1.7$ ohm cm^2. (c) Standard rate constant–potential curve assuming a constant Tafel slope of 40 mV. $D_{Cl^-} = 5 \times 10^{-6}$ cm s^{-1}, $D_{Cl_2} = 7 \times 10^{-6}$ cm s^{-1}, $E^0 = 1050$ mV SCE, and $R = 1.7$ ohm cm^2. (d) Common experimental and calculated current–potential curve using the parameters of Fig. 13(c). The broken curve refers to the calculated "reversible" curve.

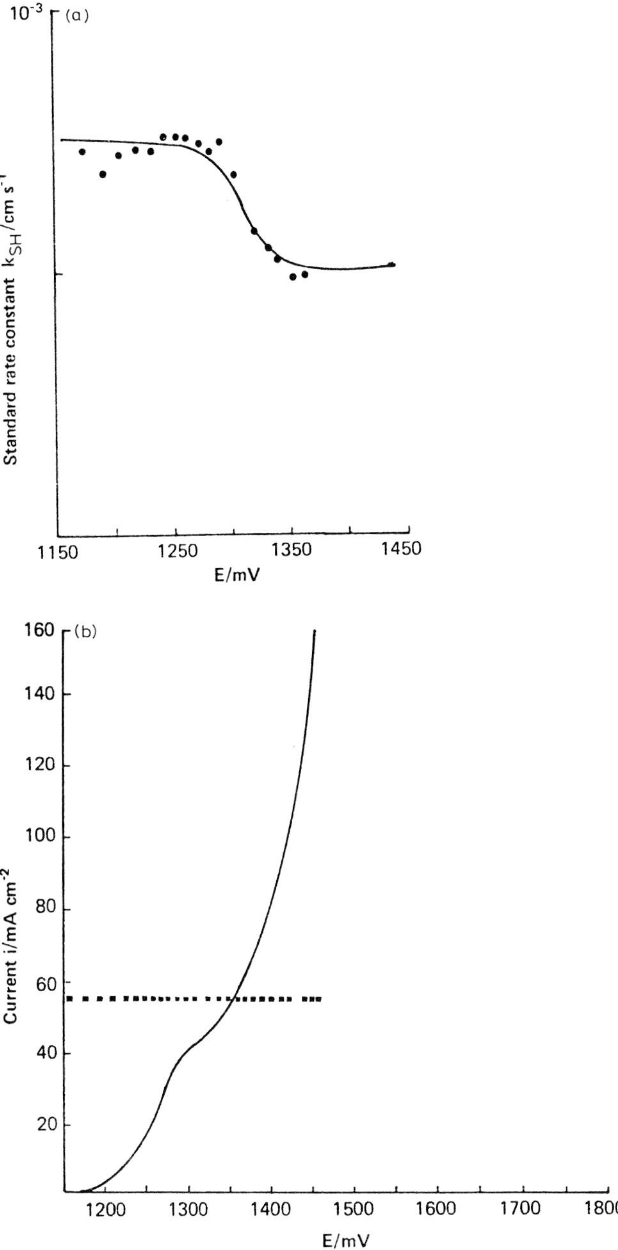

Fig. 14. Rotating (45 Hz) ruthenium dioxide/titanium dioxide electrode (35% w/w ruthenium dioxide) in 0.1 M NaCl solution. (a) Standard rate constant–potential curve for the chloride oxidation reaction [reaction (1)] assuming a constant Tafel slope of 70 mV, $D_{Cl^-} = 5 \times 10^{-6}$ cm s^{-1}, $D_{Cl_2} = 7 \times 10^{-6}$ cm s^{-1}, $E_1^0 = 1050$ mV SCE, and $R = 2.2$ ohm cm^2. The characteristics of the oxygen evolution reaction [reaction (2)] with a Tafel slope of 200 mV were chosen to be $k_2 = 1 \times 10^{-8}$ cm s^{-1}, $E_2^0 = 1257$ mV SCE and $D_{O_2} = 1 \times 10^{-5}$ cm^2 s^{-1}. (b) Common experimental and calculated current–potential curve using parameters of Fig. 14(a). The broken curve refers to the calculated "reversible" curve.

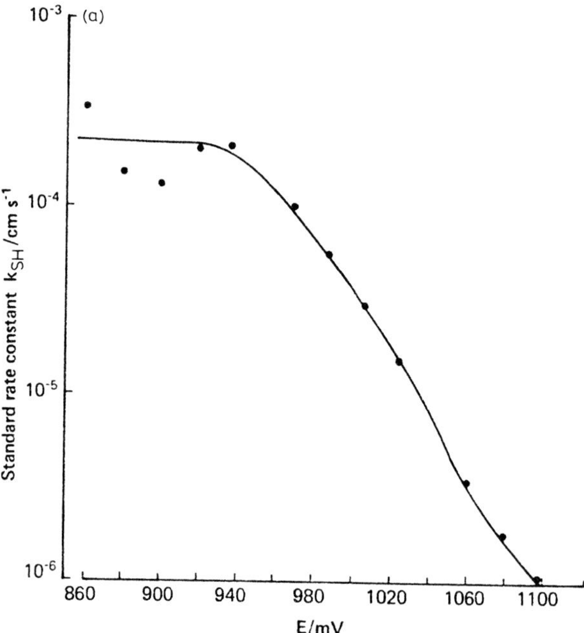

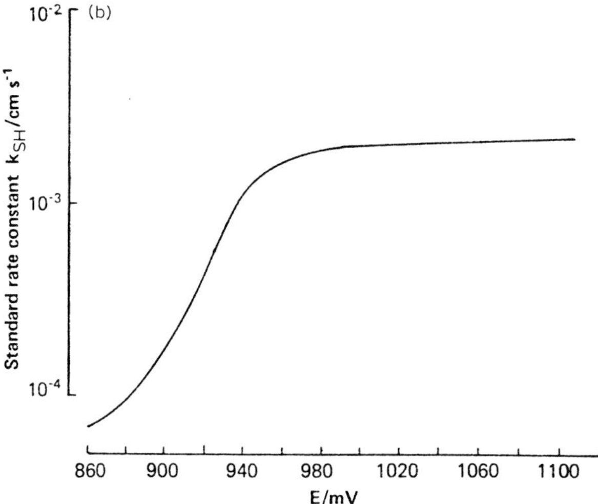

Fig. 15. Rotating ruthenium dioxide/titanium dioxide electrode (35% w/w ruthenium dioxide) in 3×10^{-2} M KBr + 1 M $NaClO_4$ solution. (a) Standard rate constant–potential curve assuming a constant Tafel slope of 40 mV. $D_{Br^-} = 9.6 \times 10^{-6}$ cm s^{-1}, $D_{Br_2} = 9.2 \times 10^{-6}$ cm s^{-1}, $E^0 = 895$ mV SCE, and $R = 0.8$ ohm cm^2. (b) Standard rate constant–potential curve assuming a constant Tafel slope of 120 mV. $D_{Br^-} = 9.6 \times 10^{-6}$ cm s^{-1}, $D_{Br_2} = 9.2 \times 10^{-6}$ cm s^{-1}, $E^0 = 895$ mV SCE, and $R = 0.8$ ohm cm^2. (c) Common experimental and calculated current–potential curve using the parameters of Fig. 15(a). The square symbols refer to the calculated "reversible" curve. (d) Double layer capacity curve.

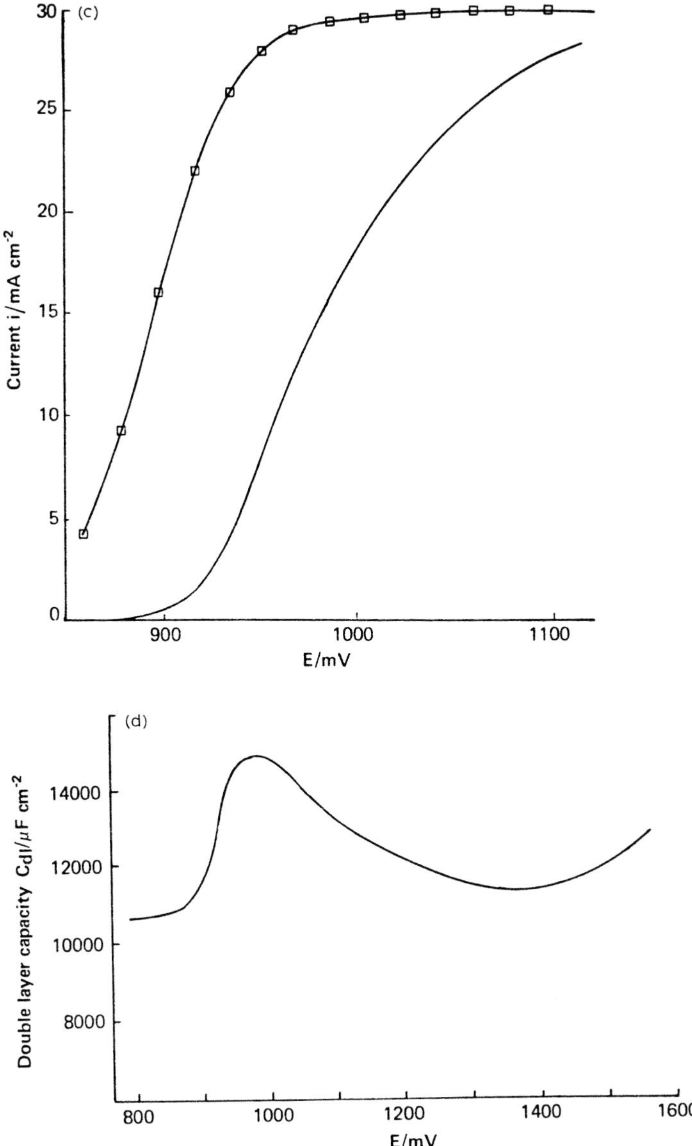

Fig. 15. Continued

For any complexing situation, the first question is: by changing the concentration of the complexing agent [chloride ion in the case of reaction (1)], is the change in current given by eqn. (26) changed sufficiently to decide which complex is involved in the electron exchange reaction? The calculated concentration of the various complexes containing palladium and

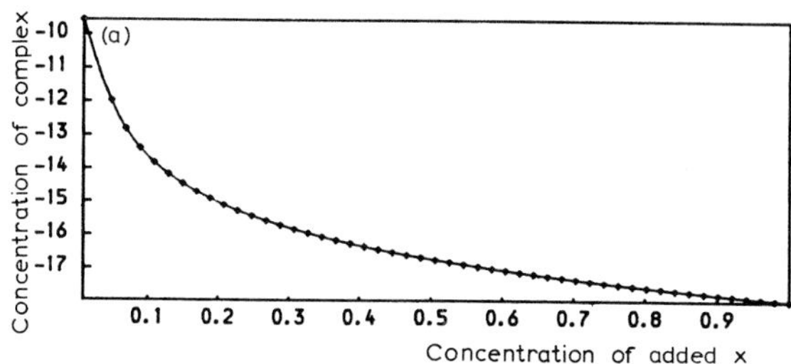

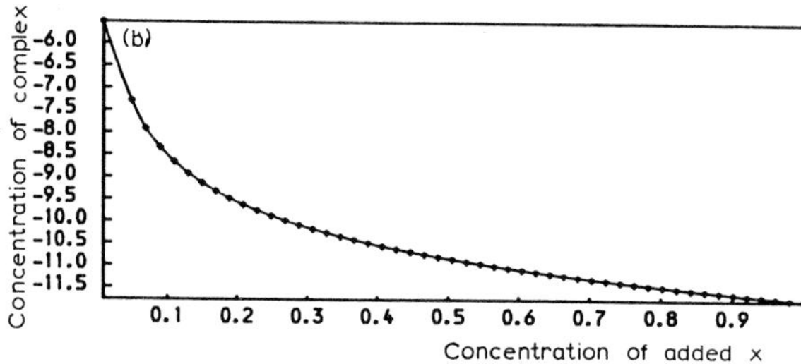

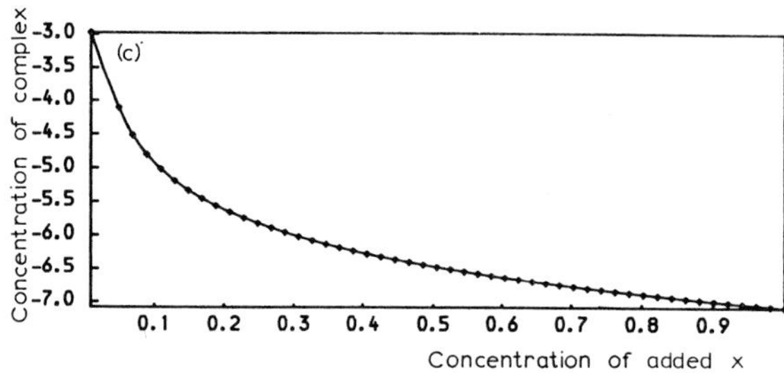

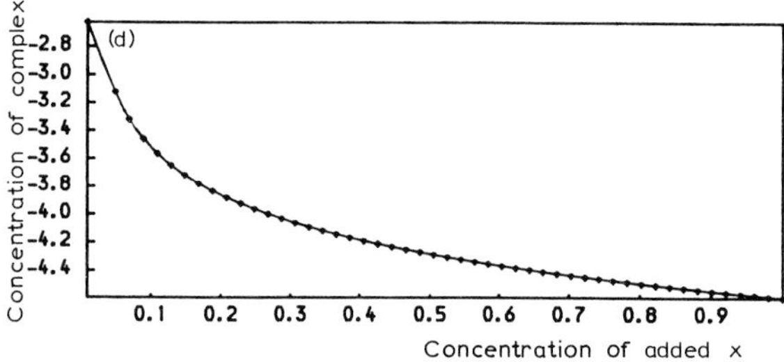

Fig. 16. The calculated molar amount of the various complexes in a solution containing 10^{-2} M $PdCl_2$ and HCl solution. The amount of HCl is given on the x-axis. The stability constants used were: $K_1 = 6.2$, $K_2 = 4.7$, $K_3 = 2.5$, $K_4 = 2.6$, and $K_{hydrol} = 13$. (a) Pd^{2+} ion concentration as a function of HCl concentration; (b) $(PdCl)^+$ ion concentration as a function of HCl concentration; (c) $PdCl_2$ ion concentration as a function of HCl concentration; (d) $(PdCl_3)^-$ ion concentration as a function of HCl concentration; (e) $(PdCl_4)^{2-}$ ion concentration as a function of HCl concentration; and (f) $(PdOH)^+$ ion concentration as a function of HCl concentration.

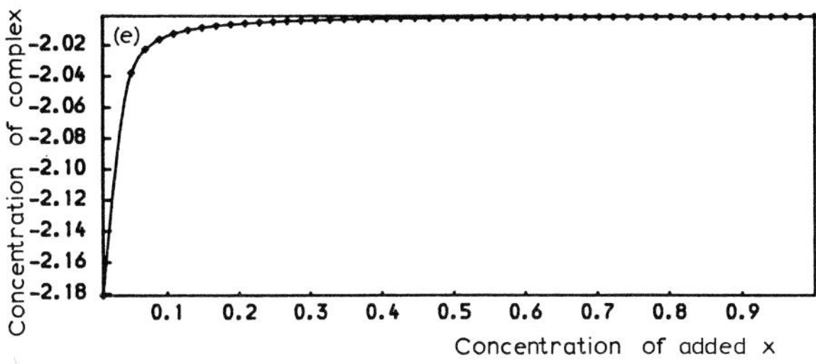

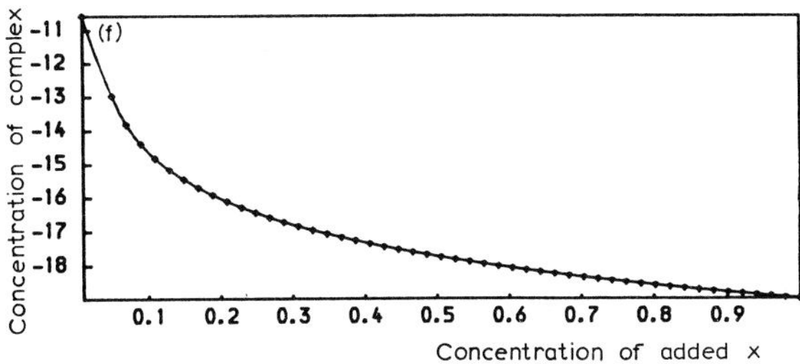

Fig. 16. Continued

chloride is shown in Fig. 16 as a function of the chloride ion concentration. It is clear that, on changing the concentration of chloride from 0.1 to 1.0 M, say, each of the palladium chloro complexes changes in a characteristic way. The concentrations shown in Fig. 16 were calculated by solving the set of equations

$$K_1[Pd^{2+}][Cl^-] - [PdCl^+] = 0 \tag{72}$$

$$K_2[PdCl^+][Cl^-] - [PdCl_2] = 0 \tag{73}$$

$$K_3[PdCl_2][Cl^-] - [PdCl_3^-] = 0 \tag{74}$$

$$K_4[PdCl_3^-][Cl^-] - [PdCl_4^{2-}] = 0 \tag{75}$$

$$K_{hydrol}[Pd^{2+}][OH^-] - [PdOH^+] = 0 \tag{76}$$

There is some uncertainty in these diagrams as the listed [29] stability constants are appropriate for infinite dilution and not the 1 M, say, ionic strength that is often convenient. However, the relative changes should be reasonably accurate. The measured current–potential and impedance–potential data should reflect the data given in Fig. 16, depending on which complex is involved in the rate-determining step.

There are a number of ways in which this problem could be tackled, but they are all governed by the number of parameters which control the reaction.

In the case of palladium, the measured equilibrium potential as a function of the chloride ion concentration (hydrochloric acid) at constant ionic strength (perchloric acid) together with the calculated concentrations of the various complexes using stability constant data leads to the standard poten-

TABLE 1

Calculated values of E^0 from eqn. (77)

Reaction	Standard potential (SCE, mol cm^{-3})/mV
$Pd/PdCl_4$	5
$Pd/PdCl_3$	172
$Pd/PdCl_2$	337
Pd/PdCl	575
Pd/Pd^{2+}	845

tials given in Table 1. These are calculated using the Nernst equation in the form

$$E_e = (E^0)_{Pd/PdCl_4^{2-}} + \frac{RT}{2F} \ln \frac{[PdCl_4^{2-}]}{[Cl^-]^4} \tag{77}$$

for each of the complexes.

The procedure used here is to interpret steady-state current–potential and impedance–potential data obtained for two concentrations of chloride ion. In the data analysis, each complex in turn is assumed to be involved in the rate-determining step.

The results of treating some experimental current–potential and impedance–potential data are shown in Figs. 17–19.

Two solutions, 1 M hydrochloric acid + 0.01 M $PdCl_2$ (A) and 0.1 M hydrochloric acid + 0.91 M perchloric acid + 0.01 M $PdCl_2$ (B), have been analysed in Fig. 17(a) and (b) by impedance–potential measurements, to give standard rate constant–potential, double layer capacity–potential, and current–potential curves for a particular reaction mechanism, in this case for the $PdCl^+$ complex

$$\begin{array}{ccccccccc} PdCl_4^{2-} & \rightleftharpoons & PdCl_3^- & \rightleftharpoons & PdCl_2 & \rightleftharpoons & PdCl^+ & \rightleftharpoons & Pd^{2+} \\ & & & & & & k_f \upharpoonleft\downharpoonright k_b & & \\ & & & & & & Pd + Cl^- - 2e & & \end{array} \tag{78}$$

The graphs [Fig. 17(a) and (b)] show the large increase in standard rate constant as the potential goes negative, suggesting that the palladium electrode is much more active for the deposition reaction at potentials less than about 100 mV. This effect is also reflected in Fig. 17(c) and (d) in which the double layer capacity–potential curves are reproduced. These show that the double layer capacity sharply increases with negative potential. The main reason for this effect is, undoubtably, an area increase as palladium metal is deposited. Figure 17(e) and (f) show the associated log current–potential curves (corrected for ohmic resistance). These curves are also reproduced by calculation from the measured impedance–potential curves.

Similar graphs can be produced for the two solutions 1 M hydrochloric acid (C) and 0.1 M hydrochloric acid + 0.9 M perchloric acid (D), which do not contain palladium ions in solution. The results are shown in Fig. 18(a) and (b) as the standard rate constant potential curves, where the data have again been analysed for the PdCl complex using reaction scheme (78). Figures 17(b) and 18(a) confirm the involvement of PdCl; however the standard rate constant is orders of magnitude less than at the cathodic potentials in the presence of palladium ions in the solution. A further complication is that the Tafel slope appears to depend on the chloride ion concentration, and also the double layer capacity, shown in Fig. 17(c) and (d), tends to fall as the potential goes more positive. Experience indicates that, for simple dissolution kinetics, the double layer capacity rises with potential due to surface area rise. The measured current–potential (corrected for ohmic effects) curves are shown in Figs. 17(b) and 18(e). Again, these are also consistent with the impedance–potential measurements.

For interest, Figs. 18(b) and 19(a) show standard rate constant–potential curves obtained by analysing the same current–potential and impedance–potential data, but assuming that $PdCl_2$ is the electroactive species, i.e. that the deposition–dissolution reactions occurs by the reaction scheme

$$\begin{array}{c} PdCl_4^{2-} \rightleftharpoons PdCl_3^- \rightleftharpoons PdCl_2 \rightleftharpoons PdCl^+ \rightleftharpoons Pd^{2+} \\ k_f \downarrow\uparrow k_b \\ Pd + 2\,Cl^- - 2\,e \end{array} \qquad (79)$$

The conclusions to be drawn from the work are as follows.

(a) Investigations involving current–potential measurements or linear potential sweep measurements are not sufficient for investigating metal deposition–dissolution reactions.

(b) For palladium during deposition, large changes in area occur, depending on potential, which make investigation difficult. A tentative suggestion is that the double layer capacity can be used to correct for the surface area.

(c) For palladium during dissolution, the surface seems to be protected against surface area change. This effect is probably due to the presence of a palladium oxide during dissolution.

(d) The large difference in standard rate constant between the potentials at which anodic dissolution takes place and at which deposition takes place make interpretation difficult. However, the closest correspondence between experiment and theory is certainly given by a reaction mechanism which involves one of the lower chloride-containing complexes, probably PdCl (as here) or possibly $PdCl_2$.

17. Corrosion of dental amalgams

This has been investigated as a model for the behaviour of materials in biological environments. In this investigation [30], various amalgams containing zinc, copper, tin, silver, and mercury have been investigated. The electrolyte used is 2/3 strength Ringer's solution. The philosophy has been

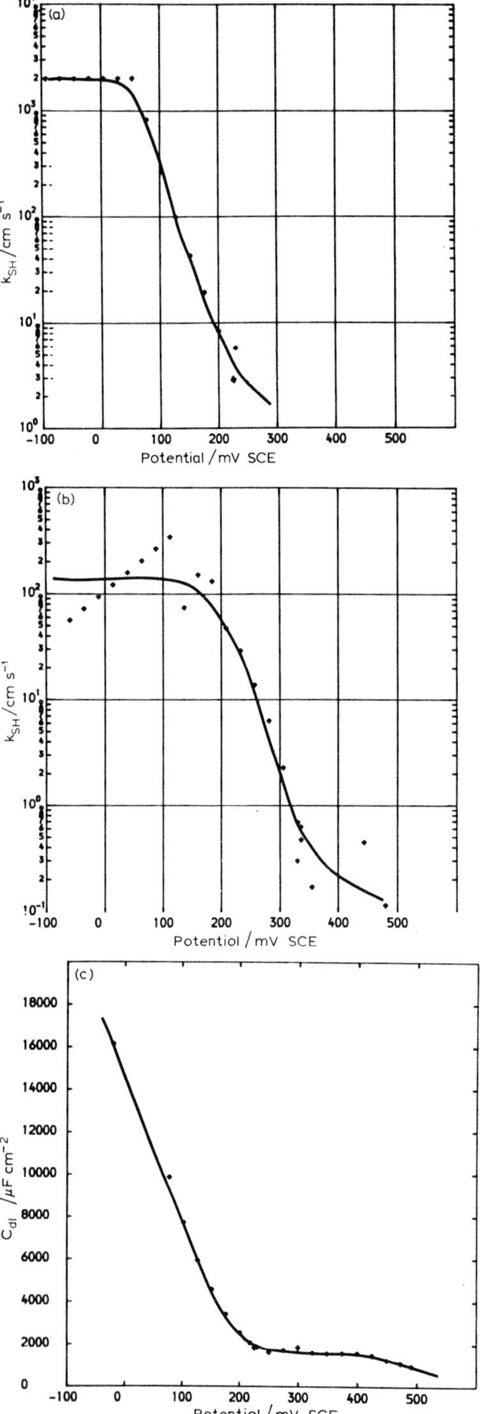

(a)
k_{SH} /cm s^{-1}
Potential / mV SCE
(b)
k_{SH} /cm s^{-1}
Potentiol / mV SCE
(c)
C_{dl} /μF cm^{-2}
Potentiol / mV SCE

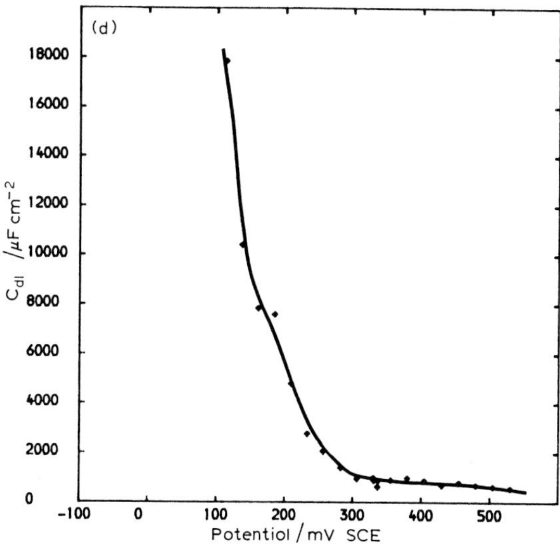

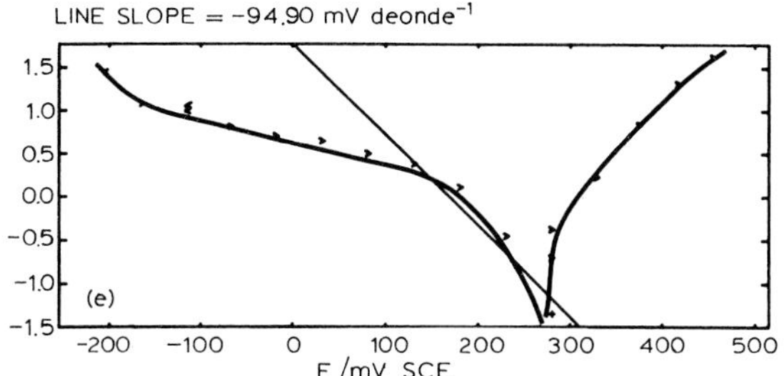

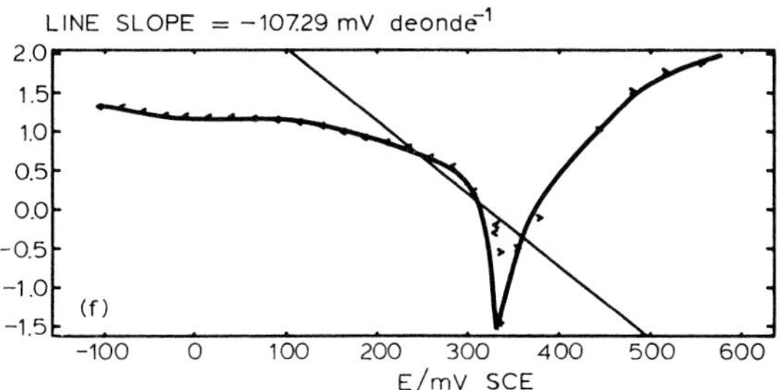

Fig. 17. Analysis of current–potential and impedance–potential data for the active deposition–dissolution of palladium in solutions of 1 M hydrochloric acid + 0.01 M $PdCl_2$ (A) and 0.1 M hydrochloric acid + 0.91 M perchloric acid + 0.01 M $PdCl_2$ (B). (a) Standard rate constant–potential curve calculated according to the reaction scheme (78) using experimental data obtained for palladium in solution A with the parameters b_a = 220 mV, b_c = 60 mV, and E^0 = 575 mV. (b) Standard rate constant–potential curve calculated according to the reaction scheme (78), using experimental data obtained for palladium in solution B with the parameters b_a = 220 mV, b_c = 60 mV, and E^0 = 575 mV. (c) Double layer capacity–potential curve for solution A. (d) Double layer capacity–potential curve for solution B. (e) Current–potential curve for solution A. (f) Current–potential curve for solution B.

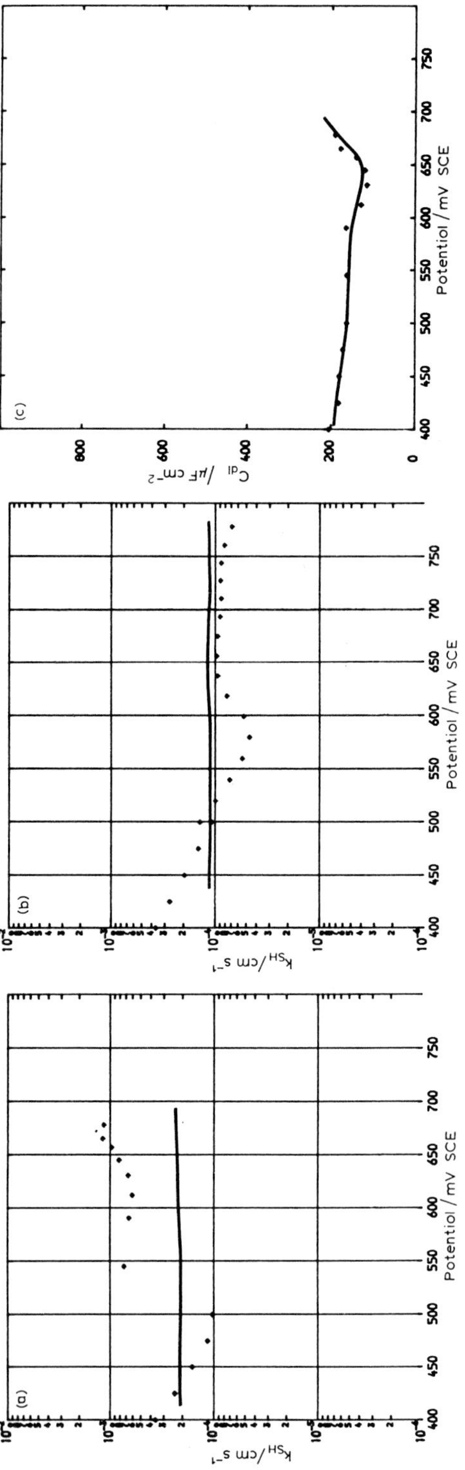
(a)
(b)
(c)
k_SH/cm s^-1
C_dl /μF cm^-2
Potentiol / mV SCE
400 450 500 550 600 650 700 750
0 200 400 600 800

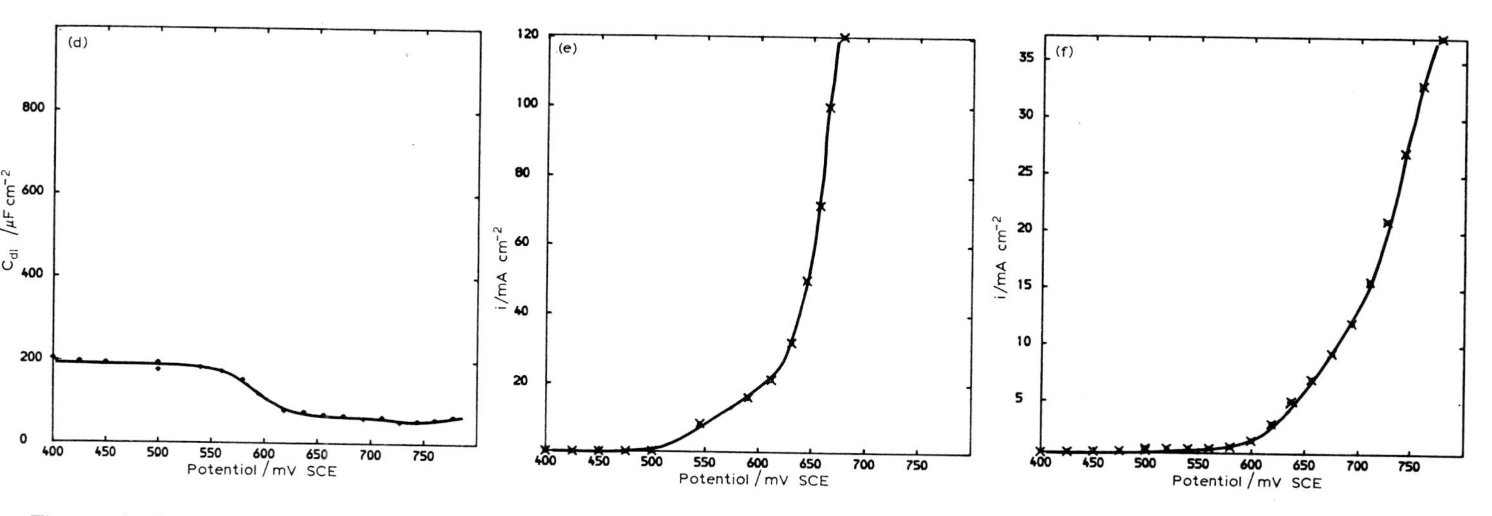

Fig. 18. Analysis of current–potential and impedance–potential data for the active deposition–dissolution of palladium in solutions of 1 M hydrochloric acid (C) and 0.1 M hydrochloric acid + 0.9 M perchloric acid (D). (a) Standard rate constant–potential curve calculated according to the reaction scheme (78) using experimental data obtained for palladium in solution C with the parameters b_a = 220 mV, b_c = 70 mV, and E^0 = 575 mV. (b) Standard rate constant–potential curve calculated according to the reaction scheme (78) for experimental data obtained for palladium in solution D with the parameters b_a = 220 mV, b_c = 140 mV, and E^0 = 575 mV. (c) Double layer capacity–potential curve for solution C. (d) Double layer capacity–potential curve for solution D. (e) Current–potential curve for solution C. (f) Current–potential curve for solution D.

References p. 497

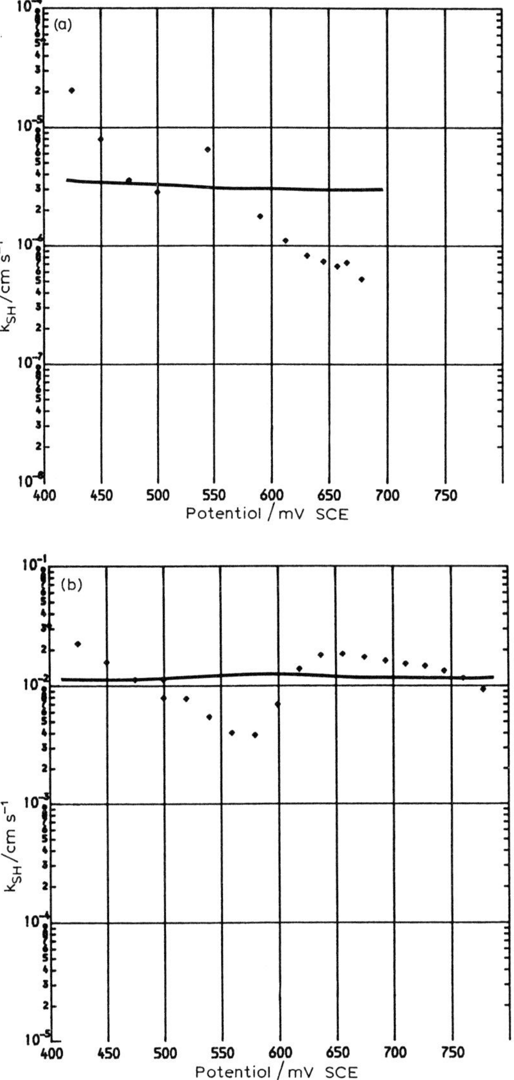

Fig. 19. Further analysis of current–potential and impedance–potential data for the active deposition–dissolution of palladium in solutions of 1 M hydrochloric acid (C) and 0.1 M hydrochloric acid + 0.09 M perchloric acid (D). (a) Standard rate constant–potential curve calculated according to the reaction scheme (79) for experimental data obtained for palladium in solution C with the parameters $b_a = 220\,\text{mV}$, $b_c = 70\,\text{mV}$, and $E^0 = 337\,\text{mV}$. (b) Standard rate constant–potential curve calculated according to the reaction scheme (79) for experimental data obtained for palladium in solution D with the parameters $b_a = 220\,\text{mV}$, $b_c = 140\,\text{mV}$, and $E^0 = 337\,\text{mV}$.

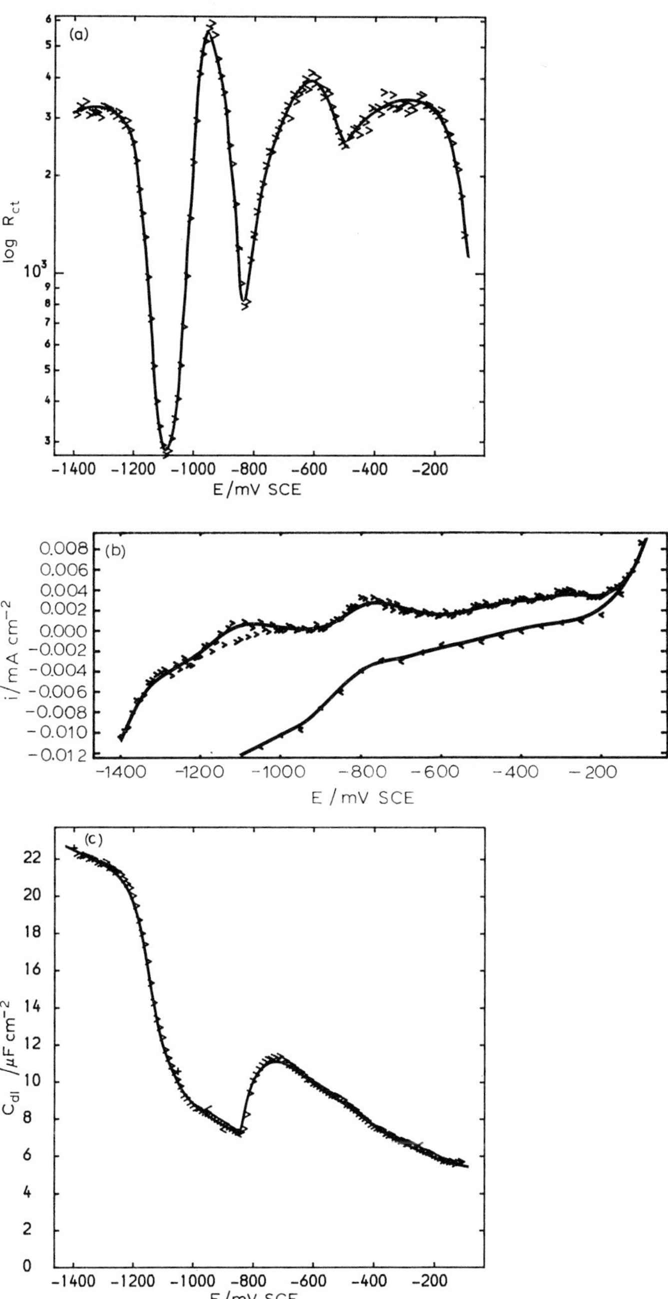

Fig. 20. Typical result of measuring the characteristics of dissolution of a dental amalgam (Dispersalloy, electrode 37 HD) in 2/3 strength Ringer's solution. The measurements are almost in the steady state at 60 s per potential point. (a) Charge transfer–potential curve; (b) Current–potential curve, also showing the return potential sweep; and (c) double layer capacity–potential curve.

References p. 497

to run steady-state measurements of state current–potential and high-frequency impedance–potential and analyse for the usual charge transfer–potential, double layer capacity–potential, Warburg coefficient (if necessary)–potential, and the corrected current–potential curves. These "spectra" have been acquired for the amalgams and the pure components of the amalgams in the presence and absence of dissolved oxygen. It has been necessary in this work because of the low currents, high impedances, and the long time constants for the formation of the surface films to measure over long times and use a specially developed potentiostat. An example of the type of results are shown in Fig. 20, where a set of curves for Dispersalloy are shown. The charge transfer–potential, Fig. 20(a), curve shows a series of minima due to the dissolution and passivation (from negative potentials) of Zn and Sn to Sn^{2+} and Sn to Sn^{4+} and then, finally, at the most positive potential the dissolution of the copper component of the alloy in the 2/3 Ringer's solution. These features echo the current–potential curve shown in Fig. 20(b), but with enhanced sensitivity. The corresponding double layer capacity–potential curve is shown in Fig. 20(c). It is not yet possible to synthesise these curves from those of the pure components, but work is proceeding [30].

18. Conclusions

The application of computers in a serious way to problems in electrode kinetics and electrochemistry has barely started. It will need a tremendous effort to bring into the subject the advances that have occurred in other branches of physical science. Specifically, subjects which need to be introduced into electrochemistry, in order to make full use of data that can now be obtained, are those of artificial intelligence and database organisation.

List of symbols

C_A^b	bulk concentration
C_A^s	surface concentration
C_T	total concentration of complexes
$[MX_n]_s$	surface concentration
D_A	diffusion coefficient
E^0	standard potential
E_e	equilibrium potential
f	flux
i	current density
j	square root of -1
k_{SH}	standard rate constant at E^0

k_f	forward rate constant
k_b, k'_b	backward rate constant
p	Laplace transform variable
Z	impedance of the interface
$Z(\omega)$	measured impedance
R_{ct}	charge transfer resistance
σ	Warburg coefficient
ω	frequency of a.c.
δ	diffusion layer thickness

References

1 Opportunities in Chemistry, National Academy Press, Washington, DC, 1985.
2 Chem. Eng. News, Jan (1984).
3 R.E. Dessy, A/C Interface, Anal. Chem., (1984–1986).
4 Chem. Eng. News, Aug (1985).
5 Science, April (1985).
6 J.O'M. Bockris, B.E. Conway, E. Yeager and R. White (Eds.), Comprehensive Treatise of Electrochemistry, Vol. 8, Plenum Press, New York, 1984.
7 J.S. Mattson, H.B. Mark and H.C. Macdonald Jr. (Eds.), Electrochemistry, Calculations, Simulation, and Instrumentation Computers in Chemistry and Instrumentation, Vol. 2, Dekker, New York, 1972.
8 D.C. Johnson, Anal. Chem., 54 (1982) 9R.
9 D.C. Johnson, M.D. Ryan and G.S. Wilson, Anal. Chem., 56 (1984) 7R.
10 J.G. Osteryoung and R.A. Osteryoung, Anal. Chem., 57 (1985) 101A.
11 M.W. Kendig, U. Bertocci and J.E. Strutt (Eds.), Computer Aided Acquisition and Analysis of Corrosion Data, The Electrochemical Society, New York, 1985.
12 A. Carrick, Computers and Instrumentation, Heyden, London, 1979.
13 J. Heinze, Angew. Chem., 23 (1984) 831.
14 M. Sargent, III and R.L. Shoemaker, The IBM Personal Computer from the Inside Out, Addison-Wesley, New York, 1984.
15 J.W. Cooper, The Minicomputer in the Laboratory: with Examples Using the PDP-11, Wiley-Interscience, New York, 2nd edn., 1985.
16 J.A. Harrison, Electrochim. Acta, 29 (1984) 703.
17 J.A. Harrison, Electrochim. Acta, 27 (1982) 1113.
18 M. Sluyters-Rehbach and J.H. Sluyters, in E. Yeager, J.O'M. Bockris, B.E. Conway and S. Sarangapani (Eds.), Comprehensive Treatise of Electrochemistry, Vol. 9, Plenum Press, 1984.
19 S.K. Rangarajan, J. Electroanal. Chem., 55 (1974) 297, 329, 337, 363.
20 E.J. Kelly, in J.O'M. Bockris, B.E. Conway and R.E. White (Eds.), Modern Aspects of Electrochemistry, Vol. 14, Plenum Press, New York, 1982.
21 J.A. Harrison, R.E. Plimley and J. Tyrovolas, J. Electroanal. Chem., 225 (1987) 139.
22 J.A. Harrison and D.E. Williams, Electrochim. Acta, 31 (1986) 1063.
23 S. Trasatti, Electrochim. Acta, 29 (1984) 1503.
24 L.I. Krishtalik, Electrochim. Acta, 26 (1981) 329.
25 J.S. Robinson (Ed.), Chlorine Production Processes, Chemical Technology Review No. 185, Noyes Data Corp, NJ, 1981.
26 C. Jackson (Ed.), Modern Chlor-Alkali Technology, Vol. 2, Ellis Horwood, Chichester, 1983.
27 J.A. Harrison and S.D. Hermijanto, J. Electroanal. Chem., 225 (1987) 159.
28 I.L. Cooper and J.A. Harrison, Electrochim. Acta, 29 (1984) 1147.
29 L.G. Sillen and A.E. Martell (Eds.), Stability Constants, Special Publication 17, The Chemical Society, London 1964.
30 J.A. Harrison and M.J. Gross, J. Appl. Electrochem., submitted for publication.

Index

A

a.c. voltammetry, at channel electrodes, 198
aluminium corrosion, studied by ellipsometry, 442
4-aminobenzonitrile reduction, ex-situ ESR study, 309
amperometric sensor, 142
anodic stripping voltammetry, at channel electrodes, 204, 205
anthracene reduction, in dry acetonitrile, 161
Auger electron spectroscopy, 1, 106, 107

B

backward implicit method (BIM), 184, 190
–, advantages of, 185
–, in investigation of Singh and Dutt approximation, 190, 191
band bending, at semi-conductor electrodes, 358
bismuth oxide electrodes, 136

C

calcium carbonate dissolution, 268
–, amperometric detection, 271, 279, 282
–, calcite dissolution, at pH 3, 279
–, –, in aqueous maleic acid, 283
–, –, in aqueous polymaleic acid, 281
–, inhibition of, 271, 283, 287
–, pit formation at crystal surface, 280, 283
–, study at channel electrode, 268
–, study by rotating disc technique, 269
CE reactions, 174, 218, 253
channel electrodes (CE), 139, 140, 173, 176, 317
–, a.c. voltammetry at, 198
–, anodic stripping voltammetry, 204, 205
–, calcite dissolution, at pH 3, 279
–, –, in aqueous maleic acid solution, 283
–, –, inhibition by polymaleic acid, 281
–, cell design and fabrication, 220
–, CE reactions, 218
–, chronoamperometry at, 200, 264
–, comparison with rotating disc electrode, 177
–, convective–diffusion equations, 179
–, coordinate systems at, 177
–, coupled homogeneous reactions, 206
–, ECE/DISP1 mechanism transport limited currents, 206, 338
–, ECE reactions, 206
–, –, Singh and Dutt approximation, 207
–, –, theory of steady-state voltammetry, 207
–, edge effects, 186, 187
–, flow systems, 222
–, DISP1 reactions, 206
–,–, Singh and Dutt approximation, 210
–, –, theory of steady-state voltammetry, 209
–, double channel electrodes, 225, 226
–, –, anodic metal dissolution, 236
–, double tubular electrodes, 225
–, ECE/DISP1 reactions discrimination, 211, 213
–, EC reactions, 217
–, EC′ reactions, 219
–, ESR studies, 222
–, ionic solid dissolution study, 268
–, –, amperometric detection, 271, 279
–, –, dissolution at finite rate, 272
–, –, mass transport controlled dissolution, 271
–, –, shielding experiments, 271, 279, 282
–, Levich equation, 140, 181, 183, 184, 185, 186, 318
–, –, Lévêque approximation, 181
–, linear sweep and cyclic voltammetry at, 196
–, mass transport to, 174, 181
–, –, Levich equation, 181
–, –, numerical methods, 184
–, –, Singh and Dutt approximation, 188

–, –, turbulent flow, 244
–, non-uniform accessibility, 177, 178
–, pitting corrosion study, 256
–, –, experimental investigation, 264
–, –, theoretical models of, 256
–, pulsed flow voltammetry at, 206
–, Singh and Dutt approximation, 179
–, –, prediction of linear sweep and cyclic voltammetric response, 198
–, steady-state voltammetry at, 194
–, theoretical treatment of problems at, 179
–, thin layer cell behaviour, 186
–, velocity profile, 181
chronoamperometry, at channel electrodes, 200, 264
chronopotentiometry, at channel electrodes, 200
collection efficiency, double channel electrodes, 226, 227, 228
–, rotating ring–disc electrode, 133, 178
–, rotating disc gap and ring parameters, 133
–, –, EC reactions, 229, 231
–, –, EC′ reactions, 229, 255
–, wall jet electrode, 142
computers, calculation of current–potential curves, 460, 463, 466
–, calculation of impedance–potential curves, 461, 464, 467
–, on-line accumulation of data, 453, 455
–, –, chloride and bromide oxidation on RuO_2/TiO_2 electrodes, 477
–, –, dental amalgam corrosion, 489
–, –, dissolution and passivation of titanium, 470
–, –, palladium chloro complexes reduction, 481
–, –, stainless steel corrosion, 475
–, processing of data, 455, 459
Cottrell equation, 152, 201
cross-sectional transmission electron microscopy (XTEM), 448
cyclic voltammetry, at in-vivo microelectrode, 151

D

data, on-line accumulation of, 453, 455
–, –, experimental strategies, 456
–, processing, 455, 459
differential contrast interferometry, 280, 284, 287
diffusion coefficient, measurement by rotation speed step experiment, 138, 139
diffusion layer, rotating disc, 129, 130
DISP1 reactions, 160, 206, 322, 336, 338
–, reduction of fluorescein in aqueous solution, 213
DISP2 reactions, 160
double channel electrodes, 225, 226
–, anodic metal dissolution, 236
–, collection efficiency at, 226, 227, 228
–, –, EC reactions, 229, 231
–, –, EC′ reactions, 229, 255
double electrodes, 129, 133
double tubular electrodes, 225

E

E reactions, 459
–, comparison of experimental and computer calculated theory, 468
–, current–potential curve, computer calculation of, 460
–, impedance–potential curve, computer calculation of, 461
–, with complexing reactant, current–potential curve, computer calculation of, 463
–, –, impedance–potential curve, computer calculation of, 464
EC reactions, 174, 217, 229, 231, 321
ECE/DISP1 reactions, 159, 176, 336
–, discrimination at channel electrode, 211, 213, 338
–, discrimination at microelectrodes, 160
ECE reactions, 159, 206, 320, 322, 329, 336
–, anthracene reduction in acetonitrile, 161
EC′ reactions, 219, 229, 255
EC_2 reactions, 313
EE reactions, 464
–, current–potential curve, computer calculation of, 466
–, impedance–potential curve, computer calculation of, 467
EELS, 117
electrochemical ESR, 139, 297
–, historical development, 305
electron spin resonance (ESR), 129, 297, 298
–, *g* factor, 298, 299
–, hyperfine coupling, 300
–, instrumentation, 305
–, linewidth, 303
–, spin trapping, 346
–, theory, 298
electroreflectance, at semiconductors, 385
–, computation of theoretical electroreflectance spectra, 401

–, experimental configuration, 391, 392
–, experimental results, 413
–, –, GaAs, 420
–, –, GaP, 413, 415
–, –, non-aqueous solvents, 423
–, –, –, GaAs, 423
–, –, –, GaP, 425
–, –, problems, 391, 392
–, Franz–Keldysh theory, application to III/IV semiconductors, 405
–, –, a.c. amplitude effect, 409
–, –, d.c. potential effect, 406
–, –, donor density effect, 407
–, –, inhomogeneous electric field effect, 405
–, –, optical cross-section effect, 409
–, –, thermal broadening effect, 408
–, low field theories, 402
–, theories, 392, 393
ellipsometry, 427
–, adsorbed biological materials, 444, 446
–, aluminium corrosion, 441, 442
–, complex layer profiling, 448
–, complex refractive index, 430
–, –, film thickness, 432
–, Fresnel equations, 430, 450
–, ideal 3-phase model, 434, 438
–, instrumentation, 432
–, intensity modulating spectrometer, 432, 434
–, nearly perfect 3-phase model, 439
–, non-uniform and rough films, 441, 442
–, nulling modulating spectrometer, 432, 433
–, spectroscopic, 446, 448
–, theory, 427
–, –, phase of refracted light, 428
–, –, refractive index, 430
–, thin films, 444
enzyme electrodes, 133
ESR, *see* electron spin resonance
etch pits, formation in calcite dissolution, 280, 283
ethylene adsorption on gold, SERS study, 96
EXAFS, 105
ex-situ electrochemical ESR, 297, 302
–, Albery tubular electrode, 222, 310
–, reduction of 4-aminobenzonitrile, 309

F

finite difference methods, 184
fluorescence quenching, by adsorption at SERS active surface, 83
Fresnel equations, 430, 450
FTIR, *see* infrared spectroscopy, Fourier transform techniques
FT-IRRAS, 62

H

Hale transformation, 134
homovanillic acid, in-vivo release by microelectrode stimulation, 151
hydrodynamic flow, 310, 317
hydrodynamic electrodes, 129, 317

I

inner Helmholtz plane (IHP), 91
in-situ electrochemical ESR, 297, 310
–, Adams' cell, 306, 307
–, Albery's semi-annular tubular electrode, 322
–, Allendoerfer's coaxial electrode, 314
–, Allendoerfer and Carroll coaxial flow cell, 316
–, Bard and Goldberg cell, 312
–, –, study of homogeneous kinetics in, 313
–, Bond's low-temperature cell, 313
–, Cauquis grid electrode, 310
–, Compton and Coles channel electrode cell, 220, 317, 318, 323
–, –, dicyanobenzene reduction in acetonitrile, 329
–, –, ECE/DISP1 reactions, fluorescein reduction, 336
–, –, polymer coated electrodes, 339, 340, 341
–, –, –, moleuclar motion within, 343, 344
–, –, study of homogeneous kinetics, 319, 323, 329, 332, 334, 336
–, –, –, ESR detection efficiency, 319
–, –, –, steady-state measurements, 319, 320
–, –, –, transient measurements, 321, 338
–, –, triphenylacetic acid oxidation, 330
–, –, triphenylmethanol oxidation, 334
–, Compton and Waller coaxial in-situ cell, 323
–, –, angular sensitivity profile, 326
–, –, hydrodynamics, 326
–, –, steady-state ESR behaviour, 328
–, –, vertical sensitivity profile, 325
–, Kastening strip electrode, 311
–, Maki and Geske cell, 297, 305
–, perchlorate oxidation, 306, 312
–, spin-labelled electrode, 344
–, spin trapping, 346

infrared reflection adsorption spectroscopy (IRRAS), 61
–, FT-IRRAS, 62
infrared spectroscopy (IR), attenuated total reflection (ATR), 13, 15, 70
–, –, corrosion, 17
–, –, double layer, 20
–, –, electropolymerisation, 19
–, –, ion radical intermediates, 15
–, –, molecular adsorption, 16
–, electrically modulated IR (EMIRS), 25, 26
–, –, applications, 27
–, –, adsorbed hydrogen, 36
–, –, adsorbed ethylene, 39
–, –, adsorption, 28, 39
–, –, CO_2 reduction, 31, 32
–, –, electron transfer in thin layer, 28
–, –, electro-oxidation of small molecules, 29
–, –, methanol oxidation, 30, 31
–, external reflection, 25
–, Fourier transform techniques (FTIR), 14, 17, 42, 43
–, –, investigations of double layer, 48
–, –, study of adsorption, 51, 58
–, –, study of ion radical intermediates, 55
–, internal reflection, 2, 10
–, –, theory, 10
–, modulated specular reflection, 25
–, polarization modulation techniques, 61, 64
–, sensitivity, 1
–, solvent absorption, 1
–, specular external reflection, 2, 6
–, –, theory, 6
–, subtractively normalised interfacial FTIR (SNIFTIRS), 47, 48, 51
–, –, CO_2 reduction, 60
–, transient techniques, 67
–, –, pyrrole oxidation, 72
–, transmission, 2, 24
interferometry, 42, 43
internal photoemission, at oxide-covered electrodes, 360, 361, 375
–, bismuth oxide film, 375
–, Pb/PbO interface, 377
ion selective electrodes, 150
IR, *see* infrared spectroscopy
iron dissolution, studied at double channel electrode, 240
IRRAS, *see* infrared reflection absorption spectroscopy

K

Kolbe oxidation, 330
–, triphenylacetic acid oxidation, 330

L

laser Raman spectroscopy, photocurrent generation, 354
lead corrosion, study by photocurrent spectroscopy, 372
LEED, 105, 107
Levich equation, channel electrode, 140, 318
–, coaxial electrode, 326
–, rotating disc electrode, 130
–, tubular electrode, 322
lithium deposition, study at microelectrode, 167

M

Michelson interferometer, 42, 43
microband electrodes, 155, 157, 158
microcylinder electrodes, 155
microelectrodes, 149
–, applications, 159
–, current at finite disc electrode, 153
–, design and construction, 155, 157
–, ECE reactions at, 160
–, ECE/DISP1 discrimination, 160
–, electronic equipment, 157
–, finite size effects, 152
–, homogeneous kinetics study, 154, 159
–, in-vivo studies, 150
–, –, normal pulse voltammetry, 151
–, nucleation and phase growth at, 169, 170
–, ohmic drop at, 158, 163, 167
–, resistive media, 163, 165, 166
–, transient studies, 167
–, –, lithium deposition, 167
microring electrodes, 155
microtubular electrodes, 244
–, current–potential relationship for steady-state voltammetry at, 251
–, CE reactions, 253
–, coupled homogeneous kinetics, 253
–, EC′ reactions, 255
–, electrode kinetics, 253
–, turbulent flow at, 244
modified electrodes, 133
–, polymer coated, in-situ ESR studies, 339
–, Raman spectroscopy of, 99
–, spin-labelled electrode, 344
Mott–Schottky relationship, 360, 385

N

normal pulse voltammetry, in-vivo microelectrode, 151

O

ohmic drop, 149, 307, 316
optically transparent electrodes, use in photocurrent spectroscopy, 365, 368
optical rotating disc electrode, 146, 147
outer Helmholtz plane (OHP), 91
oxide electrodes, 133

P

packed bed electrode, 142
packed bed wall jet electrode, 131, 132, 142
perchlorate oxidation, ESR study, 306, 312
photocolloidal systems, 147
photocurrent conversion efficiency, 359, 361, 362, 363
–, determination of, 365
–, Gartner equation, 359
photocurrent spectroscopy, 353, 363
–, anodic oxidation of iron, 369
–, application to semiconductors and insulators, 353
–, Bi_2S_3 films, 365, 368
–, collection of photogenerated carriers, 357
–, derivation of absorption spectra, 366
–, determination of film thickness, 371
–, excitation of electrons in metals, 354
–, excitation of electrons in semiconductors and insulators, 356
–, experimental aspects, 364
–, interband excitation of electrons, 354
–, internal photoemission, 375
–, lead corrosion, 372
–, light absorption in crystalline lattice, 354
–, optically transparent electrodes, 365
–, organometallic films on metal electrodes, 381
–, PbO films on lead electrodes, 371
–, photocurrent generation in semiconductor films of finite thickness, 360
–, photosensitivity of anodic films on metals 363, 364
–, p-type surface films, 378
–, sensitivity of spectrometer, 364
–, TiO_2 films on titanium, 371
photoelectrochemistry, 101, 146, 364
–, quantum yields at metal electrodes, 355
photogalvanic cells, 147
photon absorption in solids, selection rules, 356
photovoltages, 353, 356
pitting corrosion, study at channel electrodes, 256
–, –, iron, 264
–, –, stainless steel, 265
polymer-coated electrodes, 339, 343
–, poly(methylthiophene), 19
–, poly(*N*-vinylcarbazole), 340
–, poly(pyrrole), 341
potential of zero charge (PZC), 91
potentiometric electrode, 136, 138
pulsed anodic stripping voltammetry, 205
pulsed flow voltammetry, 206
pyrrole oxidation, study by transient IR, 72

R

Raman spectroscopy, *see also* surface enhanced Raman spectroscopy, 1, 2, 79
–, electrochemical cells, 85, 86, 87
–, FTIR methods, 79
–, resonance Raman sensitivity enhancement, 80
–, chemical species adsorbed at metal electrodes, 94
–, modified electrodes, 99
–, non-aqueous systems, 100
–, semiconductor electrodes, 101
reflectance spectroscopy, 427
rotating disc electrode, 129
–, CE reactions, 174
–, diffusion layer at, 129, 130
–, EC reactions, 174, 175
–, mass transport to, 174
–, rotation speed step measurement of diffusion coefficient, 138, 139
–, sinusoidal modulation of rotation speed, 206
–, uniform accessibility, 129, 177, 178
–, velocity pattern at, 129, 130
rotating ring–disc electrode, 129, 133
–, analytical solutions, 134
–, collection efficiency at, 133, 178
–, pH transients, 136
–, numerical solution, 134
–, simulation programs, 134
–, transients, 136

S

Schottky barrier, 357
semi-annular tubular electrode, 322
semiconductor electrodes, a.c. impedance studies, 389, 390
–, depletion layer, 385
–, electroreflectance at, 385

–, flat band potential, 385
–, internal field and potential, study by a.c. impedance, 388
–, –, study by measurement of photovoltages, 391
–, potential distribution at, 385, 388
–, Raman spectroscopy of, 101
–, space charge region in, during photoexcitation, 358
SERS, *see* surface enhanced Raman spectroscopy
SERRS, *see* surface enhanced resonance Raman spectroscopy
SIMS, 1
SiO_2 on silicon, studied by ellipsometry, 434
Singh and Dutt approximation, channel electrodes, 179
SNIFTIRS, *see* infrared spectroscopy
spectroscopic ellipsometry, 446, 448
spin trapping, 347
surface enhanced Raman spectroscopy, 2, 81
–, colloidal particles, 94
–, electrode potential dependence, 91
–, electrode surface activation for, 88
–, enhancement by oxidation–reduction cycles, 89
–, fluorescence quenching by adsorption at SERS active surface, 83
–, limitations, 2
–, non-aqueous systems, 100
–, adsorbed alkenes, 98
–, adsorbed benzenes and substituted benzenes, 97
–, adsorbed ethylene, 96
–, adsorbed pyridine, 81
surface enhanced resonance Raman spectroscopy, 85, 102
surface plasmon polaritons (SPP), 20, 21

T

Tafel analysis, 327
thin layer cell, limiting current at, 328
thionine-coated electrodes, 99, 136, 137
tin oxide electrodes, 146
tin oxide films, study by photocurrent spectroscopy, 371
tube electrode, 129, 132, 176
–, design and fabrication, 221
–, ESR use, 222
–, Levich equation, 133, 322
–, mass transport to under turbulent flow regime, 244, 247
tubular electrodes, *see* tube electrode

U

ultra-high vacuum techniques (UHV), 105
–, experimental techniques, 106
–, –, direct transfer systems, 107, 110
–, –, glove box transfer, 112
–, –, parallel experiments, 113, 114
–, –, electrode transfer in and out of an electrolyte, immersion and emersion, 115
–, adsorbed hydrogen on platinum, 118
–, iodine adsorption on platinum, 123
–, oxidised platinum electrodes, 122
–, silver deposition on platinum, 122
UHV, *see* ultra-high vacuum techniques

W

wall jet electrodes, collection efficiency, 142
–, colloidal deposition at, 142, 143, 145
–, non-uniform accessibility, 144
–, packed bed wall jet electrode, 131, 132, 142

X

XPS, 1, 106
X-ray diffraction, 105